# Survey of the World's Aphids

# Survey of the World's Aphids

V.F. Eastop & D. Hille Ris Lambers

Dr. W. Junk b.v., Publishers, The Hague 1976

Authors' addresses

V.F. Eastop
British Museum (Natural History)
Cromwell Road
LONDON SW7 5BD
England

D. Hille Ris Lambers
Center for Aphid Research
BENNEKOM
The Netherlands



ISBN 90 6193 561 X
Dr. W. Junk b.v., Publishers, The Hague 1976
Cover design Max Velthuijs

## CONTENTS

# INTRODUCTION

A general survey of the aphids of the world has been tried only once before, 58 years ago, by Wilson & Vickery (1918). They catalogued the species but not the genera. Börner (1930) listed - and partly keyed - most of the aphid genera. In 1952 he published a very useful catalogue of Central European aphids. Patch's (1938) Food-plant Catalogue provides much information about the then known aphids of the world. Besides there are a number of recent publications on regional aphid fauna's systematizing information from that area.

We tried something different. The present work is not merely a list of described genera and species. An attempt was made to classify the known species, in order to facilitate future revisions, to stabilize nomenclature, and to make it easier to avoid homonymy in the future. We included synonymy, and for our purpose re-examined as many species as we had or could borrow.

No authentic material is known of a large number of species. Many such species were identified from the descriptions, but many others had to be left in the genus in which they were described, which does not necessarily mean that we believe they belong there.

We did not indicate the authors of new combinations or new synonymy. In a few cases of homonymy we asked living authors to provide new names, that are published in our book. Once or twice we gave new names.

Inevitably there are both errors and omissions. We shall be grateful for corrections and additions, particularly if supported with material. The initials BM or HRL after a valid name indicate that there is material in the collection of the British Museum (Natural History), London, or the Hille Ris Lambers collection at Bennekom, Netherlands. Such initials after a synonym in italics indicate that type material of that synonym is available.

The book consists of two parts: I. Aphid Genera of the World with their Species, (p. 7), II. Index by species-group Names and infra-subspecific Names, (p. 463).

## STRUCTURE OF THE SURVEY

1. *Genus-group names (genera and subgenera)*
Genus-group names, whether valid or not, are listed in one alphabetical series. References, abbreviated according to the 4th edition of the World List of the Scientific Periodicals, are given. The type species in the original combination of generic and specific name as given by the author who cited the type is stated, if necessary followed by its oldest available name. If the genus-group name is a junior synonym, its senior synonym follows "=". A list of junior synonyms follows, in italics.

2. *Species-group names (specific and subspecific names) and infra-subspecific names (form or variety).*
Subspecific names and infra-subspecific names are only listed under the species to which they belong. Specific names, whether valid or not, are listed in alphabetical order both under those genera in which they were originally described, and under those in which we think they belong. The presence of parentheses around author(s) plus date indicates transferred species, and the genus from which they were transferred follows in parentheses. Junior synonyms are listed, in italics and in alphabetical order, only below those combinations of genus- and specific name we consider valid. If a generic name in italics between parentheses follows, the synonym was transferred from another genus.

Example: "Aphis persicae Kaltenbach, 1843 = Brachycaudus(Appelia)schwartzi Börner, 1931", but no synonymy is found below Brachycaudus schwartzi, nor below Appelia schwartzi, but only below "Brachycaudus (Appelia) schwartzi (Börner, 1931)" where the synonym is listed as "persicae Kaltenbach, 1843 (Aphis)" in italics.

If a taxon was originally described in a subgenus, it is primarily not listed under the subgeneric name in capitals of the general list, but only under its generic name plus subgeneric name. The same method was followed if a taxon was transferred by us to a subgenus, another subgenus, or the same subgenus of another genus.

Example: "Dactynotus (Uromelan) stachydis Bozhko, (1957) 1961 to Uroleucon (Uromelan)" can primarily only be found below a caption "subgen. Uromelan Mordvilko, 1914" in normal print in an alphabetical list of species originally described in Dactynotus (Uromelan) that follows the list of species originally described in (the nominal subgenus) Dactynotus Rafinesque, 1817. From there we transferred it directly to Uroleucon (Uromelan) where it appears in an alphabetical list headed by a caption "Subgen. Uromelan Mordvilko, 1914" after the species placed in Uroleucon Mordvilko, 1914 (in capitals). It is found there as "stachydis (Bozhko, (1957) 1961) (Dactynotus (Uromelan))".

A "?" preceding a species name indicates our doubt abouts its generic or synonymic position, or its synonymy.

3. *Dates of publication.*
Sometimes author's names are followed by two dates. If there is an "&" between the two dates, the species was twice validly described by the same author as a new taxon, under the same name. If the second date is between parentheses the relevant publication was distributed at a later date than the second date printed on or in the publication. If the first date is between parentheses, the name was first used without a valid description, later validly described.

4. *Invalid combinations.*
In the past many species-group names have been used in invalid secondary combinations with generic names. Such combinations usually have not been listed. However, we made some exceptions. When the author of a new genus cited a secondary combination as the type species, that combination is listed, valid or not. Besides, we listed, as secondary combinations, a few taxa of great economic importance that frequently occur in literature on economic entomology, but which differ from those we accept.

Example: Aphis persicae Sulzer, 1776 is also listed as Myzus persicae (Sulzer) and Myzodes persicae (Sulzer), and in each case transferred to Myzus (Nectarosiphon) persicae (Sulzer, 1776), where the secondary combinations do not occur among the junior synonyms. Aphis pisi Kaltenbach, 1843 also appears as Macrosiphum pisi (Kaltenbach), which combination is not listed among the junior synonyms of Acyrthosiphon pisum (Harris, 1776).

5. *Bibliographical references to species-group and lower taxa.*
We did not give references to species-group, and infra-subspecific taxa. Had we included such information, the book would have been very much longer and even more expensive. However, most original descriptions can be traced rather easily via the extensive bibliographies recently published by Szelegiewicz (1969), Sharma (1969, 1971, 1972) and Smith (1972). The earlier descriptions can be traced via Wilson & Vickery (1918) and Patch (1938), those of nearly all European species through Börner (1952).

6. *Fossil genera, Adelgidae, and Phylloxeridae.*
Fossil genera and genera of the families Adelgidae and Phylloxeridae are listed in the same alphabetical order as the other aphid genera, but their species have not been listed. We have insufficient experience with these taxa. Heie (1967) listed the species of fossil aphids. Carter (1971, appendix 2) gave a check-list of the Adelgidae of the world.

7. *Misidentifications.*
Misidentifications have not been considered except when they affected the use of generic names.

## ACKNOWLEDGEMENTS

Nearly all contemporary colleagues have given invaluable help by the gift or loan of specimens, often types. Unfortunately we cannot mention them all by name. Dr. W.W. Sampson, Lafayette, California donated his aphid collection, including types, to the second author. Prof. M. Martelli, Milano, solved the riddle of the way G. del Guercio's paper of 1908 Sulla sistema e biologia del Fillosserini was published, and enabled one of us to decode the G. del Guercio collection and to remount many of the formerly useless slides. Thanks to the cooperation of Drs Louise Russell. Washington (D.C.) and W.D. Fronk, Fort Collins, Colorado, type material of nearly all the species described by Gillette and Palmer could be cleared, re-mounted and re-examined, Dr. N. Akramovsky, Yerevan, U.S.S.R. helped with photocopies of papers by V.N. Rusanova, and with the translation of some papers from the Azerbaidjan language. Drs. W.H. Paik, Suwon, Korea; H. Higuchi, Oyama City; M. Miyazaki, Nishinasu; and M. Sorin, Kuratayama, Japan, helped with translations from the Japanese. Ir. J. Gut, Wageningen, Netherlands, translated Russian and Ukrainian papers. Miss P. Gilbert, Mrs. W.C. van Beekhoff-Wolleswinkel, and Miss. J. Williamson were of great help in tracing references, and with the preparation of the manuscript.

## BIBLIOGRAPHY

Börner, C., 1952/3, Europae centralis Aphides, *Mitt. thüring. bot. Ges.* 4(3): 1-484 (1952) + 485-488 (1953).

Carter, C.I., 1971, Conifer Woolly Aphids (Adelgidae) in Britain, *Bull. Forestry Commn, Lond.* 42: 51pp.

Heie, O., 1967, Studies on Fossil Aphids (Homoptera: Aphidoidea), *Spolia zool. Mus. haun.* 26: 274pp.

Patch, E.M., 1938, Food-plant Catalogue of the Aphids of the World, *Bull. Me agric. Exp. Stn* 393: 35-431.

Sharma, M.L., 1969-1972, Bibliography of Aphidoidea, 3v. Editions Paulines Sherbrooke, Quebec (v.I: 293pp. (1969); II: 221pp. (1971); III: 207pp. (1972).)

Smith, C.F., 1972, Bibliography of the Aphididae of the World, *Tech. Bull. N. Carol. agric. Exp. Stn* 216: 717pp.

Szelegiewicz, H., 1969, Mszyce-Szkodniki roslin wektory chorob wirusowych i producenci spadzi-Bibliografia. (Aphids- pests of plants, vectors of virus diseases and the cause of production losses. Bibliography) *Kom. ochr. Rosl. Polskiej Akad. Nauk*, 250 pp.

Wilson, H.F. & Vickery, R.A., 1918, A species list of the Aphididae of the World and their recorded Food plants, *Trans. Wis. Acad. Sci. Arts. Lett.* 19(1): 22-355.

## ABBREVIATIONS USED

f. = forma
ms = manuscript
nom. nov. = nomen novum
q.v. = quod vide (which see) following the generic name of the accepted combination.
subg. = subgenus
subgen. = subgenus
subsp. = subspecies
var. = variety

# PART I. APHID GENERA OF THE WORLD WITH THEIR SPECIES

ABAMELEKIA del Guercio, 1906 (1905)
Redia 3: 364
Type species Abamelekia lazarewi del Guercio, 1906 = Aphis chinensis Bell, 1851
= Schlechtendalia Lichtenstein, 1883

A. lazarewi del Guercio, 1906 = Schlechtendalia chinensis (Bell, 1851)

ABSINTHAPHIS Paik, 1972
Illustrated Encyclopedia of Fauna and Flora of Korea, Seoul 13 (Insecta 5): 116 legend to figure.
Type species Absinthaphis koraiensis Paik, 1972
*Absinthaphis* Remaudière ex Stary & Remaudière, 1973
Here used as subg. of Aphis Linnaeus, 1758

A. koraiensis Paik, 1972 to Aphis (Absinthaphis)

ABSINTHAPHIS Remaudière ex Stary & Remaudière, 1973
Entomophaga 18: 288, described as subg. of Protaphis Börner, 1952
Type species Cryptosiphum cinae Nevsky, 1928
= Absinthaphis Paik, 1972

ABURA Matsumura, 1917
J. Coll. Agric. Tohoku imp. Univ. 7: 407
Type species Abura momocola Matsumura, 1917

A. momocola Matsumura, 1917 nomen dubium ?

ACANTHAPHIS del Guercio, 1908
Sulla sistema e biologia dei Fillosserini, privately published, p. 156
Purportedly reprinted from Boll. Soc. ent. ital. 38: 153-188 + 3 plates, but not published in that journal.
Type species Phylloxera corticalis Kaltenbach, 1867
= Moritziella Börner, 1908
Phylloxeridae

ACANTHAPHIS Matsumura, 1918 nec del Guercio, 1908
Trans. Sapporo nat. Hist. Soc. 7: 15
Type species Acanthaphis rubi Matsumura, 1918
= Matsumuraja Schumacher, 1921

A. rubi Matsumura, 1918 to Matsumuraja Schumacher, 1921
zeni Tseng & Tao, 1938 = Matsumuraja rubifoliae (Takahashi, 1931)

ACANTHOCALLIS Matsumura, 1917
J. Coll. Agric. Tohoku imp. Univ, 5: 368
Type species Acanthocallis quercicola Matsumura, 1917
*Arakawana* Matsumura, 1917
Here used as subg. of Tuberculatus Mordvilko, 1894

A. quercicola Matsumura, 1917 to Tuberculatus (Acanthocallis)

ACANTHOCHERMES Kollar, 1848
Sber. Akad. Wiss. Wien 1: 191
Type species Acanthochermes quercus Kollar, 1848
Phylloxeridae

ACANTHULIPES Börner, 1952
Mitt. thüring. bot. Ges. 4 (3): 132
Type species Alphitoaphis carpathica Knechtel & Manolache, 1942 = Xenomyzus corticis Aizenberg, 1935
= Xenomyzus Aizenberg, 1935

ACAUDA Shinji, 1924 lapsus pro Acaudus van der Goot, 1913

A. etadorii Shinji, 1924 lapsus pro itadori Shinji, 1924 (Acaudus)
sanguisorbae Shinji, 1924 = Macchiatiella itadori (Shinji, 1924)

ACAUDELLA Nevsky, 1929
Zool. Anz. 82: 211
Type species Acaudella puchovi Nevsky, 1929

A. puchovi Nevsky, 1929
rubida Börner, 1939 type of Pseudacaudella Börner, 1944 q.v.

ACAUDINUM Börner, 1930
Arch. klassif. phylogen. Ent. 1: 132
Type species Aphis centaureae Koch, 1854 (not preoccupied by the unavailable name Aphis centaureae Ph. F. Gmelin, 1758)

A. beheni Remaudière & Davatchi, 1959 paratypes BM HRL
centaureae (Koch, 1854) (Aphis) (nec Ph. F. Gmelin, 1758, invalid) BM HRL
*dolychosiphon* Mordvilko, 1928 (*Anuraphis*)
*scabiosae* Hille Ris Lambers, 1959
*scabiosae* Koch, 1854 (*Aphis*) in litteris synonym of centaureae, nec Aphis scabiosae Scopoli, 1763; Schrank, 1801

A. longisetosum Holman, 1970 paratypes BM HRL
scabiosae Hille Ris Lambers, 1959 = centaureae (Koch, 1854)

ACAUDUS van der Goot, 1913
Tijdschr. Ent. 56: 97
Type species Acaudus lychnidis (Linnaeus, 1758) = Aphis lychnidis Linnaeus, 1758
*Brachycaudina* Börner, 1950
*Neobrachycaudus* Narzikulov, 1965
*Prunaphis* Shaposhnikov, 1964
Here used as subg. of Brachycaudus van der Goot, 1913

A. almatinus Nevsky, 1951 to Brachycaudus (Acaudus)
bipapillata Theobald, 1923 (type of Neoacaudus Theobald, 1927) = Brachycaudus helichrysi (Kaltenbach, 1843)
calami Theobald, 1923 = Schizaphis rotundiventris (Signoret, 1860)
convolvuli Nevsky, 1951 = ? Brachycaudus (Thuleaphis) amygdalinus (Schouteden, 1905)
etadorii Shinji, 1924 (Acauda) = Macchiatiella itadori (Shinji, 1924)
jozankeanus Hori & Matsumura ex Hori, 1927 = Macchiatiella itadori (Shinji, 1924)
klugkisti Börner, 1942 to Brachycaudus (Acaudus)
rhamni Hori, 1927 = Macchiatiella itadori (Shinji, 1924)
sanguisorbae Shinji, 1924 (Acauda) = Macchiatiella itadori (Shinji, 1924)

ACONITAPHIS Ivanovskaja ex Ivanovskaja & Ostanin, 1971
In Cherepanov, A.I., New and little-known species of Siberian Fauna, Academy of Sciences, Novosibirsk 5: 14.
Type species Aconitaphis salebrosus Ivanovskaja ex Ivanovskaja & Ostanin, 1971

A. salebrosus Ivanovskaja ex Ivanovskaja & Ostanin, 1971

ACROSIPHUM Shinji, 1922 lapsus pro Tuberosiphum Shinji, 1922

ACUTICAUDA Hille Ris Lambers, 1956
Boll. Lab. Ent. agr. Filippo Silvestri 14: 292
Type species Aphis asterensis Gillette & Palmer, 1929

A. asterensis (Gillette & Palmer, 1929) (Aphis) HRL
erigerontis Leclant & Remaudière, 1967 paratypes BM HRL
solidaginifoliae (Williams, 1911) (Aphis) BM HRL

ACUTOSIPHON R.C. Basu, A.K. Ghosh & RayChaudhuri, 1970
Proc. zool. Soc. Calcutta 23: 84
Type species Acutosiphon obliquoris R.C. Basu, A.K. Ghosh & RayChaudhuri, 1970

A. obliquoris R.C. Basu, A.K. Ghosh & RayChaudhuri, 1970 HRL

ACYRTHOSIPHON Mordvilko, 1914
Faune Russie, Ins. Hém. 1 (1): 75
Type species Aphis pisi Kaltenbach, 1843 = A. pisum Harris, 1776
*Hottesina* Börner, 1950
*Lactucobium* Hille Ris Lambers, 1947
*Macchiatiella* del Guercio, 1917 nec 1909
*? Macrocaudus* Shinji, 1930
*Mirotarsus* Börner, 1939
*Tenuisiphon* Mordvilko, 1948
*Tlja* Mordvilko, 1914

A. anthyllidis Börner, 1950 = loti (Theobald, 1912)
astragali Eastop, 1971 types BM HRL
*astragali* Narzikulov, 1972
astragali (Narzikulov ex Narzikulov & Umarov, 1972 nomen nudum) Narzikulov, 1972 = astragali Eastop, 1971
auctum (Walker, 1849) (Aphis) type BM HRL
*shawi* Stroyan, 1957 type BM
*silenicola* Hille Ris Lambers, 1955 HRL
aurlandicum Heikinheimo, 1966
berkemiae Shinji, 1941 to ? Macrosiphoniella del Guercio, 1911
bidenticola Smith, 1960 paratypes BM HRL
bidentis Eastop, 1953 type BM HRL
boreale Hille Ris Lambers, 1952 BM HRL
brachysiphon Hille Ris Lambers, 1952 BM HRL
brevicorne Hille Ris Lambers, 1960 BM HRL
brevis Richards, 1963 = pseudodirhodum (Patch, 1919)
calvulum Ossiannilsson, 1958
caraganae (Cholodkovsky, 1907) (Siphonophora) BM HRL
caraganae subsp. occidentale Hille Ris Lambers, 1947 BM HRL
caraganae subsp. tadzhikistanicum Narzikulov, 1951
catharinae Nevsky, 1926 to Amphorophora Buckton, 1876
circifoliae Shinji, 1935 = Aulacorthum cirsicola (Takahashi, 1923)
cirsifoliae Shinji, 1935 vide circifoliae Shinji, 1935
codonopsis Miyazaki, 1971 = Aulacorthum (Neomyzus) taiwanum subsp. codonopsis (Miyazaki, 1971)

A. crepidis Holman & Szelegiewicz, 1975
cyatheae Holman, 1974 HRL
cyparissiae (Koch, 1855) ( Siphonophora) BM HRL
cyparissiae subsp. propinquum Mordvilko, 1914 BM HRL
cyparissiae subsp. turkestanicum Nevsky, 1929
daturae Rusanova, 1943 nomen nudum
dauricum Szelegiewicz, 1963 HRL
dubium Mordvilko, 1914 = gossypii Mordvilko, 1914
elaeocarpi Tao, 1963 to Sinomegoura Takahashi, 1960
emeljanovi Mordvilko, 1914
emeri (Hille Ris Lambers) Stroyan, 1955 nomen nudum
ericetorum Hille Ris Lambers, 1959 BM HRL
euphorbiae Börner, 1949 HRL
*thracicus* Tashev, 1962 HRL
euphorbiae subsp. neerlandicum Hille Ris Lambers, 1947 = neerlandicum Hille Ris Lambers, 1947
evodiae (Takahashi, 1929) (Macrosiphum)
fragariaevescae Nevsky, 1951
*vescae* Nevsky ex Narzikulov & Umarov, 1969
genistae Mordvilko, 1914
geranicola Hille Ris Lambers, 1935 = malvae (Mosley, 1841)
ghanii Eastop, 1971 types BM HRL
gossypicola Shinji, 1936
gossypii Mordvilko, 1914 type of Tenuisiphon Mordvilko, 1948 BM HRL
*dubium* Mordvilko, 1914
*gossypii* subsp. *paczoskii* Mordvilko, 1914
*sesbaniae* David, 1956
*skrjabini* Mordvilko, 1914
? *umarovi* Narzikulov, 1972
gossypii subsp. paczoskii Mordvilko, 1914 = gossypii Mordvilko, 1914
graminearum Mordvilko, 1919 to Sitobion Mordvilko, 1914
hissaricum Umarov, 1966
hoffmanni Takahashi, 1937 to Aulacophoroides Tao, nomen novum
ignotum Mordvilko, 1914 BM HRL
*spiraeae* Narzikulov, 1957
*spiraeae* Rupais, 1961
*spiraeae* subsp. *canescentis* Chakrabarti & RayChaudhuri, 1974 (*Acyrthosiphon* (*Metopolophium*))
*spireae* A.K. Ghosh, Chakrabarti, Chowdhuri & RayChaudhuri, 1969 (*Acyrthosiphon* (*Metopolophium*))

A. ilka Mordvilko, 1914 BM
kamtshatkanum Mordvilko, 1914
knechteli (Börner, 1950) (Metopolophium) BM HRL
kondoi Shinji, 1938 BM HRL
kuwanai Takahashi, 1933 to Aulacorthum Mordvilko, 1914
lactucae (Passerini, 1860) (Siphonophora) BM HRL
*barri* Essig, 1949 (*Macrosiphum*) paratype BM
*lactucarium* Börner, 1913 (*Macrosiphum*)
*scariolae* Nevsky, 1929
lambersi Leclant & Remaudière, 1974 to Acyrthosiphon (Xanthomyzus)
loti (Theobald, 1913) ( Macrosiphum) types BM HRL
*anthyllidis* Börner, 1950 HRL
*gracilipes* Börner, 1950 (*Metopolophium*)
macrosiphum (Wilson, 1912) (Illinoia) BM HRL
malvae (Mosley, 1841) (Aphis) BM HRL
*bosqi* Blanchard, 1932 (*Macrosiphum*)
*cornelli* Patch, 1926 (*Macrosiphum*)
*erigeroniella* Theobald, 1926 (*Myzus*) BM
*geranii* Kaltenbach, 1862 (*Aphis*)
*geranicola* Hille Ris Lambers, 1935 HRL
*pelargonii* Kaltenbach, 1843 (*Aphis*)
*sodalis* Walker, 1848 (*Aphis*) type BM
*zerozalphum* Knowlton, 1935 (*Macrosiphum*)
malvae subsp. agrimoniae (Börner, 1940) (Aulacorthum) BM HRL
malvae subsp. poterii Prior & Stroyan, 1964 BM HRL
malvae subsp. potha Börner, (1943) 1950 (Acyrthosiphon (Metopolophium))
malvae subsp. rogersii (Theobald, 1913) (Macrosiphum) BM HRL
moltshanovi Mordvilko, 1914 BM HRL
mordvilkoi Nevsky, 1928
navozovi Mordvilko, 1914
neerlandicum Hille Ris Lambers, 1947 BM HRL
*euphorbiae* subsp. *neerlandicum* Hille Ris Lambers, 1947 HRL
nigripes Hille Ris Lambers, 1935 BM HRL
*superba* Börner, 1950 (*Hottesina*)
nigripes subsp. blattnyi Pintera ex Szelegiewicz, 1968 HRL
nigripes subsp. peucedani (Bozhko, 1959) (Hottesina) HRL
*nigripes* subsp. *peucedani* Szelegiewicz, 1967 HRL
nigripes subsp. peucedani Szelegiewicz, 1967 = nigripes subsp. peucedani Bozhko, 1959
norvegicum Mordvilko, 1914

A. onobrychidis (rectification of onobrychis Boyer de Fonscolombe, 1841) (Aphis) = pisum (Harris, 1776)
onobrychis subsp. galegae Börner, 1952 = pisum (Harris, 1776)
orientale Mordvilko, 1914
pamiricum Nevsky, 1929
papaverinum Nevsky ex Pek, 1957: 176
parvum Börner, 1950 BM HRL
pedicularis Richards, 1972 to ? Metopolophium Mordvilko, 1914
pelargonii (Kaltenbach, 1843) (Aphis) = malvae (Mosley, 1841)
pentatrichopus Hille Ris Lambers, 1974 BM HRL
phaseoli Chakrabarti, A.K. Ghosh & RayChaudhuri, 1971
? phaseoli (Shinji, 1930) (Macrocaudus)
photiniae Takahashi, 1936 type of Sinomegoura Takahashi, 1960 q.v.
pisi (Kaltenbach, 1843) (Aphis) = pisum (Harris, 1776)
pisi subsp. brevicaudatum Takahashi, 1965 = pisum (Harris, 1776)
pisi subsp. turanicum Mordvilko, 1914 = pisum (Harris, 1776)
pisi subsp. ussuriense Mordvilko, 1914 = pisum (Harris, 1776)
pisum (Harris, 1776) (Aphis) BM HRL
*basalis* Walker, 1848 (*Aphis*) type BM
*corydalis* Oestlund, 1886 (*Siphonophora*)
*destructor* Johnson, 1900 (*Nectarophora*)
*lathyri* Mosley, 1841 (*Aphis*)
*onobrychis* Boyer de Fonscolombe, 1841 (*Aphis*)
*onobrychis* subsp. *galegae* Börner, 1952
*pisi* Curtis, 1860 (*Aphis*)
*pisi* Kaltenbach, 1843 (*Aphis*)
*pisi* subsp. *brevicaudatum* Takahashi, 1965
*pisi* subsp. *turanicum* Mordvilko, 1914
*pisi* subsp. *ussuriense* Mordvilko, 1914
*promedicaginis* del Guercio, 1930 (*Anuraphis (Macchiatiella)*)
*spartii* Koch, 1855 (*Siphonophora*)
*spartii* subsp. *nigricantis* Börner, 1952
*theobaldii* Davis, 1915 (*Macrosiphum*)
*trifolii* del Guercio, 1917 (*Macchiatiella*)
*trifolii* Pergande, 1904 (*Macrosiphum*)
*trifolii* Theobald, 1913 (*Macrosiphum*)
pisum subsp. ononis (Koch, 1855) (Siphonophora) BM HRL
*ononis* Ferrari, 1872 (*Siphonophora*)
porrifolii (Börner, 1950) (Aulacorthum) HRL
pseudodirhodum (Patch, 1919) (Macrosiphum) BM HRL
*brevis* Richards, 1963 HRL

A. purshiae (Palmer, 1938) (Macrosiphum) BM HRL
rhododendri Takahashi, 1937 to Sinomegoura Takahashi, 1960
rjabushinskii Mordvilko, 1914 (errore Acyrthosiphon (Microlophium) sibiricum subsp. rjabushinskiji Mordvilko, 1914, page 81. Corrected in Addenda on page 5; described as Acyrthosiphon rjabushinskii Mordvilko, 1919 page 244 = Microlophium rjabushinskii (Mordvilko, 1914)
rosaefoliae Takahashi, 1931 = Rhodobium porosum (Sanderson, 1900)
rubi (Kaltenbach, 1843) (Aphis) to Amphorophora Buckton, 1876
rubi subsp. amurense Mordvilko, 1919 = Amphorophora amurensis (Mordvilko, 1919)
rubi Narzikulov, 1957 HRL
*? rubi* subsp. *elliptici* Stroyan & Nagaich, 1964 BM
rubi subsp. elliptici Stroyan & Nagaich, 1964 = ? rubi Narzikulov, 1957
rumicis Narzikulov (1963 nomen nudum) ex Narzikulov & Umarov, 1969 HRL
salviae Nevsky, 1929 to Kakimia Hottes & Frison, 1931
scariolae Nevsky, 1929 (type of Lactucobium Hille Ris Lambers, 1947) = lactucae (Passerini, 1860)
sesbaniae David, 1956 = gossypii Mordvilko, 1914
shawi Stroyan, 1957 = auctum (Walker, 1849)
shinanonum Miyazaki, 1971 HRL
silenicola Hille Ris Lambers, 1955 = auctum (Walker, 1849)
skrjabini Mordvilko, 1914 = gossypii Mordvilko, 1914
soldatovi Mordvilko, 1914 HRL (?)
soldatovi subsp. tadzhikistanicum Narzikulov & Umarov, 1969
sophorae Narzikulov & Umarov, 1969 (Acyrthosiphon (Metopolophium))
spartii subsp. nigricantis Börner, 1952 = pisum (Harris, 1776)
spiraeae Narzikulov, 1957 = ignotum Mordvilko, 1914
spiraeae Rupais, 1961 = ignotum Mordvilko, 1914
spireaellae Umarov, 1964 = ignotum Mordvilko, 1914
svalbardicum Heikinheimo, 1968
taiheisanum (Takahashi, 1935) (Macrosiphum) type of Neoacyrthosiphon Tao, 1963 q.v.
thracicum Tashev, 1962 = euphorbiae Börner, 1940
titovi Mordvilko, 1932 nomen nudum
tutigula (Hottes, 1933) (Adactynus) BM HRL
umarovi (Narzikulov ex Narzikulov & Umarov, 1972 nomen nudum) Narzikulov, 1972 = ? gossypii Mordvilko, 1914
vandenboschi Hille Ris Lambers, 1974 HRL
vasiljevi Mordvilko, 1915
vescae Nevsky ex Narzikulov & Umarov, 1969 p. 172 = fragariaevescae Nevsky, 1951

A. wasintae (Hottes, 1933) (Adactynus) BM HRL
watanabei Miyazaki, 1971 to Aulacorthum Mordvilko, 1914

Subgen. Amphorophora Buckton, 1876

A. (Amphorophora) rubi subsp. amurense Mordvilko, 1919 = Amphorophora amurensis (Mordvilko, 1919)
(Amphorophora) rubi subsp. zhuravlevi Mordvilko, 1919 = ? Amphorophora rubi (Kaltenbach, 1843)

Subgen. Aulacorthum Mordvilko, 1914

A. (Aulacorthum) sclerodorsi Kumar & Burkhardt, 1971 to Aulacorthum Mordvilko, 1914
(Aulacorthum) sensoriatus David, Narayanan & Rajasingh, 1971 to Aulacorthum Mordvilko, 1914
(Aulacorthum) spinacaudatum Kumar & Burkhardt, 1971 to Aulacorthum Mordvilko, 1914

Subgen. Liporrhinus Borner, 1939

A. (Liporrhinus) chelidonii (Kaltenbach, 1843) (Aphis) BM HRL

Subgen. Metopolophium Mordvilko, 1914

A. (Metopolophium) arctogenicolens Richards, 1964
(Metopolophium) chandrani David & Narayanan, 1968 to Metopolophium Mordvilko, 1914
(Metopolophium) graminearum Mordvilko, 1919 to Sitobion Mordvilko, 1914
(Metopolophium) longicaudatum David & Hameed, 1975 to Metopolophium Mordvilko, 1914
(Metopolophium) potha Börner ex Franz, 1943 nomen nudum = Acyrthosiphon malvae subsp. potha Börner, 1950
(Metopolophium) simlaensis Chakrabarti & RayChaudhuri ex Chakrabarti, A.K. Ghosh & RayChaudhuri, 1914 to Metopolophium Mordvilko, 1914
(Metopolophium) sophorae Narzikulov & Umarov, 1969 to Acyrthosiphon Mordvilko, 1914
(Metopolophium) spiraeae Narzikulov, 1957 = Acyrthosiphon ignotum Mordvilko, 1914
(Metopolophium) spiraeae subsp. canescentis Chakrabarti & RayChaudhuri, 1974 = Acyrthosiphon ignotum Mordvilko, 1914
(Metopolophium) spireae A.K. Ghosh, Chakrabarti, Chowdhuri & RayChaudhuri, 1969 = Acyrthosiphon ignotum Mordvilko, 1914

Subgen. Microlophium Mordilko, 1914

A. (Microlophium) ranunculi Mordvilko, 1914 key = Aulacorthum solani (Kaltenbach, 1843)
(Microlophium) sibiricum Mordvilko, 1914 to Microlophium Mordvilko, 1914
(Microlophium) sibiricum subsp. kirgiz Mordvilko, 1914 = Microlophium sibiricum (Mordvilko, 1914)
(Microlophium) urticae subsp. meridionale Mordvilko, 1914 = Microlophium carnosum (Buckton, 1876)

Subgen. Xanthomyzus Narzikulov, 1966

A. (Xanthomyzus) glaucii (Narzikulov, 1966) (Xanthomyzus) HRL
(Xanthomyzus) lambersi Leclant & Remaudière, 1974(Acyrthosiphon) BM HRL

ADACTYNUS Rafinesque, 1818
Am. mon. Mag. Crit. Rev.3(1): 18, described as subg. of Aphis Linnaeus, 1758
Type species Aphis pteris-aquilinoides Rafinesque, 1817, invalid
Invalid

A. katonkae Hottes, 1933 to Uroleucon (Lambersius)
kiowanepus Hottes, 1933 to Macrosiphum Passerini, 1860
niwanista Hottes, 1933 to Sitobion Mordvilko, 1914
tutigula Hottes, 1933 to Acyrthosiphon Mordvilko, 1914
wasintae Hottes, 1933 to Acyrthosiphon Mordvilko, 1914

ADELGES Vallot, 1836
C. r. Acad. Sci. Paris 3: 72
Mem. Acad. Dijon 1836: 224
Type species Adelges laricis Vallot, 1836
Adelgidae

AEDISIPHUM lapsus pro Oedisiphum van der Goot, 1917

AFGHANAPHIS Takahashi, 1966
Results Kyoto Univ. Sc. Exp. Karakorum & Hindukush 8: 267
Type species Afghanaphis ulmi Takahashi, 1966 = Schizoneura lanuginosa Hartig, 1839
= Schizoneura Hartig, 1839

A. ulmi Takahashi, 1966 = Eriosoma (Schizoneura) lanuginosum (Hartig, 1839)

AGRIOAPHIS Walker, 1870
Zoologist (2) 5: 2000
Type species Aphis myricae Kaltenbach, 1843
= Myzocallis Passerini, 1860

A. colyricola Shinji, 1935 = Pterocallis (Mesocallis) pteleae (Matsumura, 1919)
hashibamii Shinji, 1935 = Pterocallis (Mesocallis) pteleae (Matsumura, 1919)
moriokae Shinji, 1935 = Pterocallis (Recticallis) alnijaponicae (Matsumura, 1919)
sasacola Shinji, 1935 = Takecallis sasae (Matsumura, 1917)

AICEONA Takahashi, 1921
Spec. Rep. Formosa agric. Exp. Stn 20: 85
Type species Aiceona actinodaphnis Takahashi, 1921

A. actinodaphni Takahashi, 1921; page 85 emended to actinodaphis in 'Additions and Corrections' BM
japonica Takahashi, 1960 BM HRL
litseae A.N. Basu & Hille Ris Lambers, 1968 = titabarensis (RayChaudhuri & A.K. Ghosh, 1964)
longisetosa M.R. Ghosh & RayChaudhuri, 1973 = pseudosugii David, Sekhon & Bindra, 1970
malayana Takahashi, 1960 paratypes BM HRL
osugii Takahashi, 1924
pallida A.K. Ghosh & RayChaudhuri, 1973 HRL
paraosugii A.K. Ghosh, M.R. Ghosh & RayChaudhuri, 1971
pseudosugii David, Sekhon & Bindra, 1970 HRL
*longisetosa* M.R. Ghosh & RayChaudhuri, 1973
retipennis David, Rajasingh & Narayanan, 1970 HRL
robustiseta M.R. Ghosh & RayChaudhuri, 1973 HRL
siamensis Takahashi, 1941
tanakai Takahashi, 1963
titabarensis (RayChaudhuri & A.K. Ghosh, 1964) (Lachnus) BM HRL
*litseae* A.N. Basu & Hille Ris Lambers, 1968 BM HRL

AIXAPHIS Heie, 1970
Entomologica scand. 1: 115
Type species Tetraneura oligocenica N. Theobald. 1937
Fossil. Oligocene

AKKAIA Takahashi, 1919
Insect Wld 23: 439
Type species Akkaia polygoni Takahashi, 1919

A. bengalensis A.N. Basu, 1968 HRL
kagoshimana Suenaga, 1934 = taiwana Takahashi, 1935
neopolygoni M.R. Ghosh & RayChaudhuri, 1971 HRL
odaiensis Takahashi, 1961
polygoni Takahashi, 1919 BM HRL
*enkianthae* Shinji, 1924 (*Cavariella*)
*enkianthai* Shinji, 1923 (*Phorodon*)
taiwana Takahashi, 1933 BM HRL
*kagoshimana* Suenaga, 1934
umefoliae Shinji, 1924 = Myzus cerasi subsp. umefoliae (Shinji, 1924)

ALEUROCANTHUS (Aleyrodidae)

A. palmae Ghesquière, 1934 = to Cerataphis Lichtenstein, 1882

ALEURODAPHIS van der Goot, 1917
Contrib. Faune Indes Néerl. 1 (3): 239
Type species Aleurodaphis blumeae van der Goot, 1917

A. asteris Takahashi & Sorin, 1958 BM HRL
blumeae van der Goot, 1917 HRL
*japonica* Takahashi, 1923 (*Astegopteryx*)
*nobukii* Shinji, 1923
mikaniae Takahashi, 1925
nobukii Shinji, 1923 = blumeae van der Goot, 1917

ALEUROSIPHON Takahashi, 1966
Trans. Am. ent. Soc. 92: 527, described as subg. of Aphis Linneaus, 1758
Type species Aphis smilacifoliae Takahashi, 1921

A. smilacifoliae (Takahashi, 1921) (Aphis) BM HRL

ALHAMBRA Gomez-Menor, 1958
Boln. E. Soc. esp. Hist. nat. (Biol.) 55: 404
Type species Brachyunguis carthami Das, 1918
= Protaphis Börner, 1952

ALLAPHIS Mordvilko, 1921
Izv. sev. oblast. Sta. Zashch. Rast. Vredit. 3: 57
Type species Callaphis caricicola Mordvilko (1908 nomen nudum) 1914
= Thripsaphis Gillette, 1917

A. caricis Mordvilko, 1921 (type of Trichocallis Börner, 1930) to Thripsaphis (Trichocallis)
caricis subsp. amurensis Mordvilko, 1921 to Thripsaphis (Trichocallis)

ALLARCTAPHIS Börner, 1949
Beitr. tax. Zool. 1: 54
Type species Chaitophorus nassonowi Mordvilko, 1895
= Chaitophorus Koch, 1854

ALLOAMBRIA Richards, 1966
Can. Ent. 98: 756
Type species Alloambria caudata Richards, 1966
Fossil, cretaceous Canadian amber

ALLOCOTAPHIS Börner, 1950
Neue europäische Blattlausarten, privately published, p. 4
Type species Neanuraphis quaestionis Börner, 1942
*Malaphis* Shaposhnikov, 1951

A. quaestionis (Börner, 1942) (Neanuraphis) BM HRL
*magna* Shaposhnikov, 1951 (*Malaphis*) HRL

ALLOMYZUS Takahashi, 1965
Insecta matsum. 28: 31
Without species

ALLOTHORACAPHIS Takahashi, 1958
Insecta matsum. 22: 8
Type species Thoracaphis piyananensis Takahashi, 1935

A. piyananensis (Takahashi, 1935) (Thoracaphis) BM HRL

ALLOTRICHOSIPHUM Takahashi, 1962
Trans. Shikoku ent. Soc. 7: 70
Type species Trichosiphum kashicola Kurisaki, 1920

A. assamense RayChaudhuri, M.R. Ghosh, M. Banerjee & A.K. Ghosh, 1973
kashicola (Kurisaki, 1920) (Trichosiphum) paratype HRL
*kyushuensis* Tao, 1960 (*Paratrichosiphum*)

ALOEPHAGUS Essig, 1950
Pan-Pacif. Ent. 26: 22
Type species Aloephagus myersi Essig, 1950

A. myersi Essig, 1950 paratype BM HRL

ALPHITOAPHIS Hottes, 1926
Proc. biol. Soc. Wash. 39: 116
Type species Aphis lonicericola Williams, 1911

A. carpathica Knechtel & Manolache, 1941 (type of Acanthulipes Börner, 1952) =
Trichosiphonaphis (Xenomyzus) corticis (Aizenberg, 1935)
lonicericola (Williams, 1911) (Aphis) BM HRL

AMALANCON Scudder, 1890
U. S. Geol. Survey Territories 13: 270
Type species Amalancon lutosus Scudder, 1890 = Siphonophoroides antiqua Buckton, 1883
= Siphonophoroides Buckton, 1883, fossil

AMBARAPHIS Richards, 1966
Can. Ent. 98: 752
Type species Ambaraphis costalis Richards, 1966
Fossil, Cretaceous Canadian amber

AMEGOSIPHON Narzikulov, 1958
Trudȳ Akad. Nauk. tadzh. SSR 89: 21
Type species Rhopalosiphoninus (?) platicaudus Narzikulov, 1953
*Elbourzaphis* Remaudière & Davatchi, 1959

A. platicaudum (Narzikulov, 1953) (Rhopalosiphoninus (?)) BM HRL
*behboudii* Remaudière & Davatchi, 1959 (*Elbourzaphis*) paratypes BM HRL

AMELANCHIERIA Shaposhnikov, 1950
Ént. Obozr. 31: 224, described as subg. of Nearctaphis Shaposhnikov, 1950
Type species Aphis sensoriata Gillette & Bragg, 1918
= Nearctaphis Shaposhnikov, 1950

AMMIAPHIS Börner, (1944) 1952
In Brohmer, P., Fauna von Deutschland, ed. 5, Leipzig, p. 216, in key, without species (1944)

Mitt. thüring. bot. Ges. 4 (3): 115 (1952)
Type species Aphis sii Koch, 1855

A. sii (Koch, 1855) (Aphis) BM HRL

AMPHICERCIDUS Oestlund, 1922
Rep. Minn. St. Ent. 19: 126
Type species Aphis pulverulens Gillette, 1911
*Melanosiphum* Shinji, 1942

A. flocculosus (Gillette & Palmer, 1929) (Anuraphis) BM HRL
indicus Hille Ris Lambers & A.N. Basu, 1966 BM HRL
japonicus (Hori, 1927) (Anuraphis) BM HRL
*lonicericola* Shinji, 1942 (*Melanosiphum*)
laniger (Takahashi, 1927) (Anuraphis) BM ?
maxsoni Palmer, 1936 to Cedoaphis Oestlund, 1922
pulverulens (Gillette, 1911) (Aphis) BM HRL
tuberculatus David, Narayanan & Rajasingh, 1971 HRL

AMPHORINOPHORA MacGillivray, 1958
Temminckia 10: 26, 36, described as subg. of Masonaphis Hille Ris Lambers, 1939
Type species Amphorophora crystleae Smith & Knowlton, 1939
Here used as subg. of Illinoia Wilson, 1910

AMPHOROPHORA Buckton, 1876
Monograph of the British Aphides, London 1: 187
Type species Amphorophora ampullata Buckton, 1876
*Eunectarosiphon* del Quercio, 1913
*Rhopalosiphum* van der Goot, 1913 nec Koch, 1854

A. accidentalis Knowlton, 1929 to ? Hyperomyzus (Neonasonovia)
agathonica Hottes, 1950 paratypes BM HRL
alleni Knowlton & Fronk, 1942 = urtica Essig, 1942
alni Mason, 1925 to Illinoia Wilson, 1910
ampullata Buckton, 1876 (type of Rhopalosiphum van der Goot, 1913 nec Koch, 1854) types BM HRL
*? dryopteridis* Matsumura, 1918 (Megoura)
*? shidae* Shinji, 1922
ampullata subsp. bengalensis Hille Ris Lambers & Basu, 1966 BM HRL
ampullata subsp. laingi Mason, 1925 BM HRL

A. amurensis (Mordvilko, 1919) BM HRL
*rubi* subsp. *amurense* Mordvilko, 1919 (*Acyrthosiphon*) and (*Acyrthosiphon (Amphorophora)*)
*rubiphaga* Takahashi, 1961
aridus Knowlton, 1929 = Illinoia grindeliae (Williams, 1911)
arnicae Glendenning, 1926 = Illinoia (Oestlundia) davidsoni (Mason, 1925)
arnicae subsp. thatcheri Knowlton & Allen, 1937 = Illinoia (Oestlundia) davidsoni (Mason, 1925)
azaleae Mason, 1925 (type of Ericobium MacGillivray, 1958) to Illinoia Wilson, 1910
bartholomewi Essig, 1942 = Illinoia (Amphorinophora) crystleae subsp. bartholomewi (Essig, 1942)
bonnevilla Knowlton & Allen, 1937 to ? Macrosiphum Passerini, 1860
borealis Mason, 1925 to Illinoia Wilson, 1910
bournevilla Knowlton & Allen, lapsus pro bonnevilla Knowlton & Allen, 1937
braggi Mason, 1925 = ? Hyperomyzus (Neonasonovia) nabali (Oestlund, 1886)
brevitarsis Gillette & Palmer, 1933 to Illinoia Wilson, 1910
catharinae (Nevsky, 1928) (Acyrthosiphon) HRL
ceanothi Bartholomew, 1932 to Illinoia Wilson, 1910
cicutae Shinji, 1917 = Wahlgreniella nervata (Gillette, 1908)
commelinensis Smith, 1960 to Utamphorophora Knowlton, 1947
cosmopolitanus Mason, 1925 = Hyperomyzus lactucae (Linnaeus, 1758)
crataegi Tissot, 1932 = Utamphorophora crataegi (Monell, 1889)
cryptopteris Matsumura, lapsus pro dryopteridis (Matsumura, 1918) (Megoura)
crystleae Smith & Knowlton, 1939 (type of Amphorinophora MacGillivray, 1958) to Illinoia (Amphorinophora)
davidsoni Mason, 1925 to Illinoia (Oestlundia)
digitalisii Theobald, 1928 = rubi (Kaltenbach, 1843)
essigwanai Mason, 1925 = Indomegoura indica (van der Goot, 1916)
evansii Theobald, 1923 = Microlophium carnosum (Buckton, 1876)
filipendulae Miyazaki, 1971 BM HRL
forbesi Richards, 1959 paratypes BM HRL
formosana Takahashi, 1923 type of Aulacophora Tao, 1963; and of Aulacophoroides Tao, nomen novum q.v.
fronki Knowlton, 1945 = Hyperomyzus (Neonasonovia) nabali subsp. fronki (Knowlton, 1945)
gei (Börner, 1939) (Nectarosiphon) BM HRL
*franzi* Börner, 1942 (*Hyperomyzus*)
geranii Gillette & Palmer, 1929 paratypes BM HRL
goldamaryae Knowlton, 1938 to Illinoia Wilson, 1910

A. halli Knowlton, 1927 = Wahlgreniella nervata (Gillette, 1908)
hayhursti Mason, 1925 = Hyperomyzus (Neonasonovia) nabali (Oestlund, 1886)
henryi Balachowsky & Cairaschi, 1941 = Wahlgreniella nervata subsp. arbuti (Davidson, 1910)
hieracioides Theobald, 1926 = Hyperomyzus (Neonasonovia) picridis (Börner & Blunck, 1916)
howardii Wilson, 1911 to Glabromyzus Richards, 1960
ichigo Shinji, 1922 to ? Rhopalosiphoninus Baker, 1920
janesi Knowlton, 1938 = Wahlgreniella nervata (Gillette, 1908)
kalmiaflora Tissot & Pepper, 1944 = Illinoia azaleae subsp. kalmiaflora (Tissot & Pepper, 1944)
katoi Takahashi, 1925 to Micromyzus van der Goot, 1917
katsurae Shinji, 1930 = Aulacorthum cercidiphylli (Matsumura, 1918)
kesocqua Hottes, 1950
lactucae (Linnaeus, 1758) (Aphis) to Hyperomyzus Börner, 1933
laingi Mason, 1925 = ampullata subsp. laingi Mason, 1925
lathyri Shinji, 1921 = Megoura crassicauda Mordvilko, 1919
latysiphon Davidson, 1912 type of Rhopalosiphoninus Baker, 1920 q.v.
ledi Wahlgren, 1938 (type of Testataphis Börner, 1952) to Neoamphorophora Mason, 1924
lilicola Shinji, 1933 = Indomegoura indica (van der Goot, 1916)
lonicericola Takahashi, 1921 type of Neorhopalomyzus Tao, 1963 q.v.
malvicola Shinji, 1933 = Aulacorthum magnoliae (Essig & Kuwana, 1918)
maxima Mason, 1925 to Illinoia (Oestlundia)
minima Mason, 1925 = Capitophorus hippophaes (Walker, 1852)
mitchelli Mason, 1925 = Hyperomyzus (Neonasonovia) nabali (Oestlund, 1886)
nebularis Hottes & Frison, 1931 lapsus pro nebulosa Hottes & Frison, 1931
nebulosa Hottes & Frison, 1931 = Glabromyzus howardii (Wilson, 1911)
nigra J.M. Baker, 1934 to Hyperomyzus (Neonasonovia)
nigricornis Knowlton, 1927 to Hyperomyzus (Neonasonovia)
occidentalis Essig, 1942 = Macrosiphum osmaroniae (Wilson, 1912)
osborni Knowlton, 1942 = Hyperomyzus (Neonasonovia) ribiellus (Davis, 1919)
pacifica Hill, 1968 paratypes BM HRL
pallida Mason, 1925 to Illinoia Wilson, 1910
parviflori Hill, 1958 paratypes BM HRL
patchiae Essig, 1942 = Illinoia goldamaryae (Knowlton, 1938)
pawtincae Hottes, 1934 BM HRL
pergandei Mason, 1925 = Hyperomyzus (Neonasonovia) nabali (Oestlund, 1886)
peruviana Essig, 1953
petiolaris Knowlton & Allen, 1935 to Hyperomyzus (Neonasonovia)

A. phacelia Essig, 1942 to Illinoia Wilson, 1910
phyllanthi Takahashi, 1937 to Sitobion Mordvilko, 1914
reticulata Mason, 1925 to Illinoia Wilson, 1910
rhododendronia Mason, 1925 = Illinoia azaleae subsp. rhododendronia (Mason, 1925)
rhokalaza Tissot & Pepper, 1944 to Illinoia (Masonaphis)
rossi Hottes & Frison, 1931 paratypes BM HRL
rubi (Kaltenbach, 1843) (Aphis) BM HRL
*conjuncta* Walker, 1848 (*Aphis*)
*cynoglossi* Walker, 1848 (*Aphis*)
*digitalisii* Theobald, 1928
*fragariella* Theobald, 1905 (*Siphonophora*)
*rubi* subsp. *zhuravlevi* Mordvilko, 1919 (*Acyrthosiphon (Amphorophora)*)
rubi subsp. idaei (Börner, 1939) (Nectarosiphon) BM HRL
*idaei* Börner, 1939 (*Nectarosiphon*)
rubicumberlandi Knowlton & Allen, 1937 paratypes BM HRL
rubifragariella Theobald lapsus pro Siphonophora fragariella Theobald, 1905
rubiphaga Takahashi, 1961 = amurensis Mordvilko, 1919
rubitoxica Knowlton, 1934 BM HRL
scabripes Miyazaki, 1968 BM HRL
sensoriata Mason, 1923 BM HRL
shidae Shinji, 1922 = ampullata Buckton, 1876
singularis Hottes & Frison, 1931 to Utamphorophora Knowlton, 1947
sonchicola Shinji, 1941 = Hyperomyzus lactucae (Linnaeus, 1758)
sonchifoliae Takahashi, 1923 = Hyperomyzus carduellinus (Theobald, 1915)
stachydis Heikinheimo, 1955 (Amphorophora (Ampullosiphon)) to Cryptomyzus (Ampullosiphon)
stachyophila Hille Ris Lambers, 1966 types BM HRL
stellariae Strand, 1925 to Myzus (Nectarosiphon)
stolonis Robinson, 1974 paratype BM HRL
subterrans Wilson, 1915 to Rhopalosiphoninus Baker, 1920
takahashii Mason, 1925 to Aulacorthum Mordvilko, 1914
tigwatensa Hottes, 1933 paratypes BM HRL
tolmiea Essig, 1942 to Macrosiphum Passerini, 1860
triticum Theobald, 1923 = Hyperomyzus lactucae (Linnaeus, 1758)
tuberculaceps Essig, 1942 to Macrosiphum Passerini, 1860
urtica Essig, 1942 paratypes BM HRL
*alleni* Knowlton & Fronk, 1942
utahensis Knowlton & Allen, 1945 = Illinoia (Oestlundia) davidsoni (Mason, 1925)

vaccinii Mason, 1925 = Illinoia azaleae (Mason, 1925)
vagans (van der Goot, 1917) (Rhopalosiphum)
viburni Takahashi, 1925 to Wahlgreniella Hille Ris Lambers, 1949
vicicola Shinji, 1941 = Megoura crassicauda Mordvilko, 1919
wahnaga Hottes, 1952 to Illinoia Wilson, 1910
wikstroemiae Mamet, 1939 to Sitobion Mordvilko, 1914
yomenae Shinji, 1931, ? lapsus pro Macrosiphum yomenae Shinji, 1922

Subgen. Ampullosiphon Heikinheimo, 1955

A. (Ampullosiphon) stachydis Heikinheimo, 1955 type of Ampullosiphon Heikinheimo, 1955 to Cryptomyzus (Ampullosiphon)

Subgen. Galiaphis Ossiannilsson, 1954

A. (Galiaphis) annae Ossiannilsson, 1954 BM HRL
(Galiaphis) cryptotaeniae Takahashi, 1965 HRL
(Galiaphis) japonica Takahashi, 1965 HRL

AMPHOROSIPHON Hille Ris Lambers, 1949
Temminckia 8: 242, described as subg. of Amphorophora Buckton, 1876
Type species Delphiniobium pulmonariae Börner, 1942

A. pulmonariae (Börner, 1942) (Delphiniobium) BM HRL

AMPULLOSIPHON Heikinheimo, 1955
Suom. hyönt. Aikak. 21: 5, described as subg. of Amphorophora Buckton, 1876
Type species Amphorophora (Amphorosiphon) stachydis Heikinheimo, 1955
Here used as subg. of Cryptomyzus Oestlund, 1922

AMRHOPALOSIPHUNICUS Shinji, 1933
Kontyû 7: 270
Type species Rhopalosiphum tiliae Matsumura, 1918
= Rhopalosiphoninus Baker, 1920

AMYCLA Koch, 1857
Die Pflanzenläuse Aphiden, Nürnberg, p. 301
Type species Amycla fuscifrons Koch, 1857 = Aphis ulmi Linnaeus, 1758
= Tetraneura Hartig, 1841

A. albicornis Koch, 1857 = Smynthurodes betae Westwood, 1849
fuscicornis Koch, 1857 to Pemphigus Hartig, 1839
fuscifrons Koch, 1857 = Tetraneura ulmi (Linnaeus, 1758)

AMYZUS Hille Ris Lambers, 1946
In Alta, H. & Docters van Leeuwen, W.M., Gallenboek, Amsterdam, p. 156
Type species Amyzus ligustri (Mosley, 1841) = Aphis ligustri Mosley, 1841
= Nectarosiphon Schouteden, 1901

ANACAUDUS lapsus pro Anocaudus A.K. Ghosh, Chakrabarti, Chowdhuri & RayChaudhuri, 1969

ANAMESON Mordvilko, (1914) 1919
Faune Russie, Ins. Hém. 1 (1): 63 (in key, without species); 1 (2): 336 (1919)
Type species Anameson kamtshaticum Mordvilko, 1919
= Sitobion Mordvilko, 1914

A. kamtshaticum Mordvilko, 1919 to Sitobion Mordvilko, 1914

ANAULACORTHUM A.K. Ghosh & RayChaudhuri, 1973 (1972)
Proc. zool. Soc. Calcutta 25: 93, described as subg. of Aulacorthum Mordvilko, 1914
Type species Aulacorthum (Anaulacorthum) fagopyri A.K. Ghosh & RayChaudhuri, 1973

A. fagopyri A.K. Ghosh & RayChaudhuri, 1972 (Aulacorthum (Anaulacorthum)) HRL

ANCONATUS Buckton, 1883
Monograph of the British Aphides, London 4: 177
Type species Anconatus dorsuosus Buckton, 1883
Fossil, Miocene

ANIFERELLA Richards, 1966
Can. Ent. 98: 759
Type species Aniferella bostoni Richards, 1966
Fossil, Cretaceous Canadian amber

ANIRAPHIS Mordvilko, 1933 lapsus pro Anuraphis del Guercio, 1907

ANISOPHLEBA Koch, 1857
Die Pflanzenläuse Aphiden, Nürnberg, p. 320
Type species Anisophleba hamadryas Koch, 1857 = Adelges laricis Vallot, 1836
= Adelges Vallot, 1836
Adelgidae

ANNAJA Börner, 1952
Mitt. thüring. bot. Ges. 4 (3): 101, described as subg. of Yezabura Matsumura, 1917
Type species Yezabura brancoi Börner, 1950
= Dysaphis Börner, 1931

AMNRAPHIS Mimeur, 1936 lapsus pro Anuraphis del Guercio, 1907

ANOCAUDUS A.K. Ghosh, Chakrabarti, Chowdhuri & RayChaudhuri, 1969
Orient. Insects 3: 328
Type species Anocaudus taxus A.K. Ghosh, Chakrabarti, Chowdhuri & RayChaudhuri, 1969
= Prociphilus Koch, 1857

A. taxus A.K. Ghosh, Chakrabarti, Chowdhuri & RayChaudhuri, 1969 to Prociphilus Koch, 1857

ANOCHETIUM Wood-Baker, 1943
Proc. R. Ir. Acad. (B) 49: 140
Type species Anochetium nondescriptum Wood-Baker, 1943 = Laingia psammae Theobald, 1922
= Laingia Theobald, 1922

A. nondescriptum Wood-Baker, 1943 = Laingia psammae Theobald, 1922

ANOECIA Koch, 1857
Die Pflanzenläuse Aphiden, Nürnberg, p. 275
Type species Aphis corni Fabricius, 1775
*Neanoecia* Börner, 1950
*Subanoecia* Börner, 1950

A. agrostidis Börner, 1950 = corni (Fabricius, 1775)
caricis Pergande, (vide Mordvilko, 1931: 877)

A. corni (Fabricius, 1775) (Aphis) BM HRL
*agrostidis* Börner, 1950
*corni* Hartig, 1841 (*Schizoneura*)
*disculigera* Börner, 1950
*graminis* del Guercio, 1895 (*Schizoneura*)
*obscura* Walker, 1852 (*Schizoneura*) BM
cornicola (Walsh, 1863) (Eriosoma) BM
*eleusinis* Thomas, 1878 (*Rhizobius*)
*fungicola* Walsh, 1863 (*Eriosoma ?*)
*panicola* Thomas, 1879 (*Schizoneura*)
*querci* auct. nec Fitch, 1859 (*Eriosoma*)
cornimaris Bozhko, 1957
disculigera Börner, 1950 = corni (Fabricius, 1775) BM
fulviabdominalis (Sasaki, 1899) (Schizoneura) BM HRL
furcata (Theobald, 1913) (Forda) BM HRL
graminis Gillette & Palmer, 1924 BM HRL
haupti Börner, 1950 HRL
karatanei Sasaki, 1936 to Nipponaphis Pergande, 1906
krizusi (Börner, 1950) (Neanoecia)
major Börner, 1950 BM HRL
mimeuri Börner, 1950
mirae Narzikulov, 1968
nemoralis Börner, 1950 BM HRL
oenotherae Wilson, 1911 BM HRL
onigurumii Shinji, 1923 (type of Tuberocorpus Shinji, 1929) to Kurisakia Takahashi, 1924
? panici (Thomas, 1878) (Tychea)
piri Essig & Kuwana, 1918 = Nippolachnus piri Matsumura, 1917
pskovica Mordvilko, 1916 (type of Paranoecia Zwölfer, 1957) to Anoecia (Paranoecia)
querci (Fitch, 1859) (Eriosoma) to Stegophylla Oestlund, 1922 but the name has often been applied to A. cornicola (Walsh, 1863)
setariae Gillette & Palmer, 1924 BM HRL
stipae Mamontova, 1968 (Anoecia (Neanoecia))
vagans (Koch, 1856) (Schizoneura) BM HRL

*cerealium* Szanislo, 1880/1 (*Schizoneura*)
*kochii* Lichtenstein, 1885 (*Schizoneura*)
*? radicicola* Mokrzecky, 1903 (*Schizoneura*)
*rossica* Cholodkovsky, 1897 (*Colopha*)
*venusta* Passerini, 1860 (*Schizoneura*)
*viridis* Börner & Blunck, 1916

viridis Börner & Blunck, 1916 = vagans (Koch, 1856)
willcocksi Theobald, 1915 BM HRL
zirnitsi Mordvilko, 1931 HRL

Subgen. Neanoecia Börner, 1950

A. (Neanoecia) stipae Mamontova, 1968 to Anoecia Koch, 1857

Subgen. Paranoecia Zwölfer, 1957

A. (Paranoecia) pskovica Mordvilko, 1916 (Anoecia) BM HRL

ANOMALAPHIS Baker, 1920
Bull. U.S. Dept. Agric. 826: 32
Type species Anomalaphis comperei Pergande ex Baker, 1920

A. casimiri Carver, 1971 paratype BM HRL
comperei Pergande ex Baker, 1920 BM HRL

ANOMALOSIPHUM Takahashi, 1934
Stylops 3: 54
Type species Anomalosiphum pithecolobii Takahashi, 1934

A. indigoferae A.K. Ghosh, M.R. Ghosh & RayChaudhuri, 1971
pithecolobii Takahashi, 1934 BM
takahashii Tao, 1947

ANTALUS Adams, 1965
Proc. R. ent. Soc. Lond. (B) 34: 83
Type species Antalus albatus Adams, 1965

A. aeschynomenidis van Harten & Ilharco, 1972 HRL
albatus Adams, 1965 BM HRL
humulariae van Harten & Ilharco, 1971 BM HRL

ANTHEMIDAPHIS Tashev, 1967
Dokl. bulg. Akad. Nauk. 20(10): 1069
Type species Anthemidaphis oligommata Tashev, 1967

A. oligommata Tashev, 1967 HRL

ANTHRACOSIPHON Hille Ris Lambers, 1947
Temminckia 7: 195
Type species Anthracosiphon hertae Hille Ris Lambers, 1947

A. galii Mamontova, 1961 to Linosiphon Börner, 1950
hertae Hille Ris Lambers, 1947 cotypes BM HRL

ANTHRACOSIPHONIELLA A.N. Basu, 1969
Orient. Insects 3: 169
Type species Anthracosiphoniella maculatum A.N. Basu, 1969

A. maculata A.N. Basu, 1969 HRL

ANTIQUAPHIS Heie, 1967
Spol. zool. Mus. haun. 26: 88
Type species Antiquaphis robusta Heie, 1967
= Schizoneurites Cockerell, 1915
Fossil

ANURAPHIS del Guercio, 1907
Redia 4: 190
Type species Aphis pyri Koch, 1854 nec Boyer de Fonscolombe, 1841 = Aphis farfarae Koch, 1854

A. abrotaniella Theobald, 1919 = Brachycaudus helichrysi (Kaltenbach, 1843)
aconiti Mordvilko, 1928 to Brachycaudus (Acaudus)
aegyptiaca Hall, 1926 to Brachycaudus (Thuleaphis)
aisenstadti Rusanova, 1943 nomen nudum
aleshevi Rusanova, 1943 nomen nudum
ammobii Hori, 1929 = Brachycaudus helichrysi (Kaltenbach, 1843)
apiifolia Theobald, 1923 to Dysaphis Börner, 1931
appelii Börner, 1925 = Dysaphis (Pomaphis) aucupariae (Buckton, 1879)
artemirhizus Shinji, 1924 = Sappaphis piri Matsumura, 1918
artemisiae Takahashi, 1923 type of Micraphis Takahashi, 1931 q.v.
arundinariae Tissot, 1933 to Rhopalosiphum Koch, 1854
brevisiphon del Guercio, 1930 = Brachycaudus helichrysi (Kaltenbach, 1843) (fundatrix)
cantauriella Theobald lapsus pro centauriella
capparidis Nevsky, 1929
cardui (Linnaeus, 1758) (Aphis) to Brachycaudus (Acaudus)
cardui Shinji, 1941 = Brachycaudus (Acaudus) cardui (Linnaeus, 1758)
cardui subsp. turanica Mordvilko ex Nevsky, 1929 = Brachycaudus (Acaudus) cardui (Linnaeus, 1758)

A. cardui var. inconstans Rusanova, 1943 nomen nudum
catonii Hille Ris Lambers, 1935 BM HRL
centauriella Theobald, 1921 = Brachycaudus helichrysi (Kaltenbach, 1843)
cerasicola Mordvilko ex Nevsky, 1929: 722 to ? Brachycaudus (Acaudus)
cerasicola var. egregia Rusanova, 1943 nomen nudum
chrysothamnicola Gillette & Palmer, 1929 to Aphis (Zyxaphis)
chrysothamnicola var. infrequens Knowlton, 1929 = Aphis infrequens (Knowlton, 1929)
cinerariae Theobald, 1923 = Brachycaudus helichrysi (Kaltenbach, 1843)
controversus Rusanova, 1943 nomen nudum
cortusae Nevsky, 1929
crathaegi del Guercio, 1930 errore pro Aphis crataegi Kaltenbach, 1843
crathaegiphila del Guercio, 1930 to Dysaphis Börner, 1931
crepii Rusanova, 1943 nomen nudum
cyani Theobald, 1923 = Brachycaudus helichrysi (Kaltenbach, 1843)
cynariella Theobald, 1924 = Aphis fabae Scopoli, 1763
debilicornis Gillette & Palmer, 1929 to Aphis Linnaeus, 1758
distincta Mordvilko, 1928 to Brachycaudus van der Goot, 1913
dolychosiphon Mordvilko, 1928 = Acaudinum centaureae (Koch, 1854)
elaeagni Rusanova, 1943 nomen nudum
elpatjevskyi Rusanova, 1943 nomen nudum
emicis Mimeur, 1935 to Dysaphis Börner, 1931
ephedrae Nevsky, 1929 type of Ephedraphis Hille Ris Lambers, 1959 q.v.
erratica del Guercio, 1917 = Aphis citricola van der Goot, 1912
farfarae (Koch, 1854) (Aphis) BM HRL
*kochi* del Guercio, 1930
*kochi* Schouteden, 1903 (*Aphis*)
*pyri* Koch, 1854 (*Aphis*) nec Boyer de Fonscolombe, 1841; Hartig, 1841
*pyriella* Theobald, 1929 types BM
*? vacillans* Walker, 1849 (*Aphis*)
farfarae subsp. dianae Shaposhnikov, 1974
fasciata del Guercio, 1920 = Brachycaudus helichrysi (Kaltenbach, 1843)
ferganica Mordvilko ex Nevsky, 1929 nomen nudum
filaginea del Guercio, 1911 to Aphis (Protaphis)
filaginis var. anthemidis del Guercio, 1930 = Brachycaudus helichrysi (Kaltenbach, 1843)
filifoliae Gillette & Palmer, 1928 to Aphis (Zyxaphis)
flavescens del Guercio, 1930 = Brachycaudus helichrysi (Kaltenbach, 1843)
flavicephala del Guercio, 1930 = Brachycaudus (Acaudus) cardui (Linnaeus, 1758)

A. flocculosa Gillette & Palmer, 1929 to Amphicercidus Oestlund, 1922
floris Monzen, 1934
foeniculus Theobald, 1923 to Dysaphis Borner, 1931
formosana Takahashi, 1927 to Aphis (Protaphis)
glaucifolia Theobald, 1923 = Brachycaudus helichrysi (Kaltenbach, 1843)
gutierrezis Pack & Knowlton, 1929 to Aphis Linnaeus, 1758
helichrysi (Kaltenbach, 1843) (Aphis) to Brachycaudus van der Goot, 1913
helichrysi subsp. asiatica Nevsky, 1929 = Brachycaudus (Thuleaphis) amygdalinus (Schouteden, 1905)
helichrysi subsp. asiatica var. nigriventris Nevsky, 1929 = Brachycaudus helichrysi (Kaltenbach, 1843)
heliotropii Rusanova, 1943 nomen nudum
ilexis Moritsu, 1958 = Ryoichitakahashia prunifoltae (Shinji, 1924)
insititiella del Guercio, 1930 = Brachycaudus helichrysi (Kaltenbach, 1843)
iteae Tissot, 1932 to Aphis Linnaeus, 1758
japonica Hori, 1927 to Amphicercidus Oestlund, 1922
katsurae Shinji, vide Shiraki, 1952: 63
kochi del Guercio, 1930 (perhaps reference to Aphis kochi Schouteden, 1905) = Anuraphis farfarae (Koch, 1854)
lanigera Takahashi, 1927 to Amphicercidus Oestlund, 1922
lithospermi Rusanova, 1943 nomen nudum
longicauda Baker, 1920 to Aphis Linnaeus, 1758
madronae Essig, 1926 to Aphis Linnaeus, 1758
magnituberculata Rusanova, 1943 nomen nudum
maidi-radicis (Forbes, 1891) (Aphis) to Aphis Linnaeus, 1758
malibejli Rusanova, 1943 nomen nudum
masseei Theobald, 1927 = Brachycaudus (Acaudus) persicae (Passerini, 1860)
melampyri del Guercio, 1911 = Brachycaudus (Acaudus) lychnidis group
menthae del Guercio, 1930 = Aphis affinis del Guercio, 1911
menthaecola Nevsky, 1929 = ? Brachycaudus helichrysi (Kaltenbach, 1843)
minima Tissot, 1932 to Aphis Linnaeus, 1758
minutissima Gillette & Palmer, 1928 to Aphis (Zyxaphis)
multituberculata Rusanova, 1942 to Pterocomma Buckton, 1879 teste Shaposhnikov, 1965: 295
mume Hori, 1927 to ? Brachycaudus van der Goot, 1913
mumecola Shinji, 1930 = Brachycaudus helichrysi (Kaltenbach, 1843)
nashi Shinji, 1944 = Sappaphis piri Matsumura, 1918
ornata Gillette & Palmer, 1929 to Aphis Linnaeus, 1758
oxiana Nevsky, 1929 = Brachycaudus (Appelia) prunicola (Kaltenbach, 1843)

A. oxyacanthae del Guercio, 1930 = Dysaphis (Pomaphis) pyri (Boyer de Fonscolombe, 1841)
padi (Linnaeus, 1758) (Aphis) records were usually based on Brachycaudus helichrysi (Kaltenbach, 1843) although Aphis padi Linnaeus comes in Rhopalosiphum Koch, 1854
pectinatae del Guercio, 1930 = Brachycaudus (Acaudus) cardui (Linnaeus, 1758)
persicae var. compositarum Rusanova, 1943 nomen nudum
persicae var. praecipuus Rusanova, 1943 nomen nudum
persicae-niger (E.F. Smith, 1890) (Aphis) = Brachycaudus persicae (Passerini, 1860)
petherbridgei Theobald, 1929 = Brachycaudus (Acaudus) cardui (Linnaeus, 1758)
phlomicola Nevsky, 1929 to Roepkea Hille Ris Lambers, 1935
piricola Okamoto & Takahashi, 1927 = Sappaphis piri Matsumura, 1918
poae del Guercio, 1916 = Brachycaudus helichrysi (Kaltenbach, 1843)
populi del Guercio, 1911 to Brachycaudus (Acaudus)
projacobeae del Guercio, 1930 = Brachycaudus (Acaudus) cardui (Linnaeus, 1758)
prolappae del Guercio, 1930 = Dysaphis lappae (Koch, 1854)
pruni (Réaumur) del Guercio, 1930 = Brachycaudus helichrysi (Kaltenbach, 1843)
pruniavium Nevsky, 1929 = Brachycaudus helichrysi (Kaltenbach, 1843)
pruniavium var. ajvasovi Rusanova, 1943 nomen nudum
prunicola (Kaltenbach, 1843) (Aphis) to Brachycaudus (Appelia) but the name has also been applied to Brachycaudus (Acaudus) persicae (Passerini, 1860)
prunicola subsp. suppleta Rusanova, 1943 nomen nudum
prunifex Theobald, 1926 = Brachycaudus (Appelia) prunicola (Kaltenbach, 1843)
pruniphila var. cefaliflava del Guercio, 1930 = Brachycaudus (Acaudus) cardui (Linnaeus, 1758)
pyraria del Guercio, 1930 nomen nudum
pyriella Theobald, 1929 = farfarae (Koch, 1854)
pyrilaseri Shaposhnikov, 1950 BM HRL
rosae Baker lapsus pro roseus
roseus Baker, 1921 = Dysaphis (Pomaphis) plantaginea (Passerini, 1860)
rumecicola Hori, 1927 to Dysaphis Börner, 1931
rumicella Theobald, 1924 = Dysaphis radicola (Mordvilko, 1897)
rumicicola Hori, 1927 = Dysaphis rumecicola (Hori, 1927)
schwartzi (Börner, 1931) (Appelia) to Brachycaudus (Appelia)
senecio-radicis Gillette & Palmer, 1929 to Aphis Linnaeus, 1758

A sherardiae Theobald, 1926 = Brachycaudus helichrysi (Kaltenbach, 1843)
siciliensis Theobald, 1927 = Dysaphis crataegi subsp. siciliensis (Theobald, 1927)
subterranea (Walker, 1852) (Aphis) lectotype BM HRL
*depilosa* Nevsky, 1951 (*Yezabura*)
*heraclei* Koch, 1857 (*Aphis*)
tasashaari Rusanova, 1945 nomen nudum
tulipae (Boyer de Fonscolombe, 1841) (Aphis) to Dysaphis Börner, 1931
valerianae Takahashi, 1932 = Aphis patriniae Takahashi, 1966
verbascivorus Rusanova, 1943 nomen nudum
verbascivorus var. discors Rusanova, 1943 nomen nudum
viridana Nevsky, 1929 to Brachycaudus van der Goot, 1913
viridescens del Guercio, (1917) 1930 to ? Aphis Linnaeus, 1758
warei Theobald, 1923 = Brachycaudus helichrysi var. warei (Theobald, 1923)
xanthii del Guercio, 1913 = Brachycaudus helichrysi (Kaltenbach, 1843)

Subgen. Clavisiphon del Guercio, 1930

A. (Clavisiphon) distincta del Guercio, 1930 nomen dubium
(Clavisiphon) elegans del Guercio, 1930 (type of Clavisiphon del Guercio, 1930; and of Carazzia del Guercio, 1930) = Eucarazzia elegans (Ferrari, 1872)

Subgen. Macchiatiella del Guercio, 1917 nec 1909

A. (Macchiatiella) crathaegaria del Guercio, 1930 to Dysaphis Börner, 1931
(Macchiatiella) crathaegi errore pro Aphis crataegi Kaltenbach, 1843
(Macchiatiella) depressa del Guercio, 1930 = Brachycaudus (Acaudus) persicae (Passerini, 1860)
(Macchiatiella) flavicephala del Guercio, 1930 = Brachycaudus (Acaudus) cardui (Linnaeus, 1758)
(Macchiatiella) hirta del Guercio, 1930 = Dysaphis (Pomaphis) pyri (Boyer de Fonscolombe, 1841)
(Macchiatiella) medicaginea del Guercio, 1930 = Aphis craccivora Koch, 1854
(Macchiatiella) menthae del guercio, 1930 = Aphis affinis del Guercio, 1911
(Macchiatiella) oblonga del Guercio, 1930 = Brachycaudus (Acaudus) persicae (Passerini, 1860)
(Macchiatiella) pectinata del Guercio, 1930 = Brachycaudus (Acaudus cardui (Linnaeus, 1758)
(Macchiatiella) philippaeae errore pro phelipaeae Passerini, 1879 (Aphis)
(Macchiatiella) projacobeae del Guercio, 1930 = Brachycaudus (Acaudus) cardui (Linnaeus, 1758)

A. (Macchiatiella) prolappae del Guercio, 1930 = Dysaphis lappae (Koch, 1854)
(Macchiatiella) promedicaginis del Guercio, 1930 = Acyrthosiphon pisum (Harris, 1776)
(Macchiatiella) pruniphila del Guercio, 1930 = Brachycaudus (Acaudus) cardui (Linnaeus, 1758)
(Macchiatiella) pruniphila var. cefaliflava del Guercio, 1930 = Brachycaudus (Acaudus) cardui (Linnaeus, 1758)
(Macchiatiella) rumicifila errore pro rumiciphila del Guercio, 1930
(Macchiatiella) rumiciphila del Guercio, 1930 = Dysaphis radicola (Mordvilko, 1897)
(Macchiatiella) senecii del Guercio, 1930 = Brachycaudus (Acaudus) cardui (Linnaeus, 1758)
(Macchiatiella) tanaceticola del Guercio, 1930 = ? Coloradoa tanacetina (Walker, 1850)
(Macchiatiella) trifolii del Guercio, 1917 = Acyrthosiphon pisum (Harris, 1776)
(Macchiatiella) viridescens del Guercio, 1930 to ? Aphis Linnaeus, 1758

ANURIELLA del Guercio, 1920 (1921)
Redia 14: 113, see footnotes on pp. 136 & 210 for date of publication
Type species Anuriella dorsolineata del Guercio, 1920
Nomen dubium

A. dorsolineata del Guercio, 1920 nomen dubium

ANUROMYZUS Shaposhnikov, 1959
Ént. Obozr. 38: 156, described as subg. of Dysaphis Börner, 1931
Type species Dysaphis (Anuromyzus) cotoneasteris Shaposhnikov, 1959

A. cotoneasteris (Shaposhnikov, 1959) (Dysaphis (Anuromyzus)) BM HRL

AORISON Mordvilko, 1914·
Faune Russie, Ins. Hém. 1: 72, without species

APATHAPHIS Börner, 1952
Mitt. thüring. bot. Ges. 4 (3): 73, described as subg. of Aphis Linnaeus, 1758
Type species Aphis clematidis Koch, 1854
= Aphis Linnaeus, 1758

APHANOSTIGMA Börner, 1909
Mitt. biol. BundAnst. Ld. u. Forstw. 8: 61
Type species Phylloxera piri Cholodkovsky, 1903
*Cinacium* Kishida, 1924
Phylloxeridae

APHANTAPHIS Scudder, 1890
U.S. Geol. Survey Territories 13: 253
Type species Aphantaphis exsuca Scudder, 1890
Fossil

APHANUS Gistel, 1837 (nec Laporte de Castelnau, 1832)
Faunus, Munich, 1: 111
Type species Chermes lapidarius Fabricius, 1803
= Gisteliella Strand, 1928, nomen dubium
? Aphidoidea

APHIDELLA mis-spelling for Aphidiella

APHIDIELLA Theobald, 1923
Entomologists' mon Mag. 59: 105, spelled Aphidella
Type species Aphidiella secretocauda Theobald, 1923 = Aphis avenae Fabricius, 1775
= Sitobion Mordvilko, 1914

A. secretocauda Theobald, 1923 = Sitobion avenae (Fabricius, 1775)

APHIDIOIDES Motschulsky, 1857 (1856)
Études ent. 5: 29
Type species Aphidioides succifera Motschulsky, 1856
Fossil

APHIDOPSIS Scudder, 1890
U.S. Geol. Survey of the Territories 13: 260
Type species Aphidopsis margarum Scudder, 1890
Fossil

APHIDOUNGUIS Takahashi, 1963
Kontyû 31: 164-165
Type species Aphidounguis mali Takahashi, 1963 = Watabura nishiyae Matsumura, 1917
= Watabura Matsumura, 1917

A. mali Takahashi, 1963 = Watabura nishiyae Matsumura, 1917

APHIDULA Nevsky, 1929
Zool. Anz. 82: 208
Type species Aphidula althaeae Nevsky, 1929 = Aphis davletshinae Hille Ris Lambers, 1966
= Aphis Linnaeus, 1758

A. althaeae Nevsky, 1929 = Aphis davletshinae Hille Ris Lambers, 1966
flava Nevsky, 1929 = Aphis gossypii Glover, 1877

APHIDULUM Handlirsch, 1939
Ann. Naturh. Mus. Wien 49: 163
Type species Aphidulum pusillum Handlirsch, 1939
Fossil, not an aphid, vide Heie, 1967: 195

APHIDURA Hille Ris Lambers, 1956
Boll. Lab. Ent. agr. Filippo Silvestri 14: 293
Type species Aphidura ornata Hille Ris Lambers, 1956
*Cerasomyzus* Narzikulov, 1958

A. bharatia David, Sekhon & Bindra, 1970 HRL
bozhkoae (Narzikulov, 1958) (Myzus (Cerasomyzus)) BM HRL
delmasi Remaudière & Leclant, 1965 paratype BM HRL
gypsophilae Mamontova, 1963 HRL
mingens Pintera, 1970 paratype BM HRL
ornata Hille Ris Lambers, 1956 cotypes BM HRL
ornatella Narzikulov & Winkler, 1960
pannonica Szelegiewicz, 1967 paratype BM HRL
pannonica subsp. cretacea Mamonteva, 1968
picta Hille Ris Lambers, 1956 HRL
pujoli (Gomez-Menor, 1950) (Myzaphis) HRL

APHIDUROMYZUS Umarov & Ibramova, 1967
Dokl. Akad. Nauk. tadzhik 10 (3): 58
Type species Aphiduromyzus rosae Umarov & Ibramova, 1967

A. rosae Umarov & Ibramova, 1967 HRL

APHIOIDES Passerini, 1860 nec Rondani, 1848
Gli Afidi, Parma, p. 28 footnote
Type species Aphis populea Kaltenbach, 1843
= Pterocomma Buckton, 1879

APHIOIDES Rondani, 1848
Nuovi Ann. Sci. nat. Bologna, (2) 8: 439
Type species Aphis bursaria Linnaeus, 1758
= Pemphigus Hartig, 1839

APHIS Linnaeus, 1758
Syst. nat. ed. 10: 451
Type species Aphis sambuci Linnaeus, 1758
*Apathaphis* Börner, 1952
*Aphidula* Nevsky, 1929
*Asiataphis* Narzikulov, 1970
*Cerosipha* del Guercio, 1900
*Chaitophoroides* Mordvilko, 1908
*Comaphis* Börner, 1940
*Debilisiphon* Shaposhnikov, 1950
*Doralida* Börner, 1950
*Doralina* Börner, 1940
*Doralis* auctt. nec Leach, 1827
*Leucosiphon* Börner, 1952
*Longirostrina* Kumar & Burkhardt, 1971
*Longirostris* Kumar & Burkhardt, 1970 nec S.W.D., 1836
*Medoralis* Börner, 1952
*Microsiphon* del Guercio, 1907
*Papillaphis* Börner, 1952
*Pergandeida* Schouteden, 1903
*Tuberculaphis* Börner, 1952
*Uraphis* del Guercio, 1907
*Wapuna* Hottes & Wehrle, 1951

A. aba Miller, 1937
abbreviata Patch, 1912 = nasturtii Kaltenbach, 1843
abietaria Walker, 1852 = fabae Scopoli, 1763
abietina Walker, 1849 type of Elatobium Mordvilko, 1914 q.v. and of Neomyzaphis Theobald, 1926
abietis Kirby, 1818 probably not intended as new species but as transfer of Chermes abietis Linnaeus, 1758 to Aphis
abietis Walker, 1848 = Cinara pilicornis (Hartig, 1841)
abnobae von Engelberg, 1811, not an aphid, = Javesella pellucida (Fabricius, 1794)
abrotani Walker, 1852 to Macrosiphoniella del Guercio, 1911

A. absinthii Linnaeus, 1758 to Macrosiphoniella del Guercio, 1911
acaenovinae Eastop, 1961 types BM HRL
acanthi Schrank, 1801 = fabae subsp. cirsiiacanthoidis Scopoli, 1763
acanthoidis (Börner, 1940) (Cerosipha) BM HRL
acanthopanaci Matsumura, 1917 BM
acaroides Rafinesque, 1818 invalid
acericola Walker, 1848 to Periphyllus van der Hoeven, 1863
aceriella Theobald, 1929 = viburni Scopoli, 1763
acerina Walker, 1848 to Drepanosiphum Koch, 1855
aceris Ph. F. Gmelin, 1758 nomen oblitum
aceris Linnaeus, 1761 to Periphyllus van der Hoeven, 1863
aceris platanoides Stewart, 1802 nomen nudum
aceris pseudoplatani Mosley, 1841 = Drepanosiphum platanoidis (Schrank, 1801)
acetosae Linnaeus, 1761 BM HRL
acetosella Theobald, 1918 = nasturtii Kaltenbach, 1843
achilleae Fabricius, 1777 nomen dubium
achinifoliae Bertels, 1973 lapsus pro schinifoliae E.E. Blanchard, 1939
achyranthi Theobald, 1929 BM HRL
acori Theobald, 1923 = Schizaphis rotundiventris (Signoret, 1860)
acrita Smith, 1940
addita Walker, 1849 = fabae Scopoli, 1763
adducta Walker, 1849 = fabae Scopoli, 1763
adivae Shiraki, 1952 = Toxoptera odinae (van der Goot, 1917)
adjecta Walker, 1849 = Brachycaudus helichrysi (Kaltenbach, 1843)
adjucta Walker lapsus pro adjuta
adjuta Walker, 1848 = Sitobion avenae (Fabricius, 1775)
adjuvans Walker, 1848 to Aspidaphis Gillette, 1917
adscita Walker, 1848 = Brachycaudus helichrysi (Kaltenbach, 1843)
adusta Zehntner, 1897 = Rhopalosiphum maidis (Fitch, 1856)
advena Walker, 1849 = fabae Scopoli, 1763
aegeris Shinji lapsus pro aeglis
aeglis Shinji, 1922 = Toxoptera citricidus (Kirkaldy, 1907)
aegopodii Scopoli, 1763 to Cavariella del Guercio, 1911
aegopodii podagrariae Stewart, 1802 nomen nudum
affinis del Guercio, 1911 BM HRL
    *menthae* del Guercio, 1930 (*Anuraphis*)
affinis var. gardeniae del Guercio, 1913 = gossypii Glover, 1877
africana Theobald, 1914 = Rhopalosiphum maidis (Fitch, 1856)
agastachyos Hille Ris Lambers, 1974 HRL

A. agathona Hottes, 1950 to Braggia Gillette & Palmer, 1929
ageratoidis Oestlund, 1886 = coreopsidis (Thomas, 1878)
alba Ratzeburg, 1844 = Kaltenbachiella pallida (Haliday, 1838)
albella Nevsky, 1951
albipes Oestlund, 1887 (type of Aphthargelia Hottes, 1958; and of Thargelia Oestlund, 1922) = Aphthargelia symphoricarpi (Thomas, 1878)
alchemillae Börner, 1940 (Doralina)
alemedensis Clarke, 1903 = Brachycaudus (Acaudus) cardui (Linnaeus, 1758)
aliena Theobald, 1915 BM
allii Lichtenstein, 1884 nomen nudum
alni Fabricius, 1781 (type of Pterocallis Passerini, 1860; and of Subcallipterus Mordvilko, 1894) = Pterocallis alni (de Geer, 1773)
alni de Geer, 1773 to Pterocallis Passerini, 1860
alni Schrank, 1801 nec de Geer, 1773 = Glyphina schrankiana Börner, 1950
alstroemeriae Essig, 1953 paratype BM
alterna Walker, 1849 nomen dubium
althaea Harris, 1776 nomen dubium
althaeae Nevsky, 1929 = davletshinae Hille Ris Lambers, 1966
althaeae subsp. afghanica Narzikulov & Umarov, 1971 = davletshinae subsp. afghanica Narzikulov & Umarov, 1971
amaranthi Holman, 1974 HRL
ambrosia Rafinesque, 1818 invalid
amenticola Kaltenbach, 1874 nomen dubium
amerinae Hartig, 1841 to Plocamaphis Oestlund, 1922
amica Walker, 1848 = Macrosiphoniella pulvera (Walker, 1848)
ampelophila del Guercio, 1913 = illinoisensis Shimer, 1866
amygdali Blanchard, 1840 to Hyalopterus Koch, 1854
amygdali Buckton, 1879 = Brachycaudus (Appelia) schwartzi (Börner, 1931)
amygdalinus Schouteden, 1905 (type of Brevicaudus Shaposhnikov, 1964) to Brachycaudus (Thuleaphis)
amygdalipersicae Mosley, 1841 = Hyalopterus amygdali (Blanchard, 1840)
andreinii del Guercio, 1917 nomen dubium
angelicae Koch, 1854 type of Dysaphis Börner, 1931 q.v.
annuae Oestlund, 1886 = Rhopalosiphum padi (Linnaeus, 1758)
annulata Hartig, 1841 to Tuberculatus (Tuberculoides)
annulipes Rafinesque, 1818 invalid
antennata Kaltenbach, 1843 type of Monaphis Walker, 1870 q.v.; and of Bradyaphis Mordvilko, 1894
antherici Holman, 1966 paratypes BM HRL
anthrisci Kaltenbach, 1843 to Semiaphis van der Goot, 1913

A. aparines Fabricius, 1775 = fabae Scopoli, 1763
aparines Kaltenbach, 1843 = Myzus cerasi (Fabricius, 1775)
aparinis Blanchard, 1840 = fabae Scopoli, 1763
apigraveolens Essig, 1941 cotypes BM
apii Theobald, 1925 = fabae Scopoli, 1763
apocyni Koch, 1854 = fabae Scopoli, 1763
apposita Walker, 1850 = Brachycaudus helichrysi (Kaltenbach, 1843)
aquatica Jackson, 1908 = Rhopalosiphum nymphaeae (Linnaeus, 1761)
aquilegia-canadensis Rafinesque, 1817 invalid
aguilegiae flava Kittel, 1827, invalid
aguilegiae nigra Kittel, 1827, invalid
arabis-mollis Rafinesque, 1818 invalid
araliaeradicis (Strom, 1938) (Cerosipha)
aralia-hispida Rafinesque, 1817 invalid
aranaeiformis Germar & Berendt, 1856 to Pseudamphorophora Heie, 1967
arbuti Ferrari, 1872 HRL
archangelicae Scopoli, 1763 to Cavariella del Guercio, 1911
argentinae-radicis Gillette & Palmer, 1932 to Nearctaphis Shaposhnikov, 1950
armata Hausmann, 1802 BM HRL
armoraciae Cowen, 1895 BM HRL
artemifoliae Shinji, 1922
artemisiae Boyer de Fonscolombe, 1841 to Macrosiphoniella del Guercio, 1911
artemisiae Ph. F. Gmelin, 1758 nomen oblitum
artemisiae Goeze, 1778
artemisiae Passerini, 1860 = Cryptosiphum artemisiae Buckton, 1879
artemisiae Takahashi ex Shinji, 1935 = kurosawai Takahashi, 1921
artemisicola Williams, 1911 (type of Artemisaphis Knowlton & Roberts, 1947) to Obtusicauda (Artemisaphis)
artemisifoliae Shinji, 1922 = kurosawai Takahashi, 1921
arum (ascribed to van der Goot ex Das, 1918: 204) Patch, 1938 lapsus pro Aphis papilionacearum van der Goot, 1918 ?
arundinariae Takahashi, 1937 to Melanaphis van der Goot, 1917
arundinis Fabricius, 1775 = ? Diuraphis (Holcaphis) calamagrostis (Ossiannilsson, 1959)
arundis Fabricius lapsus pro arundinis Fabricius, 1775
ascita Walker, 1852 nomen dubium
asclepiadis Fitch, 1851 BM HRL
asperulae Walker, 1848 = Myzus cerasi (Fabricius, 1775)
assidua Walker, 1849 nomen dubium
assueta Walker, 1849 nomen dubium

A. asterensis Gillette & Palmer, 1929 type of Acuticauda Hille Ris Lambers, 1956 q.v.
astericola Tissot, 1932
asteris Walker, 1849 (type of Asterobium Hille Ris Lambers, 1938) to Macrosiphoniella (Asterobium)
astilbes Matsumura, 1917
astragali Ossiannilsson, 1959 (Aphis (Pergandeida)) paratype BM HRL
astragalicola Holman & Szelegiewicz, 1971 HRL
astragalina Hille Ris Lambers, 1974 HRL
atomaria Walker, 1849 = Macrosiphoniella pulvera (Walker, 1848)
atriplicis Fabricius, 1775 nec Linnaeus, 1758 = fabae Scopoli, 1763
atriplicis Linnaeus, 1761 to Hayhurstia del Guercio, 1917
atromaculata Hille Ris Lambers, 1974 HRL
atronitens Cockerell, 1903 = craccivora Koch, 1854
atuberculata Hille Ris Lambers, 1935 HRL
aubletia Sanborn, 1904
aucta Walker, 1849 to Acyrthosiphon Mordvilko, 1914
aucupariae Buckton, 1879 to Dysaphis (Pomaphis)
audax Hille Ris Lambers, 1947 = triglochinis Theobald, 1926
aurantii Boyer de Fonscolombe, 1841 to Toxoptera Koch, 1856
austriaca Hille Ris Lambers, 1950 HRL
avellanae Blanchard, 1840 = Myzocallis coryli (Goeze, 1778)
avellanae Schrank, 1801 type of Corylobium Mordvilko, 1914 q.v.
avenae Fabricius, 1775 to Sitobion Mordvilko, 1914
avenae sativae Schrank, 1801 = Rhopalosiphum padi (Linnaeus, 1758)
baccharicola Hille Ris Lambers, 1974 HRL
bakeri Cowen, 1895 type of Nearctaphis Shaposhnikov, 1950 q.v.
ballotae Passerini, 1860 to Brachycaudus (Acaudus)
balloticola Szelegiewicz, 1968 nom. nov. pro ballotae auctt. nec Passerini, 1860 HRL
balsamines Kaltenbach, 1862 to Impatientinum Mordvilko, 1914
balsamitae O.F. Müller, 1776 nomen dubium
bambusae Fullaway, 1910 type of Melanaphis van der Goot, 1917 q.v.
bartsiae Walker, 1849 = Brachycaudus helichrysi (Kaltenbach, 1843)
basalis Walker, 1848 = Acyrthosiphon pisum (Harris, 1776)
bauhiniae Theobald, 1918 = Aphis gossypii (Glover, 1877)
bazzii E.E. Blanchard, 1923 = fabae subsp. solanella Theobald, 1913
beccabungae Koch, 1855 = frangulae subsp. beccabungae Koch, 1855
beccabungae subsp. turanica Nevsky, 1929
*veronicae* Nevsky, 1929 nec Walker, 1848

beccarii del Guercio, 1917 = craccivora Koch, 1854
bella Walsh, 1863 (type of Lineomyzocallis Richards, 1965) to Myzocallis (Lineomyzocallis)
bellidis Mosley, 1841 nomen dubium
bellis Buckton, 1879 = Brachycaudus helichrysi (Kaltenbach, 1843)
bellula Walker, 1849 nomen dubium
berberidis Fitch, 1851 = Liosomaphis berberidis (Kaltenbach, 1843)
berberidis Kaltenbach, 1843 type of Liosomaphis Walker, 1868 q.v.
berlinskii Huculak, 1968 BM HRL
berteroae Szelegiewicz, 1966 paratypes BM HRL
betulae Ph. F. Gmelin, 1758 nomen oblitum
betulae Linnaeus, 1758 to Glyphina Koch, 1856. Nominally, as betulae L. of Walker, the type of Euceraphis Walker, 1870
betulae alni Rossi, 1790 = Pterocallis alni (de Geer, 1773)
betulaecolens Fitch, 1851 (as Callipterus betulaecolens Fitch type of Siphonocallis del Guercio, 1914) to Calaphis Walsh, 1863
betularia Kaltenbach, 1843 (type of Callipterinella van der Goot, 1913) = Callipterinella tuberculata (von Heyden, 1837)
betulicola Kaltenbach, 1843 (type of Kallistaphis Kirkaldy, 1905 and of Neocallipterus van der Goot, 1915) to Calaphis Walsh, 1863
betulina Walker, 1852 nomen dubium
bicolor Haldeman, 1844 = pomi de Geer, 1773
bicolor Koch, 1855 (type of Hydaphias Börner, 1930) = Hydaphias helvetica Hille Ris Lambers, 1947
bidentis Theobald, 1929 = citricola van der Goot, 1912
bifrons Walker, 1848 nomen dubium
bituberculata Wilson, 1914 = Hysteroneura setariae (Thomas, 1878)
boerneri Franssen, 1927 = hederae f. boerneri Franssen, 1927
bonnevillensis Knowlton, 1928 to Brachyunguis Das, 1918
borealis Curtis, 1828 = ? Cinara confinis (Koch, 1856)
bougainvilleae Börner ex Bodenheimer, 1930 & 1937 nomen nudum
brachysiphon Narzikulov, 1964 = nom. nov. pro Aphis clematidis Nevsky, 1929 nec Koch, 1854 = ? vitalbae Ferrari, 1872
brassicae Ph. F. Gmelin, 1758 nomen oblitum
brassicae Linnaeus, 1857 type of Brevicoryne van der Goot, 1915 q.v.
brassica-napus Rafinesque, 1818 invalid
brevifurca Monell, 1879 nomen nudum
brevis Sanderson, 1902 = Nearctaphis crataegifoliae (Fitch, 1851)
breviseta Holman, 1966 paratypes BM HRL

A. brevisiphona Theobald, 1913 = fabae Scopoli, 1763
brevitarsis Szelegiewicz, 1963
brohmeri Börner, 1952 BM HRL
brunellae Schouteden, 1903 BM HRL
brunnea Ferrari, 1872 HRL
bryophyllae Shinji, 1922 = gossypii Glover, 1877
buddleiae Theobald, 1918 = verbasci Schrank, 1801
bufo Walker, 1848 to Iziphya Nevsky, 1929
bulleri Robinson & Rojanavongse, 1976 (paratype) BM HRL
bumeliae Schrank, 1801 type of Prociphilus Koch, 1857 q.v.
bupleuri (Börner, 1932) (Doralis) BM HRL
bupleuri-sensoriata Bozhko, 1959
bursaria Linnaeus, 1758 type of Pemphigus Hartig, 1839 q.v.; and of Aphioides Rondani, 1848
butomi Schrank, 1801 = Rhopalosiphum nymphaeae (Linnaeus, 1761)
buxtoni Theobald, 1920 = sambuci Linnaeus, 1758
cacaliasteris Hille Ris Lambers, 1947 BM HRL
cacaliasteris subsp. helvetica Hille Ris Lambers, 1947 HRL
*haroi* Nieto Nafria, 1974 HRL
*senecionis* Börner, 1940 (*Doralis*)
cactae Hereman, 1840 nomen dubium
cadiva Walker, 1849 nomen dubium
calaminthae (Börner, errore 1940; 1952, p. 222) (Doralina)
calendulicola Monell, 1879
caliginosa Hottes & Frison, 1931 paratypes BM
callipterus Hartig, 1841 to Callipterinella van der Goot, 1913
callunae Theobald, 1915 type BM HRL
calotropidis del Guercio, 1916 = nerii Boyer de Fonscolombe, 1841
camelliae Kaltenbach, 1843 = Toxoptera aurantii (Boyer de Fonscolombe, 1841)
camellicola del Guercio, 1913 nomen dubium
campanulae Kaltenbach, 1843 to Uroleucon (Uromelan)
campanula-riparia Rafinesque, 1817 invalid
canae Williams, 1911 to Aphis (Zyxaphis)
canberrae Eastop, 1961 type of Casimira Eastop, 1966 q.v.
candicans Fitch, 1877 nomen nudum ? vide Thomas in Monell & Thomas, 1878 and Monell, 1879: 26.
candicans Passerini, 1879 to Dysaphis Börner, 1931
candicans Thomas, 1877/8 nomen nudum
capreae Fabricius, 1775 (type of Corynosiphon Mordvilko, 1914) = Cavariella pastinacae (Linnaeus, 1758)
caprifoliae Iglisch, 1975

A. capsellae Kaltenbach, 1843 member of the gossypii complex q.v.
capsellae Koch, 1854 = Brachycaudus (Acaudus) cardui Linnaeus, 1758
carbocolor Gillette, 1907 = rumicis Linnaeus, 1758
cardaminae Shinji, 1922 nomen dubium
cardaminensis Shinji, 1935 lapsus or irregular emendation of cardaminae ?
carduella Walsh, 1963 BM
cardui Ph. F. Gmelin, 1857 nomen oblitum
cardui Linnaeus, 1758 (type of Prunaphis Shaposhnikov, 1964 and of Neobrachycaudus Narzikulov, 1965) to Brachycaudus (Acaudus)
cardui var. naumburgensis Franssen, 1927 = fabae subsp. cirsiiacanthoidis Scopoli, 1763
carduina Walker, 1850 type of Capitophorus van der Goot, 1913 q.v.
cari Essig, 1917
caricis Schouteden, 1906 to Schizaphis (Paraschizaphis)
caroliboerneri (Remaudière, 1952) (Cerosipha) paratypes BM HRL
carotae Koch, 1854 (type of Semiaphis van der Goot, 1913) = Semiaphis dauci (Fabricius, 1775)
caryae Harris, 1841 type of Longistigma Wilson, 1909 q.v. and of Davisia del Guercio, 1909
caryella Fitch, 1855 type of Monellia Oestlund, 1887 q.v.
caryella var. costalis Fitch, 1855 = Monellia costalis (Fitch, 1855)
cassiae Hereman, 1840 nomen dubium
castanea Koch, 1854 = fabae subsp. cirsiiacanthoides Scopoli, 1763
castanea-vesca Haldeman, 1844 = Calaphis castaneae (Fitch, 1856)
catalpae Mamontova, 1950 BM paratypes HRL
cathartica del Guercio, 1909 = nasturtii Kaltenbach, 1848
ceanothi Clarke, 1903 cotypes BM HRL
*ceanothi-hirsuti* Essig, 1911
ceanothi-hirsuti Essig, 1911 = ceanothi Clarke, 1903
celastrii Matsumura, 1917 BM HRL
celtis Shinji, 1922 to Toxoptera Koch, 1856
centaureae Ph. F. Gmelin, 1758 nomen oblitum
centaureae Koch, 1854 type of Acaudinum Börner, 1930 q.v.
cephalanthi Thomas, 1878 BM HRL
cephalicola Cowen, 1895 = Nearctaphis bakeri (Cowen, 1895)
cerasi Fabricius, 1775 type of Myzus Passerini, 1860 q.v. and of Myzoides van der Goot, 1913
cerasi O.F. Müller, 1776 = Myzus cerasi (Fabricius, 1775)
cerasicolens Fitch, 1851 = Euceraphis punctipennis (Zetterstedt, 1828)
cerasifoliae Fitch, 1855 to Rhopalosiphum Koch, 1854

A. cerasina Walker, 1850 nomen dubium
cerastii (Börner, 1950) (Doralina)
cerastii Kaltenbach, 1846 to Brachycolus Buckton, 1879
ceratoniae Lichtenstein, 1884 nomen nudum
cercocarpi Gillette & Palmer, 1929 metatypes BM HRL
cerealis Kaltenbach, 1843 = Sitobion avenae (Fabricius, 1775)
certa Walker, 1849 to Myzus (Nectarosiphon)
chaerophylli Koch, 1854 = fabae Scopoli, 1763
chamomillae Koch, 1854 = Brachycaudus (Acaudus) cardui (Linnaeus, 1758)
chelidonii Kaltenbach, 1843 (type of Liporrhinus Börner, 1939) to Acyrthosiphon (Liporrhinus)
chenophyllum-canadense Rafinesque, 1817 invalid
chenopodii Cowen, 1895 = Hayhurstia atriplicis (Linnaeus, 1761)
chenopodii Schrank, 1801 = Hayhurstia atriplicis (Linnaeus, 1761)
chenopodium-canadense Rafinesque, 1817 invalid
chetansapa Hottes & Frison, 1931 paratypes BM HRL
chilopsidi Davletschina, 1964
chinensis Bell, 1851 type of Schlechtendalia Lichtenstein, 1883 q.v.
chipetae Hottes, 1933 = Cedoaphis incognita Hottes & Frison, 1931
chloris Koch, 1854 BM HRL
chloris subsp. citrina Nevsky, 1929 = citrina Nevsky, 1929
chloroides Nevsky, 1929 = gossypii Glover, 1877
chrysanthemi Koch, 1854 = Brachycaudus (Acaudus) cardui (Linnaeus, 1758)
chrysanthemi Walker, 1849 = Brachycaudus helichrysi (Kaltenbach, 1843)
chrysanthemicola Williams, 1911 = Coloradoa rufomaculata (Wilson, 1908)
chrysothamni Wilson, 1915 to Aphis (Zyxaphis)
cichorea Figari, 1864 (see Hall, 1926: 48)
cichorii Koch, 1854 errore pro intybi Koch, 1854
circezandis Fitch, 1870 = gossypii Glover, 1877, suppressed name, Russell, 1968
cirsii Linnaeus, 1758 to Uroleucon Mordvilko, 1914
cirsii-acanthoidis Scopoli, 1763 = fabae subsp. cirsiiacanthoidis Scopoli, 1763
cirsiioleracei (Börner, 1932) (Cerosipha) HRL
cirsina Ferrari, 1872 = fabae subsp. cirsiiacanthoidis Scopoli, 1763
cistata Walker, misprinted name placed in Dryobius by Buckton, 1881: 78 = Cinara costata (Zetterstedt, 1828)
cisti del Guercio, 1911
cisti Lichtenstein, 1884 nomen nudum
cisti Walker, 1852 nomen dubium

A. cisticola Leclant & Remaudière ex Remaudière & Leclant, 1972 HRL
cistiella Theobald, 1923 = craccivora Koch, 1854
citri Ashmead, 1887 = ? gossypii Glover, 1877
citricola van der Goot, 1912 BM lectoparatype HRL
*bidentis* Theobald, 1929
? *croomiae* Shinji, 1922
*deutziae* Shinji, 1922
*erratica* del Guercio, 1917 (*Anuraphis*)
*malvoides* van der Goot, 1917
? *mitsubae* Shinji, 1922
*nigricauda* van der Goot, 1917
*pirifoliae* Shinji, 1922
*pseudopomi* Bertels, 1973
*pseudopomi* E.E. Blanchard, 1939
*spiraecola* Patch, 1914
*viburnicolens* Swain, 1919
citricola del Guercio, 1917 = craccivora Koch, 1854
citrina Nevsky, 1929
*chloris* subsp. *citrina* Nevsky, 1929
citrulli Ashmead, 1882 = gossypii Glover, 1877
clematidis Koch, 1854 type of Apathaphis Börner, 1952 BM HRL
clematidis subsp. simalensis Kumar & Burkhardt, 1970 BM
clematidis errore Nevsky, 1929 nec Koch, 1854 = brachysiphon Narzikulov, 1964 nomen novum = ? vitalbae Ferrari, 1872
clerodendri Matsumura, 1917 BM HRL
clerodendri subsp. amamiana Takahashi, 1966
cliftonensis Stroyan, 1964 holotype BM HRL
clinopodii Passerini, 1861
clydesmithi Stroyan, 1970
cnici Schrank, 1801 = Brachycaudus (Acaudus) cardui (Linnaeus, 1758)
cnici Williams, 1910 = Bipersona ochrocentri (Cockerell, 1903)
coffeae Nietner, 1861 = Toxoptera aurantii (Boyer de Fonscolombe, 1841)
cognatella Jones, 1943 = euonymi Fabricius, 1775
collega Walker, 1848 = Macrosiphoniella pulvera (Walker, 1848)
colocasiae Matsumura, 1917 = gossypii Glover, 1877
comes Walker, 1848 to Clethrobius Mordvilko, 1928
commelinae Shinji, 1922 = gossypii Glover, 1877
commensalis Stroyan, 1952 BM HRL
*steinbergi* Shaposhnikov, 1952 HRL
commoda Walker, 1848 nomen dubium

A. comodoensis Narzikulov, 1964
comosa (Börner, 1950) (Doralida) HRL
compositae Theobald, 1915 = ? fabae subsp. solanella Theobald, 1913
compositellae Theobald, 1924 to Nasonovia Mordvilko, 1914
confusa Walker, 1849 BM HRL
*? conspersa* Walker, 1849
*ochropus* Koch, 1854
*scabiosae* Schrank, 1801 nec Scopoli, 1763
conjuncta Walker, 1848 = Amphorophora rubi (Kaltenbach, 1843)
consolidae Passerini, 1864 = symphyti Schrank, 1801
consona Walker, 1849 nomen dubium
consors Walker, 1848 = Myzus (Nectarosiphon) persicae (Sulzer, 1776)
conspersa Walker, 1849 ? = confusa Walker, 1849
consueta Walker, 1848 = Sitobion avenae (Fabricius, 1775)
consumpta, Walker, 1849 = Brachycaudus helichrysi (Kaltenbach, 1843)
contermina Walker, 1849 = Lipaphis erysimi (Kaltenbach, 1843)
convecta Walker, 1849 = Brachycaudus helichrysi (Kaltenbach, 1843)
conviva Walker, 1849 = Brachycaudus (Kaltenbach, 1843)
convolvuli Kaltenbach, 1843 (type of Macrosiphum del Guercio, 1900 nec Passerini, 1860; and of Nectarosiphon Schouteden, 1901) = Myzus (Nectarosiphon) persicae (Sulzer, 1776)
convolvulicola Ferrari, 1872 = ? gossypii Glover, 1877
cooki Essig, 1911 = Rhopalosiphum maidis (Fitch, 1856)
coprosmae Laing ex Tillyard, 1926 BM HRL
coreopsidis (Thomas, 1878) (Siphonophora) BM HRL
*ageratoidis* Oestlund, 1886
*frondosae* Oestlund, 1886
coriaria Börner, 1952 = lantanae Koch, 1854
corni Fabricius, 1775 type of Anoecia Koch, 1857 q.v.
cornifila del Guercio, 1911 nomen dubium
cornifoliae Fitch, 1851 BM HRL
cornus-stricta Rafinesque, 1818 invalid
coronillae Ferrari, 1872 BM HRL
*scaliai* del Guercio, 1915
*trifolii* Börner, 1950 (*Doralida*) nec Aphis trifolii Oestlund, 1887
coronillae subsp. arenaria Hoffmann, 1968
*scaliai* subsp. *arenaria* Hoffmann, 1968
coronopifoliae Bartholomew, 1932
coryli Goeze, 1778 type of Myzocallis Passerini, 1860 q.v.
coryli Mosley, 1841 = Corylobium avellanae (Schrank, 1801)

A. costata Zetterstedt, 1828 to Cinara Curtis, 1835
coweni Palmer, 1938 BM HRL
*veratri* Cowen, 1895 nec Walker, 1852
craccae Linnaeus, 1758 BM HRL
*viciae* Fabricius, 1781
craccivora Koch, 1854 BM HRL
*atronitens* Cockerell, 1903
*beccarii* del Guercio, 1917
*cistiella* Theobald, 1923
*citricola* del Guercio, 1917
*dolichi* Montrouzier, 1861
*hordei* del Guercio, 1913
*isabellina* del Guercio, 1917
*kyberi* Hottes, 1930
*laburni* of extra-european authors nec Kaltenbach, 1843
*leguminosae* Theobald, 1915
*loti* Kaltenbach, 1862
*loti* subsp. *gollmicki* Börner, 1952 *(Pergandeida (Doralida))*
*? medicaginea* del Guercio, 1930 *(Anuraphis (Macchiatiella))*
*medicaginis* auctt. prior to 1950 nec Koch, 1854
*meliloti* Börner, 1939 *(Doralis)*
*mimosae* Ferrari, 1872
*? onobrychidis* Goureau, 1863
*oxalina* Theobald, 1925 type BM
*papilionacearum* van der Goot, 1918
*robiniae* Macchiati, 1885
*salsolae* Börner, 1940 *(Doralina)*
*? salviae* Walker, 1852 BM HRL
craccivora subsp. pseudacaciae Takahashi, 1966 HRL
crataegaria Walker, 1850 to Ovatus van der Goot, 1913
crataegella Theobald, 1912 = Rhopalosiphum insertum (Walker, 1849) although nominally nomen novum pro Aphis crataegi Buckton, 1879 nec Kaltenbach, 1843.
crataegi Buckton, 1879 = pomi de Geer, 1773
crataegi Kaltenbach, 1843 to Dysaphis Börner, 1931
crataegi Tullgren, 1919 lapsus pro Prociphilus crataegi Tullgren, 1909
crataegifoliae Fitch, 1851 (type of Fitchiella Shaposhnikov, 1950) to Nearctaphis Shaposhnikov, 1950
crataegifoliae Shinji, 1922 incertae sedis name preoccupied

A. crataegina Walker ex Theobald, 1926, p. 281 = Ovatus crataegarius (Walker, 1850)
crataegus-coccinea Rafinesque, 1818 invalid
crepidis (Börner, 1940) (Cerosipha) BM HRL
crinosa Paik, 1969
*ligustri* Shinji, 1923 (*Pterocomma*) nec Aphis ligustri Mosley, 1841; Kaltenbach, 1843
crispi Davletschina, 1964 = ? rhoicola Hille Ris Lambers, 1956
crithmi Buckton, 1886 to Dysaphis Börner, 1931
croomiae Shinji, 1922 = ? citricola van der Goot, 1912
crypta Pack & Knowlton, 1929 to Aphis (Zyxaphis)
cucubali Ph. F. Gmelin, 1758 nomen oblitum
cucubali Passerini, 1863 to Brachycolus Buckton, 1879
cucumeris Forbes, 1882 = gossypii Glover, 1877
cucurbitae Buckton, 1879 = gossypii Glover, 1877
cuscutae Davis, 1919 BM
cydoniae Boisduval, 1867 = pomi de Geer, 1773
cymbalariae Schouteden, 1900 = Myzus (Nectarosiphon) persicae (Sulzer, 1776)
cynarae Theobald, 1915 to Dysaphis Börner, 1931
cynoglossi Lichtenstein, 1884
cynoglossi Walker, 1848 = Amphorophora rubi (Kaltenbach, 1843)
cyperi Walker, 1848 to Thripsaphis (Trichocallis)
cytisorum Hartig, 1841 BM HRL
*laburni* Kaltenbach, 1843
dahliae Mosley, 1841 = fabae Scopoli, 1763
dallmani Theobald, 1924 = Sitobion fragariae (Walker, 1848)
dasiphorae Szelegiewicz & Holman, 1971 HRL
dauci Fabricius, 1775 (type of Doralis Leach, 1827) to Semiaphis van der Goot, 1913
davidsoniella Theobald, 1927 = rumicis Linnaeus, 1857
davisi Patch, 1917 = maculatae Oestlund, 1887
davisiana del Guercio, 1913 = sambuci Linnaeus, 1857
davletshinae Hille Ris Lambers, 1966 HRL
*althaeae* Nevsky, 1929 (*Aphidula*)
*althaeae* Nevsky, 1929 nec althaea Harris, 1776
davletshinae subsp. afghanica Narzikulov & Umarov, 1971
*althaeae* subsp. *afghanica* Narzikulov & Umarov, 1971
debilicornis (Gillette & Palmer, 1929) (Anuraphis) BM HRL
decepta Hottes & Frison, 1931 paratypes BM HRL
delicatula Heer, 1853 (Aphis ?) Fossil, correct genus unknown

A. deposita Walker, 1848 = Myzus (Nectarosiphon) persicae (Sulzer, 1776)
derelicta Walker, 1849 = Myzus (Nectarosiphon) persicae (Sulzer, 1776)
despecta Walker, 1849 nomen dubium
detracta Walker, 1849 = Brachycaudus helichrysi (Kaltenbach, 1843)
deutziae Shinji, 1922 = citricola van der Goot, 1912
devecta Walker, 1849 to Dysaphis Börner, 1931
dianthi Schrank, 1801 = Myzus (Nectarosiphon) persicae (Sulzer, 1776)
diervilla-lutea Rafinesque, 1817 invalid
diminuta Walker, 1850 = ? Brachycaudus helichrysi (Walker, 1843)
diospyri Thomas, 1879
diphaga Walker, 1852 BM HRL
*scorodoniae* del Guercio, 1911
diplepha Rafinesque, 1818 invalid
dipsaci Schrank, 1801 = Macrosiphum rosae (Linnaeus, 1758)
dirhoda Walker, (1848) 1849 type of Metopolophium Mordvilko, 1914 q.v. and of Goidanichiellum Martelli, 1950
discolor Burmeister, 1836 = ? Euceraphis punctipennis (Zetterstedt, 1828)
discolor Haldeman, 1844 preoccupied
discrepans Koch, 1857 nomen nudum in litteris
dispar Walker, 1848 nomen dubium
dissita Walker, 1849 nomen dubium
diversa Walker, 1848 nomen dubium
dolichi Montrouzier, 1861 = craccivora Koch, 1854
dolichii Signoret, 1885
donacis Passerini, 1862 to Melanaphis van der Goot, 1917
dragocephalus Bozhko, date ?
droserae Takahashi, 1921 BM
dryophila Schrank, 1801 (type of Vacuna Passerini, 1860 nec von Heyden, 1837) to Thelaxes Westwood, 1840
dubia Brodie, 1845 fossil, ? to Diptera
dubia Curtis, 1842 = Myzus (Nectarosiphon) persicae (Sulzer, 1776)
durantae Theobald, 1918 = punicae Passerini, 1863
durranti Das, 1918 = punicae Passerini, 1863
ecballii Rusanova, (1943 nomen nudum) 1948
edentula Buckton, 1879 = Rhopalosiphum insertum (Walker, 1849)
egens Walker, 1852 nomen dubium
egomae Shinji, 1922
egressa Walker, 1849 = Myzus (Nectarosiphon) persicae (Sulzer, 1776)
elatinoidei Nevsky, 1929
elegantula Szelegiewicz, 1963 HRL

A. epilobiaria Theobald, 1927 (Walker, m.s.) BM
epilobii Kaltenbach, 1843 BM HRL
*triphaga* Walker, 1852
*vetusta* Hottes, 1927
*virgata* del Guercio, 1911
epilobii Kittel, 1827 invalid
epilobiina Walker, 1849 = praeterita Walker, 1849
epipactis Theobald, 1927 = ilicis Kaltenbach, 1843
equiseticola Ossiannilsson, 1964 BM HRL
erecta del Guercio, 1911 = fabae Scopoli, 1763
ericae Walker, 1852 (Hardy m.s.) nomen dubium
erigeron-canadense Rafinesque, 1818 invalid
erigeron-philadelphicum Rafinesque, 1817 invalid
erigeron-strigosum Rafinesque, 1818 invalid
erigerontis Holman, 1966 to Aphis (Protaphis)
eriobotryae Schouteden, 1905 = pomi de Geer, 1773
eriogoni Cowen, 1895 to Braggia Gillette & Palmer, 1929
eriophori Walker, 1848 to Ceruraphis Börner, 1926
eryngii E.E. Blanchard, 1923 = fabae subsp. eryngii E.E. Blanchard, 1923
eryngiiglomerata Bozhko, 1963 HRL
erysimi Kaltenbach, 1843 type of Lipaphis Mordvilko, 1928 q.v.
esulae (Börner, 1940) (Pergandeida) HRL
etiolata Stroyan, 1952 types BM HRL
eugeniae van der Goot, 1917 BM HRL
*hardyi* Eastop, 1966 types BM HRL
euonymi Fabricius, 1775 BM HRL
*cognatella* Jones, 1943
euonymi auctt. prior to 1950, nec Fabricius, 1775 = fabae subsp. solanella Theobald, 1914
eupatorii Oestlund, 1886 = nostras Hottes, 1930
eupatorii Passerini, 1863
euphorbiae Kaltenbach, 1843 BM HRL
euphorbiae subsp. tirucallis Hille Ris Lambers, 1954 = tirucallis Hille Ris Lambers, 1954
euphorbiae Walker, 1849 nomen dubium
euphrasiae Walker, 1849 = Myzus cerasi (Fabricius, 1775)
euryae Shinji, 1922 nomen dubium
excelsioris Dahlbom, 1851 nomen dubium
explorata Rusanova, 1948
*paliuri* var. *exploratus* Rusanova, 1943

A. exsors Rusanova (1943 nomen nudum) 1948
extranea Walker, 1849 nomen nudum
exul Walker, 1849 = sambuci Linnaeus, 1758
fabae Blanchard, 1840 = fabae Scopoli, 1763
fabae Scopoli, 1763 BM HRL
*abietaria* Walker, 1852 type BM
*addita* Walker, 1849
*adducta* Walker, 1849 lectotype BM
*advena* Walker, 1849 type BM
*aparines* Fabricius, 1775
*aparinis* Blanchard, 1840
*apii* Theobald, 1925 types BM
*apocyni* Koch, 1854
*atriplicis* Fabricius, 1775 nec Linnaeus, 1758
*brevisiphona* Theobald, 1913 types BM
*chaerophylli* Koch, 1854
*cynariella* Theobald, 1924 *(Anuraphis)*
*dahliae* Mosley, 1841
*erecta* del Guercio, 1911
*fabae* Blanchard, 1840
*fumariae* Blanchard, 1840 partim
*hortensis* Fabricius, 1781
*indistincta* Walker, 1849
*inducta* Walker, 1849 type BM
*nerii* Kaltenbach, 1843 nec Boyer de Fonscolombe, 1841
*phlomoidea* del Guercio, 1911
*polyanthis* Passerini, 1863 nec J.F. Gmelin, 1790
*roseum* Macchiati, 1881 *(Myzus)*
*rubra* Macchiati, 1884 *(Myzus)*
*rubrum* del Guercio, 1900 *(Myzus)*
*rumicis* auctt. prior to 1930 nec Linnaeus
*silybi* Passerini, 1861
*sinensis* del Guercio, 1900
*thlaspeos* Schrank, 1801
*translata* Walker, 1849 type BM
*tuberosae* Boyer de Fonscolombe, 1841
*valerianina* del Guercio, 1911
*watsoni* Theobald, 1929

A. fabae subsp. cirsiiacanthoidis Scopoli, 1763 BM HRL
*acanthi* Schrank, 1801
*cardui* var. *naumburgensis* Franssen, 1927
*castanea* Koch, 1854
*cirsina* Ferrari, 1872
? *neoreticulata* Theobald, 1927 BM
? *philadelphi* Börner, 1921
? *reticulata* Theobald, 1922 nec Wilson, 1915
*serratulae* Schrank, 1801
fabae subsp. eryngii E.E. Blanchard, 1923 BM
*eryngii* E.E. Blanchard, 1923
fabae subsp. mordvilkoi Börner & Janisch, 1922
fabae subsp. solanella Theobald, 1914 types BM HRL
*bazzii* E.E. Blanchard, 1923
? *compositae* Theobald, 1915 types BM
*dusmeti* Gomez Menor, 1950 partim
*euonymi* auctt. prior to 1950 nec Fabricius, 1775
*insularis* E.E. Blanchard, 1923
*solanella* Theobald, 1914
*solanophilus* E.E. Blanchard, 1923
fagi Linnaeus, 1767 type of Phyllaphis Koch, 1857 q.v.
falcariae Rusanova, 1948
falcarii Rusanova, 1943 nomen nudum
falki Hoffmann, 1972 = pseudocomosa Stroyan, 1972
familiaris Walker, 1848 = Brachycaudus helichrysi (Kaltenbach, 1843)
farfarae Koch, 1854 to Anuraphis del Guercio, 1907
farinosa Gmelin, 1790 type of Leucosiphon Börner, 1952 BM HRL
*furcula* Zetterstedt, 1838
*neosaliceti* E.E. Blanchard, 1939 nec Shinji & Kondo, 1938
*neosaliceti* Bertels, 1973 nec Shinji & Kondo, 1938
*saliceti* Kaltenbach, 1843
*salicicola* Thomas, 1878 *(Siphonophora)*
*spectabilis* Ferrari, 1872
farinosa subsp. yanagicola Matsumura, 1917
*neosaliceti* Shinji & Kondo, 1938
? *neosalici* Shinji, 1938
feminea Hottes, 1930 (Aphis (Anuraphis)) BM HRL
*tuberculata* Patch, 1914 nec von Heyden, 1837
ferruginea-striata Essig, 1941 = Dysaphis apiifolia (Theobald, 1923)

A. ficicola Takahashi, 1921 = Toxoptera odinae (van der Goot, 1917)
ficus Theobald, 1918 BM HRL = ? gossypii Glover, 1877
filaginis Boyer de Fonscolombe, 1841 = Pemphigus populinigrae (Schrank, 1801)
filaginis Lichtenstein, 1884 nomen nudum
filipendulae Matsumura, 1917
fitchii Sanderson, 1902 = Rhopalosiphum insertum (Walker, 1849)
flava Nevsky, 1929 = gossypii Glover, 1877
flava (Nevsky, 1929) (Aphidula) = gossypii Glover, 1877
flaveola Walker, 1840 nomen dubium
flavida Ivanovskaja, 1959 to Brachyunguis Das, 1918
flipandulae Matsumura lapsus pro filipendulae Matsumura, 1917
floridanae Tissot, 1932
floris-rapae Curtis, 1842 = Brevicoryne brassicae (Linnaeus, 1758)
foliorum ulmi de Geer, 1773 invalid trinominal = Eriosoma (Schizoneura) ulmi (Linnaeus, 1758)
folsomii Davis, 1908 cotypes BM HRL
*parthenocissi* Williams, 1911
forbesi Weed, 1889 BM HRL
formosana Takahashi, 1921 to Melanaphis van der Goot, 1917
foveolata del Guercio, 1916 = nerii Boyer de Fonscolombe, 1841
fragariae Walker, 1848 to Sitobion Mordvilko, 1919
frangulae Kaltenbach, 1845 member of the gossypii complex q.v.
*gallicae* del Guercio, 1911
frangulae Koch, 1855 (type of Doralina Börner, 1940) = frangulae Kaltenbach, 1845
frangulae subsp. beccabungae Koch, 1855 see gossypii complex
*beccabungae* Koch, 1855
frangulae subsp. testacea Thomas, 1968 see gossypii complex
franzi Holman, 1975 HRL
fraserae Gillette & Palmer, 1929 metatypes BM HRL
fraterna Strom, 1938 to Brevicoryne van der Goot, 1915
fraxini Fabricius, 1777 to Prociphilus Koch, 1857
fraxini Geoffroy, 1762 invalid
frequens Walker, 1848 to Diuraphis (Holcaphis)
frigidae Oestlund, 1886 type of Epameibaphis Oestlund, 1922 q.v.
frondosae Oestlund, 1886 = coreopsidis Thomas, 1878
fukii Shinji, 1922 BM HRL
*fukis* Monzen, 1927 (*Myzus*)
*petasiticola* Takahashi, 1924
fulva Walker, m.s. see Doncaster, 1961: 76

A. fumanae Remaudière & Leclant, 1972 HRL
fumariae Blanchard, 1840 partim = fabae Scopoli, 1763
fumipennella Fitch, 1855 to Melanocallis Oestlund, 1922
funesta Hottes & Frison, 1931 paratypes BM
funitecta (Börner, 1950) (Brachysiphum) BM HRL
furcata Patch, 1914 = ? Rhopalosiphum cerasifoliae (Fitch, 1855)
furcipes Rafinesque, 1817 invalid
furcula Zetterstedt, 1838 = farinosa Gmelin, 1790
fusca Geoffroy, 1762 = Stomaphis quercus (Linnaeus, 1758)
fusciclava Rafinesque, 1817 invalid type of Cladoxus Rafinesque, 1818, not an aphid.
fuscipennis Zetterstedt, 1840 = Symydobius oblongus (von Heyden, 1837)
galeopsidis Kaltenbach, 1843 (type of Myzella Börner, 1930) to Cryptomyzus Oestlund, 1922
galiae Ivanovskaja, 1959
galii Kaltenbach, 1843 = galiiscabri Schrank, 1801
galiiscabri Schrank, 1801 BM HRL
*galii* Kaltenbach, 1843
gallarum Gmelin, 1790 = Tetraneura ulmi (Linnaeus, 1758)
gallarum Kaltenbach, 1856 = Cryptosiphum artemisiae Buckton, 1879
gallarum abietis de Geer, 1773 = Adelges (Sacchiphantes) abietis (Linnaeus, 1758)
gallarum ulmi de Geer, 1773 = Tetraneura ulmi (Linnaeus, 1758)
gallicae del Guercio, 1911 = frangulae Kaltenbach, 1845
genistae Kaltenbach, 1843 (type of Uraphis del Guercio, 1907) = genistae Scopoli 1763 + kaltenbachi Hille Ris Lambers, 1955
genistae Scopoli, 1763 BM HRL
*genistae* Kaltenbach, 1843 partim
gentianae (Börner, 1940) (Doralina) BM HRL
geranii Hereman, 1840 nomen nudum
geranii Kaltenbach, 1862 = Acyrthosiphon malvae (Mosley, 1841)
gerardiae (Thomas, 1897) (Siphonophora) BM HRL
gerardianae Mordvilko, 1929 BM HRL
gibbosa Rafinesque, 1818 invalid
gillettei Cowen, 1895 = helianthi Monell, 1879
githaginella Theobald, 1927 = nasturtii Kaltenbach, 1843
gladioli Felt, 1908 = Dysaphis tulipae (Boyer de Fonscolombe, 1841)
glandulosa Kaltenbach, 1846 type of Pleotrichophorus Börner, 1930 q.v.
glareosae Bozhko, (1959) 1963 (Aphis (Pergandeida)) BM HRL
glechomae Walker, 1848 nomen dubium
glyceriae Kaltenbach, 1843 type of Sipha Passerini, 1860 q.v.

A. glycines Matsumura, 1917 BM HRL
*justiceae* Shinji, 1922
gnaphalii Shinji, 1922 nomen dubium
gnaphalii Walker, 1849 = Sitobion avenae (Fabricius, 1775)
gomphoricarpi van der Goot, 1912 = nerii Boyer de Fonscolombe, 1841
gossypii Glover, 1877 is a member of a group of interfertile species and subspecies not yet fully understood but including:-
capsellae Kaltenbach, 1843 BM HRL
frangulae Kaltenbach, 1845 BM HRL
*? mamontovae* Davletshina, 1964
*rhamni* Kaltenbach, 1843 nec Boyer de Fonscolombe, 1841
frangulae subsp. beccabungae Koch, 1855 BM HRL
frangulae subsp. testacea Thomas, 1968 BM
gossypii Glover, 1877 BM HRL
*affinis* var. *gardeniae* del Guercio, 1913
*aurantii* var. *limonii* (del Guercio, 1917) (*Toxoptera*)
*bauhiniae* Theobald, 1918
*bryophyllae* Shinji, 1922
*chloroides* Nevsky, 1929
*circezandis* Fitch, 1870, suppressed Russell, 1968
*? citri* Ashmead, 1887
*citrulli* Ashmead, 1882
*colocasiae* Matsumura, 1917
*commelinae* Shinji, 1922
*commelinae* Shinji, 1924 (Cerosipha)
*? convolvulicola* Ferrari, 1872
*cucumeris* Forbes, 1882
*cucurbiti* Buckton, 1879 type BM
*? ficus* Theobald, 1918 type BM
*flava* Nevsky, 1929
*flava* Nevsky, 1929 (*Aphidula*)
*gossypii* var. *callicarpae* Takahashi, 1921
*gossypii* var. *viridula* Nevsky, 1929
*hederella* Theobald, 1915 type BM
*helianthi* del Guercio, 1916
*? heliotropii* Macchiati, 1885
*hibiscifoliae* Shinji, 1922
*inugomae* Shinji, 1922
*leonuri* Takahashi, 1921 (*Toxoptera*)
*ligustriella* Theobald, 1914 type BM

*? lilicola* Williams, 1911
*malvacearum* van der Goot ex Das, 1918
*malvoides* Das, 1918 nec van der Goot, 1917
*? minuta* Wilson, 1911
*monardae* Oestlund, 1887
*? oxalis* Macchiati, 1884
*parvus* Theobald, 1915
*? perillae* Shinji, 1922
*pomonella* Theobald, 1916 type BM
*pruniella* Theobald, 1918 type BM
*shirakii* Takahashi, 1921
*solanina* Passerini, 1863
*tectonae* van der Goot, 1917
*tridacis* Theobald, 1929
*vitifoliae* Shinji, 1922

A. gossypii var. callicarpae Takahashi, 1921 = gossypii Glover, 1877
gossypii var. lutea Nevsky, 1929
gossypii var. obscura Nevsky, 1929
gossypii var. viridula Nevsky, 1929 = gossypii Glover, 1877
gracilis Walker, 1852 = Hyalopterus pruni (Geoffroy, 1762)
graminum Rondani, (1847) 1852 type of Schizaphis Börner, 1931 q.v.
granaria Kirby, 1897 (type of Sitobion Mordvilko, 1914) = Sitobion avenae Fabricius, 1775)
grandiflorae Holman, 1961 nomen nudum
grandis Juchnevitch, (1968) 1970
gregalis Knowlton, 1928 cotype BM HRL
*nigragregalis* Knowlton, 1935 paratype BM
grosmannae (Börner, 1952) (Toxopterina (Tuberculaphis)) HRL
grossulariae Kaltenbach, 1843 BM HRL
*penicillata* Buckton, 1879
gurnetensis Cockerell, 1921 Fossil, Oligocene
gutierrezis (Pack & Knowlton, 1929) (Anuraphis)
hamamelidis Pepper, 1950 BM HRL
hardyi Eastop, 1966 = eugeniae van der Goot, 1917
haroi Nieto Nafria, 1974 = cacaliasteris subsp. helvetica Hille Ris Lambers, 1947
healyi Cottier, 1953 BM HRL
hederae Kaltenbach, 1843 BM HRL
hederae f. boerneri Franssen, 1927 BM HRL
hederae f. pseudohederae Theobald, 1927 BM HRL

A. hederella Theobald, 1915 = gossypii Glover, 1877
hederiphaga Takahashi, 1966 cotypes BM HRL
helianthemi Ferrari, 1872 BM HRL
helianthemi subsp. obscura Bozhko (1957) 1961 (Aphis (Cerosipha))
helianthi del Guercio, 1917 = gossypii Glover, 1877
helianthi Monell, 1879 BM HRL
*gillettei* Cowen, 1895
*oxybaphi* Oestlund, 1887
*phenax* Cockerell, 1903 (*Myzus*)
*? sorensoni* Knowlton, 1928
*yuccae* Cowen, 1895
*yuccicola* Wilson, 1911
helichrysi Kaltenbach, 1843 to Brachycaudus van der Goot, 1913
heliotropii Macchiati, 1885 = ? gossypii Glover, 1877
heraclei Cowen, 1895 = heraclella Davis, 1919
heraclei Koch, 1856 = Anuraphis subterranea (Walker, 1852)
heraclella Davis, 1919 BM HRL
*heraclei* Cowen, 1895 nec Koch, 1854
heraclicola Shinji, 1939
hermistonii Wilson, 1915 to Aphis (Zyxaphis)
herniariae Mamontova, 1963 HRL?
hibernaculorum Boyer de Fonscolombe, 1841 nomen dubium
hibiscifoliae Shinji, 1922 = gossypii Glover, 1877
hieracii Kaltenbach, 1843 nec Schrank, 1801 (type of Submacrosiphum Hille Ris Lambers, 1931) = Nasonovia nigra (Hille Ris Lambers, 1931)
hieracii Schrank, 1801 BM HRL
hieracium-paniculatum Rafinesque, 1818 invalid, type of Dactynotus Rafinesque, 1818
hieracium-venosum Rafinesque, 1817 invalid
hiltoni Essig, 1922
hippophaes Walker, 1852 to Capitophorus van der Goot, 1913
hirsuta Germar & Berendt, 1856 type of Palaeosiphon Heie, 1967 fossil q.v.
hirticornis Walker, 1848 to Periphyllus van der Hoeven, 1863
hispanica Hille Ris Lambers, 1959 HRL
holci Ferrari, 1872 = Rhopalosiphum padi (Linnaeus, 1758)
holci Hardy, 1850 nomen nudum
holci Hardy of Hille Ris Lambers, 1939 (type of Holcaphis Hille Ris Lambers, 1939) = Diuraphis (Holcaphis) holci (Hille Ris Lambers, 1956)
hordei del Guercio, 1913 = craccivora Koch, 1854
hordei Kyber, 1815 = Sitobion avense (Fabricius, 1775)

A. horii Takahashi, 1923 BM HRL
*euscaphis* Paik, 1965 (*Sappaphis*) BM HRL
*sambuci* subsp. *horii* Takahashi, 1923
hortensis Fabricius, 1781 = fabae Scopoli, 1763
houghtonensis Troop, 1906 to Kakimia Hottes & Frison, 1931
humilis Walker, 1852 to Hyalopteroides Theobald, 1916
humuli Schrank, 1801 type of Phorodon Passerini, 1860 q.v.
humuli (Tseng & Tao, 1938) (Cerosipha) BM
hyoscyami Kittel, 1827 invalid
hyperici Monell, 1879 to Brachysiphum van der Goot, 1913
hypochoeridis (Börner, 1940) (Cerosipha) BM HRL
hypochoeridis Fabricius, 1779 to Uroleucon Mordvilko, 1914
hypochoeridis Hille Ris Lambers, lapsus pro hypochoeridis Börner, 1940
ichigo Shinji, 1922 BM HRL
ichigocola Shinji, 1924 BM HRL
idaei van der Goot, 1912 BM HRL
*mordvilkiana* Dobrovljansky, 1913
ilicicola Boisduval, 1867 = Lachnus roboris (Linnaeus, 1758)
ilicis Kaltenbach, 1943 BM HRL
*epipactis* Theobald, 1927 types BM
illata Walker, 1849 nomen dubium
illinoisensis Shimer, 1866 BM HRL
*ampelophila* del Guercio, 1913
*viticola* Thomas, 1878 (*Siphonophora*)
impacta Walker, 1849: xxxv nomen dubium
impacta Walker, 1849: xlviii nomen dubium
impatiensae Shinji, 1924 type of Eumyzus Shinji, 1929 q.v.
impatientis Thomas, 1878 HRL
impingens Walker, 1852 = Glyphina betulae (Linnaeus, 1758)
incerta Nevsky, 1929
incerta Walker, 1849 = ? Aulacorthum solani (Kaltenbach, 1843)
inculta Walker, 1849 nomen dubium
incumbens Walker, 1849 = Brachycaudus helichrysi (Kaltenbach, 1843)
indecisa Walker, 1849 = Aulacorthum solani (Kaltenbach, 1843)
indigoferae Shinji, 1922
indistincta Walker, 1849 = fabae Scopoli, 1763
inedita Rusanova, (1943 nomen nudum) 1948
infrequens (Knowlton, 1929)
*chrysothamnicola* subsp. *infrequens* Knowlton, 1929 (*Anuraphis*)
infuscata Koch, 1854 = Rhopalosiphum nymphaeae (Linnaeus, 1761)

A. inhaerens Walker, 1852 = Betulaphis quadrituberculata (Kaltenbach, 1843)
inserta Walker, 1849 to Rhopalosiphum Koch, 1854
insessa Walker, 1849 = Brachycaudus helichrysi (Kaltenbach, 1843)
insita Walker, 1849 to Ovatus van der Goot, 1913
insita Walker, 1852 = Brachycaudus (Acaudus) cardui (Linnaeus, 1758)
insititiae Koch, 1854 = Brachycaudus helichrysi (Kaltenbach, 1843)
insons Hottes, 1930
instabilis Buckton, 1897 = Brachycaudus (Acaudus) cardui (Linnaeus, 1758)
insularis Blanchard, 1923 = fabae subsp. solanella Theobald, 1914
insularis Hille Ris Lambers, 1959 nec E.E. Blanchard, 1923 = longirostrata Hille Ris Lambers, 1966
internata Walker, 1849 nomen dubium
intybi Koch, 1855 BM HRL
inugomae Shinji, 1922 = gossypii Glover, 1877
inulae Walker, 1849 (type of Ovatoides Börner, 1939) to Ovatus (Ovatoides)
iridis del Guercio, 1900 = Dysaphis tulipae (Boyer de Fonscolombe, 1841)
iridis Franssen, 1928 = newtoni Theobald, 1927
isabellina del Guercio, 1917 = craccivora Koch, 1854
isatidis Boyer de Fonscolombe, 1841 = Brevicoryne brassicae (Linnaeus, 1758)
iteae (Tissot, 1932) (Anuraphis) BM
jaceae Linnaeus, 1758 (type of Uromelan Mordvilko, 1914) to Uroleucon (Uromelan)
jacobaeae Schrank, 1801 BM HRL
*senecionis* Koch, 1854
*senecionisdoriae* Bozhko, 1953
jacobea-balsamita Rafinesque, 1818 invalid
jani Ferrari, 1872
janischi (Börner, 1940) (Doralis) HRL
japonica Essig & Kuwaṇa, 1918 = Ovatus malisuctus (Matsumura, 1918)
juglandina Walker, 1848 nomen dubium
juglandis Blanchard, 1840 nomen dubium
juglandis Frisch, 1734 prebinominal, type of Callaphis Walker, 1870 q.v.; and of Callipterinola Strand, 1928; and of Callipterus Koch, 1855 nec Agassiz, 1846; and of Panaphis Kirkaldy, 1904; and of Ptychodes Buckton, 1881
juglandis Goeze, 1778 to Callaphis Walker, 1870
junci Kaltenbach, 1875 = Ceruraphis eriophori (Walker, 1848)
juniperi de Geer, 1773 (type of Cupressobium Börner, 1940) to Cinara Curtis, 1835
jurineae Bozhko, 1953 HRL
justiceae Shinji, 1922 = glycines Matsumura, 1917

A. kachena Hottes, 1934 to Nearctaphis Shaposhnikov, 1950
kalopanacis (Hori, 1927) (Pergandeida)
kaltenbachi Hille Ris Lambers, 1955 BM HRL
*genistae* Kaltenbach, 1843 partim
*ononidis* Schouteden, 1903 (*Pergandeida*) nec Kaltenbach, 1845 (Aphis)
*schoutedeni* Hille Ris Lambers, 1956 nec Kirkaldy, 1906
kambashirae Shinji, 1922 nomen dubium
klimeschi (Börner, 1950) (Doralida) BM HRL
knowltoni Hottes & Frison, 1931 BM HRL
kochi Schouteden, 1903 (nomen novum pro pyri Koch, 1854 nec Boyer de Fonscolombe, 1841) = Anuraphis farfarae (Koch, 1854)
kogomecola Matsumura, 1917
kurosawai Takahashi, 1921 BM HRL
*artemisiae* Takahashi ex Shinji, 1935 nec Boyer de Fonscolombe, 1841; Passerini, 1860.
*artemisifoliae* Shinji, 1922
kyberi Hottes, 1930 (nomen novum pro hordei del Guercio, 1913 nec Kyber, 1815) = craccivora Koch, 1854
laburni Kaltenbach, 1843 = cytisorum Hartig, 1841
laburni var. labiatarum Rusanova, 1943 nomen nudum
laciniariae Gillette & Palmer, 1929 paratype BM HRL
lactucae Linnaeus, 1758 type of Hyperomyzus Börner, 1933 q.v.
lactucae Shinji, 1922
lagerstroemiae Lichtenstein, 1884 nomen nudum
lambersi (Börner, 1940) (Doralina) BM HRL
lamii Koch, 1854 to Brachycaudus (Acaudus)
lamiorum (Börner, 1950) (Doralina) BM HRL
lanata Salisbury, 1816 = Eriosoma lanigerum (Hausmann, 1802)
lanata Zetterstedt, 1840 = Asiphum tremulae (Linnaeus, 1761)
lanigera Hausmann, 1802 (type of Myzoxylus Blot, 1831) to Eriosoma Leach, 1818
lantanae Koch, 1854 (type of Chaitophoroides Mordvilko, 1909 (1908)), BM HRL
*coriaria* Börner, 1952
*setacea* Hille Ris Lambers, 1935 (*Doralis*)
lantanaella Theobald, 1925 = Ceruraphis eriophori (Walker, 1848)
lapathi Börner & Blunck, 1916 = Dysaphis radicola (Mordvilko, 1897)
lappae Koch, 1854 to Dysaphis Börner, 1931
lappae f. cirsii Passerini, 1874 = ? Dysaphis lappae subsp. cirsii (Börner, 1950)

A. largiflua Menge, 1856 fossil
laricis Walker, 1848 (type of Cinaria Borner, 1939) = Cinara laricis (Hartig, 1839
lata Walker, 1850 = Brachycaudus (Acaudus) cardui (Linnaeus, 1758)
lateralis Walker, 1848 = Brachycaudus (Acaudus) cardui (Linnaeus, 1758)
lathyri Mosley, 1841 = Acyrthosiphon pisum (Harris, 1776)
lavaterae Kittel, 1827 invalid
leguminosae Theobald, 1915 = craccivora Koch, 1854
lentiginis Buckton, 1879 = Dysaphis (Pomaphis) plantaginea (Passerini, 1860)
leonopodii errore pro leontopodii
leontodoniella Theobald, 1915 = Dysaphis radicola (Mordvilko, 1897)
leontodontis (Börner, 1950) (Doralina) BM HRL
leontopodii Schouteden, 1903 = Brachycaudus helichrysi (Kaltenbach, 1843)
leonulii Shinji, 1935
leptorhyncha David, Sekhon & Bindra, 1970 BM HRL
leucanthemi Scopoli, 1763 = Brachycaudus (Acaudus) cardui (Linnaeus, 1758)
lichtensteini Leclant & Remaudière ex Remaudière & Leclant, 1972 HRL
ligustici Fabricius, 1781 nomen dubium
ligustici scotici Stewart, 1817 nomen nudum
ligustri Kaltenbach, 1843 = Myzus (Nectarosiphon) ligustri (Mosley, 1841)
ligustri Mosley, 1841 (type of Amyzus Hille Ris Lambers, 1946) to Myzus (Nectarosiphon)
ligustriella Theobald, 1914 = gossypii Glover, 1877
liliago F.P. Müller, 1968 paratypes BM HRL
lilicola Williams, 1911 = ? gossypii Glover, 1877
lilii Lichtenstein, 1884 nomen nudum
limonii Contarini, 1847 to Staticobium Mordvilko, 1914
limonii Walker, 1852 = Staticobium staticis (Theobald, 1923)
linariae Lichtenstein, 1884 nomen nudum
linguae Opmanis, 1928 = nasturtii Kaltenbach, 1843
lini (Bozhko, 1959) (Brachyunguis (Parabrachyunguis)) HRL
lini Holman, 1966 paratypes BM HRL
lithospermi Wilson, 1915
littoralis Walker, 1848 to Sipha Passerini, 1860
longicauda (Baker, 1920) (Anuraphis) fundatrix cotype BM
longicaudatus Millière, 1853 fossil
longicornis Menge, 1856 fossil
longini Huculak, 1968 HRL
longipes Dufour, 1833 (type of Pterochlorus Passerini, 1850) = Lachnus roboris (Linnaeus, 1758)

A. longirostrata Hille Ris Lambers, 1966 HRL
*insularis* Hille Ris Lambers, 1959 nec E.E. Blanchard, 1923 HRL
*longirostris* Börner, 1950 (*Cerosipha*) nec Aphis longirostris Fabricius, 1776
longirostris Fabricius, 1787 to Stomaphis Walker, 1870
longisetosa A.N. Basu, 1969 HRL
*ruborum* subsp. *longisetosa* A.N. Basu, 1969 HRL
longituba Hille Ris Lambers, 1966 HRL
lonicerae Boyer de Fonscolombe, 1841 = Hyadaphis foeniculi (Passerini, 1860)
lonicerae Monell, 1879 (type of Gypsoaphis Oestlund, 1923) = Gypsoaphis oestlundi Hottes, 1930
lonicerae Mosley, 1841 = Hyadaphis foeniculi (Passerini, 1860)
lonicerae Siebold, 1839 (type of Judenkoa Hille Ris Lambers, 1949) to Rhopalomyzus (Judenkoa)
lonicericola Williams, 1911 type of Alphitoaphis Hottes, 1926 q.v.
loti Kaltenbach, 1862 (type of Doralida Börner, 1950) = craccivora Koch, 1854
lotiradicis Stroyan, 1972 HRL
lugentis Williams, 1911 BM HRL
*nyctalis* Hottes & Frison, 1931 paratypes BM
lupini Gillette & Palmer, 1929 paratypes BM HRL
lupinehansoni Knowlton, 1935
lupuli Rusanova, (1943 nomen nudum) 1948
luridus Hottes & Frison, 1931 = saniculae Williams, 1911
lutescens Monell, 1879 = nerii Boyer de Fonscolombe, 1841
luzulae Kaltenbach, 1862 = Ceruraphis eriophori (Walker, 1848)
lychnidis Linnaeus, 1758 (as Acaudus = type of Acaudus van der Goot, 1913) to Brachycaudus (Acaudus)
lychnidis dioicae Cederhjelm, 1798 = Brachycaudus (Acaudus) lychnidis (Linnaeus, 1758)
lycopicola (Shinji, 1941) (Cerosipha)
lycopsidis Walker, 1846 = Sitobion avenae (Fabricius, 1775)
lythri Schrank, 1801 to Myzus Passerini, 1860
macrostyla Heer, 1856 fossil
mactata Walker, 1849 = Rhopalosiphum insertum (Walker, 1849)
maculata von Heyden, 1837 to Pterocallis Passerini, 1860
maculatae Oestlund, 1887 BM HRL
*davisi* Patch, 1917 BM
maculella Fitch, 1855 to Monellia Oestlund, 1887
madronae (Essig, 1926) (Anuraphis) BM
magnipennis Haliday, see Lichtenstein, 1884: 28
magnoliae Macchiati, 1884

A. magnopilosa Nevsky, 1929
mahaleb Koch, 1854 = Myzus lythri (Schrank, 1801)
maidiradicis Forbes, 1891 BM HRL
*? menthae-radicis* Cowen, 1895
maidis Fitch, 1856 to Rhopalosiphum Koch, 1854
mali Fabricius, 1775 = pomi de Geer, 1773
mali var. bivincta Fitch, 1855
mali var. fulviventris Fitch, 1855
mali var. immaculata Fitch, 1855
mali var. nigricollis Fitch, 1855
mali var. nigriventris Fitch, 1855
mali var. obsoleta Fitch, 1855
mali var. pallidicornis Fitch, 1855
mali var. tergata Fitch, 1855
mali var. thoracica Fitch, 1855
mali var. triseriata Fitch, 1855
malifoliae Fitch, 1855 = Nearctaphis crataegifoliae (Fitch, 1851)
malvacearum van der Goot ex Das, 1918 (nomen novum pro malvoides Das, 1918 nec van der Goot, 1917) = gossypii Glover, 1877
malvae Koch, 1854 = umbrella (Börner, 1950)
malvae Mosley, 1841 to Acyrthosiphon Mordvilko, 1914
malvoides Das, 1918 = gossypii Glover, 1877
malvoides van der Goot, 1917 = citricola van der Goot, 1912
mammulata Gimingham & Hille Ris Lambers, 1949 BM HRL
*? rhamnicola* Mamontova, 1953 types HRL
mamontovae Davletshina, 1964 = ? frangulae Kaltenbach, 1845
manitobensis Robinson & Rojanavongse, 1976 paratype BM HRL
marginella Fitch, 1855 = Monellia costalis (Ftich, 1855)
marginipennis Haldeman, 1844 nomen dubium
marthae Essig, 1953 paratypes BM HRL
marutae Oestlund, 1886 = Brachycaudus helichrysi (Kaltenbach, 1843)
masoni Richards, 1963 paratype BM HRL
mathiolae Theobald, 1918 = nasturtii Kaltenbach, 1843
mathiolellae Theobald, 1918 = Lipaphis erysimi (Kaltenbach, 1843)
maydis Fitch lapsus pro maidis
mayeri Gmelin, 1790 nom. nov. pro tanacenti Mayer, 1779
medicaginis Koch, 1854 BM HRL
meijigusae Shinji, 1922
melampyrum-latifolium Rafinesque, 1817 invalid
melissae Walker, 1852 = Ovatus crataegarius (Walker, 1850)
menthae Walker, 1852 = Ovatus crataegarius (Walker, 1850)

A. menthae-radicis Cowen, 1895 = ? maidiradicis Forbes, 1891
middletonii Thomas, 1879 BM
? *swezeyi* Fullaway, 1910
millefolii de Geer, 1773 (type of Dielcysmura Mordvilko, 1914) to Macrosiphoniella del Guercio, 1911
mimosae Ferrari, 1872 = craccivora Koch, 1854
mimuli Oestlund, 1887 HRL
minima (Tissot, 1932) (Anuraphis) BM
minuta Wilson, 1911 = ? gossypii Glover, 1877
mirifica (Börner, 1950) (Doralina) BM HRL
miscanthi Takahashi, 1921 = Melanaphis formosana (Takahashi, 1921)
mitsubae Shinji, 1922 = ? citricola van der Goot, 1912
mizutakarashi Shinji, 1939
mokulen Shinji, 1922 = Toxoptera odinae (van der Goot, 1917)
molluginis (Börner, 1950) (Doralina) BM HRL
molluginis Koch, 1854 = Myzus cerasi (Fabricius, 1775)
momifoliae Shinji, 1924 = Elatobium momii (Shinji, 1922)
momii Shinji, 1922 to Elatobium Mordvilko, 1914
monardae Oestlund, 1887 = gossypii Glover, 1877
mongolica Szelegiewicz, 1963
montana Rafinesque, 1814 invalid
montanicola Hille Ris Lambers, 1950 HRL
morae Kittel, 1827 invalid
mordvilkiana Dobrovljansky, 1913 = idaei van der Goot, 1912
mordwilkoi Börner & Janisch, 1922 = fabae subsp. mordwilkoi Börner & Janisch, 1922
mori Clarke, 1903
morletti Heer, 1853 fossil
mulini Hille Ris Lambers, 1974 HRL
mulinicola Hille Ris Lambers, 1974 HRL
mumfordi Takahashi, 1935
musae Schouteden, 1906 to Rhopalosiphum Koch, 1854
myopori Macchiati, 1882
myosotidis Koch, 1854 (type of Brachycaudus van der Goot, 1913) = Brachycaudus helichrysi (Kaltenbach, 1843)
myricae Kaltenbach, 1843 (type of Agrioaphis Walker, 1870) to Myzocallis Passerini, 1860
myrmecaria Boisduval, 1867 = ? Smynthurodes betae Westwood, 1849
napelli Schrank, 1801 (type of Brachycaudina Börner, 1930) to Brachycaudus (Acaudus)
narzikulovi Szelegiewicz, 1963

A. nasturtii Kaltenbach, 1843 BM HRL
*abbreviata* Patch, 1912 BM
*acetosella* Theobald, 1918 BM
*cathartica* del Guercio, 1909
*githaginella* Theobald, 1927 types BM
*linguae* Opmanis, 1928
*mathiolae* Theobald, 1918 types BM
*neopolygoni* Theobald, 1927 presumed type BM
*pedicularis* Buckton, 1879 lectotype BM
? *plantaginifolii* Nevsky, 1929
*polygoni* van der Goot, 1912 nec Walker, 1848; Macchiati, 1885
*rhamni* auctt. nec Boyer de Fonscolombe, 1841
*transiens* Walker, 1849
? *zizyphi* Theobald, 1922 type BM
neilliae Oestlund, 1887 BM HRL
nelsonensis Cottier, 1953 BM
neogillettei Palmer, 1938 BM HRL
neomexicana (W.P. Cockerell & T.D.A. Cockerell, 1901) (Myzus) BM HRL
*ribigillettei* Knowlton & Allen, 1939 paratypes BM HRL
*ribis* Sanborn, 1904 nec Linnaeus, 1758, O.F. Müller, 1776
*sanborni* Patch, 1914
*tonahasa* Hottes, 1950
neopolygoni Theobald, 1927 = nasturtii Kaltenbach, 1843
neopolygoni Shinji, 1939
neoreticulata Theobald, 1927 = ? fabae subsp. cirsiiacanthoidis Scopoli, 1763
*reticulata* Theobald, 1922 nec Wilson, 1915
neosaliceti Bertels, 1973 lapsus pro neosaliceti E.E. Blanchard, 1939
neosaliceti E.E. Blanchard, 1939 = farinosa Gmelin, 1790
neosaliceti Shinji & Kondo, 1938 = farinosa subsp. yanagicola Matsumura, 1917
neosalici Shinji, 1938 = ? farinosa subsp. yanagicola Matsumura, 1917
neospiraeae Takahashi, 1966 HRL
nepetae Kaltenbach, 1843 BM HRL
neriastri Boisduval, 1867 = nerii Boyer de Fonscolombe, 1841
nerii Boyer de Fonscolombe, 1841 BM HRL
*asclepiadis* Passerini, 1853 (*Myzus*) nec Fitch, 1851
*calotropidis* del Guercio, 1916
*foveolata* del Guercio, 1916
*gomphoricarpi* van der Goot, 1912
*leptadeniae* Vuillet & Vuillet, 1914 (*Siphonophora*)

*lutescens* Monell, 1879
*neriastri* Boisduval, 1867
*nerii* de Stefani Perez, 1901 (*Cryptosiphum*)
*nigripes* Theobald, 1914 types BM
*paolii* del Guercio, 1916
A. nerii Kaltenbach, 1843 = fabae Scopoli, 1763
nervosa Zetterstedt, 1840 = Pterocomma salicis (Linnaeus, 1758)
newtoni Theobald, 1927 types BM HRL
*iridis* Franssen, 1928 nec del Guercio, 1900
nigra Theobald, 1916 = Dysaphis ranunculi (Kaltenbach, 1843)
nigra (Wilson, 1911) (Pergandeida)
nigragregalis Knowlton, 1935 = gregalis Knowlton, 1928
nigratibialis Robinson ex Robinson & Chen, 1969 paratype BM HRL
nigricans van der Goot, 1917 = Toxoptera citricidus (Kirkaldy, 1907)
nigricauda van der Goot, 1917 = citricola van der Goot, 1912
nigripes Theobald, 1914 = nerii Boyer de Fonscolombe, 1841
nigritarsis von Heyden, 1837 (type of Callipteroides Mordvilko, 1909) = Euceraphis punctipennis (Zetterstedt, 1828)
nigro-rufa Walker, 1848 nomen dubium
nivalis Hille Ris Lambers, 1960 HRL
nociva Walker, 1849 = Brachycaudus helichrysi (Kaltenbach, 1843)
nostras Hottes, 1930
*eupatorii* Oestlund, 1886 nec Passerini, 1863
nuda pini DeGeer, 1773 (invalid) = Cinara pini (Linnaeus, 1758)
nutricata Walker, 1849 nomen dubium
nyctalis Hottes & Frison, 1931 = lugentis Williams, 1911
nymphaeae Ph. F. Gmelin, 1758 nomen oblitum
nymphaeae Linnaeus, 1761 (type of Rhopalosiphum Koch, 1854; and of Siphocoryne Passerini, 1860 nec 1863; and of Siphonaphis van der Goot, 1915) to Rhopalosiphum Koch, 1854
oblonga von Heyden, 1837 type of Symydobius Mordvilko, 1894 q.v.
obnoxia Mordvilko, 1916 = ? Rhopalosiphum maidis (Fitch, 1856)
ochrocentri Cockerell, 1903 to Bipersona Hottes, 1926
ochropus Koch, 1854 = confusa Walker, 1849
octotuberculata Mamontova, 1955 = schneideri (Börner, 1940)
odorikonis Matsumura, 1917 BM HRL
oenanthis Lichtenstein, 1884 nomen nudum
oenotherae Oestlund, 1887 BM HRL
oenotherae var. rufa Gillette, 1927
oestlundi Gillette, 1927 BM HRL
ogilviei Theobald, 1928 BM

A. onobrychidis Goureau, 1863 = ? craccivora Koch, 1854
onobrychis Boyer de Fonscolombe, 1841 = Acyrthosiphon pisum (Harris, 1776)
ononidis Kaltenbach, 1846 type of Therioaphis Walker, 1870 q.v.
onopordi Schrank, 1801 = Brachycaudus (Acaudus) cardui (Linnaeus, 1758)
opima Buckton, 1879 = Brachycaudus (Acaudus) cardui (Linnaeus, 1758)
opuli Sulzer, 1776 = viburni Scopoli, 1763
oreaster Rafinesque, 1818 invalid
oregonensis Wilson, 1915 to Aphis (Zyxaphis)
origani Passerini, 1860 BM HRL
ornata (Gillette & Palmer, 1929) (Anuraphis) metatype BM HRL
orobanches Passerini, 1879 nomen dubium
ovina Walker, M.S. = praeterita Walker, 1849
oxalina Theobald, 1925 = craccivora Koch, 1854
oxalis Macchiati, 1884 = ? gossypii Glover, 1877
oxyacanthae Schrank, 1801 nomen dubium
oxybaphi Oestlund, 1887 = helianthi Monell, 1879
padi Linnaeus, 1758 to Rhopalosiphum Koch, 1854
pahanensis Takahashi, 1950 to Melanaphis van der Goot, 1917
palans Walker, 1852 nomen dubium
paliuri Lichtenstein, 1884 nomen nudum
paliuri var. exploratus Rusanova, 1943 nomen nudum = exploratus Rusanova, 1948
pallescens Heer, 1853 fossil
pallida Walker, 1848 = Aulacorthum solani (Kaltenbach, 1843)
pallipes Hartig, 1841 to Lachnus Burmeister, 1835
palmae Baehr, 1908 = Cerataphis lataniae (Boisduval, 1867)
palmerae Remaudière, 1954 nom. nov. pro Anuraphis senecio-radicis Gillette & Palmer, 1929 = Aphis senecioradicis (Gillette & Palmer, 1929)
paludicola Hille Ris Lambers, 1959 HRL
*palustris* Börner, 1940 (*Pergandeida*) nec Aphis palustris Theobald, 1929 HRL
palustris Theobald, 1929 (type of Euschizaphis Hille Ris Lambers, 1947) to Schizaphis (Euschizaphis)
paolii del Guercio, 1916 = nerii Boyer de Fonscolombe, 1841
papaveris Fabricius, 1781 = fabae Scopoli, 1763
papaveris var. buxi del Guercio, 1911 = Toxoptera aurantii (Boyer de Fonscolombe, 1841)
papilionacearum van der Goot, 1918 = craccivora Koch, 1854
paralios Hille Ris Lambers ex Ilharco, 1974 nomen nudum
parietariae Lichtenstein, 1884 nomen nudum

A. parietariae Theobald, 1922 BM HRL
    *parietariella* Börner, 1952 (*Toxopterina (Tuberculaphis)*)
  parthenocissi Williams, 1911 = folsomi Davis, 1908
  particeps Walker, 1848 = Myzus (Nectarosiphon) persicae (Sulzer, 1776)
  parvus Theobald, 1915 = gossypii Glover, 1877
  passeriniana (del Guercio, 1900) (Cerosipha) BM HRL
  pastinacae Linnaeus, 1758 type of Cavariella del Guercio, 1911 q.v.
  patagonica E.E. Blanchard, 1944 BM HRL
  patriniae Takahashi, 1966 BM HRL
    *valerianae* Takahashi, 1932 (*Anuraphis*) nec Cowen, 1895 (Aphis)
  pavlovskii (Narzikulov, 1974) (Chomaphis)
  pawneepae Hottes, 1934 paratype BM HRL
  pectinatae Nördlinger, 1880 (type of Buchneria Börner, 1952) to Cinara Curtis, 1835
  pedicularis Buckton, 1879 = nasturtii Kaltenbach, 1843
  pelargonii Kaltenbach, 1843 = Acyrthosiphon malvae (Mosley, 1841)
  penicillata Buckton, 1879 = grossulariae Kaltenbach, 1843
  penstemonicola Gillette & Palmer, 1929 metatypes BM HRL
  penstemonis Williams, 1911
  perforata Signoret, 1867 = Periphyllus acericola (Walker, 1848)
  perillae Shinji, 1922 = ? gossypii Glover, 1877
  persequens Walker, 1852 to Macrosiphoniella del Guercio, 1911
  persicae Boyer de Fonscolombe, 1841 = Brachycaudus (Appelia) schwartzi (Börner, 1931)
  persicae Kaltenbach, 1843 = Brachycaudus (Appelia) schwartzi (Börner, 1931)
  persicae Koch, 1857 = Brachycaudus (Appelia) schwartzi (Börner, 1931)
  persicae Morren, 1836 = Myzus (Nectarosiphon) persicae (Sulzer, 1776)
  persicae Sulzer, 1776 (type of Rhopalosiphum Passerini, 1860 nec Koch, 1854) to Myzus (Nectarosiphon)
  persicaecola Boisduval, 1867 = Brachycaudus (Acaudus) persicae (Passerini, 1860)
  persicae-niger E.F. Smith, 1890 = Brachycaudus (Acaudus) persicae (Passerini, 1860)
  persicariae Hartig, 1841 partim = Hyalopterus amygdali (Blanchard, 1840)
  persicarum Boisduval, 1867 = ? Myzus amygdalinus (Nevsky, 1928)
  persicophila Rndn. in scheda, Passerini, 1860 = Myzus (Nectarosiphon) persicae (Sulzer, 1776)
  persola Walker, 1848 = ? Myzus (Nectarosiphon) persicae (Sulzer, 1776)
  persorbens Walker, 1849 = Brachycaudus helichrysi (Kaltenbach, 1843)
  pertinax Hille Ris Lambers, lapsus pro audax = triglochinis Theobald, 1926

A. petasiticola Takahashi, 1924 = fukii Shinji, 1922
petasitidis Buckton, 1879 = Brachycaudus helichrysi (Kaltenbach, 1843)
phaceliae Gillette & Palmer, 1929 HRL
pheidolei Theobald, 1916 = Melanaphis sacchari (Zehntner, 1897)
phelipaeae Passerini, 1879 = ? Brachycaudus (Acaudus) cardui (Linnaeus, 1758)
philadelphi Börner, 1921 = ? fabae subsp. cirsiiacanthoidis Scopoli, 1763
phlomoidea del Guercio, 1911 = fabae Scopoli, 1763
phragmitidicola Oestlund, 1886 = Hyalopterus pruni (Geoffroy, 1762)
piceae Panzer, 1801 (type of Mecinaria Börner, 1949) to Cinara Curtis, 1835
picridicola Holman, 1966 to Aphis (Protaphis)
picridis Fabricius, 1775 to Uroleucon Mordvilko, 1914
picta Walker, 1849 = sambuci Linnaeus, 1758
pilicornis Hartig, 1841 to Cinara Curtis, 1835
pilosa Haldeman, 1844 = ? Pterocomma salicis (Linnaeus, 1758)
pilosa Walker, 1849 = ? Cryptaphis poae (Hardy, 1850)
pilosa Zetterstedt, 1840 nomen dubium in Cinara pini (Linnaeus) group
pilosellae (Börner, 1952) (Cerosipha) BM HRL
*hieracii* Börner, 1940 (*Cerosipha*) nec Aphis hieracii Schrank, 1801
*pilosellae* Hille Ris Lambers, 1956 HRL
pilosellae Hille Ris Lambers, 1956 = pilosellae Börner, 1952
pilosicauda Gillette & Palmer, 1932 paratype BM HRL
pimpinellae Kaltenbach, 1843 to Semiaphis van der Goot, 1913
pineti Fabricius, 1781 to Schizolachnus Mordvilko, 1909
pini Ph. F. Gmelin, 1758 nomen oblitum
pini Linnaeus, 1758 type of Cinara Curtis, 1835 q.v.
pini subsp. maritimae Dufour, 1833 = Cinara maritimae (Dufour, 1833)
pinicolens Fitch, 1851 = ? Euceraphis mucida (Fitch, 1856)
piniphila Ratzeburg, 1844 to Cinara Curtis, 1835
piperis Kittel, 1827 invalid
pirifoliae Shinji, 1922 = citricola van der Goot, 1912
pisi Curtis, 1860 = Acyrthosiphon pisum (Harris, 1776)
pisi Kaltenbach, 1843 (type of Acyrthosiphon Mordvilko, 1914) = Acyrthosiphon pisum (Harris, 1776)
pistaciae Ph. F. Gmelin, 1758 nomen oblitum
pistaciae Linnaeus, 1767 type of Baizongia Rondani, 1848 q.v.
pistacinacae Ph. F. Gmelin lapsus pro pistaciae
pisum Harris, 1776 to Acyrthosiphon Mordvilko, 1914
piutapa Hottes & Wehrle, 1951 = Xerophilaphis tetrapteralis (Cockerell, 1902)
plana Brodie, 1845, fossil, ? to Heteroptera
plantaginifolii Nevsky, 1929 = ? nasturtii Kaltenbach, 1843

A. plantaginis Goeze, 1778 BM HRL
*plantaginis* Schrank, 1801
plantaginis Schrank, 1801 = plantaginis Goeze, 1778
plantarum-aquaticum Fabricius, 1794 = Rhopalosiphum nymphaeae (Linnaeus, 1761)
plantarum-aquaticum Ph. F. Gmelin, 1758 nomen oblitum
platanicola Walker, 1848 lapsus pro platani Kaltenbach, 1843, originally described in Lachnus.
platanoidis Hartig, 1841 = Drepanosiphum platanoidis (Schrank, 1801)
platanoidis Schrank, 1801 type of Drepanosiphum Koch, 1855 q.v.
platylobii Carver & White, 1970 paratype BM HRL
poae Hardy, 1850 to Cryptaphis Hille Ris Lambers, 1947
podagrariae Schrank, 1801 BM HRL
polanisiae Oestlund, 1886
polanisia-graveolens Rafinesque, 1818 invalid
polaris Curtis, 1828 lapsus pro borealis
pollinosa Walker, 1849 BM HRL
*farinosus* del Guercio, 1913 (*Cladobius*) HRL
*pollinaria* Börner, 1952 (*Cerosipha (Uraphis)*)
polyanthis J.F. Gmelin, 1790 nomen dubium
polyanthis Passerini, 1863 (nec J.F. Gmelin, 1790) = fabae Scopoli, 1763
polyanthis Sulzer, 1776, no such thing, based on 'polyanth. tuberosum Sulzer, 1776'
polygala-senega Rafinesque, 1818 invalid
polygonacea Matsumura, 1917 HRL
polygonata (Nevsky, 1929) (Pergandeida) BM HRL
*? avicularis* Hille Ris Lambers, 1931 (*Doralis*) HRL
*polygoni* Macchiati, 1885 nec Walker, 1848
polygoni van der Goot, 1912 = nasturtii Kaltenbach, 1843
polygoni Macchiati, 1885 = polygonata (Nevsky, 1929)
polygoni Shinji, 1935 nomen dubium
polygoni Walker, 1848 = Aspidaphis adjuvans (Walker, 1848)
pomi de Geer, 1773 type of Medoralis Börner, 1952 BM HRL
*bicolor* Haldeman, 1844
*crataegi* Buckton, 1879 nec Kaltenbach, 1843
*cydoniae* Boisduval, 1867
*eriobotryae* Schouteden, 1905
*mali* Fabricius, 1775
pomi Vallot, 1802 invalid
pomonella Theobald, 1916 = gossypii Glover, 1877
popovi Mordvilko, 1932

A. populea Kaltenbach, 1843 (type of Cladobius Koch, 1856; and of Aphioides Passerini, 1860; and of Aristaphis Kirkaldy, 1905) to Pterocomma Buckton, 1879
populeti Panzer, 1801 to Chaitophorus Koch, 1854
populi Linnaeus, 1758 to Pachypappa Koch, 1856
populi-albae Boyer de Fonscolombe, 1841 to Chaitophorus Koch, 1854
populifoliae Fitch, 1851 to Pterocomma Buckton, 1879
populi-nigrae Ph. F. Gmelin, 1758 nomen oblitum
populi-nigrae Schrank, 1801 to Pemphigus Hartig, 1839
populus-grandidentata Rafinesque, 1818 invalid
populus-trepida Rafinesque, 1818 invalid
portulacae Shinji, 1935 nomen dubium
potentillae Nevsky, 1929
potentillae Walker, 1850 to Chaetosiphon (Pentatrichopus)
praeterita Walker, 1849 BM HRL
*epilobiina* Walker, 1849
*ovina* Walker M.S.
proffti (Börner, 1942) (Doralina) BM HRL
prunaria Walker, 1850 = Rhopalosiphum nymphaeae (Linnaeus, 1761)
pruni Fabricius, 1775 (type of Hyalopterus Koch, 1854) = Hyalopterus pruni (Geoffroy, 1762)
pruni Geoffroy, 1762 to Hyalopterus Koch, 1854
pruni Koch, 1854 = Brachycaudus (Acaudus) cardui (Linnaeus, 1758)
pruni Scopoli, 1763 = Phorodon humuli (Schrank, 1801)
pruni cerasi Stewart, 1802 nomen nudum
prunicola Kaltenbach, 1843 to Brachycaudus (Appelia) (but the name has also been applied to Brachycaudus (Acaudus) persicae (Passerini, 1860)
prunicoleus Ashmead, 1881 = Hysteroneura setariae (Thomas, 1878)
pruniella Theobald, 1918 = gossypii Glover, 1877
prunifoliae Fitch, 1855 = Rhopalosiphum padi (Linnaeus, 1758)
pruni mahaleb Boyer de Fonscolombe, 1841 = Myzus lythri (Schrank, 1801)
prunina Walker, 1848 = Brachycaudus helichrysi (Kaltenbach, 1843)
prunorum Dobrovljansky, 1913 = Rhopalosiphum nymphaeae (Linnaeus, 1761)
prunus Shinji, 1922 = Brachycaudus helichrysi (Kaltenbach, 1843)
psammophila Szelegiewicz, 1967 paratypes BM HRL
pseudeuphorbiae Hille Ris Lambers, 1948 BM HRL
pseudoavenae Patch, 1917 = Rhopalosiphum padi (Linnaeus, 1758)
pseudobrassicae Davis, 1914 = Lipaphis erysimi (Kaltenbach, 1843)
pseudocardui Theobald, 1915 to Aphis (Protaphis)

A. pseudocomosa Stroyan, 1972 BM HRL
    *falki* Hoffmann, 1972
pseudocytisorum Hille Ris Lambers, 1967 HRL
pseudohederae Theobald, 1927 = hederae f. pseudohederae Theobald, 1922
pseudohieracii Theobald, 1912 nomen dubium
pseudopaludicola Tashev, 1965 HRL
pseudopomi Bertels, 1973 lapsus pro pseudopomi E.E. Blanchard, 1939
pseudopomi E.E. Blanchard, 1939 = citricola van der Goot, 1912
pseudopulchella E.E. Blanchard, 1944
pseudovalerianae Gillette & Palmer, 1932 HRL
pteris-aquilinoides Rafinesque, 1917 invalid, type of Adactynus Rafinesque, 1818 invalid
pulchella Hottes & Frison, 1931 paratypes BM
pulegii del Guercio, 1911 HRL
pulsatillae Ossiannilsson, 1959 (Aphis (Cerosipha)) paratypes BM HRL
pulsatillicola Holman, 1966 paratypes BM HRL
pulvera Walker, 1848 to Macrosiphoniella del Guercio, 1911
pulverulens Gillette, 1911 type of Amphicercidus Oestlund, 1922 q.v.
punctatella Fitch, 1855 nomen dubium
punctatella auctt. nec Fitch = Tuberculatus tuberculatus (Richards, 1965)
punctipennis Zetterstedt, 1828 to Euceraphis Walker, 1870
punicae Passerini, 1863 BM HRL
    *durantae* Theobald, 1918 types BM
    *durranti* Das, 1918
    *punicella* Theobald, 1915 type BM
punicae Shinji, 1922
punicella Theobald, 1915 = punicae Passerini, 1863
pyrastri Boisduval, 1867 = Melanaphis pyraria (Passerini, 1861)
pyri Boyer de Fonscolombe, 1841 (type of Pomaphis Börner, 1939) to Dysaphis (Pomaphis)
pyri Hartig, 1841 = Dysaphis (Pomaphis) plantaginea (Passerini, 1860)
pyri Kittel, 1827 invalid
pyri Koch, 1854 (type of Anuraphis del Guercio, 1907) = Anuraphis farfarae (Koch, 1854)
pyri Vallot, 1802 invalid
quadrituberculata Kaltenbach, 1843 to Betulaphis Glendenning, 1926
quaerens Walker, 1849 = Cryptomyzus galeopsidis (Kaltenbach, 1843)
quercea Kaltenbach, 1843 type of Tuberculatus Mordvilko, 1894 q.v.
quercifoliae Walsh, 1863 to Drepanosiphum Koch, 1855
quercus Kaltenbach, 1843 (type of Tuberculoides van der Goot, 1913; and of

Neotuberculatus van der Goot, 1915) = Tuberculatus (Tuberculoides) annulatus (Hartig, 1841)

A. quercus Linnaeus, 1758 type of Stomaphis Walker, 1870 q.v.

quercus-monticula Haldeman, 1844 nomen dubium

radicicola Schouteden, 1906 emendation of radicola Mordvilko, 1897 = Dysaphis radicola (Mordvilko, 1897)

radicola Mordvilko, 1897 to Dysaphis Börner, 1931

radicum Boyer de Fonscolombe, 1841 = Tetraneura ulmi (Linnaeus, 1758)

radicum Craveri, 1915 nomen dubium

radicum Kirby, 1818 nomen nudum

raji (Kumar & Burkhardt, 1970) (Longirostris)

ramona Swain, 1918 BM HRL

ranunculi Kaltenbach, 1843 to Dysaphis Börner, 1931

ranunculina Walker, 1852 type of Tubaphis Hille Ris Lambers, 1947 q.v.

rapae Curtis, 1842 = Myzus (Nectarosiphon) persicae (Sulzer, 1776)

raphani Schrank, 1801 = Brevicoryne brassicae (Linnaeus, 1758)

rectolactens Menge, 1856 fossil

reducta Walker, 1848 = Macrosiphoniella pulvera (Walker, 1848)

redundans Walker, 1849 = Myzus (Nectarosiphon) persicae (Sulzer, 1776)

relata Walker, 1849 nomen dubium

remaudierei Börner, 1952 (Aphis (Doralis))

reticulata Theobald, 1922 nec Wilson, 1915 = neoreticulata = ? fabae subsp. cirsiiacanthoidis Scopoli, 1763

reticulata Wilson, 1915

rhamnellae Shinji, 1922

rhamni Boyer de Fonscolombe, 1841 type of Macchiatiella del Guercio, 1909 q.v. nec 1917; Koch, 1854 and many later authors applied the name to Aphis nasturtii Kaltenbach, 1843

rhamni Kaltenbach, 1843 = frangulae Kaltenbach, 1845

rhamnicola Mamontova, 1953 BM, types HRL = ? mammulata Gimingham & Hille Ris Lambers, 1949

rhamniphila S.K. David, Narayanan & Rajasingh, 1971 HRL

rhei Koch, 1854 = rumicis Linnaeus, 1758

rheicola Nevsky, 1951

rhodryas Rafinesque, 1818 invalid

rhoicola Hille Ris Lambers, 1956 HRL

*? crispi* Davletschina, 1964

ribicola Kaltenbach, 1843 (type of Nasonovia Mordvilko, 1914) = Nasonovia ribisnigri (Mosley, 1841)

ribiensis Gillette & Palmer, 1929 BM HRL

A. ribigillettei Knowlton & Allen, 1939 = neomexicana (W.P. Cockerell & T.D.A. Cockerell, 1901)
ribis Linnaeus, 1758 type of Cryptomyzus Oestlund, 1922 q.v.
ribis O.F. Müller, 1776 = Cryptomyzus ribis (Linnaeus, 1758)
ribis Sanborn, 1904 = neomexicana (W.P. Cockerell & T.D.A. Cockerell, 1901)
ribis-nigri Mosley, 1841 to Nasonovia Mordvilko, 1914
ripariae Oestlund, 1886
robiniae Macchiati, 1885 = craccivora Koch, 1854
roboris Linnaeus, 1758 (type of Dryaphis Kirkaldy, 1904 and Dryobius Koch 1855 nec Le Conte, 1850; and of Pteroclorus Rondani, 1843) to Lachnus Burmeister, 1835
robusta Walker, 1849 nomen dubium
rociadae Cockerell, 1903 to Brachycaudus (Acaudus)
roepkei (Hille Ris Lambers, 1931) (Cerosipha) HRL
roripae (Palmer, 1938) (Cerosipha) HRL
rosae Linnaeus, 1758 type of Macrosiphum Passerini, 1860 q.v.; and of Nectarophora Oestlund, 1877; and of Siphonophora Koch, 1855
rosarum Kaltenbach, 1843 type of Myzaphis van der Goot, 1913 q.v.
rosa-suaveolens Rafinesque, 1818 invalid
rubecula Haldeman, 1844 to ? Uroleucon Mordvilko, 1914
rubi Kaltenbach, 1843 (type of Eunectarosiphon del Guercio, 1913) to Amphorophora Buckton, 1876
rubiae Narzikulov, 1964
rubicola Haldeman errore pro rubecula
rubicola Oestlund, 1887 BM HRL
*rubiphila* Patch, 1914
rubicolens (Hori, 1919) (Cerosipha)
rubifolii (Thomas, 1879) (Sipha) BM HRL
rubiphila Patch, 1914 = rubicola Oestlund, 1887
rubiradicis Robinson, 1969 paratype BM HRL
ruborum (Börner, 1932) (Doralis) BM HRL
ruborum subsp. longisetosus A.N. Basu, 1969 = longisetosa A.N. Basu, 1969
rudbeckiae Fitch, 1851 (type of Tritogenaphis Oestlund, 1922) to Uroleucon Mordvilko, 1914
rufipes Hartig, 1841 to Pterocomma Buckton, 1879
rufomaculata Wilson, 1908 type of Coloradoa Wilson, 1910 q.v.
rufula Walker, 1849 to Schizaphis Börner, 1931
rumexicolens Patch, 1917 to Brachycaudus (Thuleaphis)
rumicis auctt. prior to 1930 nec Linnaeus = mostly Aphis fabae Scopoli, 1763
rumicis Leach, 1827 nomen nudum, used by Börner as type of Doralis Leach, 1827

A. rumicis Linnaeus, 1758 BM HRL
*carbocolor* Gillette, 1907 HRL
*davidsoniella* Theobald, 1927 type BM
*rhei* Koch, 1853
rumicis lapathi Stewart, 1802 nomen nudum
rutae Shinji, 1922 = Toxoptera odinae (van der Goot, 1917)
sacchari Zehntner, 1897 (type of Longiunguis van der Goot, 1917) to Melanaphis van der Goot, 1917
salicariae Koch, 1855* BM HRL
*corniella* Hille Ris Lambers, 1935 (*Doralis*) BM HRL
saliceti Kaltenbach, 1843 = farinosa Gmelin, 1790
salicina Zetterstedt, 1840 = Tuberolachnus salignus (Gmelin, 1790)
salicis Linnaeus, 1758 (type of Melanoxanterium Schouteden, 1901 and of Melanoxanthus Buckton, 1879) to Pterocomma Buckton, 1879
salicis O.F. Müller, 1776 = Pterocomma salicis (Linnaeus, 1758)
salicis Sulzer, 1776 = Tuberolachnus salignus (Gmelin, 1790)
salicis farinosa de Geer, 1773 invalid
salicis-farinosa Villers, 1789 invalid
salicis minor Kittel, 1827, invalid
salicivora Walker, 1848 (type of Tranaphis Walker, 1870) = Chaitophorus capreae (Mosley, 1841)
salicti Harris, 1841 = Pterocomma salicicola (Uhler, 1862)
salicti Schrank, 1801 to Chaitophorus Koch, 1854
saligna Gmelin, 1790 to Tuberolachnus Mordvilko, 1909
salviae Walker, 1852 = ? craccivora Koch, 1854
sambucaria Passerini, 1860 = sambuci Linnaeus, 1758
sambuci Linnaeus, 1758 BM HRL
*buxtoni* Theobald, 1920 types BM
*davisiana* del Guercio, 1913
*exul* Walker, 1849 probable type BM
*picta* Walker, 1849
*sambucaria* Passerini, 1860
*sambuci* O.F. Müller, 1776
*sambucifoliae* Fitch, 1851
*sambucina* Börner, 1940
*wilsoni* Laing, 1923 type BM

*Koch described his Aphis salicariae from "Weiderich". But "Weiderich" is the German name for both Lythrum salicaria, and Epilobium or Chamaenerium angustifolium. He very carefully described and figured the aphid from (Cornus stolonifera and) Epilobium angustifolium later known as Aphis corniella (Hille Ris Lambers, 1935) but derived the specific name from the wrong "Weiderich".

A. sambuci subsp. horii Takahashi, 1923 = horii Takahashi, 1923
sambuci O.F. Müller, 1776 = sambuci Linnaeus, 1758
sambucifoliae Fitch, 1851 = sambuci Linnaeus, 1758
sambucina Börner, 1940 = sambuci Linnaeus, 1758
sanborni Patch, 1914 = neomexicana (W.P. Cockerell & T.D.A. Cockerell, 1901)
sanguisorbae Schrank, 1801 BM HRL
*poterii* Börner, 1940 (*Cerosipha*)
sanguisorbae Shinji, 1935 nec Schrank, 1801 = sanguisorbicola Takahashi, 1966
sanguisorbicola Takahashi, 1966 BM HRL
*sanguisorbae* Shinji, 1935 nec Schrank, 1801
saniculae Williams, 1911 BM HRL
*luridus* Hottes & Frison, 1931 paratypes BM
sarothamni Franssen, 1928 BM HRL
sassceri Wilson, 1911
sativae Williams, 1911
scabiosae Koch, 1854 in litteris synonym of centaureae = Acaudinum centaureae (Koch, 1854)
scabiosae Schrank, 1801 = confusa Walker, 1849
scabiosae Scopoli, 1763 = Macrosiphum rosae (Linnaeus, 1758)
scaliai del Guercio, 1915 = coronillae Ferrari, 1872
scaliai subsp. arenaria Hoffmann, 1968 = coronillae subsp. arenaria Hoffmann, 1968
schilderi (Börner, 1940) (Doralina) BM HRL
schinifoliae E.E. Blanchard, 1939 BM HRL
*achinifoliae* Bertels, 1973
schneideri (Börner, 1940) (Doralina) BM HRL
*octotuberculata* Mamontova, 1955 HRL
schoutedeni Hille Ris Lambers, 1956 = kaltenbachi Hille Ris Lambers, 1955
schoutedeni Kirkaldy, 1906 = ulmariae Schrank, 1801
scirpi Kittel, 1827 invalid
scorodoniae del Guercio, 1911 = diphaga Walker, 1852
scotti Sanderson, 1905 = Hysteroneura setariae (Thomas, 1878)
secunda Walker, 1849 = ? Phorodon humuli (Schrank, 1801)
sedi Kaltenbach, 1843 BM HRL
sejuncta Walker, 1848 to Macrosiphoniella del Guercio, 1911
selini (Börner, 1940) (Doralina)
senecio Swain, 1918 = Brachycaudus helichrysi (Kaltenbach, 1843)
senecionicoides E.E. Blanchard, 1944
senecionis Koch, 1854 = jacobaeae Schrank, 1801
senecionis Williams, 1911

A. senecionisdoriae Bozhko, 1953 = jacobaeae Schrank, 1801
senecionis-fluviatilidis Bozhko, 1959 nomen nudum
senecioradicis (Gillette & Palmer, 1929) (Anuraphis) HRL
*palmerae* Remaudière, 1954
sensoriata Gillette & Bragg, 1918 (type of Amelanchieria Shaposhnikov, 1950) to Nearctaphis Shaposhnikov, 1950
serissae Shinji, 1922 BM HRL
*? serissae* Shinji, 1941 (*Cerosipha*)
serpylli Koch, 1854 BM HRL
serratulae Kaltenbach, 1843 = Uroleucon cirsii (Linnaeus, 1758)
serratulae Ph. F. Gmelin, 1758 nomen oblitum
serratulae Schrank, 1801 = fabae subsp. cirsiiacanthoidis Scopoli, 1763
seselii (Bozhko, 1959) (Debilisiphon seselli, misprint) HRL
seselii Lichtenstein, 1885 nomen nudum
setariae (Thomas, 1878) (Siphonophora) (type of Heteroneura Davis, 1919 nec Fallen, 1810) to Hysteroneura Davis, 1919
setosa Kaltenbach, 1846 to Ctenocallis Klodnitsky, 1924
sexton Westwood, 1850 = Cemonus rugifer Dahlbom, 1844, Sphecidae. Westwood's phrase 'the Aphis sexton' referring to the behaviour of the wasp, was printed as if a latin binomen was intended.
shirakii Takahashi, 1921 = gossypii Glover, 1877
sierra Essig, 1947 locotype BM
signatus Hottes & Frison, 1931 = thaspii Oestlund, 1887
sii Koch, 1855 type of Ammiaphis Börner, 1944 q.v.
silenea Ferrari, 1872 = Brachycolus cucubali (Passerini, 1863)
silenicola Holman & Szelegiewicz, 1971 HRL
silybi Passerini, 1861 = fabae Scopoli, 1763
similis Walker, 1848 = ? Brachycaudus helichrysi (Kaltenbach, 1843)
sinensis del Guercio, 1900 nomen nudum was fabae Scopoli, 1763
siphonella Essig & Kuwana, 1918 to Melanaphis van der Goot, 1917
smilacifoliae Takahashi, 1921 type of Aleurosiphon Takahashi, 1966 q.v.
socia Walker, 1848 = Brachycaudus helichrysi (Kaltenbach, 1843)
sodalis Walker, 1848 = Acyrthosiphon malvae (Mosley, 1841)
sogdiana Nevsky, 1929
solanella Theobald, 1914 = fabae subsp. solanella Theobald, 1914
solani Kaltenbach, 1843 to Aulacorthum Mordvilko, 1914
solani Kittel, 1827, invalid
solanina Passerini, 1863 = gossypii Glover, 1877
solanophilus Blanchard, 1923 = fabae subsp. solanella Theobald, 1914
solidaginifoliae Williams, 1911 to Acuticauda Hille Ris Lambers, 1955

A. solidaginis (Börner, 1950) (Doralis)
solidaginis Fabricius, 1779 to Uroleucon (Uromelan)
somei Essig & Kuwana, 1918 (type of Somaphis Shinji, 1929) = Toxoptera odinae (van der Goot, 1917)
sonchi Geoffroy, 1762 not available
sonchi Linnaeus, 1767 (type of Megalosiphum Mordvilko, 1919; and of Uroleucon Mordvilko, 1914) = Uroleucon sonchi (Linnaeus, 1767)
sonchi pruinosa Kittel, 1827 invalid
sonchi viridifurcata Kittel, 1827 invalid
sorbi Kaltenbach, 1843 (nominal type of Dentatus van der Goot, 1913 nec Gray, 1847 but van der Goot had Myzus plantagineus Passerini, 1860, not sorbi) to Dysaphis (Pomaphis)
sorensoni Knowlton, 1928 = ? helianthi Monell, 1879
sorghella Schouteden, 1906 = Melanaphis sacchari (Zehntner, 1897)
sorghi Theobald, 1904 = Melanaphis sacchari (Zehntner, 1897)
soyogo Uye, 1923 variety of Toxoptera aurantii (Boyer de Fonscolombe, 1841)
spectabilis Ferrari, 1872 = farinosa Gmelin, 1790
spinarum Hartig, 1841 = Hyalopterus pruni (Geoffroy, 1762)
spinosula Essig & Kuwana, 1918 (type of Trichosiphoniella Shinji, 1929) = Tuberocephalus sakurae (Matsumura, 1917)
spiraeae Oestlund, 1887 nomen dubium
spiraeae Schouteden, 1902 = ulmariae Schrank, 1801
spiraecola Patch, 1914 = citricola van der Goot, 1912
spiraeella Schouteden, 1903 = ulmariae Schrank, 1801
spiraephaga Müller, 1961 paratypes BM HRL
*spiraephaga* subsp. *asiatica* Narzikulov, 1964
spiraephaga subsp. asiatica Narzikulov, 1964 = spiraephaga Müller, 1961
spiraephila Patch, 1914 BM HRL
spiranthi Shinji, 1922 to ? Brachycaudus van der Goot, 1913 or Dysaphis Börner, 1931
spireaella Schouteden, lapsus pro spiraeella
stachydis Mordvilko, 1929 BM HRL
*stachydis* Börner, 1940 (*Doralina*)
steinbergi Shaposhnikov, 1952 = commensalis Stroyan, 1952
stellariae Hardy, 1850 type of Brachycolus Buckton, 1879 q.v.
steviae J.M. Baker, 1934 HRL
stranvaesiae Takahashi, 1937 BM
striata Rafinesque, 1818 invalid
stroyani Szelegiewicz, 1961 HRL
*picridis* Börner, 1950 (*Doralina*) nec Aphis picridis Fabricius, 1775

A. suberis Tavares, 1900 = Hoplocallis pictus (Ferrari, 1872)
submacula Walker, 1848 to Maculolachnus Gaumont, 1920
subnitida (Börner, 1940) (Doralina) BM HRL
subterranea Walker, 1852 to Anuraphis del Guercio, 1907
subviridis (Börner, 1940) (Cerosipha) BM HRL
succisae Holman, 1966 paratypes BM HRL
suffragans Walker, 1848 nomen dubium
sugadairensis Takahashi, 1966
sumire Moritsu, 1949 BM HRL
superabilis Walker, 1952 nomen dubium
swezeyi Fullaway, 1910 = ? middletonii Thomas, 1879
symphoricarpi Thomas, 1879 to Aphthargelia Hottes, 1958
symphyti Schrank, 1801 BM HRL
*consolidae* Passerini, 1864
symphyti var. kochiella del Guercio, 1916
tacita Huculak, 1968 BM HRL
tahasa Hottes, 1950 = Rhopalosiphum cerasifoliae (Fitch, 1855)
tahosalea Hottes & Wehrle, 1951 (Aphis (Wapuna))
tamaricis Lichtenstein, 1885 to Brachyunguis Das, 1918
tamaricis Theobald, 1918 = Brachyunguis tamaricis (Lichtenstein, 1885)
tanacenti Mayer, 1779 nomen dubium
*mayeri* Gmelin, 1790
tanacetaria Kaltenbach, 1843 to Macrosiphoniella del Guercio, 1911
tanaceti Linnaeus, 1758 to Uroleucon Mordvilko, 1914
tanaceti Linnaeus of Mordvilko, 1914 nec Linnaeus, 1758 (type of Metopeurum Mordvilko, 1914) = Metopeurum fuscoviride Stroyan, 1950
tanaceticola Kaltenbach, 1843 (type of Eurythaphis Mordvilko, 1914) = Uroleucon tanaceti (Linnaeus, 1758)
tanacetina Walker, 1850 to Coloradoa Wilson, 1910
taraxaci Kaltenbach, 1843 to Uroleucon (Uromelan)
taraxacicola (Börner, 1940) (Doralina) BM HRL
taraxacicola (Nevsky, 1951) (Cerosipha)
tashevi Szelegiewicz, 1962 paratypes BM HRL
taukogi Shinji, 1939
tavaresi del Guercio, 1908 = Toxoptera citricidus (Kirkaldy, 1907)
tectonae van der Goot, 1917 = gossypii Glover, 1877
tentans Walker, 1852 nomen dubium
tenuinervis Zetterstedt, 1840 = Symydobius oblongus (von Heyden, 1837)
tenuior Walker, 1849 = ? Cinara laricis (Hartig, 1839)
terricola Rondani, 1847 to Aphis (Protaphis)

A. tertia Walker, 1849 nomen dubium
tetradymia Knowlton, 1941 BM
tetrapteralis Cockerell, 1902 to Xerophilaphis Nevsky, 1929
tetrarhoda Walker, 1849 (type of Pentatrichopus Börner, 1930) to Chaetosiphon (Pentatrichopus)
teucrii (Börner, 1942) (Doralina) HRL
teucrii Lichtenstein, 1885 nomen nudum
thalictri Essig & Kuwana, 1918 = Longicaudus trirhodus (Walker, 1849)
thalictri Koch, 1854 type of Brachysiphum van der Goot, 1913 q.v.
thaspii Oestlund, 1887 BM
*signatus* Hottes & Frison, 1931 paratype BM
thecomae Davletschina, 1964
theobaldi Gillette & Bragg, 1918 to Cavariella del Guercio, 1911
therebinthinae Virey, 1820 nomen dubium
thermophila (Börner, 1950) (Doralina) BM HRL
thesii Holman, 1966 BM HRL
thlaspeos Schrank, 1801 = fabae Scopoli, 1763
thomasi (Börner, 1950) (Doralina) BM HRL
tiliae Linnaeus, 1758 type of Eucallipterus Schouteden, 1906 q.v.
tincta Walker, 1849 to Macrosiphum Passerini, 1860
tirucallis Hille Ris Lambers, 1954 HRL
*euphorbiae* Macchiati, 1881 (*Sipha*) nec Kaltenbach, 1843 (Aphis)
*euphorbiae* subsp. *tirucallis* Hille Ris Lambers, 1854 HRL
tomentosa Villers, 1789 = Schizolachnus pineti (Fabricius, 1781)
tomentosa pini de Geer, 1773 (invalid, type of Schizolachnus Mordvilko, 1909 (1908) = Schizolachnus pineti (Fabricius, 1781))
tonahasa Hottes, 1950 = neomexicana (W.P. Cockerell & T.D.A. Cockerell, 1901)
toriliae Rusanova, (1943 nomen nudum) 1948
tormentillae Passerini, 1879 type of Microsiphon del Guercio, 1907 BM HRL
torquens Holman, 1959 paratype BM HRL
torticauda Gillette, 1907 (type of Bipersona Hottes, 1926) = Bipersona ochrocentri (Cockerell, 1903)
tragopogonis Kaltenbach, 1843 (type of Appelia Börner, 1930) to Brachycaudus (Appelia)
transiens Walker, 1849 = nasturtii Kaltenbach, 1843
translata Walker, 1849 = fabae Scopoli 1763
transmutata Walker, 1849 nomen dubium
transparens Germar & Berendt, 1856 to Mindarus Koch, 1857. Fossil.
transposita Walker, 1849 nomen dubium

A. tremulae Linnaeus, 1761 to Asiphum Koch, 1856
tribulis Walker, 1849 type of Megourella Hille Ris Lambers, 1949 q.v.
tridacis Theobald, 1929 = gossypii Glover, 1877
tridentatae Wilson, 1915 to Pseudoepameibaphis Gillette & Palmer, 1932
trifolii Oestlund, 1887 nomen nudum
triglochinis Theobald, 1926 type BM HRL
*audax* Hille Ris Lambers, 1947 HRL
*pertinax* Hille Ris Lambers, lapsus pro audax
triphaga Walker, 1852 = epilobii Kaltenbach, 1843
tripolii Laing, 1920 type BM HRL
trirhoda Walker, 1849 type of Longicaudus van der Goot, 1913 q.v.; and of Hemiaphis Börner, 1926
tritici Fitch, 1867 based on m.s. reference to tritici Lawson
tritici Lawson, 1866 = Rhopalosiphum padi (Linnaeus, 1758)
truncata Hausmann, 1802 to Chaitophorus Koch, 1854
tshernovae Holman & Szelegiewicz, 1971 HRL
tsujii Shinji, 1922 BM
tuberculata von Heyden, 1837 (type of Procalaphis Quednau, 1954) to Callipterinella van der Goot, 1913
tuberculata Patch, 1914 = feminea Hottes, 1930
tuberosae Boyer de Fonscolombe, 1841 = fabae Scopoli, 1763
tulipae Boyer de Fonscolombe, 1841 to Dysaphis Börner, 1931
tussilaginis Walker, 1850 to Uroleucon Mordvilko, 1914
typhae Mamontova, 1959
ulicis Walker, 1870 lectotype BM HRL
ulmariae Schrank, 1801 BM HRL
*schoutedeni* Kirkaldy, 1906
*spiraeae* Schouteden, 1902 nec Oestlund, 1887
*spiraeella* Schouteden, 1903
*? ulmariella* Ossiannilsson, 1959 (Aphis (Aphidula)) BM HRL
ulmi Fabricius 1775 nec Linnaeus, 1758 (type of Mimaphidus Rondani, 1848) = Eriosoma (Schizoneura) lanuginosum (Hartig, 1841)
ulmi Ph. F. Gmelin, 1758 nomen oblitum
ulmi Linnaeus, 1758, type of Tetraneura Hartig, 1841 q.v.
ulmi (Linnaeus of Passerini) (type of Schizoneura Hartig, 1839) = Eriosoma (Schizoneura) ulmi (Linnaeus, 1758)
ulmi-campestris de Geer, 1773 = Eriosoma (Schizoneura) ulmi (Linnaeus, 1758)
umbellatarum Koch, 1854 (description) = Cavariella aegopodii Scopoli, 1763
umbelliferarum (Shaposhnikov, 1950) (Debilisiphon)

A. umbrella (Börner, 1950) (Doralina) BM HRL
*malvae* Koch, 1854 nec Mosley, 1841
unaweepiensis Hottes, 1948 paratypes BM HRL
urovaneta Hottes, 1950 to Braggia Gillette & Palmer, 1929
urticae Fabricius, 1775 = urticata J.F. Gmelin, 1790
urticae Linnaeus, 1758 = Orthezia urticae (Linnaeus), a coccid.
urticae Schrank, 1801 (type of Microlophium Mordvilko, 1914) = Microlophium carnosum (Buckton, 1876)
urticaria Kaltenbach, 1843 = urticata J.F. Gmelin, 1790
urticata Gmelin, 1790 BM HRL
*stanilandi* Laing, 1923 (*Pergandeida*)
*urticae* Fabricius, 1775 nec Linnaeus, 1758
*urticaria* Kaltenbach, 1843
utsigicola Monzen, 1929
uvaeursi Ossiannilsson, 1959 (Aphis (Pergandeida)) BM HRL
uwamizusakurae Monzen, 1929 to ? Rhopalosiphum Koch, 1854
vaccinii (Börner, 1940) (Doralis) BM HRL
vacillans Walker, 1849 = ? Anuraphis farfarae (Koch, 1854)
valdensis Brodie, 1854 type of Genaphis Handlirsch, 1907 q.v. Fossil
valerianae Cowen, 1895 BM HRL
valerianina del Guercio, 1911 = fabae Scopoli, 1763
valida Walker, 1852 nomen dubium
vallei Hille Ris Lambers & Stroyan, 1959 BM HRL
varians Patch, 1914 BM HRL
vastator Smee, 1846 = Myzus (Nectarosiphon) persicae (Sulzer, 1776)
veratri Cowen, 1895 = coweni Palmer, 1938
veratri Walker, 1852 BM HRL
veratri subsp. takagii Takahashi, 1966
verbasci Boyer de Fonscolombe, 1841 = verbasci Schrank, 1801
verbasci Schrank, 1801 BM HRL
*buddleiae* Theobald, 1918 type BM
*verbasci* Boyer de Fonscolombe, 1841
verbasci (Schrank of Nevsky, 1929) (type of Asiataphis Narzikulov, 1970) = verbasci Schrank, 1801
verbenae Macchiati, 1884 BM HRL
*verbenae* Nevsky, 1929
verbenae Nevsky, 1929 = verbenae Macchiati, 1884
verbena-hastata Rafinesque, 1818 invalid
vernoniae Thomas, 1877 BM HRL
veronicae Nevsky, 1929 = beccabungae subsp. turanica Nevsky, 1929
veronicae Walker, 1848 = Myzus cerasi (Fabricius, 1775)

A. versicolor (Börner, 1950) (Doralina) HRL
versicolor Menke, 1818 nomen dubium
verticillatae (Börner, 1940) (Doralina) BM HRL
verticolor Rafinesque, 1817 invalid
vetusta Hottes, 1927 = epilobii Kaltenbach, 1843
viburni Schrank, 1801 nec Scopoli, 1763 = Ceruraphis eriophori (Walker, 1848)
viburni Scopoli, 1763 BM HRL
*aceriella* Theobald, 1929 type BM
*opuli* Sulzer, 1776
viburniana Franssen, 1928 = Ceruraphis eriophori (Walker, 1848)
viburnicola Börner, 1916 nec Gillette, 1909 (type of Ceruraphis Börner, 1926) = Ceruraphis eriophori (Walker, 1848)
viburnicola Gillette, 1909 to Ceruraphis Börner, 1926
viburnicolens Swain, 1919 = citricola van der Goot, 1912
viburniphila Patch, 1917 BM HRL
viburnum-acerifolium Rafinesque, 1818 invalid
viburnum-opulus Rafinesque, 1818 invalid
viciae Fabricius, 1781 = craccae Linnaeus, 1758
viciae Kaltenbach, 1843 nec Fabricius, 1781 (type of Drepaniella del Guercio, 1913) = Megoura viciae Buckton, 1876
villosus Hartig, 1841 = ? Periphyllus testudinaceus (Fernie, 1852)
viminalis Boyer de Fonscolombe, 1841 (type of Tuberolachnus Mordvilko, 1909) = Tuberolachnus salignus (Gmelin, 1790)
viminalis Hartig, 1841 = Pterocomma salicis (Linnaeus, 1758)
vincae Walker, 1848 = Aulacorthum solani (Kaltenbach, 1843)
vineti Hoffmann, 1972 HRL
*coronillae* Börner, 1950 (*Doralida*) nec Aphis coronillae Ferrari, 1872
violae Schouteden, 1900 BM HRL
virgata del Guercio, 1911 = epilobii Kaltenbach, 1843
viridescens (del Guercio, 1930) (Anuraphis (Macchiatiella))
viridis Craveri, 1915 = Capitophorus elaeagni (del Guercio, 1894)
viridissima Mamontova, 1955 BM HRL
vitalbae Ferrari, 1872 BM HRL
*? brachysiphon* Narzikulov, 1964
*clematidis* errore Nevsky, 1929 nec Koch, 1854
vitellinae Hartig, 1841 = Tuberolachnus salignus (J.F. Gmelin, 1790)
vitellinae Schrank, 1801 (type of Pseudomicrella Börner, 1949) to Chaitophorus Koch, 1854
viticis Ferrari, 1872 BM HRL
vitifoliae Shinji, 1922 = gossypii Glover, 1877

A. vitis Scopoli, 1763
volutans (Börner, 1940) (Doralina) HRL
vorabonnevilla Knowlton & Smith, 1936 = Ceruraphis eriophori (Walker, 1848)
vulgaris Kyber, 1815 = Myzus (Nectarosiphon) persicae (Sulzer, 1776)
vulpiae del Guercio, 1913 = Rhopalosiphum maidis (Fitch, 1856)
wahena Hottes & Wehrle, 1950 HRL
wartenbergi (Börner, 1952) (Cerosipha) HRL
watsoni Theobald, 1929 = fabae Scopoli, 1763
wellensteini (Börner, 1950) (Doralina) BM HRL
wellensteini subsp. caucasica Bozhko, 1959
williamsi Soliman, 1927 = Bipersona ochrocentri (Cockerell, 1903)
wilsoni Laing, 1923 = sambuci Linnaeus, 1758
xanthelidis Rafinesque emendation of xanthelis
xanthelis Rafinesque, 1818 invalid
xylostei (Börner, 1950) (Doralis)
xylostei De Geer, 1773 (type of Stagona Koch, 1856) to Prociphilus (Stagona)
xylostei Schrank, 1801 nec de Geer, 1773 (type of Hyadaphis Kirkaldy, 1904; and of Siphocoryne Passerini, 1863 nec 1860) = Hyadaphis foeniculi (Passerini, 1860)
yanagicola Matsumura, 1917 = farinosa subsp. yanagicola Matsumura, 1917
yomogii Shinji, 1922
yuccae Cowen, 1895 = helianthi Monell, 1879
yuccae Lichtenstein, 1884 nomen nudum
yuccicola Wilson, 1911 = helianthi Monell, 1879
zeae Bonafous, 1835 nomen dubium
zeae Curtis, 1842 nomen nudum
zilora Hottes & Frison, 1931
zizyphi Theobald, 1922 = ? nasturtii Kaltenbach, 1843
zonassa Knowlton, 1935
zweigelti (Börner, 1940) (Pergandeida) BM HRL

Subgen. Absinthaphis Paik, 1972

A. (Absinthaphis) alba (Remaudière & Davatchi, 1959) (Protaphis) paratype BM HRL
(Absinthaphis) cinae (Nevsky, 1928) (Cryptosiphum) HRL
(Absinthaphis) hirsuta (Nevsky, 1929) (Cryptosiphum) HRL
(Absinthaphis) judenkoi (Szelegiewicz, 1959) (Brachyunguis) paratype BM HRL
(Absinthaphis) koraiensis (Paik, 1972) (Absinthaphis) HRL

A. (Absinthaphis) tashevella nomen novum* HRL
*lambersi* Tashev, 1964 (Brachyunguis) HRL

Subgen. Aleurosiphon Takahashi, 1966

A. (Aleurosiphon) smilacifoliae (Takahashi, 1921) (Aphis) to Aleurosiphon Takahashi, 1966

Subgen. Anuraphis del Guercio, 1907

A. (Anuraphis) feminea Hottes, 1930 to Aphis Linnaeus, 1758

Subgen. Aphidula Nevsky, 1929

A. (Aphidula) ulmariella Ossiannilsson, 1959 = ? Aphis ulmariae Schrank, 1801

Subgen. Cerosipha del Guercio, 1900

A. (Cerosipha) helianthemi subsp. obscura Bozhko, (1957) 1961 to Aphis Linnaeus, 1758
(Cerosipha) pulsatillae Ossiannilsson, 1959 to Aphis Linnaeus, 1758

Subgen. Doralis Leach, 1827

A. (Doralis) remaudierei Börner, 1952 to Aphis Linnaeus, 1758

Subgen. Forda von Heyden, 1837

A. (Forda) radicum Goureau, 1861 = Trama troglodytes von Heyden, 1837

Subgen. Iowana Hottes, 1954

A. (Iowana) frisoni (Hottes, 1954) (Iowana)

Subgen. Pemphigus Hartig, 1839

A. (Pemphigus) stamineus Haldeman, 1859 = Prociphilus (Paraprociphilus) tesselatus (Fitch, 1851)

Subgen. Pergandeida Schouteden, 1903

A. (Pergandeida) astragali Ossiannilsson, 1959 to Aphis Linnaeus, 1758
(Pergandeida) glareosae Bozhko, 1959 to Aphis Linnaeus, 1758
(Pergandeida) uvaeursi Ossiannilsson, 1959 to Aphis Linnaeus, 1758

* pro Brachyunguis lambersi Tashev, 1964 which when transferred to Aphis (Absinthaphis) is preoccupied by Aphis lambersi (Börner, 1940) transferred from Doralina Börner, 1940 to Aphis Linnaeus, 1758

Subgen. Protaphis Börner, 1952

A. (Protaphis) afghanica (Narzikulov & Umarov, 1972) (Brachyunguis)
(Protaphis) alexandrae (Nevsky, 1928) (Xerophilaphis)
(Protaphis) anthemiae (Ivanovskaja, 1960) (Protaphis)
(Protaphis) anthemidis (Börner, 1940) (Brachyunguis) HRL
*dudichi* Borner, 1940 (Brachyunguis)
(Protaphis) anuraphoides (Nevsky, 1928) (Xerophilaphis)
(Protaphis) artemisiae (Narzikulov, 1949) (Xerophilaphis)
(Protaphis) bimacula (Narzikulov & Umarov, 1972) (Brachyunguis)
(Protaphis) carlinae (Börner, 1940) (Brachyunguis) HRL
(Protaphis) carthami (Das, 1918) (Brachyunguis) HRL
(Protaphis) centaurea (Gomez-Menor, 1951) (Dasia) HRL
(Protaphis) chondrillae ((Mordvilko, (1932 nomen nudum) 1948)) (Xerophilaphis)
(Protaphis) deformans (Nevsky, 1929) (Xerophilaphis)
(Protaphis) delottoi (Eastop, 1958) (Brachyunguis) types BM HRL
(Protaphis) echinopis (Hille Ris Lambers, 1948) (Brachyunguis) types HRL
(Protaphis) elatior (Nevsky, 1928) (Xerophilaphis)
(Protaphis) elongata (Nevsky, 1928) (Xerophilaphis) HRL
(Protaphis) erigerontis Holman, 1966 (Aphis) paratypes BM HRL
(Protaphis) evansi (Eastop, 1958) (Brachyunguis) types BM HRL
(Protaphis) filaginea (del Guercio, 1911) (Anuraphis)
(Protaphis) formosana (Takahashi, 1927) (Anuraphis) BM
(Protaphis) funicularis (F.P. Müller, 1968) (Protaphis) paratype BM HRL
(Protaphis) hartigi (Hille Ris Lambers, 1931) (Doralis) HRL
(Protaphis) ignatii (Gomez-Menor, 1951) (Dasia) HRL
(Protaphis) nevskyi (Ivanovskaja, 1960) (Protaphis)
(Protaphis) picridicola Holman, 1966 (Aphis) paratypes BM HRL
(Protaphis) pseudocardui Theobald, 1915 (Aphis) types BM HRL
(Protaphis) rutae (Nevsky, 1928) (Xerophilaphis)
(Protaphis) scorzonerae (Mordvilko, 1937) (Xerophilaphis)
*scorzonerae* Mordvilko ex Tarbinsky & Plaviltchikov, 1948 (*Xerophilaphis*)
(Protaphis) sonchi (Narzikulov & Umarov, 1972) (Brachyunguis)
(Protaphis) striata Hille Ris Lambers, 1967 (Protaphis) HRL
(Protaphis) terraealbae Ivanovskaja, 1959 (Protaphis)
(Protaphis) terricola Rondani, 1847 (Aphis) HRL

Subgen. Uraphis del Guercio, 1907

A. (Uraphis) dianthi Mamontova, 1964 = ? Dianthaphis nigra Mamontova (described?)

Subgen. Wapuna Hottes & Wehrle, 1951

A. (Wapuna) tahosalea Hottes & Wehrle, 1951 (type of Wapuna Hottes & Wehrle, 1951) to Aphis Linnaeus, 1758

Subgen. Zyxaphis Knowlton, 1947

A. (Zyxaphis) canae Williams, 1911 (Aphis) BM HRL
(Zyxaphis) chrysothamni Wilson, 1915 (Aphis) BM HRL
(Zyxaphis) chrysothamnicola (Gillette & Palmer, 1929) (Anuraphis) metatypes BM HRL
(Zyxaphis) crypta Pack & Knowlton, 1929 (Aphis) BM
(Zyxaphis) filifoliae (Gillette & Palmer, 1928) (Anuraphis) BM HRL
(Zyxaphis) hermistonii Wilson, 1915 (Aphis) BM HRL
(Zyxaphis) minutissima (Gillette & Palmer, 1928) (Anuraphis) HRL
(Zyxaphis) oregonensis Wilson, 1915 (Aphis) BM HRL
(Zyxaphis) utahensis (Knowlton, 1947) (Zyxaphis) HRL

APHOIDES Rondani lapsus pro Aphioides

APHORODON Takahashi, 1961
Bull. Univ. Osaka Prefect. (B) 11: 2
Type species Myzus polygonifoliae Shinji, 1944
= Xenomyzus Aizenberg, 1935

A. polygoniphaga Takahashi, 1961 = Trichosiphonaphis (Xenomyzus) tade (Shinji, 1927)

APHRASTASIA Börner, 1909
Aphrastasia pectinatae (Chol.) CB. Privately published, St. Julien b. Metz, 1p.
Type species Chermes pectinatae Cholodkovsky, 1888
Here used as subg. of Adelges Vallot, 1836
Adelgidae

APHTHARGELIA Hottes, 1958
Proc. biol. Soc. Wash. 71: 43
Type species Aphis albipes Oestlund, 1887 = Aphis symphoricarpi Thomas, 1878
*Thargelia* Oestlund, 1922 nec Puengelev, 1899

A. symphoricarpi (Thomas, 1878) (Aphis) BM HRL
*albipes* Oestlund, 1887 (*Aphis*)

APLONEURA Passerini, 1863
Archo zool. Anat. Fisiol. 2: 201
Type species Tetraneura lentisci Passerini, 1856
? *Rhizoctonus* Mokrzecky, 1896
*Tycheoides* Schouteden, 1906

A. ? ampelina (Mokrzecky, 1896) (Rhizoctonus) BM HRL
juniperina Juchnevitch, 1968 = Gootiella tremulae (Tullgren, 1925)
lentisci (Passerini, 1856) (Tetraneura) BM HRL
*eragrostidis* Passerini of Schouteden, 1906 nec Passerini, 1860 (*Tychea*)
*graminis* Buckton, 1883 (*Rhizobius*) lectotype BM
? *longirostris* Mola, 1907 (*Tetranychus*)
*poae* Buckton, 1883 (*Rhizobius*) nec Thomas, 1879 lectotype BM
*poae* del Guercio, 1917 (*Neorhizobius*)
*poae* Thomas, 1879 (*Rhizobius*)
*radicum* Lichtenstein, 1879
*stramineus* del Guercio, 1917 (*Neorhizobius*)
radicum Lichtenstein, 1879 = lentisci (Passerini, 1856)
werthae Börner, 1952

APPELIA Börner, 1930
Arch. klassif. phylogen. Ent. 1: 132
Type species Aphis tragopogonis Kaltenbach, 1843
Here used as subg. of Brachycaudus van der Goot, 1913

A. schwartzi Börner, 1931 to Brachycaudus (Appelia)
tragopogonis subsp. setosus Hille Ris Lambers, 1948 to Brachycaudus (Appelia)

APPENDISETA Richards, 1965
Mem. ent. Soc. Can. 44: 75
Type species Callipterus robiniae Gillette, 1907

A. robiniae (Gillette, 1907) (Callipterus) BM HRL

APTEROIDES Mordvilko, 1929
Trudy prikl. Ent. 14: 91
lapsus pro Atheroides Haliday, 1837

ARAKAWANA Matsumura, 1917
J. Coll. Agric. Tohoku imp. Univ. 7: 375

Type species Arakawana stigmata Matsumura, 1917
= Acanthocallis Matsumura, 1917

A. stigmata Matsumura, 1917 to Tuberculatus (Acanthocallis)

ARCHICALLIS Aizenberg, 1954
Diss. biol. Sci. Lenin Univ., Moscow, p. 9
Without species

ARCHILACHNUS Buckton, 1883
Monograph of the British Aphides, London 4: 177
Type species Archilachnus pennatus Buckton, 1883
= Siphonophoroides Buckton, 1883
Fossil

ARCTAPHIS Walker, 1870
Zoologist 5: 2000
Type species: Chaitophorus populi (Linnaeus) of Koch, 1854
= Chaitophorus Koch, 1854

ARESHA Mordvilko, 1921
Izv. Sev. Oblast. Sta. Zashch. Rast. Vredit. 3: 53
Type species Aresha shelkovnikovi Mordvilko, 1921
= Rhopalosiphum Koch, 1854

A. setigera E.E. Blanchard, 1939 = Rhopalosiphum rufiabdominalis (Sasaki, 1899)
shelkovnikovi Mordvilko, 1921 = Rhopalosiphum rufiabdominalis (Sasaki, 1899)

ARIMAKIA Matsumura, 1917
J. Coll. Agric. Tohoku imp. Univ. 7: 405
Type species Arimakia araliae Matsumura, 1917
= Toxoptera Koch, 1856

A. araliae Matsumura, 1917 = Toxoptera odinae (van der Goot, 1917)
taranbonis Matsumura, 1917 = Toxoptera odinae (van der Goot, 1917)

ARISTAPHIS Kikaldy, 1905
Can. Ent. 37: 416
Type species Aphis populea Kaltenbach, 1843
= Pterocomma Buckton, 1879

ARTEMISAPHIS Knowlton & Roberts, 1947
J. Kans. ent. Soc. 20: 27
Type species Aphis artemisicola Williams, 1911
Here used as subg. of Obtusicauda Soliman, 1922

ARTHROMYZUS Börner, 1950
Neue europäische Blattlausarten, privately published, p. 12, described as subg. of Hyperomyzus Börner, 1933
Type species Rhopalosiphum staphyleae Koch, 1854, cited unambiguously by Börner, 1952: 138
= Myzosiphon Hille Ris Lambers, 1946

ASIÁTAPHIS Narzikulov, 1970
Ent. Obozr. 49: 362
Type species Aphis verbasci Schrank of Nevsky, 1929 = Aphis verbasci Schrank, 1801
= Aphis Linnaeus, 1758

ASIPHON alternative spelling of Asiphum Koch, 1856

ASIPHONAPHIS Wilson & Davis, 1919
Ent. News 30: 39
Type species Asiphonaphis pruni Wilson & Davis, 1919

A. anogis Hottes & Frison, 1931 (type of Pseudasiphonaphis Robinson, 1965) = Pseudasiphonaphis corni (Tissot, 1929)
carolinensis Knowlton, 1937 = Pseudasiphonaphis corni (Tissot, 1929)
pruni Wilson & Davis, 1919 BM HRL
utahensis Knowlton, 1937

ASIPHONELLA Theobald, 1923
Bull. Soc. ent. Egypte 1922: 76
Type species Asiphonella dactylonii Theobald, 1923

A. cynodonti (Das, 1918) (Pemphigus) BM HRL
dactylonii Theobald, 1923 types BM HRL
*graminis* E.E. Blanchard, 1944 (*Paraprociphilus*)

ASIPHUM Koch, 1856
Die Pflanzenläuse Aphiden, Nürnberg, p. 246
Type species Asiphum populi (Fabricius of) Koch, 1856 = Aphis tremulae Linnaeus,

1761; designated by Gerstaecker, 1859
*Rhizomaria* Hartig, 1857

A. ligustrinellum Koch, 1856 = Prociphilus bumeliae (Schrank, 1801)
populi Gaumont, 1923
populi (Fabricius of) Koch, 1856 = tremulae Linnaeus, 1761
rosettei Maxson, 1934 = tremulae (Linnaeus, 1761)
sacculi Gillette, 1914 to Pachypappa Koch, 1856
tremulae (Linnaeus, 1761) (Aphis) BM HRL
*lanata* Zetterstedt, 1840 (*Aphis*) nec Salisbury, 1816
*piceae* Hartig, 1857 (*Rhizomaria*)
*rosettei* Maxson, 1934

ASPIDAPHIS Gillette, 1917
Can. Ent. 49: 196
Type species Aspidaphis polygoni Gillette, 1917 = Aphis adjuvans Walker, 1848

A. adjuvans (Walker, 1848) (Aphis) lectotype BM HRL
*adjuvans* subsp. *rowei* Knowlton & Smith, 1936
*polygoni* Gillette, 1917
*polygoni* Schouteden, 1907 (*Sipha*)
*polygoni* Walker, 1848 (*Aphis*) lectotype BM
adjuvans subsp. rowei Knowlton & Smith, 1936 = adjuvans (Walker, 1848)
longicauda Richards, 1963 to Eoessigia David, Rajasingh & Narayanan, 1972
polygoni Gillette, 1917 = adjuvans (Walker, 1848)
porosiphon Börner, 1950 HRL

ASPIDAPHIUM Börner, 1939
Arb. physiol. angew. Ent. Berl. 6: 81
Type species Aspidaphium escherichi Börner, 1939

A. cuspidatae Stroyan, 1955 type BM HRL
escherichi Börner, 1939 BM HRL
*jeschkei* Börner, 1939
jeschkei Börner, 1939 = escherichi Börner, 1939
utahensis Smith & Knowlton, 1965 paratype BM HRL

ASPIDOPHORODON Verma, (1965) 1966
Sci. Cult. 31: 389 (1965)
Proc. Indian Sci. Congr. 53 (3): 358 (1965)

Indian J. Ent. 28: 507 (1966)
Type species Aspidophorodon harvensis Verma, 1966

A. harvensis Verma, 1966 HRL
salicis Miyazaki, 1971 BM HRL

ASTEGOPTERYX Karsch, 1890
Ber. dt. bot. Ges. 8: 51
Type species Astegopteryx styracophila Karsch, 1890
*Oregma* Buckton, 1893
*Trichoregma* Takahashi, 1929

A. bambusae (Buckton, 1893) (Oregma) paratypes BM HRL
*lutescens* van der Goot, 1917 (*Oregma*)
*mysorensis* David, 1956 (*Oregma*)
bambusae (Takahashi, 1935) (Trichoregma)
bambusifoliae (Takahashi, 1921) (Oregma) BM HRL
basalis (van der Goot, 1917) (Oregma) BM HRL
chinensis Tao, 1966
esakii (Takahashi, 1941) (Trichoregma)
fici Takahashi, 1923 = Reticulaphis distylii subsp. fici (Takahashi, 1923)
ficicola Takahashi, 1927
flava (Takahashi, 1950) (Trichoregma) types BM HRL
formosana (Takahashi, 1924) (Oregma) HRL
formosana subsp. neelagiriensis David, 1958 = neelagiriensis David, 1958
formosana Takahashi, 1927 type of Sinonipponaphis Tao, 1966 q.v.
fransseni Hille Ris Lambers, 1933 to Cerataphis Lichtenstein, 1882
fransseni subsp. kepongensis Takahashi, 1950 = Cerataphis fransseni (Hille Ris Lambers, 1933)
gigantea Takahashi, 1921 = Metanipponaphis cuspidatae (Essig & Kuwana, 1918)
insularis (van der Goot, 1912) (Cerataphis) BM HRL
*? mysorensis* Krishnamurti, 1930 (*Oregma*)
*? mysorensis* Theobald, 1929 (*Oregma*)
jamuritsu Takahashi, 1931 HRL
japonica Takahashi, 1923 = Aleurodaphis blumeae van der Goot, 1917
javensis Takahashi, 1921 in litteris synonym of Reticulaphis distylii (van der Goot, 1917)
kashifoliae Uye, 1924 to Xenothoracaphis Takahashi, 1958
kashiwae Uye ex Takahashi, 1958*, as Thoracaphis type of Xenothoracaphis Takahashi, 1958 q.v.
lambersi Takahashi, 1936 BM

* footnote next page

A. leeuweni Takahashi, 1936 BM
lithocarpi Takahashi, 1929 to Lithoaphis Takahashi, 1959
loranthi Tseng & Tao, 1938 to Tuberaphis Takahashi, 1933
malaccensis (Takahashi, 1950) (Trichoregma) types BM HRL
minuta (van der Goot, 1917) (Oregma) BM HRL
muiri (van der Goot, 1918) (Oregma) BM HRL
musae (Takahashi, 1925) (Trichoregma)
neelagiriensis David, 1958
*formosana* subsp. *neelagiriensis* David, 1958
nekoashi Sasaki, 1910 to Ceratovacuna Zehntner, 1897
nipae (van der Goot, 1917) (Oregma) BM HRL
pallida (van der Goot, 1917) (Oregma) HRL
pandani (Takahashi, 1935) (Trichoregma)
pseudostyracophila Shinji, 1936 ? to Ceratovacuna Zehntner, 1897
quercicola Takahashi, 1921 to Neothoracaphis Takahashi, 1958
rappardi Hille Ris Lambers, 1953 BM HRL
rhapidis (van der Goot, 1917) (Oregma) BM HRL
roepkei Hille Ris Lambers, 1932 types BM HRL
salatigensis (van der Goot, 1917) (Oregma) BM HRL
sasakii Takahashi, 1939
shitosanensis Takahashi, 1939
similis (van der Goot, 1917) (Oregma)
singaporensis (van der Goot, 1918) (Oregma) BM HRL
striata (van der Goot, 1917) (Oregma)
styraci Matsumura, 1917 to Ceratovacuna Zehntner, 1897
styracicola Takahashi, 1921 BM
styracophila Karsch, 1890 BM HRL
sumatrana Hille Ris Lambers, 1931 types BM HRL
swinhoei Takahashi, 1936 HRL
taiwana Takahashi, 1934
takahashii Strand, 1929 to Dermaphis Takahashi, 1958
takenouchii Takahashi, 1934 HRL
vandermeermohri Hille Ris Lambers, 1931 BM HRL

* Shinji, 1941: 1125 describes a Thoracaphis kashiwae Uye giving Astegopteryx kashiwae Uye, 1924, Insect World 28: 2 as the original description. That page of Insect World deals with coccids. Takahashi, 1958, Kontyû 26: 185 refers to Astegopteryx kashiwae Uye, 1924, Insect World 28:14. However, on that page Uye describes Astegopteryx kashifoliae. Earlier, 1932, Takahashi, Philip. J. Sci. 48 (1): 72 transferred Astegopteryx kashifoliae to Thoracaphis, but in 1958 l.c. used Thoracaphis kashiwae Uye, a name not used by Uye, as type of Xenothoracaphis Takahashi, 1958.

ASTEROBIUM Hille Ris Lambers, 1938
Temminckia 3: 19, described as subg. of Macrosiphoniella del Guercio, 1911
Type species Aphis asteris Walker, 1849
Here used as subg. of Macrosiphoniella del Guercio, 1911

ASTEROLECANIUM Westwood, 1897
Gardners' Chronicle (3) 12:797 nec Targioni-Tozzetti, 1868, Coccoidea.
Type species Asterolecanium orchidearum Westwood, 1897
= Cerataphis Lichtenstein, 1882

A. orchidearum Westwood, 1897 to Cerataphis Lichtenstein, 1882

ATARSAPHIS Takahashi, 1958
Kontyû 26: 181
Type species Atarsaphis quercus Takahashi, 1958 = Hamamelistes agrifoliae Ferris, 1921

A. agrifoliae (Ferris, 1921) (Hamamelistes ?) BM HRL
*quercus* Takahashi, 1958 HRL
quercus Takahashi, 1958 = agrifoliae (Ferris, 1921)

ATARSOS Gillette, 1911
Ent. News. 22: 440
Type species Atarsos grindeliae Gillette, 1911

A. grindeliae Gillette, 1911 BM HRL
orientalis Mordvilko, 1929 to Shinjia Takahashi, 1930

ATHEROIDES Haliday, (1837) 1839
In Curtis, J., A guide to an arrangement of British Insects, London, Ed. 2: 218 without description (1837)
Ann. Mag. nat. Hist. 2: 189 (1839)
Type species Atheroides serrulatus Haliday, 1839, selected by Kirkaldy, 1906
*Corealachnus* Paik, 1971

A. aplangi Pintera, 1965 = brevicornis Laing, 1920
brevicornis Laing, 1920 type BM HRL
*aplangi* Pintera, 1965 BM HRL
doncasteri Ossiannilsson, 1955 paratypes BM HRL
festucae Mordvilko, 1934 = serrulatus Haliday, 1839
hirtellus Haliday, (ex J. Curtis, 1837 nomen nudum) 1839 BM HRL
*junci* Laing, 1920 types BM
*niger* Ossiannilsson, 1954 HRL

A. junci Laing, 1920 = hirtellus Haliday, 1839
karakumi Mordvilko, 1948 HRL
*lasiagrostites* Juchnevitch, 1960
lasiagrostites Juchnevitch, 1960 = karakumi Mordvilko, 1948
niger Ossiannilsson, 1954 = hirtellus Haliday, 1839
serrulatus Haliday, (ex J. Curtis, 1837 nomen nudum) 1839 BM HRL
*aculeata* Dahl, 1912 (*Glyphina* )
*festucae* Mordvilko, 1934
*paradoxa* Theobald, 1918 (*Sipha*)
*suwonensis* Paik, 1971 (*Corealachnus*) HRL
stipae Börner, 1950 = Chaetosiphella stipae Hille Ris Lambers, 1947

AULACOPHORA Tao, 1963 nec Dejean, 1835; Jeffreys, 1882
Pl. Prot. Bull., Taiwan 5: 175
Type species Amphorophora formosana Takahashi, 1923
= Aulacophoroides Tao ex Eastop & Hille Ris Lambers, 1976

AULACOPHOROIDES Tao, nomen novum
Type species Amphorophora formosana Takahashi, 1923
*Aulacophora* Tao, 1963 nec Dejean, 1835; Jeffreys, 1882

A. formosana (Takahashi, 1923) (Amphorophora) HRL
hoffmanni (Takahashi, 1937) (Acyrthosiphon)

AULACORTHUM Mordvilko, 1914
Faune Russie, Ins. Hém., 1 (1): 52, 68
Type species nominally Aulacorthum pelargonii (Kaltenbach, 1843) but the species figured is that originally described as Aphis solani Kaltenbach, 1843: vide Hille Ris Lambers (1947) Temminckia 7: 307-308.
*Dysaulacorthum* Börner, 1939
*Melanosiphon* Börner, 1944
*Neomacrosiphum* van der Goot, 1915
*Pseudomegoura* Shinji, 1922

A. aegopodii Börner, 1939 = solani subsp. aegopodii Börner, 1939
agrimoniae Börner, 1940 = Acyrthosiphon malvae subsp. agrimoniae (Börner, 1940)
asteris Takahashi, 1965 HRL
brevicaudum Moritsu, 1958 = muradachi (Shinji, 1928)
capilanoense Robinson, 1969
*scabrosum* Richards, 1972
cercidiphylli (Matsumura, 1918) (Macrosiphum ) BM HRL
*katsurae* Shinji, 1930 (Amphorophora)

A. cirsicola (Takahashi, 1923) (Macrosiphum) BM HRL
*circifoliae* Shinji, 1935 (*Acyrthosiphon*)
*tricholobicola* Strand, 1929 (*Macrosiphum*)
clavicornis Richards, 1972 = Wahlgreniella nervata (Gillette, 1908)
cornaceae A.K. Ghosh, 1969 HRL
cylactis Börner, 1942 = solani subsp. cylactis Börner, 1942
dasi A.K. Ghosh, R.C. Basu & RayChaudhuri, 1970
dispersum van der Goot, 1917 type of Ipuka van Harten & Ilharco, 1976 q.v.
doronici Börner, 1950 = solani (Kaltenbach, 1843)
dorsatum Richards, 1967 to Sitobion Mordvilko, 1914
dryopteridis Holman, 1959 to Sitobion Mordvilko, 1914
esakii (Takahashi, 1924) (Macrosiphum) BM HRL
eumorphum E.E. Blanchard, 1922 = solani (Kaltenbach, 1843)
filicis van der Goot, 1917
flavum F.P. Müller, 1958 paratypes BM HRL
glechomae Takahashi, 1965
hirschfeldii Rusanova, 1943 nomen nudum
ibotum (Essig & Kuwana, 1918) (Macrosiphum) BM HRL
*ligustrumae* Shinji, 1927 (*Macrosiphum*)
kerriae (Shinji, 1930) (Illinoia) BM HRL
knautiae Heie, 1960 BM HRL
kuwanai (Takahashi, 1933) (Acyrthosiphon)
langei Börner, 1939 (Aulacorthum (Dysaulacorthum)) BM HRL
linderae (Shinji, 1022) (Macrosiphum)
lonicerae Hori, 1938 = Trichosiphonaphis (Xenomyzus) horii Miyazaki, 1971
loochooense Takahashi, 1939 to Staticobium Mordvilko, 1914
magnoliae (Essig & Kuwana, 1918) (Rhopalosiphum) BM HRL
*magnolifoliae* Shinji, 1941 (*Myzus*)
*malvicola* Shinji, 1933 (*Amphorophora*)
*nishikigi* Shinji, 1928 (*Macrosiphum*)
*sambuci* Matsumura, 1918 (*Rhopalosiphum*)
*sambucicola* Takahashi, 1918 (*Rhopalosiphum*)
majanthemi F.P. Müller, 1956 BM HRL
muradachi (Shinji, 1928) (Macrosiphum) BM HRL
*brevicaudum* Moritsu, 1958
nepetifolii Miyazaki, 1968 HRL
nipponicum (Essig & Kuwana, 1918) (Macrosiphum) BM HRL
*paederiae* Takahashi, 1921 (*Macrosiphum*)
palustre Hille Ris Lambers, 1947 BM HRL
phytolaccae Miyazaki, 1968 HRL

A. pirolacearum Szelegiewicz, 1967 = ? rufum Hille Ris Lambers, 1947
porrifolii Börner, 1950 to Acyrthosiphon Mordvilko, 1914
prasinum Börner, 1950 = solani (Kaltenbach, 1843)
pseudorosaefolium E.E. Blanchard, 1922 = Rhodobium porosum (Sanderson, 1900)
pterinigrum Richards, 1972 to ? Sitobion Mordvilko, 1914
rhamni M.R. Ghosh, A.K. Ghosh & RayChaudhuri, 1971
rhusifoliae Richards, 1973 to Glabromyzus Richards, 1960
rufum Hille Ris Lambers, 1947 cotypes BM HRL
*? pirolacearum* Szelegiewicz, 1967 paratypes BM HRL
scabrosum Richards, 1972 = capilanoense Robinson, 1969
scirpi van der Goot, 1917 HRL
sclerodorsi (Kumar & Burkhardt, 1971) (Acyrthosiphon (Aulacorthum)) paratype BM
sedens F.P. Müller, 1966 paratypes BM HRL
sensoriatum (David, Narayanan & Rajasingh, 1971) (Acyrthosiphon (Aulacorthum)) HRL
simplocois lapsus pro symplocois
smilacis Takahashi, 1965 = solani (Kaltenbach, 1843)
solani (Kaltenbach, 1843) (Aphis) BM HRL
*aquilegiae* Theobald, 1913 (*Macrosiphum*) types BM
*atropae* Mordvilko, 1895 (*Siphonophora*)
*aucubae* Bartholomew, 1932 (*Macrosiphum*)
*begoniae* Schouteden, 1901 (*Macrosiphum*)
*boerneri* F.P. Müller, 1952 (*Dysaulacorthum*)
*diplanterae* Koch, 1855 (*Siphonophora*)
*doronici* Börner, 1950
*duffieldii* Theobald, 1913 (*Macrosiphum*) types BM
*eumorphum* E.E. Blanchard, 1922
*gei* Theobald, 1919 (*Myzus*) types BM
*glaucii* Theobald, 1923 (*Myzus*) types BM
*hagi* Essig & Kuwana, 1918 (*Macrosiphum*)
*hagicola* Matsumura, 1917 (*Macrosiphum*)
*hederae* Theobald, 1915 (*Macrosiphum*)
*hydrocotylei* Theobald, 1925 (*Myzus*) types BM
*? incerta* Walker, 1849 (*Aphis*)
*indecisa* Walker, 1849 (*Aphis*)
*? kusaki* Shinji, 1941 (*Myzus*)
*lamii* Theobald, 1915 (*Macrosiphum*) types BM
*matsumuraeanum* Hori, 1928 (*Macrosiphum*)

*menthae* Buckton, 1876 (*Siphonophora*) types BM
*mercurialis* Theobald, 1919 (*Myzus*) types BM
*neogei* Theobald, 1926 (*Myzus*) types BM
*pallida* Walker, 1848 (*Aphis*) type BM
*pelargonii* Kaltenbach of van der Goot, 1915 (*Macrosiphum*) misidentification
*piceaellum* Theobald, 1916 (*Macrosiphum*) type BM
*polyanthi* Theobald, 1926 (*Myzus*) types BM
*prasinum* Börner, 1950
*pseudolamii* Theobald, 1926 (*Myzus*) types BM
*pseudosolani* Theobald, 1922 (*Myzus*) types BM
*ranunculi* Mordvilko, 1914 key (*Acyrthosiphon* (*Microlophium*))
*senecionis* Matsumura, 1917 (*Macrosiphum*)
*smilacis* Takahashi, 1965
*sobae* Shinji, 1922 (*Macrosiphum*)
*veronicae* del Guercio, 1900 (*Myzus*)
*veronicae* Theobald, 1913 (*Macrosiphum*) types BM
*veronicellus* Theobald, 1926 (*Myzus*) types BM
*vincae* Walker, 1848 (*Aphis*) lectotype BM

A. solani subsp. aegopodii Börner, 1939 BM HRL
solani subsp. cylactis Börner, 1942 BM HRL
*cylactis* Börner, 1942
solani subsp. orientale Hille Ris Lambers, 1949 HRL
speyeri Börner, 1939 type of Melanosiphon Börner, 1944 BM HRL
spinacaudatum (Kumar & Burkhardt, 1971) (Acyrthosiphon (Aulacorthum)) paratype BM
symplocois van der Goot, 1917 BM
syringae (Matsumura, 1918) (Macrosiphum) BM HRL
*syringae* Takahashi, 1965 BM HRL
syringae Takahashi, 1965 = syringae (Matsumura, 1918)
takahashii (Mason, 1925) (Amphorophora)
vaccinii Hille Ris Lambers, 1952 BM HRL
vaccinii subsp. parvulum Hille Ris Lambers, 1952 HRL
vandenboschi Hille Ris Lambers, 1967 paratype BM HRL
viride van der Goot, 1917 = Rhodobium porosum (Sanderson, 1900)
watanabei (Miyazaki, 1971) (Acyrthosiphon) BM HRL

Subgen. Anaulacorthum A.K. Ghosh & RayChaudhuri, 1972

A. (Anaulacorthum) fagopyri A.K. Ghosh & RayChaudhuri, 1972 to Anaulacorthum A.K. Ghosh & RayChaudhuri, 1972

Subgen. Dysaulacorthum Börner, 1939

A. (Dysaulacorthum) langei Börner, 1939 (type of Dysaulacorthum Börner, 1939)
to Aulacorthum Mordvilko, 1914

Subgen. Melanosiphum Aizenberg, 1954

A. (Melanosiphum) speyeri subsp. orientale Aizenberg, 1954 nomen nudum

Subgen. Neomyzus van der Goot, 1915

A. (Neomyzus) circumflexum (Buckton, 1876) (Siphonophora) lectotype BM HRL
*callae* Henrich, 1909 (*Siphonophora*)
*primulanum* Matsumura, 1917 (*Macrosiphum*)
*vincae* Gillette, 1908 (*Myzus*)
(Neomyzus) dendrobii A.N. Basu, 1969 HRL
(Neomyzus) dicentrae A.N. Basu, 1968 HRL
(Neomyzus) parthenocissi (Takahashi, 1965) (Myzus) BM HRL
(Neomyzus) primulum A.K. Ghosh, H. Banerjee & RayChaudhuri, 1971
(Neomyzus) taiwanum (Takahashi, 1923) (Macrosiphum) HRL
(Neomyzus) taiwanum subsp. codonopsis (Miyazaki, 1971) (Acyrthosiphon) HRL

Subgen. Perillaphis Takahashi, 1965

A. (Perillaphis) perillae (Shinji, 1924) (Macrosiphum) HRL
*perillae* Takahashi, 1924 (*Macrosiphum*)

AVICENNINA Narzikulov, 1957
Ént. Obozr. 36: 676
Type species Avicennina sogdiana Narzikulov, 1957

A. sogdiana Narzikulov, 1957

BACILLAPHIS Quednau, 1954
Mitt. biol. Reichsanst. Ld- u. Forstw. 78: 38, 47
Type species Saltusaphis ornata Theobald, 1927
= Subsaltusaphis Quednau, 1953

B. afghanica Narzikulov & Umarov, 1970 = Saltusaphis scirpus Theobald, 1915
aquatilis Ossiannilsson, 1959 to Subsaltusaphis Quednau, 1953
lambersi Quednau, 1954 to Subsaltusaphis Quednau, 1953
maritima Hille Ris Lambers, 1956 to Subsaltusaphis Quednau, 1953
pallida subsp. taurica Bozhko, 1959 to Subsaltusaphis Quednau, 1953
paniceae Quednau, 1954 to Subsaltusaphis Quednau, 1953

BAIZONGIA Rondani, 1848
Nuovi Annali Sci. nat. Bologna 9: 35
Type species Aphis pistaciae Linnaeus, 1767
*Dasia* van der Goot ex Das, 1918
*Pemphigella* Tullgren, 1909

B. oestlundi Hottes, 1949 = pistaciae (Linnaeus, 1767)
? paglianoi (Gaumont, 1930) (Pemphigella)
pistaciae (Linnaeus, 1767) (Aphis) BM HRL
*aedificator* Buckton, 1893 (*Pemphigus*) lectotype BM
*cornicularius* Passerini, 1856 (*Pemphigus*)
*corniculoides* Lichtenstein, 1880 (*Pemphigus*)
*formicina* Buckton, 1883 (*Endeis*) lectotype BM
*oestlundi* Hottes, 1949

BAIZONGIELLA E.E. Blanchard, 1944
Acta zool. lilloana 2: 44
Type species Baizongiella solanophila E.E. Blanchard, 1944 = Pemphigus canadensis del Guercio, 1913
= Pemphigus Hartig, 1839

B. solanophila E.E. Blanchard, 1944 = Pemphigus canadensis del Guercio, 1913
= ? Pemphigus populitransversus Riley, 1879

BALTICAPHIS Heie, 1967
Spolia zool. Mus. haun. 26: 160
Type species Balticaphis exsiccata Heie, 1967
Fossil, baltic amber

BALTICHAITOPHORUS Heie, 1967
Spolia zool.Mus. haun. 26: 180
Type species Baltichaitophorus jutlandicus Heie, 1967
Fossil, baltic amber

BALTICOMARAPHIS Heie, 1967
Spolia zool. Mus. haun, 26: 167
Type species Balticomaraphis latens Heie, 1967
Fossil, baltic amber

BALTICOROSTRUM Heie, 1967
Spol. zool. Mus. haun. 26: 77, described as subg. of Germaraphis Heie, 1967
Type species Germaraphis (Balticorostrum) oblonga Heie, 1967
Fossil, baltic amber

BELOCHILUM Börner, 1932
In Sorauer, P., Handbuch der Pflanzenkrankheiten 5(2): 630
Type species Belochilum inulae (Passerini) of Börner, 1932 errore Passerini, 1860 = Siphonophora inulae Ferrari, 1872
Here used as subg. of Uroleucon Mordvilko, 1914

B. inulae Börner, 1932 = Uroleucon (Belochilum) inulae (Ferrari, 1872)

BERBERIDAPHIS Narzikulov, 1960
Dokl. Akad. Nauk tadzhik. SSR 3(2): 31
Type species Liosomaphis lydiae Narzikulov, 1957

B. lydiae (Narzikulov, 1957) (Liosomaphis) HRL
nepetae Narzikulov & Mukhammediev, 1967

BERENDTAPHIS Heie, 1971
Dt. ent. Z. 18: 262
Type species Lachnus cimicoides Germar & Berendt, 1856
Fossil, baltic amber

BETACALLIS Matsumura, 1919
Trans. Sapporo nat. Hist. Soc. 7: 110
Type species Betacallis alnicolens Matsumura, 1919

B. alnicolens Matsumura, 1919 BM HRL
odaiensis Takahashi, 1961 HRL
querciphaga R.C. Basu, M.R. Ghosh & RayChaudhuri, 1974
sikkimensis R.C. Basu, M.R. Ghosh & RayChaudhuri, 1974

BETULAPHIS Glendenning, 1926
Can. Ent. 58: 96
Type species Betulaphis occidentalis Glendenning, 1926 = Aphis quadrituberculata Kaltenbach, 1843

B. arctosetis Richards, 1961 paratype BM HRL
aurea Richards, 1961 BM HRL
brevipilosa Börner, 1940 HRL
helvetica Hille Ris Lambers, 1947 BM HRL
*quadrituberculata* subsp. *helvetica* Hille Ris Lambers, 1947 HRL
*viridis* Richards, 1972
hissarica Narzikulov, 1963 HRL
japonica Takahashi, 1961 HRL
occidentalis Glendenning, 1926 = quadrituberculata (Kaltenbach, 1843)

B. pelei Hille Ris Lambers, 1952 BM HRL
quadrituberculata (Kaltenbach, 1843) (Aphis) BM HRL
*inhaerens* Walker, 1852 (*Aphis*)
*minimum* van der Goot, 1912 (*Pterocallis*)
*occidentalis* Glendenning, 1926
quadrituberculata subsp. helvetica Hille Ris Lambers, 1947 = helvetica Hille Ris Lambers, 1947
quadrituberculata subsp. intermedia Börner, 1952 HRL
viridis Richards, 1972 = helvetica Hille Ris Lambers, 1947

BICAUDELLA Rusanova, 1943
Trudy azerb. gos. Univ. (Biol.) 3(1): 43
Type species Bicaudella astragalensis Rusanova, 1943 nomen nudum
Without validly described species

B. astragalensis Rusanova, 1943 nomen nudum

BIPERSONA Hottes, 1926
Proc. biol. Soc. Wash. 39: 115
Type species Aphis torticauda Gillette, 1907 = Aphis ochrocentri Cockerell, 1903

B. hottesi Knowlton & Smith, 1936
ochrocentri (Cockerell, 1903) (Aphis) BM HRL
*cnici* Williams, 1910 (*Aphis*) nec Schrank, 1801
*torticauda* Gillette, 1907 (*Aphis*)
*williamsi* Soliman, 1927 (*Aphis*)

BITUBERCULAPHIS Rusanova, 1943
Trudy azerb. gos. Univ. (Biol.) 3(1): 34
Type species Bituberculaphis inexpectata Rusanova, 1943 nomen nudum
Without validly described species

B. inexpectata Rusanova, 1943 nomen nudum

BOERNERIA Grassi & Foa, 1908/9 nec Axelson, 1902; Willew, 1901/2
Atti Accad. naz. Lincei Rc. 17 (2): 685
Type species Phylloxera danesii Grassi & Foa, 1907
= Faoiella Börner, 1905
Phylloxeridae

BOERNERINA Bramstedt, 1940
Anz. Schädlingsk. 16: 13
Type species Boernerina depressa Bramstedt, 1940

B. alni Takahashi, 1961 HRL
depressa Bramstedt, 1940 BM HRL
variabilis Richards, 1961 BM HRL
variabilis subsp. alaskensis Hille Ris Lambers, & Hottes, 1962 paratypes BM HRL

Subgen. Boernerinella Hille Ris Lambers & Hottes, 1962

B. (Boernerinella) occidentalis Hille Ris Lambers & Hottes, 1962 type of Boernerinella Hille Ris Lambers & Hottes, 1962 paratypes BM HRL

BOERNERINELLA Hille Ris Lambers & Hottes, 1962
Ent. Ber., Amst. 22: 112, described as subg. of Boernerina Bramstedt, 1940
Type species Börnerina (Börnerinella) occidentalis Hille Ris Lambers & Hottes, 1962
Here used as subg. of Boernerina Bramstedt, 1940

BOISDUVALIA Signoret, 1868 nec Robineau-Desvoisy, 1830
Annls Soc. ent. Fr. (4) 8: 400
Type species Coccus lataniae Boisduval, 1867
= Cerataphis Lichtenstein, 1882

BOREAMYZUS Shaposhnikov, 1964
in Bei-Bienko, G.Y., Keys to the insects of the European part of the U.S.S.R. 1: 590,described as subg. of Ericaphis Börner, 1939
Type species Ericaphis latifrons (Börner, 1942) = Ovatus latifrons Börner, 1942
= Ericaphis Börner, 1939

BOZHKOJA Shaposhnikov, 1964
in Bei-Bienko, G.Y., Keys to the insects of the European part of the U.S.S.R. 1: 592, described as subg. of Brevicoryne van der Goot, 1915
Type species Brevicoryne crambe Bozhko, 1950
= Brevicoryne van der Goot, 1915

BRACHYCAUDINA Börner, 1930
Arch. klassif. phylogen. Ent. 1: 132
Type species Aphis napelli Schrank, 1801
= Acaudus van der Goot, 1913

BRACHYCAUDUS van der Goot, 1913
Tijdschr. Ent. 56: 97
Type species Aphis myosotidis Koch, 1854 = Aphis helichrysi Kaltenbach, 1843
*Neoacaudus* Theobald, 1927

B. asselbergsi Hille Ris Lambers, 1931 (nomen novum pro Aphis capsellae Koch, 1854 nec Kaltenbach, 1843) = Brachycaudus (Acaudus) cardui (Linnaeus, 1758)
bodenheimeri Börner ex Bodenheimer, 1937 nomen nudum
cardui (Linnaeus, 1758) (Aphis) to Brachycaudus (Acaudus)
cardui subsp. yosiii Takahashi, 1966 = Brachycaudus (Acaudus) cardui (Linnaeus, 1758)
cerinthis Bozhko (1957) 1961 to Brachycaudus (Appelia)
crassitibiae Nevsky, 1951
distinctus (Mordvilko, 1928) (Anuraphis)
divaricatae Shaposhnikov, 1956 to Brachycaudus (Acaudus)
divaricatellus Shaposhnikov, 1956
helichrysi (Kaltenbach, 1843) (Aphis) BM HRL
*abrotaniella* Theobald, 1919 (*Anuraphis*) types BM
*adjecta* Walker, 1849 (*Aphis*) type BM
*adscita* Walker, 1848 (*Aphis*) type BM
*ammobii* Hori, 1929 (*Anuraphis*)
*apposita* Walker, 1850 (*Aphis*) type BM
*bartsiae* Walker, 1849 (*Aphis*) lectotype BM
*bellis* Buckton, 1879 (*Aphis*) lectotype BM
*bipapillatus* Theobald, 1923 (*Acaudus*) types BM
*brevisiphon* del Guercio, 1930 (*Anuraphis*)
*cacaliae* Matsumura, 1918 ((*Siphocoryne*)
*centauriella* Theobald, 1921 (*Anuraphis*) types BM
*chrysanthemi* Walker, 1849 (*Aphis*) type BM
*cinerariae* Theobald, 1923 (*Anuraphis*) types BM
*consumpta* Walker, 1849 (*Aphis*)
*convecta* Walker, 1849 (*Aphis*)
*conviva* Walker, 1849 (*Aphis*)
*cyani* Theobald, 1923 (*Anuraphis*) types BM
*detracta* Walker, 1849 (*Aphis*) type BM
? *diminuta* Walker, 1850 (*Aphis*)
*familiaris* Walker, 1848 (*Aphis*) type BM
*fasciatus* del Guercio, 1920 (*Anuraphis*)
*filaginis* var. *anthemidis* del Guercio, 1930 (*Anuraphis*)

*flavescens* del Guercio, 1930 (*Anuraphis*)
*glaucifolia* Theobald, 1923 (*Anuraphis*) types BM
*helichrysi* subsp. *asiaticus* var. *nigriventris* Nevsky, 1929 (*Anuraphis*)
*incumbens* Walker, 1849 (*Aphis*) type BM
*insessa* Walker, 1849 (*Aphis*) type BM
*insititiae* Koch, 1854 (*Aphis*)
*insititiella* del Guercio, 1930 (*Anuraphis*)
*leontopodii* Schouteden, 1903 (*Aphis*)
*marutae* Oestlund, 1886 (*Aphis*)
? *menthaecola* Nevsky, 1929 (*Anuraphis*)
*mumecola* Shinji, 1930 (*Anuraphis*)
*myosotidis* Koch, 1854 (*Aphis*)
*nociva* Walker, 1849 (*Aphis*) type BM
*padi* auctt. nec Linnaeus, as *Anuraphis*
*persorbens* Walker, 1849 (*Aphis*) type BM
*petasitidis* Buckton, 1879 (*Aphis*) lectotype BM
*poae* del Guercio, 1916 (*Anuraphis*)
*pruni* del Guercio, 1930 (*Anuraphis*)
*pruniavium* Nevsky, 1929 (*Anuraphis*)
*prunina* Walker, 1848 (*Aphis*) types BM
*prunus* Shinji, 1922 (*Aphis*)
*senecio* Swain, 1918 (*Aphis*)
*sherardiae* Theobald, 1926 (*Anuraphis*) types BM
? *similis* Walker, 1848 (*Aphis*)
*socia* Walker, 1848 (*Aphis*) type BM
*tianshanicus* Nevsky, 1951

B. helichrysi var. warei (Theobald, 1923) (Anuraphis) types BM
helichrysi subsp. asiatica Nevsky, 1929 = Brachycaudus (Thuleaphis) amygdalinus (Schouteden, 1905)
iranicus Davatchi & Remaudière, 1953 to Brachycaudus (Acaudus)
jacobi Stroyan, 1957 to Brachycaudus (Acaudus)
lagarriguei Remaudière of Börner & Heinze, 1957 = Brachycaudus (Acaudus) mimeuri Remaudière, 1952
linariae Stroyan, 1950 to Brachycaudus (Acaudus)
lucifugus F.P. Müller, 1955 to Brachycaudus (Acaudus)
lychnicola Hille Ris Lambers, 1966 to Brachycaudus (Acaudus)
microsiphon Börner ex Bodenheimer, 1937 nomen nudum
mimeuri Remaudière, 1952 to Brachycaudus (Acaudus)
mordvilkoi Hille Ris Lambers, 1931 to Brachycaudus (Acaudus)
muchranica Abashidze, 1951 nomen nudum

B. mume (Hori, 1927) (Anuraphis)
nitidus Hille Ris Lambers, 1935 = Brachycaudus (Acaudus) persicae (Passerini, 1860)
persicae-niger (E.F. Smith, 1890) (Aphis) = Brachycaudus (Acaudus) persicae (Passerini, 1860)
plantaginis Holman & Szelegiewicz, 1975
prunarium (Nevsky) Börner ex Bodenheimer, 1930 errore pro Anuraphis pruniavium Nevsky
prunicola (Kaltenbach, 1843) (Aphis) to Brachycaudus (Appelia); name has also been applied to Brachycaudus (Acaudus) persicae (Passerini, 1860)
pruni-domesticae Nevsky, date?
salicinae Börner, 1939 BM HRL
semisubterraneus Börner, 1951 = Brachycaudus (Acaudus) persicae (Passerini, 1860)
shaposhnikovi Narzikulov, date? (in Shaposhnikov, 1956)
spiraeae Börner, 1932 BM HRL
subspinosus Nevsky, 1951 = Dysaphis microsiphon (Nevsky, 1929)
tianshanicus Nevsky, 1951 = helichrysi (Kaltenbach, 1843)
umbelliferarum Nevsky, 1951
virgatus Shaposhnikov, 1964 to Brachycaudus (Acaudus)
viridanus Nevsky, 1929 (Anuraphis)

Subgen. Acaudus van der Goot, 1913

B. (Acaudus) aconiti (Mordvilko, 1928) (Anuraphis) BM HRL
(Acaudus) almatinus (Nevsky, 1951) (Acaudus)
(Acaudus) ballotae (Passerini, 1860) (Aphis) BM HRL
(Acaudus) cardui (Linnaeus, 1758) (Aphis) BM HRL
*alemedensis* Clarke, 1903 (*Aphis*)
*asselbergsi* Hille Ris Lambers, 1931 (*Brachycaudus*)
*capsellae* Koch, 1854 (*Aphis*)
*cardui* Shinji, 1941 (*Anuraphis*)
*cardui* subsp. *turanica* Mordvilko ex Nevsky, 1929 (*Anuraphis*)
*cardui* subsp. *yosiii* Takahashi, 1966 BM HRL
*chamomillae* Koch, 1854 (*Aphis*)
*chrysanthemi* Koch, 1854 (*Aphis*)
*cnici* Schrank, 1801 (*Aphis*)
*flavicephala* del Guercio, 1930 (*Anuraphis* (*Macchiatiella*))
*insita* Walker, 1852 (*Aphis*) nec 1849 type BM
*instabilis* Buckton, 1879 (*Aphis*) lectotype BM
*lata* Walker, 1850 (*Aphis*) lectotype BM

*lateralis* Walker, 1848 (*Aphis*)
*leucanthemi* Scopoli, 1763 (*Aphis*)
*onopordi* Schrank, 1801 (*Aphis*)
*opima* Buckton, 1879 (*Aphis*) lectotype BM
? *pectinata* del Guercio, 1930 (*Anuraphis* (*Macchiatiella*))
*petherbridgei* Theobald, 1929 (*Anuraphis*)
? *phelipaeae* Passerini, 1879 (*Aphis*)
*projacobeae* del Guercio, 1930 (*Anuraphis* (*Macchiatiella*))
*pruni* Koch, 1854 (*Aphis*) nec Geoffroy, 1762; Scopoli, 1763
*pruniphila* del Guercio, 1930 (*Anuraphis* (*Macchiatiella*))
*pruniphila* var. *cefaliflava* del Guercio, 1930 (*Anuraphis*)
*senecii* del Guercio, 1930 (*Anuraphis* (*Macchiatiella*))

B. (Acaudus) cerasicola (Mordvilko ex Nevsky, 1929) (Anuraphis)
(Acaudus) divaricatae Shaposhnikov, 1956 (Brachycaudus) BM HRL
(Acaudus) iranicus Davatchi & Remaudiere, 1953 (Brachycaudus) paratypes BM HRL
(Acaudus) jacobi Stroyan, 1957 (Brachycaudus) types BM HRL
(Acaudus) klugkisti (Börner, 1942) (Acaudus) BM HRL
(Acaudus) lamii (Koch, 1854) (Aphis) BM HRL
(Acaudus) linariae Stroyan, 1950 (Brachycaudus) cotypes BM HRL
(Acaudus) lucifugus F.P. Müller, 1955 (Brachycaudus) BM HRL
(Acaudus) lychnicola Hille Ris Lambers, 1966 (Brachycaudus) BM HRL
(Acaudus) lychnidis (Linnaeus, 1758) (Aphis) BM HRL
*lychnidis dioicae* Cederhjelm, 1798 (*Aphis*)
*melampyri* del Guercio, 1911 (*Anuraphis*)
(Acaudus) malvae Shaposhnikov, 1964 (Brachycaudus (Prunaphis)) BM HRL
(Acaudus) mimeuri Remaudière, 1952 (Brachycaudus) paratypes BM HRL = ? persicae (Passerini, 1860)
*lagarriguei* Börner & Heinze, 1957 (*Brachycaudus*)
(Acaudus) mordvilkoi Hille Ris Lambers, 1931 (Brachycaudus) BM HRL
(Acaudus) napelli (Schrank, 1801) (Aphis) BM HRL
(Acaudus) persicae (Passerini, 1860) (Myzus) BM HRL
*depressa* del Guercio, 1930 (*Anuraphis* (*Macchiatiella*))
*masseei* Theobald, 1927 (*Anuraphis*)
? *mimeuri* Remaudière, 1952 (*Brachycaudus*) HRL
*nitidus* Hille Ris Lambers, 1935 (*Brachycaudus*)
*oblonga* del Guercio, 1930 (*Anuraphis* (*Macchiatiella*))
*persicaecola* Boisduval, 1867 (*Aphis*)
*persicae-niger* E.F. Smith, 1890 (*Aphis*)
*semisubterraneus* Börner, 1951 (*Brachycaudus*)

B. (Acaudus) populi (del Guercio, 1911) (Anuraphis) BM HRL
(Acaudus) rociadae (Cockerell, 1903) (Aphis) BM HRL
(Acaudus) virgatus Shaposhnikov, 1964 (Brachycaudus (Prunaphis)) HRL

Subgen. Appelia Börner, 1930
(Appelia) cerinthis Bozhko, (1957) 1961 (Brachycaudus)
(Appelia) prunicola (Kaltenbach, 1843) (Aphis) BM HRL (but the name has also been applied to Brachycaudus (Acaudus) persicae (Passerini, 1860))
*oxiana* Nevsky, 1929 (*Anuraphis*)
*prunifex* Theobald, 1926 (*Anuraphis*) types BM
(Appelia) schwartzi (Börner, 1931) (Appelia) BM HRL
*amygdali* Buckton, 1879 (*Aphis*) BM
*persicae* Boyer de Fonscolombe, 1841 (*Aphis*) nec Sulzer, 1776
*persicae* Kaltenbach, 1843 (*Aphis*) nec Sulzer, 1776
*persicae* Koch, 1854 (*Aphis*) nec Sulzer, 1776
(Appelia) tragopogonis (Kaltenbach, 1843) (Aphis) BM HRL
(Appelia) tragopogonis subsp. setosus (Hille Ris Lambers, 1948) (Appelia) BM HRL

Subgen. Mordvilkomemor Shaposhnikov, 1950

B. (Mordvilkomemor) pilosus (Mordvilko ex Nevsky, 1929) (Dentatus) BM HRL
*macrotuberculatum* Narzikulov, 1950 (*Mordvilkomemor*)

Subgen. Nevskyaphis Shaposhnikov, 1950

B. (Nevskyaphis) bicolor (Nevsky, 1929) (Dentatus) BM HRL

Subgen. Prunaphis Shaposhnikov, 1964

B. (Prunaphis) malvae Shaposhnikov, 1964 to Brachycaudus (Acaudus)
(Prunaphis) virgatus Shaposhnikov, 1964 to Brachycaudus (Acaudus)

Subgen. Thuleaphis Hille Ris Lambers, 1960

B. (Thuleaphis) acaudatus (Hille Ris Lambers, 1960) (Thuleaphis) BM HRL
(Thuleaphis) aegyptiacus (Hall, 1926) (Anuraphis) types BM HRL
(Thuleaphis) amygdalinus (Schouteden, 1905) (Aphis) BM HRL
*? convolvuli* Nevsky, 1951 (*Acaudus*)
*helichrysi* subsp. *asiatica* Nevsky, 1929 (*Anuraphis*)
(Thuleaphis) rumexicolens (Patch, 1917) (Aphis) BM HRL
(Thuleaphis) sedi (Jacob, 1964) (Thuleaphis) types BM HRL

BRACHYCOLUS Buckton, 1879
Monograph of the British Aphides, London, 2: 146
Type species Aphis stellariae Hardy, 1850

B. asparagi Mordvilko, 1929 type of Brachycoryne Aizenberg, 1935 nec Mabill, 1833; and of Brachycorynella Aizenberg, 1954 q.v.
ballii Gillette, 1908 type of Thripsaphis Gillette, 1917 q.v.
brachysiphon Richards, 1963
cerastii (Kaltenbach, 1846) (Aphis) BM HRL
cervariae Börner, 1932 to Semiaphis van der Goot, 1913
? cornicolum (Matsumura, 1918) (Siphocoryne)
cucubali (Passerini, 1863) (Aphis)* BM HRL
*melanocephalus* Buckton, 1879 (*Hyalopterus*) lectotype BM
*silenea* Ferrari, 1872 (*Aphis*)
dusmeti Gomez Menor, 1950 partim = Brevicoryne brassicae Linnaeus, 1758, the alatae are Aphis fabae subsp. solanella Theobald, 1914
gramini Takahashi, 1920 (type of Brachysiphoniella Takahashi, 1921) = Brachysiphoniella montana (van der Goot, 1917)
heraclei Takahashi, 1921 to Semiaphis van der Goot, 1913
korotnewi Mordvilko, 1901 = Diuraphis (Holcaphis) frequens (Walker, 1848)
lonicerae Shinji, 1939 = Semiaphis heraclei Takahashi, 1921
muehlei Börner, 1950 to Diuraphis Aizenberg, 1935
nodulus Richards, 1959 to Diuraphis Aizenberg, 1935
noxius Mordvilko ex Kurdjumov, 1913 type of Diuraphis Aizenberg, 1935 q.v.
singularis Börner, 1950 type of Uhlmannia Börner, 1952 q.v.
slavae Mordvilko, 1921 = ? Hyalopteroides humilis (Walker, 1852)
stellariae (Hardy, 1850) (Aphis) BM HRL
tritici Gillette, 1911 to Diuraphis (Holcaphis)
turritellus Wahlgren, 1938 to Lipaphis Mordvilko, 1928

BRACHYCORYNE Aizenberg, 1935 nec Mabill, 1833
Zap. biol. Sta Bolchevo 7-8: 157
Type species Brachycolus asparagi Mordvilko, 1929
= Brachycorynella Aizenberg, 1956

BRACHYCORYNELLA Aizenberg, 1954
Diss. biol. Sci. Lenin Univ., Moscow p. 9
Type species Brachycolus asparagi Mordvilko, 1929
*Brachycoryne* Aizenberg, 1935 nec Mabill, 1833

B. asparagi (Mordvilko, 1929) (Brachycolus) BM HRL

* Not preoccupied by Aphis cucubali Linnaeus, 1746, pre-binominal; or by Ph.F. Gmelin, 1758 nomen oblitum.

BRACHYMYZUS A.N. Basu, 1964
J. Linn. Soc. (Zool.) 45: 223
Type species Brachymyzus jasmini A.N. Basu, 1964

B. jasmini A.N. Basu, 1964 HRL

BRACHYSIPHONIELLA Takahashi, 1921
Spec. Rep. Formosa agric. Exp. stn 20: 61
Type species Brachycolus gramini Takahashi, 1920

B. montana (van der Goot, 1917) (Semiaphis) BM HRL
*graminis* Takahashi, 1920 rectification of *gramini* (*Brachycolus*)
*vandergooti* RayChaudhuri & C. Banerjee, 1974 (*Melanaphis*)

BRACHYSIPHUM van der Goot, 1913
Tijdschr. Ent. 56: 105
Type species Aphis thalictri Koch, 1854

B. funitectum Börner, 1950 to Aphis Linnaeus, 1758
hyperici (Monell, 1879) (Aphis) BM HRL
*hyperici* Thomas, 1879 (*Myzocallis*)
japonicum Takahashi, 1919 to Melanaphis van der Goot, 1917
kobachidsei Rusanova, 1943 nomen nudum
thalictri (Koch, 1854) (Aphis) BM HRL

BRACHYUNGUIS Das, 1918
Mem. Indian Mus. 6: 227
Type species Brachyunguis harmalae Das, 1918
*Parabrachyunguis* Remaudière & Davatchi, 1955

B. afghanica Narzikulov & Umarov (1971 nomen nudum) 1972 to Aphis (Protaphis)
? agrariae Bozhko, 1959 (Brachyunguis (Parabrachyunguis))
agriphylli Bozhko, reference not found, paratype HRL
anthemidis Börner, 1940 (type of Protaphis Börner, 1952) to Aphis (Protaphis)
astragali Narzikulov & Umarov, 1972 nomen nudum
bicolor Ivanovskaja, 1959
bimacula Narzikulov & Umarov (1971 nomen nudum) 1972 to Aphis (Protaphis)
bonnevillensis (Knowlton, 1928) (Aphis) BM HRL
calotropicus Menon & Pawar, 1958
camphorosmae Tashev, 1964 = Xerobion eriosomatinum Nevsky, 1929
carlinae Börner, 1940 to Aphis (Protaphis)
carthami Das, 1918 (type of Alhambra Gomez-Menor, 1958; and of Dasia Gomez-Menor, 1951 nec Gray, 1839, van der Goot, 1918) to Aphis (Protaphis)

B. cuscutae (Nevsky, 1928) (Xerophilaphis)
cynanchi (Nevsky, 1928) (Xerophilaphis)
cynanchi-acuta Bozhko, 1959
delottoi Eastop, 1958 to Aphis (Protaphis)
dudichi Börner, 1940 = Aphis (Protaphis) anthemidis (Börner, 1940)
echinopis Hille Ris Lambers, 1948 to Aphis (Protaphis)
evansi Eastop, 1956 to Aphis (Protaphis)
flavidus Ivanovskaja, 1959 (Aphis)
gomezmenori Nieto Nafria, 1974 = zygophylli Nevsky, 1929
harmalae Das, 1918 BM HRL
*pegani* Mimeur, 1935 (*Pergandeida*)
judenkoi Szelegiewicz, 1959 to Aphis (Absinthaphis)
kaussarii Remaudière & Davatchi, 1955 (Brachyunguis (Parabrachyunguis))
lambersi Tashev, 1964 = Aphis (Absinthaphis) tashevella nomen novum
letsoniae Das, 1918
lycii (Nevsky, 1928) (Xerophilaphis)
? peucedani (Nevsky, 1928) (Xerophilaphis)
plotnikovi (Nevsky, 1928) (Xerophilaphis)
shaposhnikovi Ivanovskaja, 1960
sonchi Narzikulov & Umarov, (1971 nomen nudum) 1972 to Aphis (Protaphis)
tamariciarum (Rusanova, 1943 nomen nudum) (Xerophilaphis) Ivanovskaja, 1956 = tamaricivorus (Narzikulov, 1954)
tamaricis (Lichtenstein, 1885) (Aphis) BM HRL
*tamaricifoliae* Hall, 1936 (*Pergandeida*) BM
*tamaricis* Theobald, 1918 (*Aphis*) types BM
tamaricivorus (Narzikulov, 1954) (Xerophilaphis)
*tamariciarum* Ivanovskaja, 1956
tamaricophilus (Nevsky, 1928) (Xerophilaphis)
tausaghys Nevsky, 1949 ex Narzikulov, Juchnevitch & Kan, 1971
zoijae Nevsky, 1938
zygophylli Gomez-Menor, 1951 = zygophylli Nevsky, 1929
zygophylli (Nevsky, 1929) (Xerophilaphis) HRL
*gomezmenori* Nieto Nafria, 1974
*zygophylli* Gomez-Menor, 1951

Subgen. Parabrachyunguis Remaudière & Davatchi, 1955

B. (Parabrachyunguis) agrari Bozhko, 1959 mis-spelling
(Parabrachyunguis) agrariae Bozhko, 1959 to Brachyunguis Das, 1928
(Parabrachyunguis) kaussarii Remaudière & Davatchi, 1955 (type of Parabrachyunguis Remaudière & Davatchi, 1955) to Brachyunguis Das, 1918
(Parabrachyunguis) lini Bozhko, 1959 to Aphis Linnaeus, 1750

BRADYAPHIS Mordvilko, 1894
Trav. Lab. Zool. Kab. Univ. Varsov. 1: 46
Type species Aphis antennata Kaltenbach, 1843
= Monaphis Walker, 1870

BRAGGIA Gillette & Palmer, 1929
Ann. ent. Soc. Am. 22: 28
Type species Braggia echinata Gillette & Palmer, 1929

B. agathona (Hottes, 1950) (Aphis) paratypes BM HRL
deserticola Hille Ris Lambers, 1966 HRL
deserticola subsp. thanatophila Hille Ris Lambers, 1966 HRL
echidna lapsus pro echinata
echinata Gillette & Palmer, 1929 BM HRL
eriogoni (Cowen, 1895) (Aphis) BM HRL
eriogoni subsp. atra Hille Ris Lambers, 1966 HRL
eriogoni subsp. californica Hille Ris Lambers, 1966 HRL
uncompahgrensis Hottes, 1950 paratypes BM HRL
urovaneta (Hottes, 1950) (Aphis) paratype BM HRL
urovaneta subsp. pachysiphon Hille Ris Lambers, 1966 HRL

BRASILAPHIS Mordvilko, 1930
Dokl. Akad. Nauk SSSR (A) 1930: 277
Type species Brasilaphis bondari Mordvilko, 1930

B. bondari Mordvilko, 1930 BM HRL

BREVICAUDUS Shaposhnikov, 1964
Ént. Obozr. 43: 151, described as subg. of Brachycaudus van der Goot, 1913
Type species Brevicaudus amygdalinus (Schouteden, 1905) = Aphis amygdalinus Schouteden, 1905
= Thuleaphis Hille Ris Lambers, 1960

BREVICORYNAPHIS Hille Ris Lambers, 1956
Mitt. schweiz. ent. Ges. 29: 381
Type species Brevicorynaphis schneideri Hille Ris Lambers, 1956
= Smiela Mordvilko, (1929) 1948

B. schneideri Hille Ris Lambers, 1956 to Smiela Mordvilko, 1948

BREVICORYNE van der Goot, 1915
  Beiträge zur Kenntnis der Holländischen Blattläuse, Haarlem - Berlin p. 245, attributed to Das
  Type species Aphis brassicae Linnaeus, 1758
    *Bozhkoja* Shaposhnikov, 1964

B. arctica Richards, 1963 paratype BM HRL
  barbareae Nevsky, 1929 HRL
  brassicae (Linnaeus, 1758) (Aphis) BM HRL
    *dusmeti* Gomez-Menor, 1950 partim
    *floris-rapae* Curtis, 1842 (*Aphis*)
    *isatidis* Boyer de Fonscolombe, 1841 (*Aphis*)
    *raphani* Schrank, 1801 (*Aphis*)
  buhri Börner, 1952 type of Pseudobrevicoryne Heinze, 1960 q.v.
  crambe Bozhko, 1950 (type of Bozhkoja Shaposhnikov, 1964) HRL
    *crambinistataricae* Bozhko, 1953 HRL
  crambinis-tataricae Bozhko, 1953 = crambe Bozhko, 1950
  coriandri Das, 1918 to Hyadaphis Kirkaldy, 1904
  fraterna (Strom, 1938) (Aphis) BM
    *salixutis* Smith & Knowlton, 1940 paratype BM
  salixutis Smith & Knowlton, 1940 = fraterna (Strom, 1938)
  shaposhnikovi Narzikulov, 1957

BREVICORYNELLA Nevsky, 1929
  Acta Univ. Asiae mediae 3: 21
  Type species Brevicorynella quadrimaculata Nevsky, 1929

B. quadrimaculata Nevsky, 1929

BREVITRICHOSIPHON RayChaudhuri, M.R. Ghosh, M. Banerjee & A.K. Ghosh, 1973
    Kontyû 41: 54
    Type species Brevitrichosiphon mukerjii RayChaudhuri, M.R. Ghosh, M. Banerjee & A.K. Ghosh, 1973

B. mukerjii RayChaudhuri, M.R. Ghosh, M. Banerjee & A.K. Ghosh, 1973

BROMAPHIS Amyot, 1847
  Annls Soc. ent. Fr. (2) 5: 479
  Invalid

BRYSOCRYPTA Westwood, lapsus pro Byrsocrypta Haliday, 1838

BUCHNERIA Börner, 1952
Mitt. thüring. bot. Ges. 4 (3): 41, 242
Type species Aphis pectinatae Nördlinger, 1880
= Cinara Curtis, 1835

BUCKTONIA Lichtenstein, 1896
Monographie des Pucerons du Peuplier, p. 16, described as subg. of Pemphigus Hartig, 1839
Type species Pemphigus affinis Kaltenbach, 1843
= Thecabius Koch, 1857

BURSAPHIS J.M. Baker, 1934
An. Inst. Biol. Univ. Mexico 5: 217
Type species Bursaphis solitaria J.M. Baker, 1934

B. solitaria J.M. Baker, 1934

BYRSOCRYPTA Haliday, 1838
Ann. nat. Hist. 2: 189
Type species Eriosoma ulmi-gallarum Haliday, 1838 = ? Aphis ulmi Linnaeus, 1758
Suppressed by International Commission
= Tetraneura Hartig, 1841

B. ? bordschomica Abashidze, 1951 to ? Tetraneura Hartig, 1841
hamamelidis Fitch, 1851 to Hormaphis Osten-Sacken, 1861
pallida Haliday, 1838 to Kaltenbachiella Schouteden, 1906
personata Börner, 1950 = Tetraneura ulmi (Linnaeus, 1758)
pseudobyrsa Walsh, 1863 to Pachypappa Koch, 1856
rhois Fitch, 1866 type of Melaphis Walsh, 1867 q.v.
ulmicola Fitch, 1859 type of Colopha Monell, 1887 q.v.
vagabunda Walsh, 1863 type of Mordwilkoja del Guercio, 1909 q.v.

BYRSOCRYPTOIDES Dzhibladze, 1960
Trudy Inst. Zool., Tbilisi 17: 234
Type species Byrsocryptoides zelkovae Dzhibladze, 1960
*Gharesia* Stroyan, 1963

polunini (Stroyan, 1963) (Gharesia) types BM HRL
zelkovae Dzhibladze, 1960 BM HRL
zelkovaecola Dzhibladze, 1965 BM

CACHRYPHORA Oestlund, 1922
Rep. Minn. St. Ent. 19: 132
Type species Rhopalosiphum serotinae Oestlund, 1887

C. canadensis Hille Ris Lambers, 1960 types BM HRL
imbricaria Richards, 1972 = serotinae (Oestlund, 1887)
serotinae (Oestlund, 1887) (Rhopalosiphum) BM HRL
*imbricaria* Richards, 1972

CALAPHIS Walsh, 1863 (1862)
Proc. ent. Soc. Philad. 1: 301
Type species Calaphis betulella Walsh, 1863
*Cepegillettea* Granovsky, 1928
*Kallistaphis* Kirkaldy, 1905
*Neocallipterus* van der Goot, 1915
*Siphonocallis* del Guercio, (1913) 1914

C. alni Baker, 1916 BM HRL
alnosa Pepper, 1950 paratype BM HRL
arctica Hille Ris Lambers, 1952 BM HRL
betulae Mordvilko, 1929 = flava Mordvilko, 1928
betulaecolens (Fitch, 1851) (Aphis) BM HRL
*betulaecolens* Monell, 1879 (*Callipterus*)
betulaefoliae (Granovsky, 1928) (Cepegillettea) BM HRL
betulella Walsh, 1863 BM HRL
betulicola (Kaltenbach, 1843) (Aphis) BM HRL
castaneae (Fitch, 1857) (Callipterus) BM HRL
*? castanea-vesca* Haldeman, 1844 (*Aphis*)
castaneoides Baker, 1916 HRL
coloradensis Granovsky, 1939 BM HRL
flava Mordvilko, 1928 BM HRL
*basalis* Stroyan, 1957 (*Kallistaphis*)
*betulae* Mordvilko, 1929
*granovskyi* Palmer, 1952 HRL
granovskyi Palmer, 1952 = flava Mordvilko, 1928
leonardi Quednau, 1971 paratype BM HRL
magnoliae Essig & Kuwana, 1918 type of Neocalaphis Shinji, 1927 q.v.
magnolicolens Takahashi, 1921 to Neocalaphis Shinji, 1927
minutissima Stroyan, 1953 to Callipterinella van der Goot, 1913
myricae Patch, 1923 BM HRL
nanae Tissot, 1932

C. neobetulella Quednau, 1971 BM HRL
viridipallida Palmer, 1952 BM HRL
viridis Richards, 1957

CALLAPHIS Mordvilko, (1909) 1914 nec Walker, 1870
Biol. Centralbl. 29: 102 (1909) without description
Faune Russie, Ins. Hém. 1 (1): 27 (1914), in key
Type species Callaphis caricicola Mordvilko, 1914
= Thripsaphis Gillette, 1917

C. caricicola Mordvilko, (1909) 1914 (type of Allaphis Mordvilko, 1921) to Thripsaphis Gillette, 1917
caricis Mordvilko, 1909 nomen nudum

CALLAPHIS Walker, 1870
Zoologist (2) 5: 2000
Type species Aphis juglandis Frisch, 1734 = Aphis juglandis Goeze, 1778
*Callipterinola* Strand, 1928
*Callipterus* Koch, 1855 nec Agassiz, 1846
*Panaphis* Kirkaldy, 1904
*Ptychodes* Buckton, 1881

C. juglandis (Goeze, 1778) (Aphis) BM HRL
*juglandis* Frisch, 1734 (*Aphis*) pre-binominal
nepalensis Quednau, 1973 type BM

CALLIPTERINELLA van der Goot, 1913
Tijdschr. Ent. 56: 118
Type species Aphis betularia Kaltenbach, 1843 = A. tuberculata von Heyden, 1837
*Procalaphis* Quednau, 1954

C. callipterus (Hartig, 1841) (Aphis) BM HRL
*annulatus* Koch, 1854 (*Chaitophorus*)
*betulae* Buckton, 1879 (*Chaitophorus*)
minutissima (Stroyan, 1953) (Calaphis) paratypes BM HRL
tuberculata (von Heyden, 1837) (Aphis) BM HRL
*betularia* Kaltenbach, 1843 (*Aphis*)
*tricolor* Koch, 1854 (*Chaitophorus*)

CALLIPTERINOLA Strand, 1928
Arch. Naturgesch. 92 (A 8): 47
Type species Aphis juglandis Frisch, 1734 = A. juglandis Goeze, 1778
= Callaphis Walker, 1870

CALLIPTEROIDES Mordvilko, 1909 (1908)
Ezheg. zool. Muz. 13: 377
Type species Aphis nigritarsis von Heyden, 1837 = Aphis punctipennis Zetterstedt, 1828
= Euceraphis Walker, 1870

CALLIPTERUS Koch, 1855 nec Agassiz, 1846
Die Pflanzenläuse Aphiden, Nürnberg, p. 208
Type species Aphis juglandis Frisch, 1734, selected by Passerini, 1860: 29
= Aphis juglandis Goeze, 1778
= Callaphis Walker, 1870

C. abiepinus Matsumura ex Shiraki, 1952 ? lapsus pro Nippolachnus abietinus
arundicolens Clarke, 1903 to Takecallis Matsumura, 1917
asclepiadis Monell, 1879 = Myzocallis punctata (Monell, 1879)
betulae Koch, 1855 = ? Euceraphis punctipennis (Zetterstedt, 1828)
betulaecolens Monell, 1879 (type of Siphonocallis del Guercio, (1913) 1914) = Calaphis betulaecolens (Fitch, 1851)
bicolor Koch, 1855 = Euceraphis punctipennis (Zetterstedt, 1828)
carpini Koch, 1855 to Myzocallis Passerini, 1860
caryae Monell, 1879 to Monellia Oestlund, 1887
caryaefoliae Davis, 1910 (type of Melanocallis Oestlund, 1922) = Melanocallis fumipennella (Fitch, 1855)
castaneae Buckton, 1881 = Myzocallis castanicola Baker, 1917
castaneae Fitch, 1857 to Calaphis Walsh, 1863
coryli Koch, 1855 = Myzocallis coryli (Goeze, 1778)
discolor Monell, 1879 to Myzocallis Passerini, 1860
elegans Koch, 1855 = Tinocallis platani (Kaltenbach, 1843)
flabella Gillette (errore pro Chaitophorus flabellus Sanborn, 1904) type of Caricaphis Börner, 1930
genevii Sanborn, 1904 = Therioaphis trifolii (Monell, 1879)
gigantea Cholodkovsky, 1899 type of Clethrobius Mordvilko, 1928 q.v.
hyalinus Monell, 1879 = Myzocallis punctata (Monell, 1879)
juglandicola Koch, 1855 = Chromaphis juglandicola (Kaltenbach, 1843)
mucidus Fitch, 1856 to Euceraphis Walker, 1870

C. punctatus Monell, 1879 (type of Neomyzocallis Richards, 1965) to Myzocallis Passerini, 1860
quercicola Monell, 1879 to Stegophylla Oestlund, 1922
quercifolii Thomas, 1879 = Hoplochaitophorus quercicola (Monell, 1879)
robiniae Gillette, 1907 type of Appendiseta Richards, 1965 q.v.
trifolii Monell, 1882 to Therioaphis Walker, 1870
ulmifolii Monell, 1879 to Tinocallis Matsumura, 1919
walshii Monell, 1879 to Myzocallis (Lineomyzocallis)

CAMELAPHIS Hille Ris Lambers, 1974
Boll. Zool. Agr. Bachi. (II) 11 (1972) 1974: 24, described as subg. of Tuberculatus Mordvilko, 1894
Type species Tuberculatus cornutus Richards, 1969
Here used as subg. of Tuberculatus Mordvilko, 1894

CANADAPHIS Essig, 1938
Univ. Toronto Stud. Geol. Ser. 40: 19
Type species Canadaphis carpenteri Essig, 1938
Fossil. Cretaceous

CAPITOPHORAPHIS E.E. Blanchard, 1944
Acta zool. lilloana 2: 34
Type species Capitophoraphis williamsi E.E. Blanchard, 1944 = Aphis rufomaculata Wilson, 1908
= Coloradoa Wilson, 1910

C. williamsi E.E. Blanchard, 1944 = Coloradoa rufomaculata (Wilson, 1908)

CAPITOPHORINUS Börner, 1931
Anz. Schädlingsk. 7: 129, described as subg. of Capitophorus van der Goot, 1913
Type species Capitophorus similis van der Goot, 1915
= Capitophorus van der Goot, 1913

CAPITOPHORUS van der Goot, 1913
Tijdschr. Ent. 56: 84
Type species Aphis carduina Walker, 1850
*Capitophorinus* Börner, 1931

C. acanthovillus Knowlton & Smith, 1936 to Pleotrichophorus Börner, 1930
archangelskii Nevsky, 1928 BM HRL
archangelskii subsp. afghani Narzikulov & Umarov, 1972
arctifoliae Shinji, 1924 = ? elaeagni (del Guercio, 1894)
bitrichus Knowlton & Smith, 1936 = Pleotrichophorus heterohirsutus (Gillette & Palmer, 1933)
brevinectarius Gillette & Palmer, 1933 to Pleotrichophorus Börner, 1930
bulgaricus Tashev, 1964 HRL
cannabifoliae Shinji, 1924 = Phorodon (Paraphorodon) cannabis Passerini, 1860
carduinus (Walker, 1850) (Aphis) lectotype BM HRL
*flaveolus* Walker auctt. nec Walker, 1849 (*Aphis*)
cefsmithi Knowlton, 1940 to Kakimia Hottes & Frison, 1931
chaetosiphon Nevsky, 1928 type of Chaetosiphon Mordvilko, 1914 q.v.
chaetosiphon var. tripilosus Rusanova, 1943 nomen nudum
chlorophainus Knowlton & Smith, 1936 = Pleotrichophorus elongatus (Knowlton, 1929)
chrysanthemi Theobald, 1920 to Pleotrichophorus Börner, 1930
cirsii Nevsky, 1928 = elaeagni (del Guercio, 1894)
cirsiiphagus Takahashi, 1961 HRL
codonopsidis Narzikulov, 1963 nomen nudum
corambus Hottes & Frison, 1931
cynariella Theobald, 1923 = elaeagni (del Guercio, 1894)
decampus Knowlton & Smith, 1936 to Pleotrichophorus Börner, 1930
elaeagni (del Guercio, 1894) (Myzus) BM HRL
? *arctifoliae* Shinji, 1924
*braggii* Gillette, 1908 (*Myzus*)
*carthusianus* Haviland, 1918 (*Myzus*)
*cirsii* Nevsky, 1928
*cynariella* Theobald, 1923 type BM
*elaeagni* van der Goot, 1913
*viridis* Craveri, 1915 (*Aphis*)
elaeagni van der Goot, 1913 = elaeagni (del Guercio, 1894)
elongatus Knowlton, 1929 to Pleotrichophorus Börner, 1930
eniwanus Miyazaki, 1971 HRL
essigi Hille Ris Lambers, 1953 BM HRL
feragaeus Knowlton & Smith, 1936 = Pleotrichophorus elongatus (Knowlton, 1929)
filifoliae Palmer, 1938 to Pleotrichophorus Börner, 1930
flaveola (Walker, 1849) (Aphis) = ? Aulacorthum solani (Kaltenbach, 1843) but references in Capitophorus apply to C. carduinus (Walker, 1850)

C. formosanus Takahashi, 1929 = ? Pleotrichophorus chrysanthemi (Theobald, 1920)
formosartemisiae (Takahashi, 1921) (Myzus) BM HRL
fragariae Shinji, 1924 = Chaetosiphon (Pentatrichopus) minus (Forbes, 1884)
fragarifoliae Shinji, 1941 = Chaetosiphon (Pentatrichopus) minus (Forbes, 1884)
geranii Chowdhuri, R.C. Basu, Chakrabarti & RayChaudhuri, 1969 type of Indoidiopterus Chakrabarti, A.K. Ghosh & RayChaudhuri, 1972 q.v.
gillettei Theobald, 1926 = hippophaes Walker, 1852
gnaphalodes Palmer, 1938 to Pleotrichophorus Börner, 1930
gnathalifoliae Shinji, 1924
gregarius Knowlton, 1929 to Pleotrichophorus Börner, 1930
heterohirsutus Gillette & Palmer, 1933 to Pleotrichophorus Börner, 1930
himalayensis A.K. Ghosh, M.R. Ghosh & RayChaudhuri, 1971
hippophaes (Walker, 1852) (Aphis) lectotype BM HRL
*gillettei* Theobald, 1926 probable type BM
*hippophaes* Koch, 1854 (*Rhopalosiphum*)
*minima* Mason, 1925 (*Amphorophora*)
hippophaes subsp. *indicus* A.K. Ghosh & RayChaudhuri, 1968 = indicus A.K. Ghosh & RayChaudhuri, 1968
hippophaes subsp. javanicus Hille Ris Lambers, 1953 BM HRL
hippophaes subsp. mitegoni Eastop, 1956 type BM HRL
horni Börner, 1931 BM HRL
horni subsp. gynoxantha Hille Ris Lambers, 1953 HRL
indicus A.K. Ghosh & RayChaudhuri, 1968 HRL
*hippophaes* subsp. *indicus* A.K. Ghosh & RayChaudhuri, 1968 HRL
infrequenus Knowlton & Smith, 1936 to Pleotrichophorus Börner, 1930
inulae (Passerini, 1860) (Phorodon) BM HRL
jopepperi Corpuz-Raros & Cook, 1974
longinectarius Gillette & Palmer, 1933 to Pleotrichophorus Börner, 1930
magnautensus Knowlton & Smith, 1936 to Pleotrichophorus Börner 1930
montanus Takahashi, 1931 BM HRL
oestlundi Knowlton, 1927 to Pleotrichophorus Börner, 1930
ohioensis Smith, 1940 to Pleotrichophorus Börner, 1930
pakansus Hottes & Frison, 1931 paratypes BM HRL
*vandergooti* Hille Ris Lambers, 1947 BM HRL
palmerae Knowlton, 1935 to Pleotrichophorus Börner, 1930
patonkus Hottes & Frison, 1931 to Pleotrichophorus Börner, 1930
polygoni A.K. Ghosh, M.R. Ghosh & RayChaudhuri, 1971
prunifoliae Shinji, 1924
pseudoglandulosus Palmer, 1952 to Pleotrichophorus Börner, 1930

C. pullus Gillette & Palmer, 1933 to Pleotrichophorus Börner, 1930
pycnorhysus Knowlton & Smith, 1936 to Pleotrichophorus Börner, 1930
quadritrichus Knowlton & Smith, 1936 to Pleotrichophorus Börner, 1930
rusticatus Knowlton & Smith, 1937 = Pleotrichophorus Börner, 1930
shepherdiae Gillette & Bragg, 1916 BM HRL
similis van der Goot, 1915 type of Capitophorinus Börner, 1931 BM HRL
spatulavillus Knowlton & Smith, 1936 to Pleotrichophorus Börner, 1930
stroudi Knowlton, 1948 to Pleotrichophorus Börner, 1930
takahashii Strand, 1929
tricholepidis Chakrabarti, 1976 HRL
utensis Pack & Knowlton, 1929 to Pleotrichophorus Börner, 1930
vandergooti Hille Ris Lambers, 1947 = pakansus Hottes & Frison, 1931
vernoniae A.K. Ghosh & RayChaudhuri, 1968 = Subovatomyzus leucosceptri A.N. Basu, 1964
wasatchii Knowlton, 1927 to Pleotrichophorus Börner, 1930
xanthii (Oestlund, 1886) (Siphocoryne) BM HRL
xerozoous Knowlton & Smith, 1936 to Pleotrichophorus Börner, 1930
zoomontanus Knowlton & Smith, 1936 to Pleotrichophorus Börner, 1930

CARAZZIA del Guercio, 1930
Redia 19: 500, as in litteris synonym of Clavisiphon del Guercio, 1930 by implication.
Type species Anuraphis (Clavisiphon) elegans (Ferrari, 1872) = Rhopalosiphum elegans Ferrari, 1872
= Eucarazzia del Guercio, 1921

CARICAPHIS Börner, 1930
Arch. klassif. phylogen. Ent. 1: 128
Type species "Callipterus flabella Gill" = Chaitophorus flabellus Sanborn, 1904
= Iziphya Nevsky, 1929

C. leegei (Börner, 1940) (Iziphya) (type of Juncobia Quednau, 1954) to Iziphya Nevsky, 1929

CARICOBIUM Aizenberg, 1954
Diss. biol. Sci. Lenin Univ., Moscow, p. 8
Without validly described species

C. palustre Aizenberg, 1954 nomen nudum
pilosae Aizenberg, 1954 nomen nudum
pokrowskii Aizenberg, 1954 nomen nudum

CARICOSIPHA Börner, 1939
Arb. physiol. angew. Ent. Berl. 6: 77
Type species Caricosipha paniculatae Börner, 1939

C. paniculatae Börner, 1939 BM HRL

CAROLINAIA Wilson, 1911
Can. Ent. 43: 61
Type species Carolinaia caricis Wilson, 1911

C. caricis Wilson, 1911 BM HRL
cyperi Ainslie, 1915 BM HRL
justiciae Shinji, 1924 nomen dubium
modesta Hottes, 1926 to Myzodium Börner, 1949
rhois Tissot, 1928 BM HRL
tade Shinji, 1927 to Trichosiphonaphis (Xenomyzus)

CASIMIRA Eastop, 1966
Austr. J. Zool. 14: 485
Type species Aphis canberrae Eastop, 1961

C. bhutanensis A.K. Ghosh, R.C. Basu & RayChaudhuri, 1971
canberrae (Eastop, 1961) (Aphis) types BM HRL

CATAMERGUS Oestlund, 1922
Rep. Minn. St. Ent. 19: 127
Type species Nectarophora fulvae Oestlund, 1887

C. fulvae (Oestlund, 1887) (Nectarophora) BM HRL
kickapoo (Hottes & Frison, 1931) (Macrosiphum) paratypes BM HRL

CATANEURA Scudder, 1890
U.S. Geol. Survey of the Territories 13: 245
Type species Cataneura absens Scudder, 1890
= Siphonophoroides Buckton, 1883
Fossil

CAVAHYALOPTERUS Mimeur, 1942
Bull. Soc. Sci. nat. Maroc 21: 67
Type species Cavahyalopterus graminearum Mimeur, 1942 = Brachycolus noxius Mordvilko ex Kurdjumov, 1913
= Diuraphis Aizenberg, 1935

C. graminearum Mimeur, 1942 = Diuraphis noxius (Mordvilko ex Kurdjumov, 1913)

CAVARAIELLIA Heinze, 1960
Beitr. Ent. 10: 810, described as subg. of Cavariella del Guercio, 1911
Type species Cavariella hillerislambersi Ossiannilsson, 1959 = Siphocoryne aquatica Gillette & Bragg, 1916
Here used as subg. of Cavariella del Guercio, 1911

CAVARAIELLOPSIS Heinze, 1960
Beitr. Ent. 10: 808
Type species Cavariella saxifragae Remaudière, 1959
= Cavariella del Guercio, 1911

CAVARIELLA del Guercio, 1911
Redia 7 : 323
Type species Aphis pastinacae Linnaeus, 1758
*Cavaraiellopsis* Heinze, 1960
*Corynosiphon* Mordvilko, 1914
*Metaphis* Matsumura, 1918
*Neocavariella* Shinji, 1932
*Nipposiphum* Matsumura, 1917
*Tuberculaphis* Theobald ex Das, 1918

C. aegopodii (Scopoli, 1763) (Aphis) BM HRL
*capreae* auctt. nec Fabricius, 1775 (*Aphis*)
*umbellatarum* (Koch, 1854) (*Aphis*) description, not figures
angelicae (Matsumura, 1918) (Metaphis) BM HRL
araliae Takahashi, 1921 type of Neocavariella Shinji, 1932 BM HRL
*neocapreae* Takahashi, 1921

archangelicae (Scopoli, 1763) (Aphis) BM HRL
*fusca* Macchiati, 1881 (*Toxoptera*)
*gigliolii* del Guercio, 1911
artemisiae Paik, 1965 = Coloradoa artemisicola Takahashi, 1965
aspidaphoides Hille Ris Lambers, 1969 HRL
azamii Shinji, 1930 = salicicola Matsumura, 1917
biswasi A.K. Ghosh, Basu & RayChaudhuri, 1969
borealis Hille Ris Lambers, 1952 HRL
bozhkoi Mamontova-Solucha, 1967
capreae (Fabricius, 1775) (Aphis) = pastinacae Linnaeus but the binomen Cavariella capreae usually applies to aegopodii Scopoli, 1763
cicutae (Koch, 1854) (Rhopalosiphum) BM HRL
*rutila* Mamontova, 1961 type HRL

C. digitata Hille Ris Lambers, 1969 BM HRL
enkianthae Shinji, 1924 = Akkaia polygoni Takahashi, 1919
gigliolii del Guercio, 1911 = archangelicae Scopoli, 1763
gilibertiae Takahashi, 1961 BM HRL
glauciiphaga Theobald, 1923 = theobaldi Gillette & Bragg, 1918
hendersoni Knowlton & Smith, 1936 BM HRL
heraclei Takahashi, 1961 BM HRL
hidaensis Takahashi, 1961 HRL
hillerislambersi Ossiannilsson, 1959 = Cavariella (Cavaraiellia) aquatica (Gillette & Bragg, 1916)
intermedia Hille Ris Lambers, 1969 BM HRL
japonica (Essig & Kuwana, 1918) (Siphocoryne) BM HRL
konoi Takahashi, 1939 BM HRL
*archangelicae* Oestlund, 1886 (*Siphocoryne*) nec Aphis archangelicae Scopoli, 1763
mitsubae Shinji, 1924 = salicicola Matsumura, 1917
neocapreae Takahashi, 1921 = araliae Takahashi, 1921
nigra A.N. Basu, 1964 BM HRL
nigrocaudata Takahashi, 1965
nipponica Takahashi, 1961 BM HRL
oenanthi (Shinji, 1922) (Hydronaphis) BM HRL
pastinacae (Linnaeus, 1758) (Aphis) BM HRL
*capreae* Fabricius, 1775 (*Aphis*)
*essigi* Gillette & Bragg, 1918 (*Siphocoryne*)
*rumicis* Theobald, 1925 (*Rhopalosiphum*) BM
pseudopustula Hille Ris Lambers, 1969 BM HRL
pustula Essig, 1937 HRL
rutila Mamontova, 1961 = cicutae Koch, 1854
salicicola (Matsumura, 1917) (Nipposiphum) BM HRL
*azamii* Shinji, 1930
*bicaudata* Essig & Kuwana, 1918 (*Siphocoryne*)
*mitsubae* Shinji, 1924
salicis (Monell, 1879) (Siphocoryne) BM HRL
sapporoensis Takahashi, 1961 BM HRL
saxifragae Remaudière, 1959 type of Cavaraiellopsis Heinze, 1960 paratypes BM HRL
simlaensis Chowdhuri, R.C. Basu & RayChaudhuri, 1969
takahashii Hille Ris Lambers, 1965 HRL
theobaldi (Gillette & Bragg, 1918) (Aphis) BM HRL
*glauciiphaga* Theobald, 1923

C. umbellatarum (Koch, 1854) (Aphis) = aegopodii (Scopoli, 1763) but references sometimes apply to theobaldi (Gillette & Bragg, 1918)

Subgen. Cavaraiellia Heinze, 1960

C. (Cavaraiellia) aquatica (Gillette & Bragg, 1916) (Siphocoryne) BM HRL
*hillerislambersi* Ossiannilsson, 1959 (*Cavariella*)

CEDOAPHIS Oestlund, 1922
Rep. Minn. St. Ent. 19: 127
Type species Aphis symphoricarpi Thomas of Oestlund, 1887 nec Thomas, 1879
= Cedoaphis incognita Hottes & Frison, 1931

C. incognita Hottes & Frison, 1931 BM HRL
*chipetae* Hottes, 1933 (*Aphis*) paratypes BM HRL
*symphoricarpi* Thomas of Oestlund 1887 nec Thomas, 1879
maxsoni (Palmer, 1936) (Amphicercidus) BM

CEDROBIUM Remaudière, 1954
Revue Path. vég. Ent. agric. Fr. 33: 116
Type species Cedrobium laportei Remaudière, 1954

C. laportei Remaudière, 1954 paratypes BM HRL

CEPEGILLETTEA Granovsky, 1928
Proc. ent. Soc. Wash. 30: 114
Type species Cepegillettea betulaefoliae Granovsky, 1928
= Calaphis Walsh, 1863

C. betulaefoliae Granovsky, 1928 to Calaphis Walsh, 1863

CERASOMYZUS Narzikulov, 1958
Trudy Acad. Nauk. tadzhik SSR 89: 21, described as subg. of Myzus Passerini, 1860
Type species Myzus (Cerasomyzus) bozhkoae Narzikulov, 1958
= Aphidura Hille Ris Lambers, 1956

CERATAPHIS Lichtenstein, 1882
Annls. Soc. ent. Fr. (6) 2: lxxv
Type species Coccus lataniae Boisduval, 1867
*Asterolecanium* Westwood, 1879 nec Targioni-Tozzetti, 1868
*Boisduvalia* Signoret, 1868 nec Robineau-Desvoisy, 1830

C. bambucifoliae rectified to bambusifoliae by Takahashi, 1930: 11
bambusifoliae Takahashi, 1925 BM HRL

C. betulae Mordvilko, 1901 to Hormaphis Osten-Sacken, 1861
formosana Takahashi, 1924
fransseni (Hille Ris Lambers, 1933) (Astegopteryx) BM HRL
*fransseni* subsp. *kepongensis* Takahashi, 1950 (*Astegopteryx*) paratypes BM
freycinetiae van der Goot, 1917 BM HRL
insularis van der Goot, 1912 to Astegopteryx Karsch, 1890
lataniae (Boisduval, 1867) (Coccus) BM HRL
*palmae* Baehr, 1908 (*Aphis*)
orchidearum (Westwood, 1879) (Asterolecanium) BM HRL
*brasiliensis* Hempel, 1901 (*Ceratovacuna*)
palmae (Ghesquière, 1934) (Aleurocanthus) BM HRL
*variabilis* Hille Ris Lambers, 1953 HRL
pirifoliae Shinji, 1924 = Doraphis populi (Maskell, 1898)
saccharivora Matsumura, 1917 = Ceratovacuna lanigera (Zehntner, 1897)
variabilis Hille Ris Lambers, 1953 = palmae (Ghesquière, 1934)

CERATOGLYPHINA van der Goot, 1917
Contrib. Faune Indes Néerl. 3: 237
Type species Ceratoglyphina bambusae van der Goot, 1917

C. bambusae van der Goot, 1917 BM HRL
bambusae subsp. bengalensis L.K. Ghosh, 1972 HRL

CERATOPEMPHIGIELLA Menon & Pawar, 1958
Proc. Indian Sci. Congr. 53 (3): 348
Type species Ceratopemphigiella delhieinsis Menon & Pawar, 1958

C. delhieinsis Menon & Pawar, 1958

CERATOPEMPHIGUS Schouteden, 1905
Spolia zeylan. 2: 187
Type species Ceratopemphigus zehntneri Schouteden, 1905

C. zehntneri Schouteden, 1905 syntypes BM HRL

CERATOVACUNA Zehntner, 1897
Archf. Java Suikerindust. 5: 553
Meded. Proefst. Oost-Java (N.S.) 37: 553
Type species Ceratovacuna lanigera Zehntner, 1897

C. arundinariae Takahashi, 1932 = silvestrii (Takahashi, 1927)
brasiliensis Hempel, 1901 = Cerataphis orchidearum (Westwood, 1897)

C. brevicornis Takahashi, 1940 = japonica (Takahashi, 1924)
hoffmanni Takahashi, 1936
japonica (Takahashi, 1924) (Oregma) BM HRL
*brevicornis* Takahashi, 1940
*qanici* Uye, 1924 (*Oregma*)
*sasae* Monzen, 1927 (*Oregma*)
lanigera Zehntner, 1897 BM HRL
*saccharivora* Matsumura, 1917 (*Cerataphis*)
longifera Takahashi, 1929 lapsus pro longifila
longifila (Takahashi, 1929) (Oregma) BM
nekoashi (Sasaki, 1910) (Astegopteryx) BM
? *oplismeni* Takahashi, 1924 (*Oregma*)
*oplismeni* Shinji, 1922 (*Pemphigus*)
? *orientalis* Takahashi, 1923 (*Oregma*) BM
panici (van der Goot, 1917) (Oregma) BM
pseudostyracophila (Shinji, 1936) (Astegopteryx)
silvestrii (Takahashi, 1927) (Oregma) HRL
*arundinariae* Takahashi, 1932
*subglandulosa* Hille Ris Lambers & A.N. Basu, 1966 (*Oregma*) HRL
spinulosa A.K. Ghosh & RayChaudhuri, 1972
styraci (Matsumura, 1917) (Astegopteryx)
uscare (Tao, 1964) (Oregma)

CERCIAPHIS Theobald, 1920
Bull. ent. Res. 11: 70
Type species Cerciaphis bougainvilleae Theobald, 1920 = Setaphis lutea van der Goot, 1917
= Schoutedenia Rübsaamen, 1905

C. bougainvilleae Theobald, 1920 = Schoutedenia lutea (van der Goot, 1917)
emblica Patel & Kulkarni, 1953 to Schoutedenia Rübsaamen, 1905

CERIFERELLA Carver & Martyn, 1965
Proc. R. ent. Soc. Lond. (B) 34: 38
Type species Ceriferella leucopogonis Carver & Martyn, 1965

C. dossuaria Carver & Martyn, 1965 paratypes BM HRL
leucopogonis Carver & Martyn, 1965 paratypes BM HRL

CEROSIPHA del Guercio, 1900
Nuove Relaz. R. Staz. Ent. agr. 1: 116
Type species Cerosipha passeriniana del Guercio, 1900
= Aphis Linnaeus, 1758

C. acanthoidis Börner, 1940 to Aphis Linnaeus, 1758
araliae-radicis Strom, 1938 to Aphis Linnaeus, 1758
bodenheimeri Börner MS ex Bodenheimer, 1930 & 1957 nomen nudum
californica Essig. 1944 = Rhopalosiphum rufiabdominalis (Sasaki, 1899)
caroliboerneri Remaudière, 1952 to Aphis Linnaeus, 1758
cirsii-oleracei Börner, 1932 to Aphis Linnaeus, 1758
commelinae Shinji, 1924 = Aphis gossypii Glover, 1877
crepidis Börner, 1940 to Aphis Linnaeus, 1758
cupressi Swain, 1918 type of Siphonatrophia Swain, 1918 q.v.
hieracii Börner, 1940 = Aphis pilosellae Börner, 1952
humuli Tseng & Tao, 1938 to Aphis Linnaeus, 1758
hypochoeridis Börner, 1940 to Aphis Linnaeus, 1758
longirostris Börner, 1950 = Aphis longirostrata Hille Ris Lambers, 1966
lycopicola Shinji, 1941 to Aphis Linnaeus, 1758
passeriniana del Guercio, 1900 to Aphis Linnaeus, 1758
pilosellae Börner, 1952 to Aphis Linnaeus, 1758
poterii Börner, 1940 = Aphis sanguisorbae Schrank, 1801
roepkei Hille Ris Lambers, 1931 to Aphis Linnaeus, 1758
roripae Palmer, 1938 to Aphis Linnaeus, 1758
rubicolens Hori, 1929 to Aphis Linnaeus, 1758
serissae Shinji, 1941 = Aphis serissae Shinji, 1922
subviridis Börner, 1940 to Aphis Linnaeus, 1758
taraxaci Nevsky, errore pro taraxacicola Nevsky, 1951
taraxacicola Nevsky, 1951 to Aphis Linnaeus, 1758
wartenbergi Börner, 1952 to Aphis Linnaeus, 1758

Subgen. Uraphis del Guercio, 1907

C. (Uraphis) pollinaria Börner, 1952 = Aphis pollinosa Walker, 1849

CERURAPHIS Börner, 1926
In Abderhalden, E., Handbuch der biologischen Arbeitsmethoden, Abt. IX
Teil 1, 2 Hälfte, Heft 2: 226
Type species Aphis viburnicola Börner, 1916 = Aphis eriophori Walker, 1848
*Neoceruraphis* Shaposhnikov, 1956

C. eastopi Hille Ris Lambers, 1966 types BM HRL
eriophori (Walker, 1848) (Aphis) lectotype BM HRL
*eriophori* Haliday ex Buckton, 1877 (*Hyalopterus*)
*junci* Kaltenbach, 1875 (*Aphis*)
*lantanaella* Theobald, 1925 (*Aphis*) types BM
*luzulae* Kaltenbach, 1862 (*Aphis*)
*viburni* Schrank, 1801 (*Aphis*) nec Scopoli, 1763
*viburniana* Franssen, 1928 (*Aphis*)
*viburnicola* Börner, 1916 (*Aphis*) nec Gillette, 1909
*vorabonnevilla* Knowlton & Smith, 1936 (*Aphis*)
viburnicola (Gillette, 1909) (Aphis) type of Neoceruraphis Shaposhnikov, 1956 BM HRL

CERVAPHIS van der Goot, 1917
Contrib. Faune Indes Néerl. 3: 148
Type species Cervaphis schouteniae van der Goot, 1917
*Diverosiphum* Shinji, 1922

C. cambodiensis Takahashi, 1941 = schouteniae van der Goot, 1917
echinata Hille Ris Lambers, 1956 types BM HRL
quercus Takahashi, 1918 BM HRL
*kunugii* Shinji, 1922 (*Diverosiphum*)
rappardi Hille Ris Lambers, 1956 BM HRL
schouteniae van der Goot, 1917 BM HRL
*cambodiensis* Takahashi, 1941 BM HRL

CEYLONIA Buckton, 1891
Indian Mus. Notes 2: 35
Type species Ceylonia theaecola Buckton, 1891 = Aphis aurantii Boyer de Fonscolombe, 1841
= Toxoptera Koch, 1856

C. theaecola Buckton, 1891 = Toxoptera aurantii (Boyer de Fonscolombe, 1841)

CHAETOCALLIS Börner, 1952
Mitt. thüring. bot. Ges. 4 (3): 62, as in litteris synonym of Hoplochaitophorus Granovsky, 1933
Type species Chaetocallis ruperti (Pintera, 1952) = Hoplocallis ruperti Pintera, 1952
= Hoplocallis Pintera, 1952

CHAETOGEOICA Remaudière & Tao, 1957
Revue Path. vég. Ent. agric. Fr. 36: 226
Type species Pemphigella foliodentata Tao, 1947

C. foliodentata (Tao, 1947) (Pemphigella) paratypes BM HRL

CHAETOMYZUS A.K. Ghosh & RayChaudhuri, 1962
J. Asiat. Soc. Beng. 4: 104
Type species Chaetomyzus rhododendri A.K. Ghosh & RayChaudhuri, 1962

C. rhododendri A.K. Ghosh & RayChaudhuri, 1962 HRL

CHAETOPHORELLA Börner, 1940
Neue Blattläuse aus Mitteleuropa, privately published, p. 2
Type species Ch. aceris Linnaeus nec Koch of Borner = Chaitophorus lyropictus Kessler, 1886
= Periphyllus van der Hoeven, 1863

C. fusca Börner, 1940 = Periphyllus lyropictus (Kessler, 1886)
helferi Quednau, 1954 = Periphyllus singeri (Börner, 1952)

CHAETOPHORIA Börner, 1940
Neue Blattläuse aus Mitteleuropa, privately published, p. 2
Type species Chaitophorus xanthomelas Koch, 1854 = Aphis aceris Linnaeus, 1758
= Periphyllus van der Hoeven, 1863

C. rhenana Börner, 1940 to Periphyllus van der Hoeven, 1863
singeri Börner, 1940 to Periphyllus van der Hoeven, 1863
uhlmanni Börner, 1950 = Periphyllus aceris (Linnaeus, 1758)

CHAETOPHORUS (Koch) Börner, 1931, emendation of Chaitophorus Koch, 1854

CHAETOSIPHELLA Hille Ris Lambers, 1939
Zool. Meded., Leiden, 22: 84
Type species Sipha berlesei del Guercio, 1905

C. berlesei (del Guercio, 1905) (Sipha) BM HRL
stipae Hille Ris Lambers, 1947 cotypes BM HRL
*stipae* Börner, 1950 (*Atheroides*)
*stipifolii* Bozhko, 1959
*tshernavini* subsp. *stipae* Hille Ris Lambers, 1947 HRL

C. stipifolii Bozhko, 1959 = stipae Hille Ris Lambers, 1947
tshernavini (Mordvilko, 1921) (Sipha) HRL
tshernavini subsp. stipae Hille Ris Lambers, 1947 = stipae Hille Ris Lambers, 1947

CHAETOSIPHON Mordvilko, 1914
Faune Russie, Ins. Hém. 1 (1): 71 (in key)
Type species Capitophorus chaetosiphon Nevsky, 1928, cited by Nevsky, 1929: 139

C. alpestre Hille Ris Lambers, 1953 HRL
alpestre subsp. airolensis Hille Ris Lambers, 1953 cotypes BM HRL
alpestre subsp. orientalis Shaposhnikov, 1964
chaetosiphon (Nevsky, 1928) (Capitophorus) HRL
chaetosiphon subsp. indicum Chakrabarti & A.K. Ghosh, 1970
chaetosiphon subsp. montanum Hille Ris Lambers, 1953 cotypes BM HRL
glaber David, Rajasing & Narayanan, 1971 to Chaetosiphon (Pentatrichopus)
gracilicorne David, Rajasing & Narayanan, 1971 BM HRL
janetscheki (Börner, 1949) (Pentatrichopus)
scalaris Richards, 1963 paratype BM HRL

Subgen. Pentatrichopus Börner, 1930

C. (Pentatrichopus) alpinum (Börner, 1950) (Pentatrichopus) HRL
(Pentatrichopus) coreanum (Paik, 1965) (Pentatrichopus) BM HRL
(Pentatrichopus) fragaefolii (Cockerell, 1901) (Myzus) BM HRL
*fragariae* Theobald, 1912 (*Myzus*)
(Pentatrichopus) glabrum David, Rajasing & Narayanan, 1971 (Chaetosiphon) BM HRL
(Pentatrichopus) heterotrichum Chakrabarti, A.K. Ghosh & RayChaudhuri, 1971
(Pentatrichopus) hirticorne (Takahashi, 1960) (Chaitomyzus) HRL
(Pentatrichopus) hottesi Stroyan, 1970 HRL
(Pentatrichopus) jacobi Hille Ris Lambers, 1953 BM HRL
*thomasi* subsp. *jacobi* Hille Ris Lambers, 1953 HRL
(Pentatrichopus) minus (Forbes, 1884) (Siphonophora) BM HRL
*brevipilosus* Baerg, 1922 (*Myzus*)
*fragariae* Shinji, 1924 (*Capitophorus*) nec Myzus fragariae Theobald, 1912
*fragarifoliae* Shinji, 1941 (*Capitophorus*)
(Pentatrichopus) minus f. dorsale Schaefers, 1960 BM HRL
(Pentatrichopus) potentillae (Walker, 1850) (Aphis) type BM HRL

C. (Pentatrichopus) tetrarhodum (Walker, 1849) (Aphis) type BM HRL
*neorosarum* Theobald, 1915 (*Myzus*)
*rosarum* Koch, 1855 (*Siphonophora*)
(Pentatrichopus) thomasi Hille Ris Lambers, 1953 BM HRL
*potentillae* Oestlund, 1886 (*Myzus*) nec Aphis potentillae Walker, 1850
(Pentatrichopus) thomasi subsp. jacobi Hille Ris Lambers, 1953 = jacobi Hille Ris Lambers, 1953

CHAITAPHIS Nevsky, 1928
Ent. Mitt. 17: 1917
Type species Chaitaphis tenuicauda Nevsky, 1928

C. tenuicauda Nevsky, 1928 BM HRL

CHAITOCALLIPTERUS Theobald, 1927
The Plant Lice or Aphididae of Great Britain, 2: 329
Based on a mixture of Cinara juniperi, Calaphis flava and Callipterinella calliptera, which was probably recognised before publication, but the genus was not removed from the key.
Without species

CHAITOMYZUS Takahashi, 1960
Kontyû 29: 223
Type species Chaitomyzus hirticornis Takahashi, 1960
= Pentatrichopus Börner, 1930

C. hirticornis Takahashi, 1960 to Chaetosiphon (Pentatrichopus)

CHAITOPHORAPHIS Shinji, 1923
Zool. Mag., Tokyo 35: 307
Type species Chaitophoraphis acerifloris Shinji, 1923 = Yamatocallis hirayamae Matsumura, 1917
= Yamatocallis Matsumura, 1917

C. acerifloris Shinji, 1923 (type of Dichaitophorus Shinji, 1927) = Yamatocallis hirayamae Matsumura, 1917

CHAITOPHORINELLA van der Goot, 1913
Tijdschr. Ent. 56: 112
Type species Chaitophorus testudinatus (Thornton, 1852) = Phyllophora testudinacea Fernie, 1852
= Periphyllus van der Hoeven, 1863

C. acerifoliae Takahashi, 1919 to Periphyllus van der Hoeven, 1863
koelreuteriae Takahashi, 1919 to Periphyllus van der Hoeven, 1863
kuwanaii Takahashi, 1919 to Periphyllus van der Hoeven 1863

CHAITOPHORINUS Börner, 1930
Arch. klassif. phylogen. Ent. 1: 126
Type species Chaitophorus lyropictus Kessler, 1886
= Periphyllus van der Hoeven, 1863

CHAITOPHOROIDES Mordvilko, 1909 (1908)
Ezheg. zool. Muz. 13: 382
Type species Aphis lantanae Koch, 1854
= Aphis Linnaeus, 1758

CHAITOPHORUS Koch, 1854
Die Pflanzenläuse Aphiden, Nürnberg, p. 1
Type species Aphis populi L. of Koch, selected by Gerstaecker, 1856
The species called Chaitophorus populi (L.) by Koch is not the species described by Linnaeus as Aphis populi. The International Commission on Zoological Nomenclature is ruling on the proposal that Chaitophorus leucomelas Koch, 1854 be the type species
*Allarctaphis* Börner, 1949
*Arctaphis* Walker, 1870
*Eichochaitophorus* Essig, 1912
*Micrella* Essig, 1912
*Neothomasia* Baker, 1920
*Promicrella* Börner, 1949
*Pseudomicrella* Börner, 1952
*Thomasia* Wilson, 1910 nec Poche, 1908; Rübsaamen, 1910
*Thomasiniellula* Strand, 1917
*Tranaphis* Walker, 1870

C. abditus (Hottes, 1926) (Neothomasia) HRL
abdominalis Shinji, 1924
abnormis Theobald, 1925 = leucomelas Koch, 1854
acanthipilosus Ivanovskaja, 1973
affinis Börner, 1939 = populeti (Panzer, 1801)
afganensis Narzikulov & Umarov, 1971
agropyronensis Gillette, 1911 to Sipha Passerini, 1860
albicentrus Richards, 1972 = saliciniger (Knowlton, 1927)
albitorosus Ivanovskaja, 1973

C. albus Mordvilko, 1901 = populialbae (Boyer de Fonscolombe, 1841)
albus var. albosiphon Rusanova, 1943 nomen nudum
americanus Baker, 1917 to Periphyllus van der Hoeven, 1863
annulatus Koch, 1854 = Callipterinella callipterus (Hartig, 1841)
anuraphoides Hille Ris Lambers, 1931 = vitellinae (Schrank, 1801)
artemisiae Gillette, 1911 type of Microsiphoniella Hille Ris Lambers, 1947 q.v.
austriacus Pintera, 1968 nomen nudum
balsamiferinus Hille Ris Lambers, 1960 HRL
bathiaschvili Abashidze, 1951 nomen nudum
betulae Buckton, 1879 = Callipterinella callipterus (Hartig, 1841)
betulinus van der Goot, 1912 = populeti (Panzer, 1801)
brachyunguis Börner, 1931 = nassonowi Mordvilko, 1895
brunealineatus Juchnevitch, 1968
bruneri Williams, 1911 = populicola Thomas, 1878
bucktonii del Guercio, 1894 presumably a misprint for Capitophorus in Patch, 1938
candicans (Fitch) (Aphis) 'Probably belongs to Chaitophorus', Thomas, 1878: 14
candicans Monell, 1877 nomen nudum
canens Richards, 1968 paratype BM
capreae Koch, 1854 = salicti (Schrank, 1801)
capreae (Mosley, 1841) (Cinara) BM HRL
*mariae* Mamontova, 1955 HRL
*salicivora* Walker, 1848 (*Aphis*) lectotype BM
cheni Lou, 1935
chinensis Takahashi, 1930 = saliniger (Shinji, 1927)
chrysanthemi Hille Ris Lambers, 1932 = nassonowi Mordvilko, 1895
cinereae Mamontova, 1955 = salicti (Schrank, 1801)
clarus Tseng & Tao, 1936 HRL
coleoptis Shinji, 1917 to Periphyllus van der Hoeven, 1863
coracinus Koch, 1854 to Periphyllus van der Hoeven, 1863
corax Börner, 1939 = tremulae Koch, 1854
cordatae Williams, 1911 = nigrae Oestlund, 1886
coreanus Okamoto & Takahashi, 1927 = populeti (Panzer, 1801)
crinitus Ivanovskaja, 1973
crucis (Essig, 1912) (Thomasia) cotypes BM HRL
delicata Patch, 1913 = stevensis Sanborn, 1904
diversifolia Juchnevitch, 1968 HRL?
dorocolus Matsumura, 1919 HRL
electus Rusanova, 1943 nomen nudum

C. eoessigi Hille Ris Lambers, 1966 HRL
    *longiunguis* Robinson, 1974 HRL
  essigi Gillette & Palmer, 1928 = populifolii (Essig, 1912)
  euphraticus Börner ex Bodenheimer, 1937 nomen nudum
  flabellus Sanborn, 1904 (type of Caricaphis Börner, 1930) to Iziphya Nevsky, 1929
  flavus Forbes, 1884 to Sipha Passerini, 1860
  fraxinicolus Matsumura, 1919
  furcatus Quednau MS ex Pintera, 1968 nomen nudum
  georgica Abashidze, 1951 nomen nudum
  gomesi Ilharco, 1968 paratypes BM HRL
  granulatus Koch, 1854 = Periphyllus hirticornis (Walker, 1848)
  hickelianae Mimeur, 1931 = populialbae (Boyer de Fonscolombe, 1841)
  himalayensis (Das, 1918) (Eichochaitophorus) BM HRL
  hokkaidensis Higuchi, 1972
  horii Takahashi, 1939 BM HRL
  horii subsp. beuthani Börner, 1950 (Tranaphis) BM HRL
    *beuthani* Bőrner (*Tranaphis*)
  hypogeus Schouteden, 1906 BM HRL
  inconspicuus Theobald, 1923 = populialbae (Boyer de Fonscolombe, 1841)
  indicus A.K. Ghosh, M.R. Ghosh & RayChaudhuri, 1970
  inouyei Hille Ris Lambers nomen novum pro Chaitophorus dorocola Matsumura of Higuchi, 1972 nec Matsumura, 1919 HRL
  israeleticus Hille Ris Lambers misprint of israeliticus
  israeliticus Hille Ris Lambers, 1960 HRL
    *minutus* Hille Ris Lambers, 1954 nec Sipha minuta Tissot, 1932 HRL
  japonicus Baker, 1918 = Periphyllus californiensis (Shinji, 1917)
  japonicus Essig & Kuwana, 1918 = Periphyllus kuwanaii (Takahashi, 1919)
  jaxarti Nevsky, 1929 = populeti (Panzer, 1801)
  jaxarti subsp. nigrosiphon Rusanova, 1943 nomen nudum
  jordanus Börner ex Bodenheimer, 1930 & 1937 nomen nudum = euphraticus Börner ex Bodenheimer, 1937 nomen nudum
  kapuri Hille Ris Lambers, 1966 BM HRL
  knowltoni Hille Ris Lambers, 1960 HRL
  lapponum Ossiannilsson, 1959 HRL BM
  latus Mamontova, 1955 = vitellinae (Schrank, 1801)
  laurifoliae Pintera nomen nudum ex Holman & Szelegiewicz, 1972
  leucomelas Koch, 1854 BM HRL
    *abnormis* Theobald, 1925 type BM
    *leucomelas* var. *lyratus* Ferrari, 1872
    *versicolor* Koch, 1854

C. leucomelas var. lyratus Ferrari, 1872 = leucomelas Koch, 1854
longipes Tissot, 1932 BM HRL
longisetosus Szelegiewicz, 1959 paratype BM HRL
longiunguis Robinson, 1974 = eoessigi Hille Ris Lambers, 1966
lyropictus Kessler, 1886 (type of Chaitophorinus Börner, 1930) to Periphyllus van der Hoeven, 1863
macgillivrayae Richards, 1972 = pusillus Hottes & Frison, 1931
macrostachyae (Essig, 1912) (Symdobius) cotype BM HRL
maculatus Buckton, 1899 (type of Pterocallidium Börner, 1949) = Therioaphis trifolii forma maculata (Buckton, 1899)
mariae Mamontova, 1955 = capreae (Mosley, 1841)
matsumurai Hille Ris Lambers, 1960
*salicicolus* Matsumura, 1917 nec salicicola Essig, 1911
minutus Hille Ris Lambers, 1954 = israeliticus Hille Ris Lambers, 1960
minutus (Tissot, 1932) (Sipha) BM HRL
monelli (Essig, 1912) (Micrella) cotype BM HRL
montemartinii del Guercio, 1913 to Dysaphis Börner, 1931
mordwilkoi Mamontova ex Szelegiewicz, 1960 BM HRL
mori Rusanova, 1943 nomen nudum
nassonowi Mordvilko, 1895 type of Allarctaphis Börner, 1949 BM HRL
*brachyunguis* Börner, 1931
*chrysanthemi* Hille Ris Lambers, 1932 HRL
narae Shinji, 1941
neglectus Hottes & Frison, 1931 = populifolii subsp. neglectus Hottes & Frison, 1931
negundinis Thomas, 1877 to Periphyllus van der Hoeven, 1863
niger Mordvilko, 1929 = salijaponicus subsp. niger Mordvilko, 1929
nigrae Oestlund, 1886 BM HRL
*cordatae* Williams, 1911
nigrae subsp. nigrescens Hille Ris Lambers, 1960 cotypes BM HRL
nigrae subsp. tranaphoides Hille Ris Lambers, 1960 BM HRL
nigricantis Pintera, 1968 nomen nudum
nigricantis Pintera ex Ossiannilsson, 1969 nomen nudum
nigricantis subsp. mongolicus Pintera ex Holman & Szelegiewicz, 1972 nomen nudum
nigricentrus Richards, 1972 HRL
nigritus Hille Ris Lambers, 1966 BM HRL
nodulosus Richards, 1966
nudus Richards, 1966 paratype BM HRL
pakistanicus Hille Ris Lambers, 1966 BM HRL
pallipes Richards, 1972 = saliciniger (Knowlton, 1927)

C. parvus Hille Ris Lambers, 1935 HRL
pentandrinus Ossiannilsson, 1959 paratype BM HRL
pheleodendroni Juchnevitch, 1968
populellus Gillette & Palmer, 1928 = populifolii Essig, 1912
populeti (Panzer, 1801) (Aphis) BM HRL
*affinis* Börner, 1939
*betulinus* van der Goot, 1912
*coreanus* Okamoto & Takahashi, 1927
*jaxarti* Nevsky, 1929
*populi* (Linnaeus) of Koch, 1854
*populi-sieboldi* Matsumura, 1919
*pseudotremulae* Ghulamullah, 1942
*quercicola* Juchnevitch, 1960
*sensoriatus* Mimeur, 1934
*yamanarashi* Shinji, 1941
populi (Linnaeus, 1758) (Aphis) is not a Chaitophorus but authors following Koch, 1854 applied this name to leucomelas and more rarely to other species
populi (Linnaeus) of Koch, 1854 (type of Arctaphis Walker, 1870) = populeti (Panzer, 1801)
populialbae (Boyer de Fonscolombe, 1841) (Aphis) BM HRL
*albus* Mordvilko, 1901
*hickelianae* Mimeur, 1931
*inconspicuus* Theobald, 1923 types BM
*roepkei* Börner, 1931
*saccharinus* del Guercio, 1913 (*Myzocallis*)
populicola Thomas, 1878, type of Thomasia Wilson, 1910 nec Poche, 1908; and of Thomasiniellula Strand, 1917; and of Neothomasia Baker, 1920 BM HRL
*bruneri* Williams, 1911
populicola subsp. patchae Hille Ris Lambers, 1960 cotypes BM HRL
populifoliae Oestlund (1887) Davis, 1910 nomen nudum
populifoliae( Fitch, 1851) (Aphis) is a Pterocomma but authors applied the name to populicola and stevensis
populifolii (Essig, 1912) (Eichochaitophorus) cotype BM HRL
*essigi* Gillette & Palmer, 1928
*populellus* Gillette & Palmer, 1928
populifolii subsp. neglectus Hottes & Frison, 1931 paratypes BM HRL
*neglectus* Hottes & Frison, 1931 BM
populifolii subsp. simpsoni Hille Ris Lambers, 1960 cotypes BM HRL
populi-sieboldi Matsumura, = populeti (Panzer, 1801)
pruinosae Narzikulov, 1954 HRL

C. pseudotremulae Ghulamullah, 1942 BM ? = populeti (Panzer, 1801)
pusillus Hottes & Frison, 1931 paratypes BM HRL
*macgillivrayae* Richards, 1972
pustulatus Hille Ris Lambers, 1960 cotypes BM HRL
quercicola Juchnevitch, 1960 = populeti (Panzer, 1801)
quercicola Monell, 1879 type of Hoplochaitophorus Granovsky, 1933 q.v.
quercifoliae (Fitch, 1851) (Lachnus)
ramicola (Börner, 1949) (Promicrella) BM HRL
*ramicola* subsp. *capreae* (F.P. Müller ex Börner, 1952) (*Promicrella*)
reticulata (Börner, 1950) (Pseudomicrella)
roepkei Börner, 1931 = populialbae (Boyer de Fonscolombe, 1841)
saliapterus Shinji, 1924 BM HRL
saliceti Hottes, 1930 emendation of salicti q.v.
salicicola Essig, 1911 cotype BM HRL
salicicolus Matsumura, 1917 = matsumurai Hille Ris Lambers, 1960
salicicorticis (Essig, 1912) (Symdobius) cotypes BM HRL
saliciniger (Knowlton, 1927) (Neothomasia) BM HRL
*albicentrus* Richards, 1972
*pallipes* Richards, 1972
salicis Williams, 1891 nomen nudum
salicti (Schrank, 1801) (Aphis) BM HRL
*capreae* Koch, 1854 nec Cinara capreae Mosley, 1841
*cinereae* Mamontova, 1955 HRL
*saliceti* (Schrank) Hottes, 1930 emendation
salicti subsp. incanae Pintera, 1968 nomen nudum
salicti var. alpha Rusanova, 1943 nomen nudum
salicti var. beta Rusanova, 1943 nomen nudum
salijaponicus Essig & Kuwana, 1918 BH HRL
salijaponicus subsp. niger Mordvilko, 1929 BM HRL
*jacobi* Börner, 1950 (*Pseudomicrella*)
*niger* Mordvilko, 1929
saliniger Shinji, 1929 HRL
*chinensis* Takahashi, 1930
sensoriatus Mimeur, 1934 = populeti (Panzer, 1801)
shantungensis Tseng & Tao, 1936 HRL
shaposhnikovi Mamontova ex Shaposhnikov, 1964
similis Pintera, 1868 nomen nudum
smithiae Monell, 1879 to Pterocomma Buckton, 1879
spinosus Oestlund, 1886 = Hoplochaitophorus quercicola (Monell, 1879)

C. spinulosus Koch in litteris, 1854 = Therioaphis ononidis (Kaltenbach, 1846)
stevensis Sanborn, 1904 BM HRL
*delicata* Patch, 1913
tremulae Koch, 1854 BM HRL
*corax* Börner, 1939
tricolor Koch, 1854 = Callipterinella tuberculata (von Heyden, 1837)
tridentatae Wilson, 1915 type of Flabellomicrosiphum Gillette & Palmer, 1932 q.v.
truncatus (Hausmann, 1802) (Aphis) BM HRL
utahensis (Knowlton, 1928) (Neothomasia) BM
variegatus Szelegiewicz, 1974 nomen nudum HRL
versicolor Koch, 1854 = leucomelas Koch, 1854
viminalis Monell, 1879 BM HRL
viminicola Hille Ris Lambers, 1960 BM HRL
viridis Matsumura, 1919 to Periphyllus van der Hoeven, 1863
vitellinae (Schrank, 1801) (Aphis) BM HRL
*anuraphoides* Hille Ris Lambers, 1931 HRL
*latus* Mamontova, 1955 HRL
vitellinae subsp. danubicus Pintera, 1968 nomen nudum
xanthomelas Koch, 1854 (type of Chaitophoria Börner, 1940) = Periphyllus aceris (Linnaeus, 1761)
yamanarashi Shinji, 1941 = populeti (Panzer, 1801)
yomefuri Shinji, 1922 HRL

Subgen. Allarctaphis Börner, 1949

C. (Allarctaphis) agropyricola Bozhko, 1959

CHAITOREGMA Hille Ris Lambers & A.N. Basu, 1966
Ent. Ber., Amst. 26: 15-17
Type species Oregma tattakana Takahashi, 1925

C. aderuensis (Takahashi, 1935) (Oregma)
tattakana (Takahashi, 1925) (Oregma) BM HRL
tattakana var. suishana (Takahashi, 1929) (Oregma)

CHAITOSIPHELLA lapsus pro Chaetosiphella

CHALOPHONUS Matsumura, 1917, lapsus calami pro Chaitophorus

CHATAIPHIS Nevsky, lapsus pro Chaitaphis

CHELYMORPHA Clarke, 1858 nec Dejean, 1835
The Microscope: being a popular description of ....... objects ... p. 114
Type species Chelymorpha phyllophora Clarke, 1858 = Phyllophora testudinacea Fernie, 1852
= Periphyllus van der Hoeven, 1863

C. phyllophora Clarke, 1858 = Periphyllus testudinaceus (Fernie, 1852)

CHERMAPHIS Maskell, 1884
N.Z.J. Sci. 2:
Type species Kermes pini Koch var. laevis Maskell, 1884
= Pineus Shimer, 1869
Adelgidae

CHERMES Linnaeus, 1758
Syst. Nat. (ed. 10) 1: 453
Suppressed, Bull. Zool. Nomencl. 22: 86 (1965)

C. alni Kalm, 1770 = ? Prociphilus tesselatus (Fitch, 1851)
lapidarius Fabricius, 1803 type of Aphanus Gistel, 1887 nec Laporte de Castelnau, 1832; and of Gisteliella Strand, 1926 nomen dubium
taxi Buckton, 1886 nomen dubium, not an aphid, possibly the coccid Physokermes abietis (Geoffroy)
ulmi Linnaeus, 1758 type of Schizoneura Hartig, 1837 = Eriosoma (Schizoneura) ulmi (Linnaeus, 1758)

CHEYLETUS (Acarina)

C. nigripes Mola, 1907 nomen dubium

CHILEAPHIS Essig, 1953
Proc. Calif. Acad. Sci. 28: 63
Type species Chileaphis michelbacheri Essig, 1953
Here used as subg. of Neophyllaphis Takahashi, 1920

C. michelbacheri Essig, 1953 to Neophyllaphis (Chileaphis)

CHOLODKOVSKAYA Börner, 1909
Mitt. K. biol. Anst. Ld- u. Forstw. 8: 52
Type species Chermes viridanus Cholodkovsky, 1896
Adelgidae

CHOMAPHIS Mordvilko, 1928
In Philipjev, J.M., Keys for the identification of Russian Insects, p. 204
Type species Chomaphis mira Mordvioko, 1928

C. centaureae Börner, 1950 to Dysaphis Börner, 1931
chondrillae Mordvilko, 1932 nomen nudum
cirsii Börner, 1950 = Dysaphis lappae subsp. cirsii (Börner, 1950)
mira Mordvilko, 1928 HRL
*tussilaginis* Mordvilko, 1934
pavlovskii Narzikulov, 1974 to Aphis Linnaeus, 1758
tussilaginis Mordvilko, 1934 = mira Mordvilko, 1928
verae Shaposhnikov, 1956 = Toxopterina vandergooti (Börner, 1939)

CHONDRILLOBIUM Bozhko, 1961
Trudy vsesoyuzn. ent. Obtshch. 48: 19
Type species Chondrillobium junceae Bozhko, 1961 = Hyalopteroides blattnyi Pintera, 1959

C. blattnyi (Pintera, 1959) (Hyalopteroides) BM HRL
*junceae* Bozhko, 1961 HRL
junceae Bozhko, 1961 = blattnyi (Pintera, 1959)

CHROMAPHIS Walker, 1870
Zoologist (2) 5: 2001
Type species Lachnus juglandicola Kaltenbach, 1843

C. californica (Essig, 1912) (Monellia) BM HRL
carpinicola Takahashi, 1921 to Neochromaphis Takahashi, 1921
celticolens Essig & Kuwana, 1918 = Shivaphis celti Das, 1918
coreanus Paik, 1965 to Tinocallis Matsumura, 1919
hirsutustibis Kumar & Lavigne 1970 type BM HRL
juglandicola (Kaltenbach, 1843) (Lachnus) BM HRL
*juglandicola* Koch, 1855 (*Callipterus*)
*peglandicola* Walker, 1870
nirecola Shinji, 1933 type of Chromocallis Takahashi, 1961 q.v.
peglandicola Walker, 1870 errore pro juglandicola = juglandicola (Kaltenbach, 1843)

CHROMOCALLIS Takahashi, 1961
Kontyû 29: 253
Type species Chromaphis nirecola Shinji, 1933

C. nirecola (Shinji, 1933) (Chromaphis) HRL

CHUANSICALLIS Tao, 1964
Qutly. J. Taiwan Museum 17: 217-219
Type species Chuansicallis chengtuensis Tao, 1964
C. chengtuensis Tao, 1964 BM HRL

CHUCALLIS Tao, 1964
Qutly J. Taiwan Museum 17: 221
Type species Myzocallis bambusicola Takahashi, 1921
C. bambusicola (Takahashi, 1921) (Myzocallis) BM HRL

CINACIUM Kishida, 1924
Zool. Mag., Tokyo 36: 472
Type species Cinacium iaksuiense Kishida, 1924
= Aphanostigma Börner, 1909
Phylloxeridae

CINARA Curtis, 1835
British Entomology, 12: 576
Type species Aphis pini Linnaeus, 1758
*Buchneria* Börner, 1952
*Cinarella* Börner, 1949 nec Hille Ris Lambers, 1948
*Cinarella* Hille Ris Lambers, 1948
*Cinarellia* Börner, 1952
*Cinaria* Börner, 1939
*Cinarina* Börner, 1939
*Cinaropsis* Börner, 1939
*Cupressobium* Börner, 1940
*Dilachnus* Baker, 1919 nec Fairmaire, 1896
*Dinolachnus* Börner, 1940
*Lachniella* del Guercio, 1909
*Laricaria* Börner, 1949
*Mecinaria* Börner, 1949
*Neochmosis* Laing ex Theobald, 1929
*Neocinaria* Pašek ex Pintera, 1966
*Neodimosis* Toth, 1935 errore pro Neochmosis Laing ex Theobald, 1929
*Panimerus* Laing, 1926 nec Eaton, 1913
*Pityaria* Börner, 1949
*Pseudocinara* Pasěk ex Pintera, 1966
*Subcinara* Börner, 1949

*Todolachnus* Matsumura, 1917
*Wilsonia* Baker, 1919 nec Bonaparte, 1838
C. abieticola subsp. tenuipes Chakrabarti & A.K. Ghosh ex Chakrabarti, A.K. Ghosh & RayChaudhuri, 1974 = confinis subsp. tenuipes Chakrabarti & A.K. Ghosh ex Chakrabarti, A.K. Ghosh & RayChaudhuri, 1974
abietinus (Matsumura, 1917) (Nippolachnus)
abietis (Fitch, 1851) (Lachnus)
acadiana Hottes, 1956
acutirostris Hille Ris Lambers, 1956 BM HRL
alacra Hottes & Essig, 1953
alaskana Hottes, 1964
alticola Hottes & Essig, 1954 = osborni Knowlton, 1942
americana Hottes & Essig, 1953 = arizonica (Wilson, 1919)
anzai Hottes & Essig, 1953 HRL
apacheca Hottes & Butler, 1958
apini (Gillette & Palmer, 1934) (Lachnus) BM HRL
apini var. villosa Gillette & Palmer, 1931 = villosa Gillette & Palmer, 1931
arizonica (Wilson, 1919) (Lachniella) BM HRL
*americana* Hottes & Essig, 1953*
*kinoi* Hottes & Essig, 1953*
*lassenensis* Hottes & Essig, 1953*
*sabinianus* Swain, 1919 (*Lachnus*)
*serrai* Hottes & Essig, 1953*
*? utahensis* subsp. *zoolathridii* Knowlton & Smith, 1938
atlantica (Wilson, 1919) (Lachniella) BM HRL
*carolina* Tissot, 1932 BM HRL
atra (Gillette & Palmer, 1924) (Lachnus) BM HRL
atripes Hottes, 1958
atroalbipes David, Rajasingh & Narayanan, 1970 (Cinara (Lachniella)) HRL
atrotibialis David & Rajasingh, 1968 BM HRL
*khasyae* Robinson, 1972 BM HRL
azteca Hottes & Essig, 1954
balachowskyi Remaudière, 1974 BM HRL
banksiana Pepper & Tissot, 1955 BM HRL
boerneri Hille Ris Lambers, 1956 = cuneomaculata (del Guercio, 1909)
boerneriana (Pašek, 1954) (Cinaria (Cinaropsis))
*laricis* Börner, 1942 (*Cinaropsis*) nec Lachnus laricis Hartig, 1839; Koch, 1856; Aphis laricis Walker, 1848
bonita Hottes, 1956

*We received permission to publish this synonymy found by Dr. David Voegtlin.

C. braggii (Gillette, 1917) (Lachnus) BM HRL
brauni Börner, 1940 (type of Subcinara Börner, 1949) BM HRL
brevispinosa (Gillette & Palmer, 1924) (Lachnus) BM HRL
*sclerosa* Richards, 1956 paratypes BM HRL
burrilli (Wilson, 1919) (Lachniella) BM
caliente Hottes, 1955 paratypes BM HRL
californica Hottes & Essig, 1953
caliginosa Hottes, 1961 = coloradensis (Gillette, 1917)
canadensis Hottes & Bradley, 1953 = ? cupressi (Buckton, 1881)
canatra Hottes & Bradley, 1953 BM HRL
capreae Mosley, 1841 to Chaitophorus Koch, 1854
caricifex (Cook, 1879) (Lachnus) misprint for laricifex?
carolina Tissot, 1932 = atlantica (Wilson, 1919)
caudelli (Wilson, 1919) (Lachniella)
cedri Mimeur, 1936 BM HRL
cembrae (Seitner, 1936) (Lachnus) BM HRL
*pini* var. *cembrae* Cholodkovsky, 1891 (*Lachnus*)
chamberlini Knowltoni, 1935 = ferrisi (Swain, 1918)
chibi Inouye, 1962 HRL
chinookiana Hottes, 1955 BM HRL
cistata (Buckton, 1881) (Dryobius) error pro costata (Zetterstedt, 1828) but sometimes used for stroyani Pašek, 1954*
cognita Hottes & Essig, 1954
coloradensis (Gillette, 1917) (Lachnus) BM HRL
*caliginosa* Hottes, 1961 HRL
comata (Doncaster, 1956) (Lachniella) types BM HRL
commatula Hottes & Essig, 1954
confinis (Koch, 1856) (Lachnus) BM HRL
*abieticola* Cholodkovsky, 1899 (*Lachnus*)
*abieticola* subsp. *bulgarica* Pintera, 1959 (*Todolachnus*)
*? borealis* Curtis, 1828 (*Aphis*)
*cilicica* del Guercio, 1909 (*Lachniella*)
*cilicica* var. *cecconii* del Guercio, 1909 (*Lachniella*)
*lasiocarpae* Gillette & Palmer, 1930 (*Lachnus*)
*vanduzei* Swain, 1919 (*Lachnus*)
confinis subsp. tenuipes Chakrabarti & A.K. Ghosh ex Chakrabarti, A.K. Ghosh & RayChaudhuri, 1974 HRL
*abieticola* subsp. *tenuipes* Chakrabarti & A.K. Ghosh ex Chakrabarti, A.K. Ghosh & RayChaudhuri, 1974 HRL

* Buckton described a Walker specimen on the slide of which the name costatus was indistinctly written; vide Doncaster, J.P., Bull. Br. Mus. nat. Hist: (Ent.) 28: 89

C. contortae Hottes, 1958
costata (Zetterstedt, 1828) (Aphis) BM HRL
*cistata* (Buckton) misprint for costata but name often applied to stroyani Pašek, 1954
*costata* Hartig, 1839 (*Lachnus*)
*costata* Hartig, 1841 (*Schizoneura*)
*farinosa* Cholodkovsky, 1891 (*Lachnus*)
*laricina* del Guercio, 1909 (*Lachniella*) HRL
*stigma* Curtis, 1844 (*Schizoneura*)
? *symphyti* Curtis, 1844
cronartii Tissot & Pepper, 1967 paratypes BM HRL
cuneomaculata (del Guercio, 1909) (Lachniella) BM HRL
*boerneri* Hille Ris Lambers, 1956 HRL
*laricicola* Börner, 1939 nec Lachnus laricicola Matsumura, 1917
*laricis* subsp. *cuneomaculata* del Guercio, 1909 (*Lachniella*) HRL
cupressi (Buckton, 1881) (Lachnus) types BM HRL
? *canadensis* Hottes & Bradley, 1953 HRL
*juniperi* var. *signata* del Guercio, 1909 (*Lachniella*)
*juniperinus* Mordvilko, 1895 (*Lachnus*)
*tujae* del Guercio, 1909 (*Lachniella*)
curtihirsuta Hottes & Essig, 1954 HRL
curvipes (Patch, 1912) (Lachnus) BM HRL
*utahensis* Knowlton & Smith, 1938
diabola Hottes, 1961
difficilis Hottes & Frison, 1931 = juniperivora (Wilson, 1919) teste Hottes, 1955
diversiseta Börner, 1952 = pinea (Mordvilko, 1895)
doncasteri Pašek, 1953 = laricis (Hartig, 1839)
dubia Hottes & Essig, 1954
eastopi Pintera, 1966 types BM HRL
edulis (Wilson, 1919) (Lachniella) cotypes BM HRL
edulis subsp. tanneri Knowlton, 1930 = tanneri Knowlton, 1930
engelmanniensis (Gillette & Palmer, 1924) (Lachnus) HRL
enigma Hottes & Knowlton, 1954 = obscura Bradley, 1953
escherichi (Börner, 1950) (Cinaria) BM HRL
*nudus* Mordvilko of authors
essigi Hottes, 1961
etsuhoe Inouye, 1970 = sorini Inouye, 1970
excelsae Hille Ris Lambers, 1948 = maritimae (Dufour, 1833)
ezoana Inouye, 1936 HRL

C. ferrisi (Swain, 1918) (Lachnus) BM HRL
*chamberlini* Knowlton, 1935
flexilis (Gillette & Palmer, 1924) (Lachnus) HRL
formosana (Takahashi, 1924) (Dilachnus) BM HRL
*kiangsiensis* Lou, 1935 (*Panimerus*)
fornacula Hottes, 1930 BM HRL
fresai E.E. Blanchard, 1939 BM HRL
*maui* Bradley, 1965 (*Cupressobium*)
*wacasassae* Tissot, 1945
gentneri Hottes, 1957
glabra (Gillette & Palmer, 1924) (Lachnus)
glacialis Hottes, 1964
glehna (Essig, 1915) (Lachnus) cotypes BM HRL
gracilis (Wilson, 1919) (Lachniella) BM
*tunicula* Hottes, 1958
grandis Hottes, 1956
greeni (Schouteden, 1905) (Lachnus)
guadarramae Mimeur, 1936 BM HRL
*? montanicola* Börner, 1939 (*Cinaria*)
harmonia Hottes, 1958 BM HRL
hattorii Kono & Inouye, 1930 BM HRL
*konoi* Inouye, 1941
hirsuta Hottes & Essig, 1954
hirta Hottes & Essig, 1953 = osborni Knowlton, 1942
hirticula Hottes, 1958
horii Inouye, 1941 HRL
hottesi (Gillette & Palmer, 1924) (Lachnus) BM HRL
hyperophila (Koch, 1855) (Lachnus) HRL
idahoensis Knowlton, 1935 BM HRL
*louisianensis* Boudreaux, 1949 (1948) HRL
indica Verma, 1970 BM
inscripta Hottes & Essig, 1953
intermedia (Pašek, (1952) 1954) (Cinaria) HRL
jucunda Hottes, 1958
juniperensis (Gillette & Palmer, 1925) (Lachnus) HRL
juniperi (de Geer, 1773) (Aphis) BM HRL
juniperivora (Wilson, 1919) (Lachnus) BM HRL
*difficilis* Hottes & Frison, 1931 paratype BM
kaltenbachi Hottes, 1954 = pinea (Mordvilko, 1895)
khasyae Robinson, 1972 = atrotibialis David & Rajasingh, 1968

C. kinoi Hottes & Essig, 1953 = arizonica (Wilson, 1919)
kiusa Hottes, 1957
kocheta Hottes, 1958 = nigra (Wilson, 1919)
kochi Inouye, 1939 BM HRL
kochiana (Börner, 1939) (Cinaria) BM HRL
konoi Inouye, 1941 = hattorii Kono & Inouye, 1930
kosarowi Tashev, 1962 = schimitscheki Borner, 1940
kuchea Hottes, 1958 BM
lachnirostris Hille Ris Lambers, 1966 HRL
laricicola Börner, 1939 (type of Cinarella Börner, 1949 nec Hille Ris Lambers, 1948; and of Cinarellia Börner, 1952) = cuneomaculata (del Guercio, 1909)
laricicola (Matsumura, 1917) (Lachnus) HRL
laricifex (Fitch, 1858) (Lachnus) BM HRL
laricifoliae (Wilson, 1915) (Lachnus) BM HRL
laricis (Hartig, 1839) (Lachnus) BM HRL
*doncasteri* Pašek, 1953 types BM
*laricis* Koch, 1856 (*Lachnus*) partim
*laricis* Walker, 1848 (*Aphis*) type BM
*maculosus* Cholodkovsky, 1899 (*Lachnus*)
*muravensis* Arnhart, 1927 (*Lachnus*) BM HRL
*nigrofasciata* del Guercio, 1909 (*Eulachnus*)
*nigrotuberculata* del Guercio, 1909 (*Lachniella*)
*? tenuior* Walker, 1849 (*Aphis*)
lassenensis Hottes & Essig, 1953 = arizonica (Wilson, 1919)
longipennis (Matsumura, 1917) (Lachnus) BM HRL
*momii* Shinji, 1924 (*Dilachnus*)
*quercihabitans* Takahashi, 1924 (*Dilachnus*)
longirostris (Börner, 1950) (Cinaria)
longispinosa Tissot, 1932 = pergandei (Wilson, 1919)
louisianensis Boudreaux, 1949 (1948) = idahoensis Knowlton, 1935
lunata Hottes, 1961 = melaina Boudreaux, 1949
lyalli Bradley, 1956 BM HRL
maculipes Hille Ris Lambers, 1966 BM HRL
maghrebica Mimeur, 1934 BM HRL
manitobensis Bradley, 1963 paratype BM HRL
mariana Bradley, 1957
maritimae (Dufour, 1833) (Aphis) BM HRL
*excelsae* Hille Ris Lambers, 1948 HRL
*pini* subsp. *maritimae* Dufour, 1833 (*Aphis*)

C. matsumurana Hille Ris Lambers, 1966 BM HRL
*abietis* Matsumura, 1917 (*Todolachnus*) nec Aphis abietis Walker, 1848; Lachnus abietis Fitch, 1851
medispinosa (Gillette & Palmer, 1929) (Lachnus) BM HRL
*similis* Gillette & Palmer, 1924 (*Lachnus*) nec van der Goot, 1917
melaina Boudreaux, 1949 (1948) HRL
*lunata* Hottes, 1961
metalica Hottes, 1956
minuta Hottes & Knowlton, 1954
moketa Hottes, 1957 HRL
montanensis (Wilson, 1919) (Lachniella)
montanesa Hottes, 1961
mordvilkoi (Pašek, 1954) (Cupressobium) BM ?
murrayanae (Gillette & Palmer, 1924) (Lachnus) HRL
nepticula Hottes, 1958
neubergi (Arnhart, 1930) (Lachnus) type of Pseudocinara Pašek ex Pintera, 1966 BM HRL
newelli Tissot, 1939 HRL
nigra Mamontova, 1964
nigra (Wilson, 1919) (Lachniella)
*kocheta* Hottes, 1958
nigripes Bradley, 1962 paratypes BM HRL
nigrita Hottes & Essig, 1953
nimbata Hottes, 1954 paratypes BM HRL
nitidula Hottes, 1954 BM HRL
nopporensis Inouye, 1937 HRL
nuda (Mordvilko) has been applied to both C. pini and C. escherichi. Mordvilko probably had a mixture of the two when he referred to Lachnus nudus (de Geer). He never described a Lachnus nudus Mordvilko. His material under the name Lachnus nudus is of C. escherichi (Börner, 1950) q.v.
nuda pini de Geer, 1773 (Aphis) invalid trinominal = Aphis pini Linnaeus, 1758
oblonga (del Guercio, 1909) (Lachniella)
obscura Bradley, 1953 BM HRL
*enigma* Hottes & Knowlton, 1954
occidentalis (Davidson, 1909) (Lachnus) BM HRL
onatei Hottes & Essig, 1953 = ponderosae (Williams, 1911)
ontarioensis Bradley, 1962 paratypes BM HRL
oregonensis (Wilson, 1915) (Lachnus) BM HRL
oregoni Hottes & Essig, 1953
orientalis (Takahashi, 1925) (Dilachnus) BM HRL

C. osborni Knowlton, 1942 BM HRL
*alticola* Hottes & Essig, 1954
*hirta* Hottes & Essig, 1953 HRL
osborni Tissot, 1945 (1944) nec Knowlton, 1942 = pinivora (Wilson, 1919)
osborniana Tissot, 1945 = pinivora (Wilson, 1919)
ozawai Inouye, 1970 (Cinara (Cinaropsis)) HRL
pacifica (Wilson, 1919) (Lachniella)
palaestinensis Hille Ris Lambers, 1948 BM HRL
pallidipes Hottes, 1958
parvicornis Hottes, 1958
pasheki Pintera, 1966
pectinatae (Nördlinger, 1880) (Aphis) BM HRL
*abameleki* del Guercio, 1909 (*Eulachnus*)
*macchiatii* del Guercio, 1909 (*Eulachnus*)
*martellii* Börner, 1952 (*Protolachnus*)
*pichtae* Mordvilko, 1895 (*Lachnus*)
pergandei (Wilson, 1919) (Lachniella) BM HRL
*longispinosa* Tissot, 1932
petersoni Bradley, 1963 BM HRL
piceae (Panzer, 1801) (Aphis) BM HRL
*grossus* Kaltenbach, 1846 (*Lachnus*)
*piceae* var. *paseki* Szelegiewicz, 1962
piceae var. paseki Szelegiewicz, 1962 = piceae (Panzer, 1801)
piceicola (Cholodkovsky, 1896) (Lachnus) = pilicornis (Hartig, 1841) but name often erroneously applied to stroyani Pašek, 1954
picta (del Guercio, 1909) (Lachniella)
pilicornis (Hartig, 1841) (Aphis) BM HRL
*abietis* Walker, 1848 (*Aphis*)
*flavus* Mordvilko, 1895 (*Lachnus*)
*hyalinus* Koch, 1856 (*Lachnus*)
*macrocephalus* Buckton, 1881 (*Lachnus*)
*piceicola* Cholodkovsky, 1898 (*Lachnus*)
*piceicola* var. *viridescens* Cholodkovsky, 1898 (*Lachnus*)
pilosa (Zetterstedt, 1840) (Aphis) nomen dubium in the pini group
pinata Hottes, 1955 paratypes BM
pinea (Mordvilko, 1895) (Lachnus) BM HRL
*diversiseta* Börner, 1952
*inoptis* Wilson, 1919 (*Lachniella*)
*kaltenbachi* Hottes, 1954
*pineti* (Fabricius) of Koch, 1856 nec Fabricius, 1781

C. pineti (Fabricius) (Aphis) of Koch, 1856 nec Fabricius, 1781 = pinea (Mordvilko, 1895)
pini (Linnaeus, 1758) (Aphis) BM HRL
*pineti* Hartig, 1839 (*Lachnus*)
*setosa* Börner, 1950 (*Cinaria*)
pinicola (Kaltenbach, 1843) (Lachnus) BM HRL
pinidensiflorae (Essig & Kuwana, 1918) (Lachnus) BM HRL
piniformosana (Takahashi, 1923) (Dilachnus) BM HRL
pinihabitans (Mordvilko, 1895) (Lachnus) type HRL
piniphila (Ratzeburg, 1844) (Aphis) HRL
*mingazzinii* del Guercio, 1909 (*Eulachnus*)
*pineus* var. *curtipilosa* Mordvilko, 1895 (*Lachnus*)
piniradicis Bradley ex Bradley & Wighton, 1959 paratype BM HRL
pinivora (Wilson, 1919) (Lachniella) BM HRL
*osborni* Tissot, 1945 (1944) nec Knowlton, 1942
*osborniana* Tissot, 1945
pinona Hottes, 1953 HRL
poketa Hottes, 1956
ponderosae (Williams, 1911) (Lachnus) BM HRL
*onatei* Hottes & Essig, 1953 *
pruinosa (Hartig, 1841) (Lachnus) BM HRL
*bogdanowi* Mordvilko, 1895 (*Lachnus*)
*palmerae* Gillette, 1917 (*Lachnus*) BM HRL
*pubescens* Wellenstein, 1930 (*Dilachnus*)
*radicicola* Wellenstein, 1930 (*Dilachnus*)
pseudoschwarzii Palmer, 1936 BM HRL
pseudotaxifoliae Palmer, 1952 BM HRL
pseudotsugae (Wilson, 1912) (Lachnus) BM HRL
puerca Hottes, 1954 paratypes BM HRL
pulverulens (Gillette & Palmer, 1924) (Lachnus) HRL
quercus Mosley, 1841 = Thelaxes dryophila (Schrank, 1801)
rara Bradley, 1957
rigidae Hottes, 1958
rubicunda (Wilson, 1915) (Lachnus)
russellae Pepper & Tissot, 1973 HRL
rustica Hottes, 1956
sabinae (Gillette & Palmer, 1924) (Lachnus) HRL

***We received permission to publish this synonymy found by Dr. David Voegtlin.**

C. saccharipini Hottes, 1958
saskensis Bradley, 1962
schimitscheki Börner, 1940 BM HRL
*kosarowi* Tashev, 1962 HRL
schuhi Hottes, 1957
schwarzii (Wilson, 1919) (Lachniella) BM
sclerosa Richards, 1956 = brevispinosa (Gillette & Palmer, 1924)
serrai Hottes & Essig, 1953 = arizonica (Wilson, 1919)
setulosa Hottes & Essig, 1955 BM HRL
shinjii Inouye, 1938
*fasciatus* Shinji, 1922 (*Lachnus*) nec Burmeister, 1835
sibiricae (Gillette & Palmer, 1924) (Lachnus) HRL
similis (van der Goot, 1917) (Lachnus)
sitchensis Hottes, 1958
solitaria (Gillette & Palmer, 1924) (Lachnus) BM HRL
sonata Hottes, 1955 paratypes BM
soplada Hottes, 1956
sorini Inouye, 1970 HRL
*etsuhoe* Inouye, 1970 HRL
spiculosa Bradley, 1956 BM HRL
splendens (Gillette & Palmer, 1924) (Lachnus) BM HRL
strobi (Fitch, 1851) (Eriosoma) BM HRL
stroyani Pašek, 1954 BM HRL
*cistata** of authors nec Buckton, 1881
*cistata* var. *stroyani* Pašek, 1954 (*Cinaropsis*)
*piceicola* (Cholodkovsky) auctt. nec Cholodkovsky, 1898
subterranea Bradley, 1956 BM HRL
symphyti Curtis, 1844 = ? costata (Zetterstedt, 1828)
taedae Tissot, 1932 BM HRL
taeniata (Koch, 1856) (Lachnus) HRL?
taeniatoides (Mordvilko, 1901) (Lachnus)
taiwana (Takahashi, 1925) (Dilachnus)
tanneri Knowlton, 1930 BM
*edulis* subsp. *tanneri* Knowlton, 1930
taxifolia (Swain, 1918) (Lachnus) HRL
terminalis (Gillette & Palmer, 1932) (Lachnus) HRL
thatcheri Knowlton & Smith, 1938 BM

* The name cistata (Buckton, 1881) errore pro costata Zetterstedt, 1828 has for some time been used for stroyani Pašek, 1954

C. todocola (Inouye, 1936) (Tuberolachnus) BM HRL
    *todoe* Shinji, 1941
  todoe Shinji, 1941 = todocola (Inouye, 1936)
  tonaluca Hottes & Wehrle, 1951
  tsugae Bradley, 1960 paratypes BM HRL
  tujafilina (del Guercio, 1909) (Lachniella) BM HRL
    *biotae* van der Goot, 1917 (*Lachnus*)
    *callitris* Froggatt, 1927 (*Dilachnus*)
    ? *cupresi* Gomez-Menor, 1962 (*Cinaropsis*)
    *mediterraneum* Narzikulov, 1963 (*Cupressobium*) HRL
    *pseudosabinae* Nevsky, 1929 (*Lachnus*)
    *sabinae* Nevsky, 1929 (*Lachnus*) nec Gillette & Palmer, 1924
    *thujafolia* Theobald, 1914 (*Lachniella*)
    *winonkae* Hottes, 1934 HRL
  tunicula Hottes, 1958 = gracilis (Wilson, 1919)
  utahensis Knowlton & Smith, 1938 = curvipes (Patch, 1912)
  utahensis subsp. zoolathridii Knowlton & Smith, 1938 = ? arizonica (Wilson, 1919)
  vagabunda Hottes & Essig, 1953
  vandykei (Wilson, 1919) (Lachniella) HRL
  villosa Gillette & Palmer, 1931
    *apini* var. *villosa* Gillette & Palmer, 1931
  villosa subsp. curtivillosa Hottes, 1955
  villosa subsp. parvavillosa Hottes, 1955
  wacasassae Tissot, 1945 = fresai E.E. Blanchard, 1939
  wahluca Hottes, 1952 paratypes BM HRL
  wahnaka Hottes, 1951 paratype BM HRL
  wahsugae Hottes, 1960
  wahtolca Hottes, 1953 BM HRL
  wahtolca subsp. curtiwahtolca Hottes, 1955 BM
  wanepae Hottes, 1933
  watanabei Inouye, 1970 HRL
  watsoni Tissot, 1939 BM HRL
  westi Tissot & Pepper, 1967 paratypes BM HRL
  winonkae Hottes, 1934 = tujafilina (del Guercio, 1909)
  yukona Hottes, 1964
  zoarcbursara Knowlton, 1935

Subgen. Cinaropsis Börner, 1939

C. (Cinaropsis) ozawai Inouye, 1970 to Cinara Curtis, 1835

Subgen. Lachniella del Guercio, 1909

C. (Lachniella) atroalbipes S.K. David, Rajasingh & Narayanan, 1971 to Cinara Curtis, 1835

CINARELLA Börner, 1949 nec Hille Ris Lambers, 1948
Beitr. tax. Zool. 1: 58, 59
Type species Cinara laricicola Börner, 1939 = Lachniella laricis subsp. cuneomaculata del Guercio, 1909
= Cinara Curtis, 1835

CINARELLA Hille Ris Lambers, 1948
Trans. R. ent. Soc. Lond. 99: 275, described as subg. of Cinara Curtis, 1835
Type species Lachnus pineus Mordvilko, 1895
= Cinara Curtis, 1835

CINARELLIA Börner, 1952
Mitt. thüring. bot. Ges. 4 (3): 41, described as subg. of Cinara Curtis, 1835
Type species Cinara laricicola Börner, 1939 = Lachniella laricis subsp. cuneomaculata del Guercio, 1909
= Cinara Curtis, 1835

CINARIA Börner, 1939
Arb. physiol. angew. Ent. Berl. 6: 76
Type species Aphis laricis Walker, 1848 = Lachnus laricis Hartig, 1839
= Cinara Curtis, 1835

C. escherichi Börner, 1950 (used as type of Neocinaria Pašek 1951 ms.. (ex Pintera, 1966)) to Cinara Curtis, 1835
intermedia Pasek (1952) 1954 to Cinara Curtis, 1835
kochiana Börner, 1939 (type of Laricaria Börner, 1949) to Cinara Curtis, 1835
longirostris Börner, 1950 to Cinara Curtis, 1835
montanicola Börner, 1939 = ? Cinara guadarramae Mimeur, 1936
setosa Börner, 1950 = Cinara pini (Linnaeus, 1758)

Subgen. Cinaropsis Börner, 1939

C. (Cinaropsis) boerneriana Pašek, 1954 to Cinara Curtis, 1835

CINARINA Börner, 1939
Arb. physiol. angew. Ent. Berl. 6: 76
Type species Lachnus [piceicola var. ] viridescens Cholodkovsky, 1898
= Aphis pilicornis Hartig, 1841: but Börner used the name viridescens for Cinaropsis cistata var. stroyani Pašek, 1954
= Cinara Curtis, 1835

CINAROPSIS Börner, 1939
Arb. physiol. angew. Ent. Berl. 6: 76
Type species Lachnus pinicola Kaltenbach, 1843
= Cinara Curtis, 1835

C. cistata var. stroyani Pasek, 1954 = Cinara stroyani (Pasek, 1954)
cupresi Gomez-Menor, 1962 = ? Cinara tujafilina del Guercio, 1909
laricis Börner, 1942 = Cinara boerneriana (Pasek, 1954)

CLADOBIUS Koch, 1856 nec Dejean, 1837
Die Pflanzenläuse Aphiden, Nürnberg, p. 251
Type species Aphis populea Kaltenbach, 1843
= Pterocomma Buckton, 1879

C. beulahensis Cockerell, 1904 to Pterocomma Buckton, 1879
farinosus del Guercio, 1913 = Aphis pollinosa Walker, 1849
populeus var. longirostris Mordvilko, 1909 = Pterocomma populeum (Kaltenbach, 1843)
rufulus Davidson, 1909 = Pterocomma bicolor (Oestlund, 1887)
steinheili Mordvioko, 1901 = Pterocomma rufipes (Hartig, 1841)

CLADOXUS Rafinesque, 1817
Am. Mon. Mag. Crit. Rev. 1: 361, described as subg. of Aphis Linnaeus, 1758
Type species Aphis fusciclava Rafinesque, 1817 invalid
Invalid, type not an aphid: vide Hottes, 1931

CLAVIGERUS Szepligeti, 1883
Rovaraszati Lapok 1: 4, 19 footnote
Type species Clavigerus salicis (Kaltenbach, 1843) = Aphis salicis Linnaeus, 1758
= Pterocomma Buckton, 1879

CLAVISIPHON del Guercio, 1930
Redia 19: 500, described as subg. of Anuraphis del Guercio, 1907
Type species Anuraphis (Clavisiphon) elegans (Ferrari, 1872)* = Rhopalosiphum elegans Ferrari, 1872
= Eucarazzia del Guercio, 1921

*** new type citation**

CLAVOSIPHUM Shinji, 1922
Zool. Mag., Tokyo 34: 790
Type species Clavosiphum adenocauli Shinji, 1922 = Rhopalosiphum tiliae Matsumura, 1918
= Rhopalosiphoninus Baker, 1920

C. adenocauli Shinji, 1922 = Rhopalosiphoninus tiliae (Matsumura)
sambucifoliae Shinji, 1922 = Rhopalosiphoninus hydrangeae (Matsumura, 1918)

CLETHROBIUS Mordvilko, 1928
in Philipjev, Keys for the identification of Russian insects, pp. 181, 184
Type species Clethrobius giganteus (Cholodkovsky, 1899) = Callipterus giganteus Cholodkovsky, 1899

C. betulae Mordvilko ex Tarbinsky & Plaviltchikov, 1948 = comes (Walker, 1848)
comes (Walker, 1848) (Aphis) BMHRL
*betulae* Mordvilko ex Tarbinsky & Plaviltchikov, 1948
giganteus (Cholodkovsky, 1899) (Callipterus) BM HRL

CLYPEAPHIS mis-spelling of Clypeoaphis Soliman, 1937

CLYPEOAPHIS Soliman, 1937
Entomologist's mon. Mag. 73: 181
Type species Clypeoaphis suaedae Soliman, 1937 = Longicaudus suaedae Mimeur, 1934
= Xerophilaphis Nevsky, 1928

C. stavropolensis Ivanovskaja, 1960 to Xerophilaphis Nevsky, 1928
suaedae Soliman, 1937 = Xerophilaphis suaedae (Mimeur, 1934)

CNAPHALODES Macquart ex Amyot & Serville, 1843
Hist. nat. Insectes, Hémipt. p. 594
Type species Cnaphalodes laricis Macquart, 1843 = Adelges laricis Vallot, 1836
= Adelges Vallot, 1836
Adelgidae

COCCUS (Coccoidea)

C. laricis Bouché, 1834 = ? Adelges laricis Vallot, 1836
lataniae Boisduval, 1867 type of Cerataphis Lichtenstein, 1882 q.v.; and of Boisduvalia Signoret, 1868 nec Robineau-Desvoisy, 1830

C. mali Bingley, 1803 = Eriosoma lanigerum (Hausmann, 1802)
pinicorticis Fitch, 1855 = Pineus strobus (Hartig, 1837)
strobi Baerensprung, 1849 = ? Pineus strobus (Hartig, 1837)
strobus Hartig, 1837 to Pineus Shimer, 1869
zeae-maydis Dufour, 1824 = Tetraneura ulmi (Linnaeus, 1758)

COLOPHA Monell, 1877
Can. Ent. 9: 102
Type species Byrsocrypta ulmicola Fitch, 1859
*Colophella* Börner, 1926
*Sinocolopha* Tao, 1970

C. caucasica (Dzhibladze, 1960) (Schizoneura) BM HRL
clematicola (Shinji, 1922) (Pemphigus) BM HRL
compressa (Koch, 1856) (Schizoneura) BM HRL
eragrostidis Middleton, 1878 = ulmicola (Fitch, 1859)
graminis (Takahashi, 1930 (Truncaphis) (type of Sinocolopha Tao, 1970) BM HRL
? kansugei Uye, 1924 (Eriosoma)
moriokaensis (Monzen, 1927) (Tetraneura) BM HRL
nirecola (Matsumura, 1917) (Gobaisha) BM HRL
rossica Cholodkovsky, 1897 = Anoecia vagans (Koch, 1856)
ulmicola (Fitch, 1859) (Byrsocrypta) BM HRL
*colophoidea* Howard, 1908 nomen nudum of Monell (*Tetraneura*)
*eragrostidis* Middleton, 1878
*graminis* Monell, 1882 (*Tetraneura*)
*? spicatus* Hart, 1895 (*Rhizobius*)

COLOPHELLA Börner, 1926
In Abderhalden, E., Handbuch der biologischen Arbeitsmethoden, Abt. IX, Teil 1,2 Hälfte, Heft 2: 233
Type species Tetraneura graminis Monell, 1882 = Byrsocrypta ulmicola Fitch, 1859
= Colopha Monell, 1877

COLOPHINA Börner, 1932
In Sorauer, Handbuch der Pflanzenkrankheiten 5(2): 671
Type species Eriosoma clematicola Takahashi, 1924 = Pemphigus clematis Shinji, 1922
Here used as subg. of Eriosoma Leach, 1818

COLORADOA Wilson, 1910
Ann. ent. Soc. Am. 3: 323
Type species Aphis rufomaculata Wilson, 1908
*Capitophoraphis* E.E. Blanchard, 1944
*Eurhopalosiphum* Shinji, 1942
*Lidaja* Börner, 1952
*Neaphis* Nevsky, 1929
*Stephensonia* Das, 1918

C. abrotani (Koch, 1854) (Hyalopterus) BM HRL
absinthiella Ossiannielsson, 1962 HRL
*angelicae* del Guercio, 1911 (*Siphocoryne*) see Stroyan, 1969, Bull. zool. Nomencl. 25: 174
absinthii (Lichtenstein, 1885) (Rhopalosiphum) BM HRL
*lydiae* Börner, 1932
achilleae Börner, (1939 nomen nudum) 1940 = achilleae Hille Ris Lambers, 1939
achilleae Hille Ris Lambers, 1939 BM HRL
*achilleae* Börner, (1939 nomen nudum) 1940
artemisiae (del Guercio, 1913) (Siphocoryne) BM HRL
artemisiae subsp. artemisicola Takahashi, 1965 = artemisicola Takahashi, 1965
artemisicola Takahashi, 1965 HRL
*artemisiae* Paik, 1965 (*Cavariella*) nec Siphocoryne artemisiae del Guercio, 1913
*artemisiae* subsp. *artemisicola* Takahashi, 1965 HRL
bournieri Remaudière & Leclant, 1969 BM HRL
bournieri subsp. iberica Remaudière & Leclant, 1969 HRL
brevipilosa (Ivanovskaja, 1958) (Neaphis)
campestrella Ossiannilsson, 1959 paratype BM HRL
campestris Börner, 1939 BM HRL
cydiae lapsus pro lydiae
heinzei (Börner, 1952) (Lidaja) BM HRL
heinzei subsp. paradoxa Szelegiewicz, 1969 = paradoxa Szelegiewicz, 1969
inodorella Ossiannilsson, 1959 paratypes BM HRL
kondoi (Shinji, 1942) (Eurhopalosiphum)
lydiae Börner, 1932 = absinthii (Lichtenstein, 1885)
moralesi Remaudière & Leclant, 1969 HRL
palmerae Börner, 1952 BM HRL
paradoxa Szelegiewicz, 1969 HRL
*heinzei* subsp. *paradoxa* Szelegiewicz, 1969

C. ponticae (Börner, 1940) (Neaphis) HRL
procerae (Bozhko, 1959) (Neaphis)
rufomaculata (Wilson, 1908) (Aphis) BM HRL
*chrysanthemicola* Williams, 1911 (*Aphis*)
*kiku* Hori, 1929 (*Rhopalosiphum*)
*lahorensis* Das, 1918 (*Stephensonia*)
*williamsi* E.E. Blanchard, 1944 (*Capitophoraphis*)
santolinae Hille Ris Lambers, 1948 HRL
scopariae Szelegiewicz, 1969 HRL
submissa Doncaster, 1961 types BM HRL
tadzhica Narzikulov, 1970
tadzhikistanica Narzikulov, 1974
*artemisiae* Narzikulov, 1958 (*Neaphis*)
tanacetina (Walker, 1850) (Aphis) lectotype BM HRL
? *tanaceticola* del Guercio, 1930 (*Anuraphis (Macchiatiella)*)
taurica (Mamontova, 1963) (Lidaja) HRL
viridis (Nevsky, 1929) (Neaphis) HRL
? *takahashii* Tseng & Tao, 1936 (*Micraphis*)

COMAPHIS Börner, 1940
Neue Blattläuse aus Mitteleuropa, privately published, p. 3
Type species Doralis corniella Hille Ris Lambers, 1935 = Aphis salicariae Koch, 1855
= Aphis Linnaeus, 1758

CONICAUDUS Heie, 1972
Steenstrupia 2: 255
Type species Conicaudus longipes Heie, 1972
Fossil, baltic amber

COREALACHNUS Paik, 1971
Korean Jl Entom. 1: 3
Type species Corealachnus suwonensis Paik, 1971 = Atheroides serrulatus Haliday, 1839
= Atheroides Haliday, 1837

C. suwonensis Paik, 1971 = Atheroides serrulatus Haliday, 1839

CORNAPHIS Gillette, 1913
Ann. ent. Soc. Am. 6: 491
Type species Cornaphis populi Gillette, 1913

C. populi Gillette, 1913 BM HRL

CORYLOBIUM Mordvilko, 1914
Faune Russie, Ins. Hém. 1: 71
Type species Aphis avellanae Schrank, 1801

C. avellanae (Schrank, 1801) (Aphis) BM HRL
*coryli* Mosley, 1841 (*Aphis*) nec Goeze, 1778
vandenboschi Hille Ris Lambers, 1966 to Macrosiphum (Neocorylobium)

CORYNOSIPHON Mordvilko, 1914
Faune Russie, Ins. Hém. 1: 73
Type species Aphis capreae Fabricius, 1775: only species mentioned in footnote on page 73.
= Cavariella del Guercio, 1911

COTONEASTERIA Shaposhnikov, 1964
In Bei-Bienko, Keys to the insects of the European part of the USSR 1: 584, described as subg. of Dysaphis Börner, 1931
Type species Dentatus microsiphon Nevsky, 1929
= Dysaphis Börner, 1931

CRACCIFEX Amyot, 1847
Annls Soc. ent. Fr. (2) 5: 478
Invalid

CRANAPHIS Takahashi, 1939
Philipp. J. Sci. 69: 28
Type species Monellia formosana (Takahashi, 1924) = Myzocallis formosanus Takahashi, 1924

C. arundinariae (Takahashi, 1940) (Shivaphis)
bambusicola David, Rajasingh & Narayanan, 1971 to Shivaphis Das, 1918
formosana (Takahashi, 1924) (Myzocallis)
pilosa David, Rajasingh & Narayanan, 1971 type of Subtakecallis RayChaudhuri & Pal, 1974 q.v.

CRATAEGARIA Shaposhnikov, 1964
In Bei-Bienko, Keys to the Insects of the European part of the USSR 1: 584, described as subg. of Dysaphis Börner, 1931
Type species Dysaphis crataegi (Kaltenbach, 1843) = Aphis crataegi Kaltenbach, 1843
= Dysaphis Börner, 1931

CROSAPHIS Evans, 1971
Mem. Qd. Mus. 16: 146
Type species Crosaphis anomala Evans, 1971
Fossil. Upper Triassic

CRYPTAPHIS Hille Ris Lambers, 1947
Temminckia 7: 296
Type species Cryptaphis setiger Hille Ris Lambers, 1947 = Aphis poae Hardy, 1850
*Neodecorosiphon* Heinze, 1960

C. bromi Robinson, 1967 paratypes BM HRL
geraniae Takahashi, 1961 = geranicola (Shinji, 1935)
geranicola (Shinji, 1935) (Myzus) HRL
*geraniae* Takahashi, 1961 HRL
menthae Takahashi, 1961
poae (Hardy, 1850) (Aphis) BM HRL
*muscicolens* Heinze, 1960 (*Neodecorosiphon*)
? *pilosa* Walker, 1849 (*Aphis*) nec Zetterstedt, 1840
*setiger* Hille Ris Lambers, 1947 HRL
rostrata Chakrabarti & RayChaudhuri in Chakrabarti, A.K. Ghosh & RayChaudhuri, 1974 HRL
setiger Hille Ris Lambers, 1947 = poae (Hardy, 1850)

CRYPTOMYZUS Oestlund, 1922
Rep. Minn. St. Ent. 19: 139
Type species Aphis ribis Linnaeus, 1758
*Myzella* Börner, 1930

C. alboapicalis (Theobald, 1916) (Siphocoryne) BM HRL
*ulmeri* Börner, 1922 (*Myzella*)
ballotae Hille Ris Lambers, 1953 cotypes BM HRL
behboudii Remaudière & Davatchi, 1961 paratype BM HRL
elaeagni Börner, 1950 = korschelti Börner, 1938
galeopsidis (Kaltenbach, 1843) (Aphis) BM HRL
*dispar* Patch, 1914 (*Myzus*)
*lamii* van der Goot, 1912 (*Myzus*)
*quaerens* Walker, 1849 (*Aphis*)
*scrophulariae* Buckton, 1876 (*Siphonophora*) lectotype BM
*whitei* Theobald, 1912 (*Myzus*)

C. galeopsidis subsp. citrinus Hille Ris Lambers, 1953 HRL
galeopsidis subsp. dickeri Hille Ris Lambers, 1953 HRL
heinzei Hille Ris Lambers, 1953 BM HRL
*calaminthae* Lichtenstein, 1884 (*Rhopalosiphum*) nomen nudum of Börner, 1952
korschelti Börner, 1938 BM HRL
*elaeagni* Börner, 1950
leonuri Bozhko, (1957) 1961 BM HRL
ribis (Linnaeus, 1758) (Aphis) BM HRL
*ribis* O.F. Müller, 1776 (*Aphis*)
taoi Hille Ris Lambers, 1963 BM HRL
taoi subsp. indicus A.K. Ghosh & RayChaudhuri, 1972

Subgen. Ampullosiphon Heikinheimo, 1955

C. (Ampullosiphon) stachydis (Heikinheimo, 1955) (Amphorophora (Ampullosiphon)) HRL

CRYPTOSIPHON miscorrection of Cryptosiphum

CRYPTOSIPHUM Buckton, (1875) 1879
Monograph of the British Aphides London; 1: 47 nomen nudum (1875); 2: 144 (1879)
Type species Cryptosiphum artemisiae Buckton, 1879
*Pseudolachnus* Shinji, 1922

C. artemisiae Buckton, 1879 lectotype BM HRL
*artemisiae* Passerini, 1860 (*Aphis*)
*gallarum* Kaltenbach, 1856 (*Aphis*)
*pseudogallarum* Shinji, 1941
*yomogi* Shinji, 1922 (*Pseudolachnus*)
astrachanicae Ivanovskaja, 1960
brevipilosum Börner, 1932 BM HRL
canadensis Williams, 1911 to Microsiphoniella Hille Ris Lambers, 1947
caspicae Bozhko, 1957
cinae Nevsky, 1928 (type of Absinthaphis Remaudière ex Stary & Remaudière, 1972) to Aphis (Absinthaphis)
eurotiae Mamontova, 1968 (as Cryptosiphon)
hirsutum Nevsky, 1928 to Aphis (Absinthaphis)
innokentyi Ivanovskaja, 1970
milsteppensis Rusanova, 1943 nomen nudum

C. mordvilkoi Ivanovskaja, 1960
nerii de Stefani Perez, 1901 = Aphis nerii Boyer de Fonscolombe, 1841
nevskii Börner, 1952
pseudogallarum Shinji, 1941 = artemisiae Buckton, 1879
sieversiana Ivanovskaja, 1958
silenes Buckton, 1875 nomen nudum
tahoense Davidson, 1911 = Tamalia coweni (Cockerell, 1905)

CRYPTOZIPHUM lapsus pro Cryptosiphum

CRYPTURAPHIS Silvestri, 1935
Boll. Lab. Zool. gen. agr. R. Scuola Agric. Portici 38: 290
Type species Crypturaphis grassii Silvestri, 1935

C. grassii Silvestri, 1935 cotype BM HRL

CTENOCALLIS Klodnitsky, 1924
Trudў chetvert. vses. Entomo- Fitopat- S'ezda 4 (1922): 61, described as Ctenocallis Grossheim in litt.
Type species Ctenocallis dobrovljanskyi Klodnitsky, 1924
*Gentnera* Essig, 1952
*Oniscomyzus* Börner, 1942

C. dobrovljanskyi Klodnitsky, 1924 BM HRL
*bramstedti* Börner, 1942 (*Oniscomyzus*)
israelicus Hille Ris Lambers, 1954 BM HRL
setosus (Kaltenbach, 1846) (Aphis) BM HRL
*oregona* Essig, 1952 (*Gentnera*) paratype BM HRL

CUERNAVACA J.M. Baker, 1934 nec Kirkaldy, 1913
An. Inst. Biol. Univ. Méx. 5: 210
Type species Cuernavaca mexicana J.M. Baker, 1934
= Diuraphis Aizenberg, 1935

C. mexicana J.M. Baker, 1934 to Diuraphis Aizenberg, 1935

CUPRESSOBIUM Börner, 1940
Neue Blattläuse aus Mitteleuropa, privately published, p. 1
Type species Aphis juniperi de Geer, 1773
= Cinara Curtis, 1835

C. maui Bradley, 1965 = Cinara fresai E.E. Blanchard, 1939
mediterraneum Narzikulov, 1963 = Cinara tujafilina (del Guercio, 1909)

C. mordvilkoi Pašek, 1954 to Cinara Curtis, 1835

DACTYLOSPHAERA Shimer, 1867
Proc. Acad. nat. Sci. Philad. 19: 5 (as new genus)
Type species Dactylosphaera globosum (Shimer, 1867) = Phylloxera globosa Shimer, 1867
= ? Phylloxera Boyer de Fonscolombe, 1834
Phylloxeridae

DACTYNOTUS Rafinesque, 1818
Am. mon. Mag. Crit. Rev. 3: 18, described as subg. of Aphis Linnaeus, 1758
Type species Aphis hieracium-paniculatum Rafinesque, 1818, invalid
Invalid; vide Uroleucon Mordvilko, 1914

D. astronomus Hille Ris Lambers, 1962 to Uroleucon Mordvilko, 1914
asyneumatis Holman, 1969 to Uroleucon (Uromelan)
bielawskii Szelegiewicz, 1962 to Uroleucon Mordvilko, 1914
brachysiphon Verma, 1969 nomen nudum
caligatus Richards, 1966 to Uroleucon (Lambersius)
cameronensis Takahashi, 1950 to Uroleucon (Uromelan)
campanulae subsp. longior Börner, 1950 to Uroleucon (Uromelan)
canadensis Richards, 1972 to Uroleucon (Lambersius)
carlinae Börner, 1933 to Uroleucon (Uromelan)
carthami Hille Ris Lambers, 1948 to Uroleucon (Uromelan)
chrysopsidicola Olive, 1963 to Uroleucon Mordvilko, 1914
cichorii subsp. carduicola Hille Ris Lambers, 1939 to Uroleucon Mordvilko, 1914
cichorii subsp. grossus Hille Ris Lambers, 1939 to Uroleucon Mordvilko, 1914
cichorii subsp. leontodontis Hille Ris Lambers, 1939 to Uroleucon Mordvilko, 1914
ciefi Olive, 1963 to Uroleucon Mordvilko, 1914
cirsicola Holman, 1962 to Uroleucon Mordvilko, 1914
doronici Börner, 1942 to Uroleucon (Uromelan)
ensifoliae Holman, 1965 to Uroleucon Mordvilko, 1914
fagopyri Chowdhuri, R.C. Basu, Chakrabarti & RayChaudhuri, 1969 to Uroleucon Mordvilko, 1914
glomeratae Börner, 1950 = Uroleucon (Uromelan) nigrocampanulae (Theobald, 1928)
helianthi Tao, 1963 to ? Uroleucon (Lambersius) or ? Macrosiphum Passerini, 1860

D. helianthicola Olive, 1963 to Uroleucon (Uromelan)
hieracicola Hille Ris Lambers, 1962 to Uroleucon Mordvilko, 1914
hypochoeridis Hille Ris Lambers, 1939 = Uroleucon hypochoeridis (Fabricius, 1779)
hyssopii Narzikulov & Daniarova, 1973 to Uroleucon Mordvilko, 1914
inulicola Hille Ris Lambers, 1939 to Uroleucon Mordvilko, 1914
inulicola subsp. hirticola Ossiannilsson, 1959 to Uroleucon Mordvilko, 1914
jaceae subsp. henrichi Börner, 1950 to Uroleucon (Uromelan)
jaceae subsp. scabiosae Börner ex Franz, 1949 nomen nudum = Uroleucon (Uromelan) jaceae subsp. henrichi (Börner, 1950)
jaceicola Hille Ris Lambers, 1939 to Uroleucon Mordvilko, 1914
jaceicola subsp. pasqualei Hille Ris Lambers & Stroyan, 1959 to Uroleucon Mordvilko, 1914
kashmiricus Verma, 1966 to Uroleucon Mordvilko, 1914
leonardi Olive, 1965 to Uroleucon Mordvilko, 1914
macrocaudus Tao, 1964 = Macrosiphoniella grandicauda Takahashi & Moritsu, 1963
marcatus Hille Ris Lambers, 1931 = Uroleucon cirsii (Linnaeus, 1758)
margarithae Hille Ris Lambers, 1950 = Uroleucon chondrillae (Nevsky, 1929)
minor Börner, 1940 to Uroleucon (Uromelan)
neocampanulae Takahashi, 1962 to Uroleucon (Uromelan)
nigrotibius Olive, 1963 to Uroleucon Mordvilko, 1914
nigrotuberculatus Olive, 1963 to Uroleucon Mordvilko, 1914
nodulus Richards, 1972 to Uroleucon (Lambersius)
obscuricaudatus Olive, 1965 to Uroleucon Mordvilko, 1914
ochropus Hille Ris Lambers, 1939 to Uroleucon Mordvilko, 1914
pachysiphon Börner, 1950 to Uroleucon (Uromelan)
parvotuberculatus Olive, 1963 to Uroleucon (Uromelan)
paucosensoriatus Hille Ris Lambers, 1960 to Uroleucon Mordvilko, 1914
pepperi Olive, 1965 to Uroleucon Mordvilko, 1914
picridicola Hille Ris Lambers, 1939 = Uroleucon cichorii subsp. grossum (Hille Ris Lambers, 1939)
picridiphagus Takahashi, 1962 to Uroleucon Mordvilko, 1914
pieloui Richards, 1972 = Uroleucon hieracicola (Hille Ris Lambers, 1962)
pilosellae Börner, 1933 to Uroleucon Mordvilko, 1914
pseudambrosiae Olive, 1963 to Uroleucon Mordvilko, 1914
pseudobscurus Hille Ris Lambers, 1967 to Uroleucon Mordvilko, 1914
pseudochrysanthemi Olive, 1967 to Uroleucon Mordvilko, 1914
pseudotanaceti Verma, 1969 to Uroleucon Mordvilko, 1914
ptarmicae Bozhko, 1959 nomen nudum

D. pulicariae Hille Ris Lambers, 1939 to Uroleucon Mordvilko, 1914
rapunculoides Börner (altered by Börner in 1952 to rapunculoidis), 1939 to Uroleucon (Uromelan)
reynoldensis Olive, 1965 to Uroleucon Mordvilko, 1914
richardsi Robinson, 1964 to Uroleucon (Lambersius)
russellae Hille Ris Lambers, 1960 to Uroleucon Mordvilko, 1914
saussureae Takahashi, 1962 to Uroleucon Mordvilko, 1914
similis Hille Ris Lambers, 1935 to Uroleucon (Uromelan)
solirostratus Richards, 1972 to Uroleucon Mordvilko, 1914
sonchi subsp. afghanica Narzikulov ex Narzikulov & Umarov, 1972 to Uroleucon Mordvilko, 1914
telekiae Holman, 1965 to Uroleucon Mordvilko, 1914
torajicolus Paik, 1965 = Uroleucon kikioensis (Shinji, 1942)
trachelii Börner, 1939 = Uroleucon (Uromelan) nigrocampanulae (Theobald, 1928)
tripartitus Ivanovskaja, 1971 to Uroleucon (Uromelan)
tuataiae Olive, 1963 to Uroleucon (Uromelan)
vernonicola Holman, 1974 to Uroleucon Mordvilko, 1914
wahinkae Hottes, 1933 to Kakimia Hottes & Frison, 1931
wakibae Hottes, 1934 to Fimbriaphis Richards, 1959

Subgen. Kakimia Hottes & Frison, 1931

D. (Kakimia) takalus Hottes, 1933 to Kakimia Hottes & Frison, 1931

Subgen. Lambersius Olive, 1965

D. (Lambersius) bradburyi Olive, 1965 to Uroleucon (Lambersius)
(Lambersius) brevitarsus Robinson, 1974 = Uroleucon (Lambersius zymozionense (Knowlton, 1946)
(Lambersius) crepusisiphon Olive, 1965 to Uroleucon (Lambersius)
(Lambersius) macgillivrayae Olive, 1967 to Uroleucon (Lambersius)
(Lambersius) zayasi Holman, 1974 to Uroleucon (Lambersius)

Subgen. Uromelan Mordvilko, 1914

D. (Uromelan) cephalonopli Takahashi, 1962 to Uroleucon (Uromelan)
(Uromelan) compositae subsp. evansi Eastop, 1956 to Uroleucon (Uromelan)
(Uromelan) dravidiana David, 1956 = Uroleucon (Uromelan) minutum (van der Goot, 1916)

D. (Uromelan) helenae Hille Ris Lambers, 1950 to Uroleucon (Uromelan)
(Uromelan) hieracioides Bozhko, (1957) 1961 to Uroleucon (Uromelan)
(Uromelan) jaceae subsp. aeneus Hille Ris Lambers, 1939 to Uroleucon (Uromelan)
(Uromelan) jaceae subsp. macrosiphon Hille Ris Lambers, 1939 to Uroleucon (Uromelan)
(Uromelan) jaceae subsp. reticulatus Hille Ris Lambers, 1939 to Uroleucon (Uromelan)
(Uromelan) montanivorus Mosbacher, 1959 to Uroleucon (Uromelan)
(Uromelan) riparius Stroyan, 1955 to Uroleucon (Uromelan)
(Uromelan) stachydis Bozhko, (1957) 1961 to Uroleucon (Uromelan)
(Uromelan) triphyllae Miyazaki, 1960 = Uroleucon (Uromelan) adenophorae (Matsumura, 1918)

DACTYNUS errore pro Dactynotus Rafinesque, 1818

DAKTULOSPHAIRA Shimer, 1866
Prairie Farmer for 1866: 365 lapsus pro Dactylosphaera Shimer, 1867
The articles on pp. 290 and 365 of the Prairie Farmer for 1866, and in Proc. Acad. nat. Sci. Philad. 19: 2-11, when read in conjunction indicate as the type species of Dactylosphaera: globosum (Shimer, 1867)*, vide Dactylosphaera.

DASIA Gomez-Menor, 1951 nec Gray, 1839; van der Goot, 1918
Eos, tomo extraordinaria, 1950: 98
Type species Brachyunguis carthami Das, 1918
= Alhambra Gomez-Menor, 1958 nomen novum = Protaphis Börner, 1952

D. centaurea Gomez-Menor, 1951 to Aphis (Protaphis)
ignatii Gomez-Menor, 1951 to Aphis (Protaphis)

DASIA van der Goot ex Das, 1918 nec Gray, 1839
Mem. Indian Mus. 6: 152
Type species Pemphigus aedificator Buckton, 1839 = Aphis pistaciae Linnaeus, 1767
= Baizongia Bondani, 1848

* Russell, L.M., 1975 (J. Wash. Acad. Sci. 64: 303-304) argues that Pemphigus vitifoliae is the type species of Daktulosphaira while we took the view that Shimer (1867, Proc. Acad. nat. Sci. Philad. 19: 2-11) had selected globosus as type, and had as "first reviser" clarified the obscure situation he created in the Prairie Farmer the previous year. Whatever the eventual interpretation, it may well only be of academic interest, as no convincing characters for removing either of these species from the genus Phylloxera are known.

DASYAPHIS Takahashi, 1938
Tenthredo 2: 13
Type species Tuberocorpus onigurumi Shinji, 1932 = Glyphina rhusae Shinji, 1922
*Echinaphis* Mordvilko, 1929 nec Cockerell, 1913
*Sinocallis* Tseng & Tao, 1938
*Tuberocorpus* Shinji, 1932 nec 1929

D. rhusae (Shinji, 1922) (Glyphina) BM HRL
*coreanus* Paik, 1965 (*Tuberocorpus*)
*mirabilis* Tseng & Tao, 1938 (*Sinocallis*)
*onigurumi* Shinji, 1932 (*Tuberocorpus*)
*ussuriensis* Mordvilko, 1929 (*Echinaphis*)

DAVATCHIAPHIS Remaudière, 1964
Revue Path. vég. Ent. agric. Fr. 43: 63
Type species Davatchiaphis persicus Remaudière, 1964

D. persica Remaudière, 1964 HRL

DAVIDSONIA Essig, 1912 nec Bouchard-Chantereaux, 1849
Pomona Coll. J. Ent. 4: 827
Type species Davidsonia saliciradicis Essig, 1912
= Fullawaya Essig, 1912

D. saliciradicis Essig, 1912 to Fullawaya Essig, 1912

DAVISIA del Guercio, 1909
Redia 5: 185, described as subg. of Lachnus Burmeister, 1835
Type species Aphis caryae Harris, 1841 since both the originally included species, Lachnus longistigma Monell and L. platanicola Riley are junior synonyms of Aphis caryae Harris, 1841
= Longistigma Wilson, 1909

DEBILISIPHON Shaposhnikov, 1950
Ént. Obozr. 31: 226
Type species Debilisiphon umbelliferarum Shaposhnikov, 1950
= Aphis Linnaeus, 1758

D. peucedanicarvifoliae Bozhko ex Shaposhnikov, 1964 to Aphis Linnaeus, 1758
seselli Bozhko, 1959 misprint for seselii = Aphis seselii (Bozhko, 1959)
umbelliferarum Shaposhnikov, 1950 to Aphis Linnaeus, 1758

DECOROSIPHON Börner, 1939
Arb. physiol. angew. Ent. Berl. 6: 78
Type species Decorosiphon corynothrix Börner, 1939

D. corynothrix Börner, 1939 BM HRL

DEFRACTOSIPHON Börner, 1950
Neue europäische Blattlausarten, privately published, p. 10
Type species Defractosiphon franzi Börner, 1950

D. brevisiphon Mamontova, 1968 HRL
franzi Börner, 1950
*? peucedanii* Mamontova, 1968 HRL
*? rugosus* F.P. Müller, 1973 paratype BM HRL
peucedanii Mamontova, 1968 = ? franzi Börner, 1950
rugosus F.P. Müller, 1973 = ? franzi Börner, 1950

DELPHINIOBIUM Mordvilko, 1914
Faune Russie, Ins. Hém. 1:50, 65
Type species Delphiniobium aconiti (van der Goot) = Rhopalosiphum aconiti van der Goot, 1912 = Myzus junackianus Karsch, 1887, see caption to figure 34 on page 50, and also the only species included by Mordvilko, 1928: 190

D. bogdouli Szelegiewicz, 1969
canadense (Robinson, 1968) (Kakimia) paratypes BM HRL
carpaticae Mamontova, 1966 BM
hanla Paik, 1971 HRL
junackianum (Karsch, 1887) (Myzus) BM HRL
*aconiti* van der Goot, 1912 (*Rhopalosiphum*)
junackianum subsp. sylvanae (Knechtel & Manolache, 1943) (Megoura) HRL
*aconiti* subsp. *sylvanae* Knechtel & Manolache, 1943 (*Megoura*) HRL
lycoctoni Börner, 1950 BM HRL
pulmonariae Börner, 1942 type of Amphorosiphon Hille Ris Lambers, 1949 q.v.
yezoense Miyazaki, 1971 BM HRL

DENTATUS van der Goot, 1913 nec Gray, 1847
Tijdschr. Ent. 56: 92, 146
Type species Aphis sorbi Kaltenbach, 1843 (but van der Goot had Myzus plantagineus Passerini, 1860, not Aphis sorbi).
= Pomaphis Börner, 1939

D. affinis Mordvilko, 1928 to Dysaphis Börner, 1931
bicolor Nevsky, 1929 (type of Nevskyaphis Shaposhnikov, 1950) to Brachycaudus (Nevskyaphis)
capsellae Mordvilko ex Nevsky, 1929 to Dysaphis Börner, 1931
communis Mordvilko, 1928 = Dysaphis devecta (Walker, 1849)
dodikovi Rusanova, 1943 nomen nudum
dubius Mordvilko ex Nevsky, 1929 = ? Dysaphis (Pomaphis) reaumuri (Mordvilko, 1928)
ferulae Nevsky, 1929 to Dysaphis Börner, 1931
incertus Mordvilko ex Nevsky, 1929 = Dysaphis affinis (Mordvilko, 1928)
longipilosus Mordvilko, 1918 to Dysaphis Börner, 1931
longipilosus subsp. turanicus Mordvilko, 1928 to Dysaphis Börner, 1931
malicola Mordvilko, 1928 = Dysaphis (Pomaphis) plantaginea (Passerini, 1860)
malus Nevsky, 1929 = Dysaphis (Pomaphis) pyri (Boyer de Fonscolombe, 1841)
microsiphon Nevsky, 1929 (type of Cotoneasteria Shaposhnikov, 1964) to Dysaphis Börner, 1931
pilosus Mordvilko ex Nevsky, 1929 (type of Mordvilkomemor Shaposhnikov, 1950) = Brachycaudus (Mordvilkomemor) pilosus (Mordvilko ex Nevsky, 1929)
plumbicolor Nevsky, 1929 = Dysaphis (Pomaphis) plantaginea (Passerini, 1860)
pulverinus Nevsky, 1929 to Dysaphis Börner, 1931
radicivorans Nevsky, 1929 to Dysaphis Börner, 1931
reaumuri Mordvilko, 1928 to Dysaphis (Pomaphis)

DERMAPHIS Takahashi, 1958
Insecta Matsum. 22: 9
Type species Dermaphis japonensis Takahashi, 1958

D. japonensis Takahashi, 1958 HRL
takahashii (Strand, 1929) (Astegopteryx) BM HRL

DIATOMYZUS Heie, 1970
Meddr dansk geol. Foren. 20: 163
Type species Diatomyzus eocaenius Heie, 1970
Fossil. Lower Eocene

DICHAITOPHORUS Shinji, 1927
Bull. Morioka imp. Coll. For. Agric. 11: 48
Type species Chaitophoraphis acerifloris Shinji, 1923 = Yamatocallis hirayamae Matsumura, 1917
= Yamatocallis Matsumura, 1917

DIELCYSMURA Mordvilko, 1914
Faune Russie, Ins. Hém. 1: 65
Type species Aphis millefolii de Geer, 1773
= Macrosiphoniella del Guercio, 1911

DILACHNUS Baker, 1919 nec Fairmaire, 1896
Can. Ent. 5: 253
Type species Lachniella gracilis Wilson, 1919
= Cinara Curtis, 1835

D. callitris Froggatt, 1927 = Cinara tujafilina (del Guercio, 1909)
formosanus Takahashi, 1924 to Cinara Curtis, 1835
krishni George, 1927 = Pyrolachnus pyri (Buckton, 1899)
liquidambarus Takahashi, 1925 to Longistigma Wilson, 1909
momii Shinji, 1924 = Cinara longipennis (Matsumura, 1917)
orientalis Takahashi, 1925 to Cinara Curtis, 1835
piniformosanus Takahashi, 1923 to Cinara Curtis, 1835
pubescens Wellenstein, 1930 = Cinara pruinosa (Hartig, 1841)
quercihabitans Takahashi, 1924 = Cinara longipennis (Matsumura, 1917)
radicicolus Wellenstein, 1930 = Cinara pruinosa (Hartig, 1841)
taiwanus Takahashi, 1925 to Cinara Curtis, 1835

DIMERAPHIS Becker-Migdisova, 1973
Paleont. Zh., 1973 (3): 87, described as subg. of Electrocallis Heie, 1967
Type species Electrocallis (Dimeraphis) arnoldi Becker-Migdisova, 1973
Fossil

DINIPPONAPHIS Takahashi, 1962
Bull. Univ. Osaka Prefect. B. 13: 7
Type species Nipponaphis autumnus Monzen, 1934

D. autumna (Monzen, 1934) (Nipponaphis) BM HRL
*yanonis* var. *autumna* Monzen, 1934 (*Nipponaphis*)

DINOLACHNUS Börner, 1940
Neue Blattläuse aus Mitteleuropa, privately published, p. 1
Type species Lachniella cilicica var. cecconii del Guercio, 1909 = Lachnus confinis Koch, 1856
= Cinara Curtis, 1835

DIPHORODON Börner, 1939
Arb. physiol. angew. Ent. Berl. 6: 19
Type species Phorodon cannabis Passerini, 1860
= Paraphorodon Tseng & Tao, 1938

DIPHYLLAPHIS Takahashi, 1960
Kontyû 28: 12
Type species Phyllaphis konarae (Shinji, 1924) = Phloeomyzus konarae Shinji, 1924
*Nymphaphis* Takahashi, 1960

D. alba Takahashi, 1960 HRL
konarae (Shinji, 1924) (Phloeomyzus) BM HRL
microtrema Quednau, 1971 paratypes BM
mordvilkoi (Aizenberg, 1932) (Stegophylla) BM HRL
quercus (Takahashi, 1960) (Nymphaphis) HRL

DITRICHOSIPHON RayChaudhuri, 1956
Zool. Verh., Leiden 31: 8, 10, 21, described as subg. of Eutrichosiphum Essig & Kuwana, 1918
Type species Eutrichosiphum elongatum Takahashi, 1940
Here used as subg. of Eutrichosiphum Essig & Kuwana, 1918

DIURAPHIS Aizenberg, 1935
Zap. biol. Sta Bolcheva 7-8: 157
Type species Brachycolus noxius Mordvilko ex Kurdjumov, 1913
*Cavahyalopterus* Mimeur, 1942
*Cuernavaca* J.M. Baker, 1934 nec Kirkaldy, 1913

D. mexicana (J.M. Baker, 1934) (Cuernavaca) HRL
muehlei (Börner, 1950) (Brachycolus) BM HRL
nodulus (Richards, 1959) (Brachycolus) HRL
noxia (Mordvilko ex Kurdjumov, 1913) (Brachycolus) BM HRL
*graminearum* Mimeur, 1942 (*Cavahyalopterus*)

Subgen. Holcaphis Hille Ris Lambers, 1939

D. (Holcaphis) agrostidis (Muddathir, 1965) (Holcaphis) types BM HRL
(Holcaphis) bromicola (Hille Ris Lambers, 1959) (Holcaphis) types BM HRL
(Holcaphis) calamagrostis (Ossiannilsson, 1959) (Holcaphis) paratypes BM HRL
*? arundinis* Fabricius, 1775 (*Aphis*)

D. (Holcaphis) frequens (Walker, 1848) (Aphis) type BM HRL
*korotnewi* Mordvilko, 1901 (*Brachycolus*)
(Holcaphis) holci (Hille Ris Lambers, 1956) (Holcaphis) BM HRL
*holci* Hardy of Hille Ris Lambers, 1939 (*Holcaphis*)
(Holcaphis) tritici (Gillette, 1911) (Brachycolus) BM HRL

DIVEROSIPHUM Shinji, 1922
Zool. Mag., Tokyo 34: 791
Type species Diverosiphum kunugii Shinji, 1922 = Cervaphis quercus Takahashi, 1918
= Cervaphis van der Goot, 1917

D. kunugii Shinji, 1922 = Cervaphis quercus (Takahashi, 1918)

DORALIDA Börner, 1950
Neue europäische Blattlausarten, privately published, p. 8, described as subg. of Pergandeida Schouteden, 1903
Type species Aphis loti Theobald, 1927 which is a reference to Theobald's account of Aphis loti Kaltenbach, 1862
= Aphis Linnaeus, 1758

D. comosa Börner, 1950 to Aphis Linnaeus, 1758
coronillae Börner, 1950 = Aphis vineti Hoffmann, 1972
klimeschi Börner, 1950 to Aphis Linnaeus, 1758
loti subsp. gollmicki Börner, 1952 = Aphis craccivora Koch, 1854
trifolii Börner, 1950 = Aphis coronillae Ferrari, 1872

DORALINA Börner, 1940
Neue Blattläuse aus Mitteleuropa, privately published, p. 3
Type species Aphis frangulae Koch, 1855 = Aphis frangulae Kaltenbach, 1845
= Aphis Linnaeus, 1758

D. alchemillae Börner, 1940 to Aphis Linnaeus, 1758
calaminthae Börner, (errore 1940) 1952, p. 222 to Aphis Linnaeus, 1758
cerastii Börner, 1950 to Aphis Linnaeus, 1758
gentianae Börner, 1940 to Aphis Linnaeus, 1758
lambersi Börner, 1940 to Aphis Linnaeus, 1758
lamiorum Börner, 1950 to Aphis Linnaeus, 1758
leontodontis Börner, 1950 to Aphis Linnaeus, 1758
mirifica Börner, 1950 to Aphis Linnaeus, 1758
molluginis Börner, 1950 to Aphis Linnaeus, 1758

D. picridis Börner, 1950 = Aphis stroyani Szelegiewicz, 1961
proffti Börner, 1942 to Aphis Linnaeus, 1758
salsolae Börner, 1940 = Aphis craccivora Koch, 1854
schilderi Börner, 1940 to Aphis Linnaeus, 1758
schneideri Börner, 1940 to Aphis Linnaeus, 1758
selini Börner, 1940 to Aphis Linnaeus, 1758
stachydis Börner, 1940 = Aphis stachydis Mordvilko, 1929
subnitida Börner, 1940 to Aphis Linnaeus, 1758
taraxacicola Börner, 1940 (type of Papillaphis Borner, 1952; and of Tuberculaphis Börner, 1952) to Aphis Linnaeus, 1758
teucrii Börner, 1942 to Aphis Linnaeus, 1758
thermophila Börner, 1950 to Aphis Linnaeus, 1758
thomasi Börner, 1950 to Aphis Linnaeus, 1758
umbrella Börner, 1950 to Aphis Linnaeus, 1758
versicolor Börner, 1950 to Aphis Linnaeus, 1758
verticillatae Börner, 1940 to Aphis Linnaeus, 1758
volutans Börner, 1940 to Aphis Linnaeus, 1758
wellensteini Börner, 1950 to Aphis Linnaeus, 1758

DORALIS Leach, 1827
In Risso, Hist. nat. princip. Prod. Europ. mérid. 5: 217
Type species Aphis dauci Fabricius, 1775, the only described and known species included: see China, 1961: 143. Börner, 1930: 161 invalidly selected Aphis rumicis Leach, 1827, a nomen nudum, as type, and used the genus with Aphis rumicis Linnaeus, 1758 as type species.
Suppressed by International Commission on Zoological Nomenclature, opinion 646, Bull. zool. Nomencl., 20: 31 (1963)

D. avicularis Hille Ris Lambers, 1931 = Aphis polygonata Nevsky, 1929
bupleuri Börner, 1932 to Aphis Linnaeus, 1758
corniella Hille Ris Lambers, 1935 (type of Comaphis Börner, 1940) = Aphis salicariae Koch, 1855
gambir Börner ex Schneider, 1940 nomen nudum
hartigi Hille Ris Lambers, 1931 to Aphis (Protaphis)
janischi Börner, 1940 to Aphis Linnaeus, 1758
meliloti Börner, 1939 = Aphis craccivora Koch, 1854
pini Leach, 1827 nomen nudum
ruborum Börner, 1932 to Aphis Linnaeus, 1758
rumicis Leach, 1827 nomen nudum
senecionis Borner, 1940 = Aphis cacaliasteris subsp. helvetica Hille Ris Lambers, 1947

D. setacea Hille Ris Lambers, 1935 = Aphis lantanae Koch, 1854
solidaginis Börner, 1950 to Aphis Linnaeus, 1758
ulmi Leach, 1827 nomen nudum
vaccinii Börner, 1940 to Aphis Linnaeus, 1758
xylostei Börner, 1950 to Aphis Linnaeus, 1758

DORAPHIS Matsumura & Hori ex Hori, 1929
Trans. Sapporo nat. Hist. Soc. 10: 112
Type species Doraphis populi Matsumura & Hori, 1929 = Sphaerococcus populi Maskell, 1898
*Paracerataphis* Mordvilko, 1929
populi (Maskell, 1898) (*Sphaerococcus*) BM HRL
*dorocola* Shinji, 1941 (*Thoracaphis*)
*pirifoliae* Shinji, 1924 (*Cerataphis*)
*populi* Matsumura & Hori ex Hori, 1929
*tremulae* Mordvilko, 1929 (*Paracerataphis*)
*yamanarashii* Shinji, 1941 (*Thoracaphis*)
populi Matsumura & Hori ex Hori, 1929 = populi (Maskell, 1898)

DREPANAPHIS del Guercio, 1909
Riv. Patol. veg., Padova, N.S. 4: 49
Type species Drepanosiphum acerifolii (Thomas, 1878) = Siphonophora acerifoliae Thomas, 1878
*Phymatosiphum* Davis, 1909
D. acerifoliae (Thomas, 1878) (Siphonophora) BM HRL
*allegheneyensis* Miller, 1936
allegheneyensis Miller, 1936 = acerifoliae (Thomas, 1878)
carolinensis Smith, 1941 paratype BM HRL
choanotricha Smith & Dillery, 1968 paratype BM HRL
granovskyi Smith & Knowlton, 1943 paratype BM HRL
idahoensis Smith & Dillery, 1968 paratype BM HRL
kanzensis Smith, 1941 BM HRL
keshenae Granovsky, 1931 BM HRL
knowltoni Smith & Dillery, 1968 paratype BM HRL
minutus Davis, 1910 (type of Shenahweum Hottes & Frison, 1931) to Drepanaphis (Shenahweum)
monelli (Davis, 1909) (Phymatosiphum) BM HRL
nigricans Smith, 1941 paratype BM HRL
pallida Richards, 1968

D. parva Smith, 1941 paratype BM HRL
    *? rubrum* Smith, 1941 HRL
  rubra Smith, 1941 = ? parva Smith, 1941
  sabrinae Miller, 1937 BM HRL
  saccharini Smith & Dillery, 1968 paratype BM HRL
  sauteri Takahashi, 1927 to Yamatocallis Matsumura, 1917
  simpsoni Smith, 1959 paratypes BM HRL
  spicata Smith, 1941 paratype BM HRL
  tissoti Smith, 1945 paratype BM
  tokyoensis Takahashi, 1923 to Yamatocallis Matsumura, 1917
  utahensis Smith & Knowlton, 1943 paratype BM HRL

Subgen. Shenahweum Hottes & Frison, 1931

D. (Shenahweum) minuta Davis, 1910 (Drepanaphis) paratype BM HRL

DREPANIELLA del Guercio, 1913
  Redia 9: 188
  Type species Aphis viciae Kaltenbach, 1843 = Megoura viciae Buckton, 1876
  = Megoura Buckton, 1876

DREPANOSIPHON = incorrect spelling of Drepanosiphum Koch, 1855

DREPANOSIPHONIELLA Davatchi, Hille Ris Lambers & Remaudière, 1957
  Tijdschr. Ent. 100: 125
  Type species Drepanosiphoniella aceris Davatchi, Hille Ris Lambers & Remaudière, 1957

D. aceris Davatchi, Hille Ris Lambers & Remaudière, 1957 paratypes BM HRL
  aceris subsp. fugans Remaudière & Leclant, 1972 BM HRL

DREPANOSIPHUM Koch, 1855
  Die Pflanzenläuse Aphiden, Nürnberg, p. 201
  Type species Aphis platanoides Schrank, 1801 = Aphis platanoidis Schrank, 1801

D. acerinum (Walker, 1848) (Aphis) lectotype BM HRL
    *gracilis* Börner, 1940
  aceris Koch, 1855 BM HRL
  braggii Gillette, 1907 BM HRL
  californicum Mordvilko, 1928
  dixoni Hille Ris Lambers, 1971 BM HRL
  gracilis Börner, 1940 = acerinum (Walker, 1848)

D. oregonensis Granovsky, 1939 BM HRL
*steveni* Bozhko, (1957) 1961 HRL
*zimmermanni* Börner, 1940
platanoidis (Schrank, 1801) (Aphis) BM HRL
*aceris pseudoplatani* Mosley, 1841 (*Aphis*)
*platanoidis* Hartig, 1841 (*Aphis*)
*smaragdinum* Koch, 1855
*tiliae* Koch, 1855
quercifoliae (Walsh, 1863) (Aphis)
smaragdinum Koch, 1855 = platanoidis (Schrank, 1801)
steveni Bozhko, (1957) 1961 = oregonensis Granovsky, 1939
tiliae Koch, 1855 = platanoidis (Schrank, 1801)
zimmermanni Börner, 1940 = oregonensis Granovsky, 1939

DREPANPHIS errore pro Drepanaphis

DRYAPHIS Amyot, 1847
Annls. Soc. ent. Fr. (2) 5: 481
Invalid

DRYAPHIS Kirkaldy, 1904
Entomologist 37: 279
Type species Aphis roboris Linnaeus, 1758
= Lachnus Burmeister, 1835

D. cerricola del Guercio, 1909 = Lachnus roboris (Linnaeus, 1758)
ilicina del Guercio, 1909 = Lachnus roboris (Linnaeus, 1758)
iliciphila del Guercio, 1909 to Lachnus Burmeister, 1835
minor del Guercio, 1909 = Lachnus iliciphilus (del Guercio, 1909)
roboris subsp. nigra del Guercio, 1909 = Lachnus roboris (Linnaeus, 1758)

DRYOBIUS Koch, 1855 nec Le Conte, 1850
Die Pflanzenläuse Aphiden, Nürnberg, p. 225
Type species Aphis roboris Linnaeus, 1758
= Lachnus Burmeister, 1835

amygdali van der Goot, 1912 = Pterochloroides persicae (Cholodkovsky, 1899)
cistatus Buckton, 1881 = Cinara costata (Zetterstedt, 1828)
croaticus Koch, 1855 = Lachnus roboris (Linnaeus, 1758), ♂
riparius Snellen van Vollenhoven, 1862 = Tuberolachnus salignus (Gmelin, 1790)

DRYOMYZUS Hille Ris Lambers, 1948
Trans. R. ent. Soc. Lond. 99: 285, described as subg. of Myzocallis Passerini, 1860
Type species Myzocallis (Dryomyzus) glandulosus Hille Ris Lambers, 1948
= Myzocallis Passerini, 1860

D. polychaetus David, 1969 to Myzocallis Passerini, 1860

DRYOPEIA Kirkaldy, 1904
Entomologist 37: 279
Type species Endeis bella Koch, 1857 = Aphis ulmi Linnaeus, 1758
= Tetraneura Hartig, 1841

D. hirsuta Baker, 1921 (type of Tetraneurella Hille Ris Lambers, 1970) = Tetraneura (Tetraneurella) nigriabdominalis (Sasaki, 1899)
morrisoni Baker, 1919 type of Paracolopha Hille Ris Lambers, 1966 q.v.

DUROCAPILLATA Knowlton, 1927
Ann. ent. Soc. Am. 20: 229
Type species Durocapillata utahensis Knowlton, 1927

D. utahensis Knowlton, 1927 paratypes BM HRL

DYSAPHIS Börner, 1951
Anz. Schädlingsk. 7: 9
Type species Aphis angelicae Koch, 1854 (see Stroyan, 1960: 97)
*Annaja* Börner, 1952
*Cotoneasteria* Shaposhnikov, 1964
*Crataegaria* Shaposhnikov, 1964
*Umbelliferaria* Shaposhnikov, 1964

D. affinis (Mordvilko, 1928) (Dentatus) BM HRL
*incertus* Mordvilko ex Nevsky, 1929 (*Dentatus*)
aizenbergi (Shaposhnikov, 1949) (type of Umbelliferaria Shaposhnikov, 1964) (Yezabura) HRL
albocinerea (Hille Ris Lambers, 1956) (Sappaphis (Zinia)) cotypes BM HRL
angelicae (Koch, 1854) (Aphis) BM HRL
annulata (Börner, 1950) (Yezabura)
anthrisci Börner, 1950 BM HRL
anthrisci subsp. chaerophyllina Shaposhnikov, 1961 HRL
anthrisci subsp. majkopica Shaposhnikov, 1961 HRL
apiifolia (Theobald, 1923) (Anuraphis) types BM HRL
*ferruginea-striata* Essig, 1941 (*Aphis*) HRL

D. apiifolia subsp. petroselini (Börner, 1950) (*Yezabura*) BM HRL
*inculta* subsp. *nudicaulinum* Börner, 1950 (*Yezabura*)
*inculta* subsp. *petroselini* Börner, 1950 (*Yezabura*)
atina A.K. Ghosh, R.C. Basu & RayChaudhuri, 1969 paratype BM
bonomii (Hille Ris Lambers, 1935) (Yezabura) BM HRL
brachycyclica Shaposhnikov, 1961 HRL
brancoi (Börner, 1950) (Yezabura) BM HRL
*pomaria* Shaposhnikov, 1950 (*Yezabura*)
brancoi subsp. rogersoni (Stroyan, 1955) (Sappaphis) BM HRL
bunii Shaposhnikov, 1956
candicans (Passerini, 1879) (Aphis)
capsellae (Mordvilko ex Nevsky, 1929) (Dentatus)
caucasica Shaposhnikov, 1974
centaureae (Börner, 1950) (Chomaphis)
cephalariae (Narzikulov, 1953) (Yezabura)
cephalarioides Shaposhnikov, 1956 HRL
chaerophylli (Börner, 1940) (Yezabura) BM HRL
choerophylli (Börner) emendation or lapsus for chaerophylli (Börner, 1940)
coreanus Paik, 1972 p. 357 nomen nudum?
cousiniae Narzikulov, 1967 HRL
cousiniae subsp. minor Narzikulov, 1967
crataegi (Kaltenbach, 1843) (Aphis) type of Crataegaria Shaposhnikov, 1964 BM HRL
*crataegi* subsp. *aegopodii* Börner, 1950 (*Yezabura*)
*crataegi* subsp. *anthrisci* Börner, 1950 (*Yezabura*)
*dauci* Goureau, 1867 (*Forda*)
crataegi subsp. aethusae (Börner, 1950) (Yezabura) BM HRL
crataegi subsp. kunzei (Börner, 1950) (Yezabura) BM HRL
*kunzei* Börner, 1950 (*Yezabura*)
crataegi subsp. siciliensis (Theobald, 1927) (Anuraphis) BM HRL
*siciliensis* Theobald, 1927 (*Anuraphis*)
crathaegaria (del Guercio, 1930) (Anuraphis (Macchiatiella))
crathaegiphila (del Guercio, 1930) (Anuraphis)
crithmi (Buckton, 1886) (Aphis) types BM HRL
cynarae (Theobald, 1915) (Aphis) types BM HRL
devecta (Walker, 1849) (Aphis) BM HRL
*communis* Mordvilko, 1928 (*Dentatus*)
emicis (Mimeur, 1935) (Anuraphis) BM HRL
*rheicola* Daniarova, 1971 HRL
eremuri (Narzikulov, 1954) (Yezabura)

D. eremuri subsp. balsunensis Narzikulov, 1970
ferulae (Nevsky, 1929) (Dentatus)
 ? *mirabilis* Nevsky, 1951 (*Yezabura*)
flava Shaposhnikov, 1956
foeniculus (Theobald, 1923) (Anuraphis) types BM HRL
ghanii Stroyan, 1963 = montemartinii subsp. ghanii Stroyan, 1963
hirsutissima (Börner, 1940) (Yezabura) BM HRL
inculta (Walker, 1848) (Aphis) nomen dubium, used by Börner, 1950, to replace apiifolia (Theobald, 1923)
krumbholzi F.P. Müller, 1961, paratypes BM HRL
lappae (Koch, 1854) (Aphis) BM HRL
 *prolappae* (del Guercio, 1930) (*Anuraphis (Macchiatiella)*)
lappae subsp. cirsii (Börner, 1950) (Chomaphis) BM
 ? *lappae* f. *cirsii* Passerini, 1874 (*Aphis*)
laserpitii (Börner, 1950) (Yezabura) BM HRL
lauberti (Börner, 1940) (Yezabura) BM HRL
leefmansi (Hille Ris Lambers, 1954) (Sappaphis) cotypes BM HRL
libanotidis Shaposhnikov, 1956
ligulariae (Narzikulov, 1954) (Yezabura) HRL
longipilosa (Mordvilko, 1928) (Dentatus) HRL
longipilosa subsp. turanica (Mordvilko, 1928) (Dentatus) HRL
mamontovae Shaposhnikov, 1956 = newskyi subsp. mamontovae Shaposhnikov, 1956
meridialis Shaposhnikov, 1964 HRL
 *radicola* subsp. *meridialis* Shaposhnikov, 1964 HRL
microsiphon (Nevsky, 1929) (Dentatus)
 *subspinosus* Nevsky, 1951 (*Brachycaudus*)
mirabilis (Nevsky, 1951) (Yezabura) = ? ferulae (Nevsky, 1929)
montemartinii (del Guercio, 1913) (Chaitophorus) BM
montemartinii subsp. ghanii Stroyan, 1963 type BM HRL
 *ghanii* Stroyan, 1963 type BM
multisetosa A.N. Basu, 1969 HRL
narzikulovi Shaposhnikov, 1956 HRL
neostroyani Ilharco, 1965 BM HRL
newskyi (Börner, 1940) (Neanuraphis) BM HRL
 *sphondylii* Stroyan, 1952 (*Sappaphis*) paratypes BM HRL
newskyi subsp. mamontovae Shaposhnikov, 1956
 *mamontovae* Shaposhnikov, 1956
newskyi subsp. ossiannilssoni Stroyan, 1961
 *ossiannilssoni* Stroyan, 1961

D. orientalis Mordvilko ex Shaposhnikov, 1956
ossiannilssoni Stroyan, 1961 = newskyi subsp. ossiannilssoni Stroyan, 1961
papillata (Nevsky, 1951) (Yezabura)
pavlovskyana Narzikulov, 1957 to Dysaphis (Pomaphis)
physocaulis Shaposhnikov, 1974
pseudomolli Narzikulov, 1961 (type of Neodysaphis Narzikulov, 1961) to Dysaphis (Neodysaphis)
pulverina (Nevsky, 1929) (Dentatus)
pulverina subsp. iranica Stroyan, 1972 HRL
pyraria Narzikulov, 1961
radicicola Mordvilko, Schouteden, 1906, emendation of radicola (Mordvilko, 1897)
radicivorans (Nevsky, 1929) (Dentatus) HRL
radicola (Mordvilko, 1897) (Aphis) BM HRL
*lapathi* Börner & Blunck, 1916 (*Aphis*)
*leontodoniella* Theobald, 1915 (*Aphis*)
*radicicola* Schouteden, 1906 (*Aphis*)
*rumicella* Theobald, 1924 (*Anuraphis*) BM
*rumiciphila* del Guercio, 1930 (*Anuraphis (Macchiatiella)*)
radicola subsp. meridialis Shaposhnikov, 1964 = meridialis Shaposhnikov, 1964
ranunculi (Kaltenbach, 1843) (Aphis) BM HRL
*nigra* Theobald, 1916 (*Aphis*)
*ranunculi* subsp. *bulbosi* Börner, 1950 (*Yezabura*)
rheicola Daniarova, 1971 = emicis (Mimeur, 1935)
rumecicola (Hori, 1927) (Anuraphis) HRL
sorbiarum (Narzikulov, 1954) (Yezabura)
stroyani Shaposhnikov, 1956
taisetsusana (Miyazaki, 1971) (Elatobium) HRL
tulipae (Boyer de Fonscolombe, 1841) (Aphis) BM HRL
*gladioli* Felt, 1908 (*Aphis*)
*iridis* del Guercio, 1900 (*Aphis*)
*palaestinensis* Hille Ris Lambers, 1948 (*Sappaphis*) HRL
unicauli Mukhamediev, 1964
uralensis Shaposhnikov, 1956
valentinae Shaposhnikov, 1974
vandenboschi Stroyan, 1972 HRL
zini Shaposhnikov, 1956

Subgen. Anuromyzus Shaposhnikov, 1953

D. (Anuromyzus) cotoneasteris Shaposhnikov, 1959 to Anuromyzus Shaposhnikov, 1959

Subgen. Neodysaphis Narzikulov, 1961

D. (Neodysaphis) pseudomolli Narzikulov, 1961 HRL

Subgen. Pomaphis Börner, 1939

D. (Pomaphis) ariae (Börner, 1950) (Sappaphis)
(Pomaphis) aucupariae (Buckton, 1879) (Aphis) lectotype BM HRL
*appelii* Börner, 1926 (*Anuraphis)*
(Pomaphis) brevirostris (Borner, 1950) (Myzodes (Myzodium)) BM HRL
(Pomaphis) gallica (Hille Ris Lambers, 1955) (Sappaphis) BM HRL
(Pomaphis) maritima (Hille Ris Lambers, 1955) (Sappaphis) BM HRL
(Pomaphis) maritima subsp. glabra (Hille Ris Lambers, 1955 (Sappaphis) HRL
(Pomaphis) parasorbi (Börner, 1952) (Sappaphis)
(Pomaphis) pavlovskyana Narzikulov, 1957 (Dysaphis) HRL
(Pomaphis) plantaginea (Passerini, 1860) (Myzus) BM HRL
*lentiginis* Buckton, 1879 (*Aphis*)
*mali* Ferrari, 1872 (*Myzus*)
*malicola* Mordvilko, 1928 (*Dentatus*)
*padi* del Guercio, 1930 (*Anuraphis (Macchiatiella)*)
? *plantagicola* Takahashi, 1931 (*Myzus*)
? *plantagifoliae* Shinji, 1929 (*Myzus*)
*plumbicolor* Nevsky, 1929 (*Dentatus*)
*pyri* Hartig, 1841 (*Aphis*)
*roseus* Baker, 1921 (*Anuraphis*)
(Pomaphis) plantaginis (Pašek, 1955) (Myzopsis)
(Pomaphis) pyri (Boyer de Fonscolombe, 1841) (Aphis) BM HRL
*hirta* del Guercio, 1930 (*Anuraphis (Macchiatiella)*)
*malus* Nevsky, 1929 (*Dentatus*)
*oxyacanthae* del Guercio, 1930 (*Anuraphis*)
(Pomaphis) reaumuri (Mordvilko, 1928) (Dentatus) HRL
? *dubius* Mordvilko ex Nevsky, 1929 (*Dentatus*)
(Pomaphis) sorbi (Kaltenbach, 1843) (Aphis) BM HRL

DYSAULACORTHUM Börner, 1939
Arb. physiol. angew. Ent. Berl. 6: 82, described as subg. of Aulacorthum Mordvilko, 1914
Type species Aulacorthum (Dysaulacorthum) langei Börner, 1939
= Aulacorthum Mordvilko, 1914

D. boerneri Müller, 1952 = Aulacorthum solani (Kaltenbach, 1843)
langei Börner, 1939 to Aulacorthum Mordvilko, 1914

ECHINAPHIS Cockerell, 1913
Can. Ent. 45: 229
Type species Echinaphis rohweri Cockerell, 1913
Fossil. Miocene

ECHINAPHIS Mordvilko, 1929 nec Cockerell, 1913
Trudȳ prikl. Ént. 14: 35
Type species Echinaphis ussuriensis Mordvilko, 1929 = Glyphina rhusae Shinji, 1922
= Dasyaphis Takahashi, 1938

E. ussuriensis Mordvilko, 1929 = Dasyaphis rhusae (Shinji, 1922)

EICHINAPHIS Narzikulov, 1963
Trudȳ Pamir Biol. Sta. 1: 259
Type species Eichinaphis pamirica Narzikulov, 1963

E. pamirica Narzikulov, 1963 BM HRL

EICHOCHAITOPHORUS Essig, 1912
Pomona Coll. J. Ent. 4: 715, 721
Type species Eichochaitophorus populifolii Essig, 1912
= Chaitophorus Koch, 1854

E. himalayensis Das, 1918 to Chaitophorus Koch, 1854
populifolii Essig, 1912 to Chaitophorus Koch, 1854

EICHOTOPHORUS lapsus pro Eichochaitophorus

ELATOBIUM Mordvilko, 1914
Faune Russie, Ins. Hém. 1 (1): 72
Type species Aphis abietina Walker, 1849
*Neomyzaphis* Theobald, 1926
abietinum (Walker, 1849) (Aphis) lectotype BM HRL
laricis (Rupais, 1974) (Liosomaphis)
momii (Shinji, 1922) (Aphis) BM HRL
*abietifoliae* Shinji, 1923 (*Phorodon*)
*momifoliae* Shinji, 1924 (*Aphis*)

E. piceanum (Inouye, 1939) (Neomyzaphis) HRL
taisetsusanum Miyazaki, 1971 to Dysaphis (Börner, 1931)
trochodendri Takahashi, 1960 BM HRL

ELBOURZAPHIS Remaudière & Davatchi, 1959
Revue Path. vég. Ent. agric. Fr. 38: 135
Type species Elbourzaphis behboudii Remaudière & Davatchi, 1959 = Rhopalosiphoninus (?) platicaudus Narzikulov, 1953
= Amegosiphon Narzikulov, 1958

E. behboudii Remaudière & Davatchi, 1959 = Amegosiphon platicaudus (Narzikulov, 1953)

ELECTROCALLIS Heie, 1967
Spolia zool. Mus. haun. 26: 147
Type species Electrocallis bakeri Heie, 1967
Fossil, baltic amber

ELECTROCORNIA Heie, 1972
Steenstrupia 2: 249
Type species Electrocornia antiqua Heie, 1972
Fossil, baltic amber

ELECTROMYZUS Heie, 1972
Steenstrupia 2: 250
Type species Electromyzus acutirostris Heie, 1972
Fossil, baltic amber

ELEKTRAPHIS Steffan, 1968
Zool. Jb. System. 95: 11
Type species Elektraphis polykrypta Steffan, 1968
Fossil, baltic amber, = Schizoneurites Cockerell, 1915

ENDEIS Koch, 1857 nec Philippi, 1847
Die Pflanzenläuse Aphiden, Nürnberg, p. 312
Type species Endeis bella Koch, 1857 = Aphis ulmi Linnaeus, 1758
= Tetraneura Hartig, 1841

E. bella Koch, 1857 (type of Dryopeia Kirkaldy, 1904) = Tetraneura ulmi (Linnaeus, 1758)
carnosa Buckton, 1883 to Geoica Hart, 1894
formicina Buckton, 1883 = Baizongia pistaciae (Linnaeus, 1767)
pellucida Buckton, 1883 to Geoica Hart, 1894
rosea Koch, 1857 = Tetraneura ulmi (Linnaeus, 1758)

EOESSIGIA David, Rajasingh & Narayanan, 1972
Orient. Insects 6: 35
Type species Eoessigia indica David, Rajasingh & Narayanan, 1972

E. indica David, Rajasingh & Narayanan, 1972 HRL
longicauda (Richards, 1963) (Aspidaphis) paratype BM HRL

EOMACROSIPHON Hille Ris Lambers, 1958
In MacGillivray, Temminckia 10: 24
Type species Macrosiphum nigromaculosum MacDougall, 1926

E. nigromaculosum (MacDougall, 1926) (Macrosiphum) BM HRL

EOMYZUS Takahashi, 1960
Kontyû 28: 227
Type species Myzus nipponicus Moritsu, 1949
= Micromyzodium David, 1959

E. levipes R.C. Basu & RayChaudhuri, 1974 (not seen by us)

EONAPHIS Essig, 1957
Naturaliste malgache 9: 287
Type species Eonaphis pauliani Essig, 1957

E. crotonis Quednau, 1962 paratypes BM HRL
euphorbiae Quednau, 1964 types BM
pauliani Essig, 1957 HRL

EOPINEUS Steffan, 1968
Zoologica 40 (5) (115) p. 5
Type species (chosen by Eastop & Hille Ris Lambers) Chermes strobi Ratzeburg, 1844 = Coccus strobus Hartig, 1837
= Pineus Shimer, 1869
Adelgidae

EOTRAMA HIlle Ris Lambers, 1969
Mitt. schweiz. ent. Ges. 42: 181
Type species Eotrama moerickei Hille Ris Lambers, 1969

E. moerickei Hille Ris Lambers, 1969 HRL

EPAMEIBAPHIS Oestlund, 1922
Rep. Minn. St. Ent. 19: 132
Type species Aphis frigidae Oestlund, 1886

E. atricornis Gillette & Palmer, 1933 metatype BM HRL
frigidae (Oestlund, 1886) (Aphis) BM HRL
*thornleyi* Knowlton, 1946
thornleyi Knowlton, 1946 = frigidae (Oestlund, 1886)
utahensis Knowlton & Smith, 1936 paratypes BM HRL

EPEIMAPHIS lapsus pro Epameibaphis

EPHEDRAPHIS Hille Ris Lambers, 1959
Mitt. schweiz. ent. Ges. 32: 279
Type species Anuraphis ephedrae Nevsky, 1929

E. ephedrae (Nevsky, 1929) (Anuraphis) BM HRL
ephedrae subsp. taurica Mamontova, 1963 HRL
gobica Szelegiewicz, 1963

EPIPEMPHIGUS Hille Ris Lambers, 1966
Tijdschr. Ent. 109: 204
Type species Pemphigus imaicus Cholodkovsky, 1912

E. imaicus (Cholodkovsky, 1912) (Pemphigus) BM HRL
niisimae (Matsumura, 1917) (Pemphigus) HRL

ERICAPHIS Börner, 1939
Arb. physiol. angew. Ent. Berl. 6: 80, described as subg. of Myzaphis van der Goot, 1913
Type species Myzaphis ericae Börner, 1933
*Boreamyzus* Shaposhnikov, 1964

E. empetri Ossiannilsson, 1954 = latifrons (Börner, 1942)
ericae (Börner, 1933) (Myzaphis) BM HRL
latifrons (Börner, 1942) (Ovatus) (type of Boreamyzus Shaposhnikov, 1964) BM HRL
*empetri* Ossiannilsson, 1954
*? lagarriguei* Remaudière, 1952 (*Myzodium*)
scammelli (Mason, 1942) (Myzus) HRL

ERICOBIUM MacGillivray, 1958
Temminckia 10: 12, 25, 41, described as subg. of Masonaphis Hille Ris Lambers, 1939
Type species Amphorophora azaleae Mason, 1925
= Illinoia Wilson, 1910

ERICOLOPHIUM Tao, 1963
Pl. Prot. Bull., Taiwan 5: 187
Type species Macrosiphum itoe Takahashi, 1925

E. itoe (Takahashi, 1925) (Macrosiphum)

Subgen. Neoacyrthosiphon Tao, 1963

E. (Neoacyrthosiphon) setosum Hille Ris Lambers & A.N. Basu, 1966 to Indiaphis A.N. Basu, 1969

ERIOSOMA Leach, 1818
Trans. Hort. Soc. Lond. 3: 60
Type species Eriosoma mali Leach, 1818 = Aphis lanigera Hausmann, 1802
*Myzoxylus* Blot, 1831

E. americanum (Riley, 1879) (Schizoneura) BM HRL
antennatum Mordvilko ex Shaposhnikov, 1955 nomen nudum
caryae Fitch, 1856 to Phylloxera Boyer de Fonscolombe, 1834
clematicola Takahashi, 1924 (type of Colophina Börner, 1932) = E. (Colophina) clematis (Shinji, 1922)
clematis Shinji, 1922, reference in Shinji, 1941 p. 1052, a nonentity
cornicola Walsh, 1863 (1862) to Anoecia Koch, 1857
crataegi (Oestlund, 1887) (Schizoneura) BM HRL
crataegi Stephens, 1829 nomen nudum
elsholtriae Shinji, 1935 to Kaltenbachiella Schouteden, 1906
flavum Jancke, 1930 to Eriosoma (Schizoneura)
fungicola Walsh, 1863 (1862) = Anoecia cornicola (Walsh, 1863)
gomboriense Dzhibladze, 1968 to Eriosoma (Schizoneura)
herioti Börner, 1952 BM HRL
imbricator Fitch, 1851 type of Fagiphagus Smith, 1974 q.v.
inopinatum Alfieri, 1920 = Eriosoma (Schizoneura) pyricola Baker & Davidson, 1916

E. kansugei Uye, 1924 to ? Colopha Monell, 1877
lactucae Mosley, 1841 = Pemphigus bursarius (Linnaeus, 1758)
lanigerum (Hausmann, 1802) (Aphis) BM HRL
*lanata* Salisbury, 1816 (*Aphis*)
*mali* Bingley, 1803 (*Coccus*)
*mali* Blot, 1831 (*Myzoxylus*)
*mali* Leach, 1818
laricis Mosley, 1841 = Adelges laricis Vallot, 1836, Adelgidae
longipilosum Chakrabarti & RayChaudhuri, 1975 to Eriosoma (Schizoneura)
mali Leach, 1818 = lanigerum (Hausmann, 1802)
mimicum Hottes & Frison, 1931 paratype BM
oleae Leach ex Risso, 1826 to Prociphilus Koch, 1857
phaenax Mordvilko, 1923 to Eriosoma (Schizoneura)
populi Mosley, 1841 = Pemphigus bursarius (Linnaeus, 1758)
pyri Fitch, 1851 = Prociphilus alnifoliae subsp. fitchii Baker & Davidson, 1917
pyri Westwood, 1849 nomen dubium
pyricola Baker & Davidson, 1916 to Eriosoma (Schizoneura)
querci Fitch, 1859 to Stegophylla Oestlund, 1922
quercus Mosley, 1841 nomen dubium
quercus Stephens, 1829 nomen nudum
rileyi Thomas, 1877 BM HRL
*ulmi* Riley, 1869 nec Chermes ulmi Linnaeus, 1758
rosetti Gillette, 1916 nomen nudum
strobi Fitch, 1851 to Cinara Curtis, 1835
taskhiri Ghulamullah, 1942 = Eriosoma (Schizoneura) lanuginosum (Hartig, 1841)
tessellata Fitch, 1851 to Prociphilus (Paraprociphilus)
ulmi Riley, 1869 = rileyi Thomas, 1877
ulmi-gallarum Haliday, 1838 (type of Byrsocrypta Haliday, 1839) = ? Tetraneura ulmi (Linnaeus, 1758)
ulmosedens Marchal, 1919 = Eriosoma (Schizoneura) patchiae (Börner & Blunck, 1916)
ussuriense Mordvilko ex Shaposhnikov, 1955
yangi Takahashi, 1939 to Eriosoma (Schizoneura)

Subgen. Colophina Börner, 1932

E. (Colophina) clematis (Shinji, 1922) (Pemphigus) BM HRL
*clematicola* Takahashi, 1924 (*Eriosoma*)

Subgen. Schizoneura Hartig, 1839

E. (Schizoneura) flavum Jancke, 1930 (Eriosoma) HRL
(Schizoneura) gomboriense Dzhibladze, 1968 (Eriosoma)
(Schizoneura) japonicum (Matsumura, 1917) (Schizoneura) HRL
(Schizoneura) lanuginosum (Hartig, 1839) (Schizoneura) BM HRL
*piri* Goethe, 1884 (*Schizoneura*)
*taskhiri* Ghulamullah, 1942 (*Eriosoma*)
*ulmi* Fabricius, 1775 (*Aphis*)
*ulmi* Takahashi, 1966 (*Afghanaphis*)
(Schizoneura) longipilosum Chakrabarti & RayChaudhuri, 1975 (Eriosoma) HRL
(Schizoneura) patchiae (Börner & Blunck, 1916) (Schizoneura) BM HRL
*ulmosedens* Marchal, 1919 (*Eriosoma*)
(Schizoneura) phaenax Mordvilko, 1923 (Eriosoma) BM HRL
(Schizoneura) pyricola Baker & Davidson, 1916 (Eriosoma) BM HRL
*inopinatum* Alfieri, 1920 (*Eriosoma*)
(Schizoneura) taskhiri Ghulamullah, 1942 = lanuginosum (Hartig, 1841)
(Schizoneura) ulmi (Linnaeus, 1758) (Chermes) BM HRL
*ampelorrhiza* del Guercio, 1900 (*Schizoneura*)
*fodiens* Buckton, 1881 (*Schizoneura*)
*foliorum ulmi* de Geer, 1773 (*Aphis*)
*grossulariae* Taschenberg, 1887 (*Schizoneura*)
*soror* Börner & Blunck, 1916 (*Schizoneura*)
*ulmi-campestris* de Geer, 1773 (*Aphis*)
(Schizoneura) ulmi subsp. orientale (Mordvilko ex Shaposhnikov, 1955) (Schizoneura)
(Schizoneura) yangi Takahashi, 1939 (Eriosoma) BM HRL

ESSIGELLA del Guercio, 1909
Riv. Patol. Veg., Padova, n.s. 3: 329
Type species Lachnus californicus Essig, 1909

E. agilis Hottes, 1957
braggi Hottes, 1957
californica (Essig, 1909) (Lachnus) BM HRL
claremontiana Hottes, 1957
cocheta Hottes, 1957
essigi Hottes, 1957 BM HRL
fusca Gillette & Palmer, 1924 HRL
gillettei Hottes, 1957 BM
hoerneri Gillette & Palmer, 1924 BM HRL

E. knowltoni Hottes, 1957
maculata Hottes, 1957
monelli Hottes, 1957
oregonensis Hottes, 1958
palmerae Hottes, 1957
patchae Hottes, 1957
pergandei Hottes, 1957
pineti Hottes, 1957
pini Wilson, 1919 BM HRL
robusta Hottes, 1957
swaini Hottes, 1957
wilsoni Hottes, 1957

EUAULAX Mordvilko, 1914
Faune Russie, Ins. Hém. 1: 67
Without species

EUCALLIPTERUS Schouteden, 1906
Annls Soc. ent. Belg. 50: 31
Type species Aphis tiliae Linnaeus, 1758

E. tiliae (Linnaeus, 1758) (Aphis) BM HRL
tilicola (Shinji, 1933) (Therioaphis)

EUCARAZZIA del Guercio, 1921
Redia 14: 135
Type species Eucarazzia picta del Guercio, 1921 = Rhopalosiphum elegans Ferrari, 1872
*Carazzia* del Guercio, 1930
*Clavisiphon* del Guercio, 1930

E. caucasica (Aizenberg, 1956) (Rhopalosiphoninus (Eucarazzia))
elegans (Ferrari, 1872) (Rhopalosiphum) BM HRL
*chicotei* Gomez-Menor, 1950 (*Rhopalosiphoninus*)
*elegans* del Guercio, 1930 (*Anuraphis (Clavisiphon)*)
*picta* del Guercio, 1921
*salviae* Hall, 1926 (*Rhopalosiphoninus*)
picta del Guercio, 1921 = elegans (Ferrari, 1872)

EUCERAPHIS Walker, 1870
Zoologist (2) 5: 2001
Type species Aphis betulae Linnaeus of Walker nec Linnaeus, 1758
= Aphis punctipennis Zetterstedt, 1828
*Callipteroides* Mordvilko, 1909
*Leptopteryx* Zetterstedt, 1837 nec Horsfield, 1821
*Mimocallis* Matsumura, 1919
*Quippelachnus* Oestlund, 1922

E. betulifoliae Shinji, 1922 = Hannabura alnicola Matsumura, 1917
betulijaponicae (Matsumura, 1919) (Mimocallis) HRL
brevis Baker, 1917 fundatrix of ? punctipennis (Zetterstedt, 1828)
chuansiensis Tao, 1964 type of Taoia Quednau, 1973 q.v.
deducta Baker, 1917 HRL
flava Davidson, 1912 type of Oestlundiella Granovsky, 1930 q.v.
gillettei Davidson, 1915 type of Quippelachnus Oestlund, 1922 BM HRL
indica A.K. Ghosh & RayChaudhuri, 1972 to Taoia Quednau, 1973
japonica Essig & Kuwana, 1918 = Symydobius alniarius (Matsumura, 1917)
lineata Baker, 1917 BM HRL
magnolifoliae Shinji, 1923 = Neocalaphis magnoliae (Essig & Kuwana, 1918)
mucida (Fitch, 1856) (Callipterus) BM HRL
*? pinicolens* Fitch, 1851 (*Aphis*)
ontakensis Sorin, 1970 HRL
pilosa Nevsky, 1929 HRL
punctipennis (Zetterstedt, 1828) (Aphis) BM HRL
*? betulae* Koch, 1855 (*Callipterus*)
*bicolor* Koch, 1855 (*Callipterus*)
*? brevis* Baker, 1917
*cerasicolens* Fitch, 1851 (*Aphis*)
*discolor* Burmeister, 1836 (*Aphis*)
*nigritarsis* von Heyden, 1837 (*Aphis*)
*nivalis* Zetterstedt, 1837 (*Leptopteryx*)
sitchensis Glendenning 1926 BM HRL
*variabilis* Glendenning, 1926
variabilis Glendenning, 1926 = sitchensis Glendenning, 1926

EULACHNUS del Guercio, 1909
Redia 5: 315
Type species Lachnus agilis Kaltenbach, 1843, see Bull. zool. Nom. 22: 188
*Protolachnus* Theobald, 1915

E. abameleki del Guercio, 1909 = Cinara pectinatae (Nördlinger, 1880)
agilis (Kaltenbach, 1843) (Lachnus) BM HRL
alticola Börner, 1940 HRL
americanus Takahashi, 1932
bielawskii Szelegiewicz, 1970 = thunbergii (Wilson, 1919)
bluncki Börner, 1940 = rileyi (Williams, 1910)
brevipilosus Börner, 1940 BM HRL
cembrae Börner, 1950 BM HRL
cretaceus (Mamontova, 1968) (Protolachnus)
macchiatii del Guercio, 1909 = Cinara pectinatae (Nördlinger, 1880)
mingazzinii del Guercio, 1909 = Cinara piniphila (Ratzeburg, 1844)
nigricola (Pašek, 1953) (Protolachnus) BM HRL
nigrofasciatus del Guercio, 1909 = Cinara laricis (Hartig, 1839)
pallidus Mamontova,1972
pini Takahashi, 1935 = ? thunbergii (Wilson, 1919)
piniformosanus Takahashi, 1931 = ? thunbergii (Wilson, 1919)
pumilae Inouye, 1939 HRL
rileyi (Williams, 1911) (Lachnus) BM HRL
*bluncki* Börner, 1940
rileyi subsp. tauricus Bozhko, (1957) 1961 HRL
*tauricus* Bozhko, (1957) 1961
taiwanus Takahashi, 1932 = ? thunbergii (Wilson, 1919)
tamaricis Nevsky, 1951
tauricus Bozhko, (1957) 1961 = rileyi subsp. tauricus Bozhko, (1957) 1961
thunbergii (Wilson, 1919) (Lachniella) BM HRL
*bielawskii* Szelegiewicz, 1970 HRL
*longicorpi* Shinji, 1922 (*Lachnus*)
? *pini* Takahashi, 1935
? *piniformosanus* Takahashi, 1931
? *taiwanus* Takahashi, 1932
tuberculostemmatus (Theobald, 1915) (Protolachnus) BM HRL

EUMACROSIPHUM Shinji, 1932
Isayoshi Ishii (ed.) Practical Horticulture, Tokyo, 13: 246
Type species Eumacrosiphum viciae Shinji, 1932

E. viciae Shinji, 1932

EUMYZUS Shinji, 1929
Lansania 1: 111
Type species Aphis impatiensae Shinji, 1924

E. clinopodii Takahashi, 1965
darjeelingensis R.C. Basu & RayChaudhuri, 1974
gallicola Takahashi, 1963 BM HRL
impatiensae (Shinji, 1924) (Aphis) BM HRL
*impatiensae* Monzen, 1929 (*Myzus*)

EUNECTAROSIPHON del Guercio, 1913
Redia 9: 188
Type species Aphis rubi Kaltenbach, 1843
= Amphorophora Buckton, 1876

EUPHYLLOXERA del Guercio, 1908
Sulla Sistematica e sulla Biologia dei fillosserini, privately published, p. 156. Purportedly reprinted from Boll. Soc. ent., ital. 38: 153-188 but the paper was not published in that journal.
Type species Phylloxera foveola Pergande, 1904
Phylloxeridae

EURHOPALOSIPHUM Shinji, 1942
Insect Wld 46: 98
Type species Europhalosiphum kondoi Shinji, 1942
= Coloradoa Wilson, 1910

E. kondoi Shinji, 1942 to Coloradoa Wilson, 1910

EURYTHAPHIS Mordvilko, 1914
Faune Russie, Ins. Hém. 1 (1): 66
Type species Aphis tanaceticola Kaltenbach, 1845 (= Aphis tanaceti Linnaeus, 1758), indicated by Mordvilko, 1929
= Uroleucon Mordvilko, 1914

E. tanaceticola subsp. petasitis Bozhko, 1959 = Uroleucon tussilaginis (Walker, 1850)

EUSCHIZAPHIS Hille Ris Lambers, 1947
Zoöl. Meded., Leiden 26: 324, described as subg. of Schizaphis Börner, 1931
Type species Aphis palustris Theobald, 1929
Here used as subg. of Schizaphis Börner, 1931

E. longicornis Richards, 1961 to Schizaphis (Euschizaphis)

EUTHORACAPHIS Takahashi, 1938
Tenthredo 2: 14, described as subg. of Thoracaphis van der Goot, 1917
Type species Thoracaphis umbellulariae Essig, 1932

E. cinnamoniae (Shinji, 1941) (Thoracaphis) HRL
heterotricha A.K. Ghosh & RayChaudhuri, 1973
longisetosa A.K. Ghosh & RayChaudhuri, 1973
umbellulariae (Essig, 1932) (Thoracaphis) cotype BM HRL

EUTRICHOSIPHUM Essig & Kuwana, 1918
Proc. Calif. Acad. Sci. (4) 8: 97
Type species Trichosiphum pasaniae Okajima, 1908
*Goodea* Shinji, 1922
*Holotrichosiphon* RayChaudhuri, 1956
*Neoparatrichosiphum* A.K. Ghosh & RayChaudhuri, 1962
*Paratrichosiphum* Takahashi, 1931

E. alnicola (A.N. Basu, 1968) (Paratrichosiphum) HRL
alnifoliae (Tao, 1958) (Paratrichosiphum) HRL ?
arunachali R.C. Basu, A.K. Ghosh & RayChaudhuri, 1972
assamense (A.K. Ghosh & RayChaudhuri, 1962) (Paratrichosiphum) BM HRL
*tattakanum* subsp. *assamense* A.K. Ghosh & RayChaudhuri, 1962 (*Paratrichosiphum*)
assamense A.K. Ghosh & RayChaudhuri, 1969 nec 1962
davidi RayChaudhuri, 1956 BM HRL
dubium (van der Goot, 1918) (Trichosiphum)
elongatum Takahashi, 1940 (type of Ditrichosiphon RayChaudhuri, 1956) to Eutrichosiphum (Ditrichosiphon)
flavum (Takahashi, 1941) (Paratrichosiphum) HRL
heterotrichum (RayChaudhuri, 1956) (Holotrichosiphon) HRL
khasyanum (A.K. Ghosh & RayChaudhuri, 1962) (Paratrichosiphum (Neoparatrichosiphum))
lithocarpi subsp. malayense Takahashi, 1950 = pasaniae (Okajima, 1908)
makii RayChaudhuri & Chatterjee, 1974
miniatum Takahashi of Shinji, 1941: 443 errore pro minutum Takahashi, 1923
minutum (van der Goot, 1916) (Trichosiphum)
minutum Takahashi, 1923 = parvulum Eastop & Hille Ris Lambers nomen novum
narafoliae (Shinji, 1922) (Greenidea (Goodea))
parvulum Eastop & Hille Ris Lambers nomen novum pro minutum Takahashi, 1923 nec van der Goot, 1916 BM HRL

E. pasaniae (Okajima, 1908) (Trichosiphum) BM HRL
*lithocarpae* Maki, 1918 (*Trichosiphum*)
*lithocarpi* subsp. *malayense* Takahashi, 1950
*vandergooti* RayChaudhuri, 1956 BM HRL
pseudopasaniae Szelegiewicz, 1968 BM HRL
pyri Chakrabarti, A.K. Ghosh & RayChaudhuri, 1972
quercifoliae RayChaudhuri, M.R. Ghosh, M. Banerjee & A.K. Ghosh, 1973 HRL
raychaudhurii (A.K. Ghosh, 1969) (Paratrichosiphum (Neoparatrichosiphum))
roepkei van der Goot, 1918 (Trichosiphum)
russellae (A.K. Ghosh, M.R. Ghosh & RayChaudhuri, 1971) (Holotrichosiphon)
sankari RayChaudhuri, M.R. Ghosh, M. Banerjee & A.K. Ghosh, 1973
sensoriatum (A.K. Ghosh, 1974) (Paratrichosiphum)
shiicola Takahashi, 1972 HRL
sikkimense (RayChaudhuri, M.R. Ghosh, M. Banerjee & A.K. Ghosh, 1973) (Paratrichosiphum)
sinense RayChaudhuri, 1956 type BM HRL
szechuanense Tseng & Tao, 1938
taoi A.K. Ghosh, R.C. Basu & RayChaudhuri, 1970
tattakanum (Takahashi, 1925) (Greenidea) BM HRL
vandergooti (Franssen, 1930) (Trichosiphum)
vandergooti RayChaudhuri, 1956 nec Franssen, 1930 = pasaniae (Okajima, 1908)

Subgen. Ditrichosiphon RayChaudhuri, 1956

E. (Ditrichosiphon) elongatum Takahashi, 1940 BM HRL
*javanicum* RayChaudhuri, 1956 (*Paratrichosiphum*) cotypes BM HRL

Subgen. Neotrichosiphum RayChaudhuri, M.R. Ghosh, M. Banerjee & A.K. Ghosh, 1973

E. (Neotrichosiphum) subinoyi RayChaudhuri, M.R. Ghosh, M. Banerjee & A.K. Ghosh, 1973 type of Neotrichosiphum RayChaudhuri, M.R. Ghosh, M. Banerjee & A.K. Ghosh, 1973

FAGIPHAGUS Smith, 1974
Tech. Bull. N. Carol. agric. Exp. Stn 226: 14
Type species Eriosoma imbricator Fitch, 1851

F. imbricator (Fitch, 1851) (Eriosoma) BM HRL

FERGANAPHIS Mukhamediev* nomen novum pro Sogdianella Mukhamediev, 1965 nec Schurenkova, 1939

* To be quoted as Ferganaphis Mukhamediev ex Eastop & Hille Ris Lambers, 1976

Type species Sogdianella lonicericola Mukhamediev, 1965
*Sogdianella* Mukhamediev, 1965

F. lonicericola (Mukhamediev, 1965) (Sogdianella)

FIMBRIAPHIS Richards, 1959
Can. Ent. 91: 248
Type species Fimbriaphis fimbriata Richards, 1959

F. fimbriata Richards, 1959 paratypes BM HRL
fimbriata subsp. pernettyae Prior, 1969 paratypes BM HRL
gentneri (Mason, 1947) (Macrosiphum) BM HRL
? harmstoni Knowlton, 1943 (Myzus)
lilii (Mason, 1940) (Myzus) BM HRL
scoliopi (Essig, 1936) (Macrosiphum) HRL
wakibae (Hottes, 1934) (Dactynotus)

FITCHIELLA Shaposhnikov, 1950
Ént. Obozr. 31: 223, described as subg. of Nearctaphis Shaposhnikov, 1950
Type species Aphis crataegifoliae Fitch, 1851
= Nearctaphis Shaposhnikov, 1950

FLABELLOMICROSIPHUM Gillette & Palmer, 1932
Ann. ent. Soc. Am. 25: 472
Type species Chaitophorus tridentatae Wilson, 1915

F. knowltoni Smith, 1937 paratypes BM HRL
tridentatae (Wilson, 1915) (Chaitophorus) BM HRL

FLORAPHIS Tsai & Tang, 1946
Trans. R. ent. Soc. Lond. 97: 416
Type species Floraphis meitanensis Tsai & Tsang, 1946
= Nurudea Matsumura, 1917

F. meitanensis Tsai & Tang, 1946 to Nurudea Matsumura, 1917

FOAIELLA Börner, 1909
Zool. Anz. 34: 560
Type species Phylloxera danesii Grassi & Foa, 1908
*Boerneria* Grassi & Foa, 1908/9 nec Axelson, 1902; Willew, 1901/2
Phylloxeridae

FORDA von Heyden, 1837
Ent. Beitr. Mus. senckenb. 2: 291
Type species Forda formicaria von Heyden, 1837
*Fordona* Mordvilko, 1935
*Pentaphis* Horvath, 1896
*Rhizoterus* Hartig, 1841

F. dactylidis Börner, 1950 BM HRL
*marginata* Koch sensu Mordvilko, 1921
*mordvilkoi* Börner, 1950
*sensoriata* Börner, 1950
dauci Goureau, 1867 = Dysaphis crataegi (Kaltenbach, 1843)
flavula Rohwer, 1908 = Prociphilus erigorenensis (Thomas, 1879)
follicularioides Mordvilko, 1935 = marginata Koch, 1857
formicaria von Heyden, 1837 BM HRL
*formicaria* subsp. *intermixta* Börner, 1952
*formicaria* subsp. *subnuda* Börner, 1952
*formicaria* subsp. *viridis* Mordvilko, 1935
*meridionalis* Mordvilko, 1935
? *occidentalis* Hart, 1894
*semilunaria* Passerini, 1856 (*Pemphigus*)
*vacca* Hartig, 1841 (*Rhizoterus*)
*viridana* Buckton, 1883
*viridescens* del Guercio, 1920 (*Pentaphis*)
formicaria subsp. intermixta Börner, 1952 = formicaria von Heyden, 1837
formicaria subsp. subnuda Börner, 1952 = formicaria von Heyden, 1837
formicaria subsp. viridis Mordvilko, 1935 = formicaria von Heyden, 1837
furcata Theobald, 1913 to Anoecia Koch, 1857
harukawai Tankaa, 1957 = Paracletus cimiciformis von Heyden, 1837
hexagona Theobald, 1913 = marginata Koch, 1857
hirsuta Mordvilko, 1928 BM HRL
interjecta Cockerell, 1903 = marginata Koch, 1857
kaussarii Davatchi & Remaudière, 1957 = ? orientalis George, 1920
kingii Cockerell, 1903 = marginata Koch, 1857
marginata Koch, 1857 (type of Pentaphis Horvath, 1896) BM HRL
*apuliae* del Guercio, 1920 (*Pentaphis*)
*follicularioides* Mordvilko, 1935
*follicularius* Passerini, 1861 (*Pemphigus*)
*folliculoides* Lichtenstein, 1880 (*Pemphigus*)
*hexagona* Theobald, 1913 BM
*interjecta* Cockerell, 1903

*kingii* Cockerell, 1903
*mokrzeckyi* Mordvilko, 1921
*olivacea* Rohwer, 1908
*pawlowae* Mordvilko, 1901 (*Pentaphis*)
*polonica* Mordvilko, 1921
*proximalis* Mordvilko, 1921
*pskovensis* Mordvilko, 1921
*retroflexus* Courchet, 1879 (*Pemphigus*)
*trivialis* Passerini, 1860 (*Tychea*) HRL
*wilsoni* Mordvilko, 1935

F. marginata Koch sensu Mordvilko, 1921 = dactylidis Börner, 1950
meridionalis Mordvilko, 1935 = formicaria von Heyden, 1837
mokrzeckyi Mordvilko, 1921 = marginata Koch, 1857
mordvilkoi Börner, 1950 = dactylidis Börner, 1950
natalensis Theobald, 1920 = Smynthurodes betae Westwood, 1849
occidentalis Hart, 1894 = formicaria von Heyden, 1837
olivacea Rohwer, 1908 = marginata Koch, 1857
orientalis George, 1920 BM HRL
? *kaussarii* Davatchi & Remaudière, 1957
*ussuriensis* Mordvilko, 1935
polonica Mordvilko, 1921 = marginata Koch, 1857
proximalis Mordvilko, 1921 = marginata Koch, 1857
pskovensis Mordvilko, 1921 = marginata Koch, 1857
radicum Goureau, 1861 = Trama troglodytes von Heyden, 1837
riccobonii (Stefani, 1899) (Pemphigus) BM HRL
rotunda Theobald, 1914 type BM HRL
*italica* Mordvilko, 1935 (*Fordona*)
*skorkini* Mordvilko, 1935
sensoriata Börner, 1950 = dactylidis Börner, 1950
sichangensis Remaudière & Tao, 1957 paratype BM HRL
*marginata* Tao, 1947 (*Pemphigella*) nec Forda marginata Koch, 1857
skorkini Mordvilko, 1935 = rotunda Theobald, 1914
ussuriensis Mordvilko, 1935 = orientalis George, 1920
viridana Buckton, 1883 = formicaria von Heyden, 1837
wilsoni Mordvilko, 1935 = marginata Koch, 1857

FORDONA Mordvilko, 1935
Ergebn. Fortschr. Zool. 8: 141, 198.
Type species Fordona italica Mordvilko, 1935 = Forda rotunda Theobald, 1914 = Forda von Heyden, 1837

F. italica Mordvilko, 1935 = Forda rotunda Theobald, 1914

FORMOSAPHIS Takahashi, 1925
Rep. Govt Res. Inst. Dep. Agric. Formosa 6: 52
Type species Formosaphis micheliae Takahashi, 1925

F. micheliae Takahashi, 1925 BM HRL

FRANCOA del Guercio, 1916
Redia 12: 197 (published 23. xii. 1916, see footnote on p. 277)
Type species Francoa elegans del Guercio, 1916 = Aphis rosarum Kaltenbach, 1843
= Myzaphis van der Goot, 1913

F. elegans del Guercio, 1916 = Myzaphis rosarum (Kaltenbach, 1843)

FULLAWAYA Essig, 1912
Pomona Coll. J. Ent. 4: 716
Type species Davidsonia saliciradicis Essig, 1912
*Davidsonia* Essig, 1912 nec Bouchard-Chanteroaux, 1849

F. bradleyi (Richards, 1966) (Plocamaphis) paratype BM
braggii (Gillette & Palmer, 1929) (Plocamaphis)
bulbosa (Richards, 1966) (Plocamaphis)
ontarioensis (Richards, 1966) (Plocamaphis) HRL
saliciradicis Essig, 1912 (Davidsonia) cotype BM HRL
terricola (Hottes & Frison, 1931) (Plocamaphis) paratype BM

Subgen. Pseudopterocomma MacGillivray, 1963

F. (Pseudopterocomma) hughi MacGillivray, 1963 type of Pseudopterocomma MacGillivray, 1963 q.v.

FULLAWAYELLA del Guercio, 1911
Redia 7: 462
Type species Macrosiphum kirkaldyi Fullaway, 1910 = Idiopterus nephrelepidis Davis, 1909
= Idiopterus Davis, 1909

F. formosana Takahashi, 1921 to Neotoxoptera Theobald, 1915
lonicerae Mordvilko, 1929

FUSHIA Matsumura, 1917
In Nagano, K., A collection of Essays for Mr. Yasushi Nawa, Gifu, p. 70

Type species Fushia rosea Matsumura, 1917 = Nurudeopsis yanoniella Matsumura, 1917
= Nurudea Matsumura, 1917

F. rosea Matsumura, 1917 = Nurudea yanoniella (Matsumura, 1917)

GALIAPHIS Ossiannilsson, 1954
Ent. Tidskr. 75: 123
Type species Galiaphis annae Ossiannilsson, 1954
Here used as subg. of Amphorophora Buckton, 1876

G. annae Ossiannilsson, 1954 to Amphorophora (Galiaphis)
cryptotaeniae Takahashi, 1965 to Amphorophora (Galiaphis)
japonica Takahashi, 1965 to Amphorophora (Galiaphis)

GALIOBIUM Börner, 1933
Kleine Mitteilungen über Blattläuse, privately published, p. 4
Type species Trilobaphis langei Börner, 1933
Here used as subg. of Myzus Passerini, 1860

G. galinarium Narzikulov ex Juchnevitch, 1968 to Myzus (Galiobium)

GENAPHIS Handlirsch, 1907
Die Fossilen Insekten, p. 643
Type species Aphis valdensis Brodie, 1845
Fossil, Jurassic

GENTNERA Essig, 1952
Pan-Pacif. Ent. 28: 215
Type species Gentnera oregona Essig, 1952 = Aphis setosa Kaltenbach, 1846
= Ctenocallis Klodnitsky, 1924

G. oregona Essig, 1952 = Ctenocallis setosus (Kaltenbach, 1846)

GEOICA Hart, 1894
Rep. Ill. St. Ent. 18: 101
Type species Geoica squamosa Hart, 1894 = Pemphigus utricularius Passerini, 1856
*Neoschoutedenia* Schumacher, 1923
*Pemphigetum* Mordvilko, 1928
*Schoutedenia* Mordvilko, 1912 nec Rübsaamen, 1905
*Schoutedenum* Mordvilko, 1928
*Serrataphis* van der Goot, 1917
*Trinacriella* del Guercio, 1913

G. anchusae Narzikulov, 1963
carnosa (Buckton, 1883) (Endeis) vide utricularia group
cyperi Schouteden, 1902 (type of Schoutedenia Mordvilko, 1921 nec Rübsaamen, 1905; and of Schoutedenum Mordvilko, 1928; and of Neoschoutedenia Schumacher, 1923) vide utricularia group
discreta Börner, 1952 vide utricularia group
flavescens (del Guercio, 1920) (Tetraneura)
floccosa Moreira, 1925 to Geopemphigus Hille Ris Lambers, 1933
golbachi E.E. Blanchard, 1958 = lucifuga (Zehntner, 1897)
herculana Mordvilko, 1935 = setulosa (Passerini, 1860)
horvathi Nevsky, 1929 = lucifuga (Zehntner, 1897)
lucifuga (Zehntner, 1897) (Tetraneura) BM HRL
*golbachi* E.E. Blanchard, 1958
*horvathi* Nevsky, 1929
*pseudosetulosa* Theobald, 1928 paratypes BM
*spatulata* Theobald, 1923 type BM
magnifica (del Guercio, 1913) (Trinacriella)
mimeuri (Gaumont, 1930) (Pemphigella) BM HRL
muticae (Mordvilko, 1928) (Pemphigetum) HRL
*utricularia* subsp. *rungsi* Davatchi & Remaudière, 1957 HRL
pellucida (Buckton, 1883) (Endeis) vide utricularia group
pseudosetulosa Theobald, 1928 = lucifuga (Zehntner, 1897)
setariae (Passerini, 1860) (Tychea) vide utricularia group
setulosa (Passerini, 1860) (Tychea) BM HRL
*herculana* Mordvilko, 1935
spatulata Theobald, 1923 = lucifuga (Zehntner, 1897)
squamosa Hart, 1894 vide utricularia group
utricularia (Passerini, 1856) (Pemphigus) is a member of a group of species or subspecies not yet fully understood, including:
carnosa (Buckton, 1882) (Endeis) type BM HRL
cyperi Schouteden, 1902 HRL
discreta Börner, 1952
eragrostidis (Passerini, 1860) (Tychea) type HRL
pellucida (Buckton, 1883) (Endeis) type BM HRL
setariae (Passerini, 1860) (Tychea) type HRL
squamosa Hart, 1894
utricularia (Passerini, 1856) (Pemphigus)
utriculoides (Lichtenstein, 1880) (Pemphigus)
utricularia subsp. rungsi Davatchi & Remaudière, 1957 = muticae (Mordvilko, 1928)

GEOKTAPIA Mordvilko, 1921
Izv. Sev. Oblast. Sta. Zashch. Rast. Vredit. 3: 53
Type species Geoktapia areshensis Mordvilko, 1921 = Myzus pyrarius Passerini, 1861
= Melanaphis van der Goot, 1917

G. areshensis Mordvilko, 1921 = Melanaphis pyraria (Passerini, 1861)
dashushendis Rusanova, 1943 nomen nudum

GEOPEMPHIGUS Hille Ris Lambers, 1933
Stylops 2: 197
Type species Geopemphigus surinamensis Hille Ris Lambers, 1933 = Geoica floccosa Moreira, 1925
*Xenopterygus* Smith, 1948

G. floccosus (Moreira, 1925) (Geoica) BM HRL
*ipomoiae* Smith, 1948 (*Xenopterygus*)
*surinamensis* Hille Ris Lambers, 1933
surinamensis Hille Ris Lambers, 1933 = floccosus (Moreira, 1925)

GEORGIA Wilson, 1911 nec Baird & Girard, 1853; Thomson, 1857; Bourgingorat, 1882
Can. Ent. 43: 64
Type species Georgia ulmi Wilson, 1911
= Georgiaphis Maxson & Hottes, 1926

G. gillettei Maxson & Hottes, 1926 to Georgiaphis Maxson & Hottes, 1926
ulmi Wilson, 1911 type of Georgiaphis Maxson & Hottes, 1926 q.v.

GEORGIAPHIS Maxson & Hottes, 1926
Ent. News 37: 226
Type species Georgia ulmi Wilson, 1911
*Georgia* Wilson, 1911 nec Baird & Girard, 1853; Thomson, 1857; Bourgingorat, 1882

G. gillettei (Maxson & Hottes, 1926) (Georgia) HRL
maxsoni Boudreaux, 1949 (1948)
ulmi (Wilson, 1911) (Georgia) BM HRL

GERANCHON Scudder, 1890
U.S. Geol. Survey of the Territories 13: 248
Type species Lachnus petrorum Scudder, 1877
Fossil

GERMARAPHIS Heie, 1967
Spolia zool. Mus. haun. 26: 47
Type species Lachnus dryoides Germar & Berendt, 1856
Fossil

GHARESIA Stroyan, 1963
Proc. R. ent. Soc. Lond. (B) 32: 81
Type species Gharesia polunini Stroyan, 1963
= Byrsocryptoides Dzhibladze, 1960

G. polunini Stroyan, 1963 to Byrsocryptoides Dzhibladze, 1960

GILLETTEA Börner, 1909 nec Ashmead, 1897
Zool. Anz. 34: 504
Type species Chermes cooleyi Gillette, 1907
= Gilletteella Börner, 1930
Adelgidae

GILLETTEELLA Börner, 1930
Arch. klassif. phylogen. Ent. 1: 157
Type species Chermes cooleyi Gillette, 1907
*Gillettea* Börner, 1909 nec Ashmead, 1897
Adelgidae

GISTELIELLA Strand, 1926 nomen novum pro Aphanus Gistel, 1837 nec Laporte de Castelnau, 1832
Arch. Naturgesch. 92 (A 8): 46
Type species Chermes lapidarius Fabricius, 1803
Nomen dubium, ? Aphidoidea

GLABROMYZUS Richards, 1960
Can. Ent. 92: 771
Type species Rhopalosiphum rhois Monell, 1879

G. howardii (Wilson, 1911) (Amphorophora) BM HRL
*davisi* Mordvilko, 1921 (*Rhopalosiphum*)
*nebulosa* Hottes & Frison, 1931 (*Amphorophora*)
rhois (Monell, 1879) (Rhopalosiphum) BM HRL
rhusifoliae (Richards, 1973) (Aulacorthum) BM
schlingeri Hille Ris Lambers, 1966 HRL

GLENDENNINGIA MacGillivray, 1954
Can. Ent. 86: 346
Type species Glendenningia philadelphi MacGillivray, 1954

C. philadelphi MacGillivray, 1954 paratype BM

GLOBULICAUDAPHIS Hille Ris Lambers, 1966
Tijdschr. Ent. 109: 207
Type species Globulicaudaphis pakistanica Hille Ris Lambers, 1966

G. pakistanica Hille Ris Lambers, 1966 BM HRL

GLYPHINA Koch, 1856
Die Pflanzenläuse Aphiden, Nürnberg, p. 259
Type species Vacuna betulae Kaltenbach, 1843 = Aphis betulae Linnaeus, 1758

G. aculeata Dahl, 1912 = Atheroides serrulatus Haliday, 1839
betulae (Linnaeus, 1758) (Aphis) BM HRL
*betulae* Kaltenbach, 1843 (*Vacuna*)
*betulina* Buckton, 1886 (*Thelaxes*) lectotype BM
*impingens* Walker, 1852 (*Aphis*)
eragrostidis Howard, 1908 nomen dubium
jacutensis Mordvilko, 1931/2
longiseta Richards, 1968
onigurumii Shinji, 1924 = Kurisakia onigurumi (Shinji, 1923)
pilosa Buckton, 1883 = Schizolachnus pineti (Fabricius, 1781)
pilosa Dahl, 1912 nec Buckton, 1883 = ? Laingia psammae Theobald, 1922
pterocaryae Monzen, 1927 = Kurisakia onigurumi (Shinji, 1923)
rhusae Shinji, 1922 to Dasyaphis Takahashi, 1938
saitamaensis Shinji, 1923 = Neothoracaphis quercicola (Takahashi, 1921)
schrankiana Börner, 1950 BM HRL
*alni* Schrank, 1801 (*Aphis*) nec de Geer, 1773
setosa MacGillivray, 1963 paratype BM HRL

GLYPHINAPHIS van der Goot, 1917
Contrib. Faune Indes Néerl. 1(3): 232
Type species Glyphinaphis bambusae van der Goot, 1917
*Okajimaia* Suenaga, 1933

G. bambusae van der Goot, 1917 BM HRL
*japonica* Suenaga, 1933 (*Okajimaia*)

GOBAISHA Matsumura, 1917
In Nagano, K., A collection of Essays for Mr. Yasushi Nawa, Gifu, 3: 75
Type species Gobaisha japonica Matsumura, 1917
= Kaltenbachiella Schouteden, 1906

G. japonica Matsumura, 1917 to Kaltenbachiella Schouteden, 1906
nirecola Matsumura, 1917 to Colopha Monell, 1877

GOIDANICHIELLUM Martelli, 1950
Redia 35: 314, 318
Type species Macrosiphum dirhodum (Walker, 1849) = Aphis dirhoda Walker, 1849
= Metopolophium Mordvilko, 1914

GOODEA Shinji, 1922
Zool. Mag., Tokyo 34: 731, described as subg. of Greenidea Schouteden, 1905
Type species Greenidea (Goodea) narafoliae Shinji, 1922
= Eutrichosiphum Essig & Kuwana, 1918

GOOTIELLA Tullgren, 1925
Meddn CentAnst. FörsVäs Jordbromrad., Stockh. 280: 22
Type species Gootiella tremulae Tullgren, 1925

G. alba Shaposhnikov, 1952 HRL
tremulae Tullgren, 1925 BM HRL
*juniperina* Juchnevitch, 1968 (*Aploneura*)

GREENIDEA Schouteden, 1905
Spolia zeylan. 2: 181
Type species Siphonophora artocarpi Westwood, 1890

G. aborensis A.K. Ghosh, 1974
artocarpi (Westwood, 1890) (Siphonophora) BM HRL
brideliae Takahashi, 1928 HRL
decaspermi Takahashi, 1933 HRL
distylii van der Goot, 1917 nomen novum = flacourtiae van der Goot, 1917
eugeniae Takahashi, 1941 to Greenidea (Trichosiphum) = ? formosana Maki, 1917
fici Takahashi, 1937 to Greenidea (Trichosiphum)
ficicola Takahashi, 1921 BM HRL
flacourtiae van der Goot, 1917
*distylii* van der Goot, 1917 nomen novum
himansui RayChaudhuri, M.R. Ghosh, M. Banerjee & A.M. Ghosh, 1973
hirsuta Takahashi, 1950 type BM

G. longicornis M.R. Ghosh, A.K. Ghosh & RayChaudhuri, 1971
  longirostris A.N. Basu, 1969 HRL
  luchuana Takahashi, 1930 to Mollitrichosiphum (Metatrichosiphum)
  mangiferae Takahashi, 1925 HRL
  mushana Takahashi, 1925
  myricae Takahashi, 1925 to Greenidea (Trichosiphum)
  neoficicola A.K. Ghosh, M.R. Ghosh & RayChaudhuri, 1970
  nipponica Suenaga, 1934 to Greenidea (Trichosiphum)
  photiniphaga RayChaudhuri, M.R. Ghosh, M. Banerjee & A.K. Ghosh, 1973
  psidii van der Goot, 1917 = Greenidea (Trichosiphum) formosana (Maki, 1917)
  quercifoliae Takahashi, 1921 HRL
  rappardi RayChaudhuri, 1956 paratype BM HRL
  schimae Takahashi, 1929 HRL
  sinensis RayChaudhuri, 1956 type BM HRL
  sutepensis Takahashi, 1941 HRL
  symplocosis A.K. Ghosh, R.C. Basu & RayChaudhuri, 1959
  taiwana Takahashi, 1921 to Mollitrichosiphum (Metatrichosiphum)
  tattakana Takahashi, 1925 (type of Paratrichosiphum Takahashi, 1931) to Eutrichosiphum Essig & Kuwana, 1918
  tenuicorpus (Okajima, 1908) (Trichosiphum) type of Mollitrichosiphum Suenaga, 1934 q.v.
  viticola Takahashi, 1929 (type of Paragreenidea RayChaudhuri, 1956) to Greenidea (Paragreenidea)

Subgen. Goodea Shinji, 1922

G. (Goodea) narafoliae Shinji, 1922 (type of Goodea Shinji, 1922) to Eutrichosiphum Essig & Kuwana, 1918

Subgen. Neogreenidea RayChaudhuri, M.R. Ghosh, M. Banerjee & A.K. Ghosh, 1973

G. (Neogreenidea) ayyari RayChaudhuri, M.R. Ghosh, M. Banerjee & A.K. Ghosh, 1973
  (Neogreenidea) longisetosa RayChaudhuri, M.R. Ghosh, M. Banerjee & A.K. Ghosh, 1973 type of Neogreenidea RayChaudhuri, M.R. Ghosh, M. Banerjee & A.K. Ghosh, 1973
  (Neogreenidea) querciphaga RayChaudhuri, M.R. Ghosh, M. Banerjee & A.K. Ghosh, 1973

Subgen. Paragreenidea RayChaudhuri, 1956

G. (Paragreenidea) viticola Takahashi, 1929 HRL

Subgen. Trichosiphum Pergande, 1906

G. (Trichosiphum) anonae (Pergande, 1906) (Trichosiphum) BM HRL
(Trichosiphum) bucktonis A.K. Ghosh, R.C. Basu & RayChaudhuri, 1970 HRL
(Trichosiphum) carpini Takahashi, 1963
(Trichosiphum) chiengmaiensis Robinson, 1972 paratype BM HRL
(Trichosiphum) eugeniae Takahashi, 1941 HRL = ? formosana (Maki, 1917)
(Trichosiphum) fici Takahashi, 1937 (Greenidea)
(Trichosiphum) formosana (Maki, 1917) (Trichosiphum) BM HRL
? *eugeniae* Takahashi, 1941 (*Greenidea*) HRL
*psidii* van der Goot, 1917 (*Greenidea*)
(Trichosiphum) gigantea A.K. Ghosh & RayChaudhuri, 1972
(Trichosiphum) heeri RayChaudhuri, M.R. Ghosh, M. Banerjee & A.K. Ghosh, 1973
(Trichosiphum) kuwanai (Pergande, 1906) (Trichosiphum) BM HRL
(Trichosiphum) manii A.K. Ghosh, R.C. Basu & RayChaudhuri, 1970
(Trichosiphum) myricae Takahashi, 1925 (Greenidea) HRL
(Trichosiphum) nigra (Maki, 1917) (Trichosiphum) HRL
(Trichosiphum) nipponica Suenaga, 1934 (Greenidea) BM HRL
(Trichosiphum) okajimai Suenaga, 1934 BM HRL
(Trichosiphum) prunicola A.K. Ghosh, H. Banerjee & RayChaudhuri, 1971
(Trichosiphum) schoutedeni RayChaudhuri, M.R. Ghosh, M. Banerjee & A.K. Ghosh, 1973
(Trichosiphum) sikkimensis RayChaudhuri, M.R. Ghosh, M. Banerjee & A.K. Ghosh, 1973

GREENIDEOIDA van der Goot, 1917
Contrib. Faune Indes Néerl. 1(3): 140
Type species Greenideoida elongata van der Goot, 1917

G. ceyloniae van der Goot, 1917 BM HRL
*mesuae* Takahashi, 1950 types BM HRL
ceyloniae subsp. bhalukpongensis A.K. Ghosh, H. Banerjee & RayChaudhuri, 1971
elongata van der Goot, 1917 BM HRL
*vandermeermohri* Takahashi, 1935
? gavarrai RayChaudhuri, 1972 BM (generic position doubtful!)
hannae van der Goot, 1917
lambersi A.N. Basu, 1964 HRL
mesuae Takahashi, 1950 = ceyloniae van der Goot, 1917
noonadanae Heie, 1967
vandermeermohri Takahashi, 1935 = elongata van der Goot, 1917

Subgen. Neogreenideoida RayChaudhuri, 1956

G. (Neogreenideoida) philippensis RayChaudhuri, 1956 type of Neogreenideoida
RayChaudhuri, 1956

GRYLLOPROCIPHILUS Smith & Pepper, 1968
Proc. ent. Soc. Wash, 70: 57
Type species Grylloprociphilus frosti Smith & Pepper, 1968

G. frosti Smith & Pepper, 1968 paratype BM HRL

GUERCIOJA Mordvilko, 1909 (1908)
Ezheg. zool. Muz. 13: 361
Type species Chermes populi del Guercio, 1900
= Phylloxerina Börner, 1908
Phylloxeridae

GYPSOAPHIS Oestlund, 1922
Rep. Minn. St. Ent. 19: 126
Type species Aphis lonicerae Monell, 1879 nec Siebold, 1839; Boyer de Fonscolombe, 1841; Mosley, 1841 = Gypsoaphis oestlundi Hottes, 1930

G. oestlundi Hottes, 1930 BM HRL
*lonicerae* Monell, 1879 (*Aphis*)

HALLAPHIS Doncaster, 1956
Bull. ent. Res. 47: 745
Type species Yamataphis rhodesiensis Hall, 1932

H. bamendae Eastop, 1958 type BM
ilharcoi van Harten, 1971 paratypes BM HRL
malawi van Harten, 1971
nigeriensis Eastop, 1958 types BM HRL
rhodesiensis (Hall, 1932) (Yamataphis) types BM HRL

HALMODAPHIS Mordvilko, 1914
Faune Russie, Ins. Hém. 1 (1): 66
Without species

HAMADRYAPHIS Kirkaldy, 1904
Entomologist 37: 279
Type species Pemphigus spirothecae Passerini, 1860 = spyrothecae Passerini, 1856
= Pemphigus Hartig, 1839

HAMAMELISTES Shimer, 1867
Trans. Am. ent. Soc. 1: 284
Type species Hamamelistes spinosus Shimer, 1867
*Mansakia* Matsumura, 1917
*Tetraphis* Horvath, 1896

H. agrifoliae Ferris, 1921 to Atarsaphis Takahashi, 1958
betulae Tullgren, 1909 = betulinus (Horvath, 1896) ? not new species, misidentification
betulinus (Horvath, 1896) (Tetraphis) BM HRL
*betulae* Tullgren, 1909
*? betulifoliae* Shinji, 1924 (*Thoracaphis*)
*cristafoliae* Monzen, 1954
*? shirakabae* Monzen, 1927
*tullgreni* de Meijere, 1912
betulinus subsp. miyabei (Matsumura, 1917) (Mansakia)
*? kagamii* Monzen, 1929 (*Mansakia*)
cornus Shimer, 1867 = Hormaphis hamamelidis (Fitch, 1851)
cristafoliae Monzen, 1954 = betulinus (Horvath, 1896)
gibberi Monzen, 1954 to Hormaphis Osten-Sacken, 1861
gibberi biological race grossae Monzen, 1954 to Hormaphis Osten-Sacken, 1861
shirakabae Monzen, 1927 = ? betulinus (Horvath, 1896)
spinosus Shimer, 1867 BM HRL
*papyraceae* Oestlund, 1887 (*Hormaphis*)
tuberculatus Takahashi, 1933 to Quernaphis Takahashi, 1958
tullgreni de Meijere, 1912= betulinus (Horvath, 1896)

HANNABURA Matsumura, 1917
J. Coll. Agric. Hokkaido imp. Univ. 7: 377
Type species Hannabura alnicola Matsumura, 1917

H. alnicola Matsumura, 1917 HRL
*betulifoliae* Shinji, 1922 (*Euceraphis*)

HAPLONEURA lapsus pro Aploneura Passerini, nec Haploneura Löw, 1850

HAYHURSTIA del Guercio, 1917
Redia 12: 208
Type species Hayhurstia deformans del Guercio, 1917 = Aphis atriplicis Linnaeus, 1761

H. aizenbergi Narzikulov, 1957 to Hyadaphis Kirkaldy, 1904
atriplicis (Linnaeus, 1761) (Aphis) BM HRL
*chenopodii* Cowen, 1895 (*Aphis*)
*chenopodii* Schrank, 1801 (*Aphis*)
*deformans* del Guercio, 1917
*mercurialis* Balachowsky & Cairaschi, 1941 (*Pergandeida*)
atriplicis subsp. chenopodii Mamontova, 1963
camphorosmae Hille Ris Lambers, 1959 cotypes BM HRL
deformans del Guercio, 1917 = atriplicis (Linnaeus, 1761)
tataricae Aizenberg, 1935 (type of Neohayhurstia Aizenberg, (1954) 1956 to Hyadaphis Kirkaldy, 1904

HAYHURSTIA Mordvilko, 1921 nec del Guercio, 1917
Izv. Sev. Oblast. Sta. Zashch. Rast. Vredit. 3: 45
Type species Hyalopterus dactylidis Hayhurst, 1909 = Aphis humilis Walker, 1852
= Hyalopteroides Theobald, 1916

HELOSIPHON Leclant, 1969
Annls Soc. ent. Fr. n.s. 5: 429
Type species Helosiphon eryngii Leclant, 1969

H. eryngii Leclant, 1969 paratypes BM HRL

HEMIAPHIS Börner, 1926
In Abderhalden, E., Handbuch der biologischen Arbeitsmethoden, Abt. IX, Teil 1, 2 Hälfte, Heft 2: 226
Type species Aphis trirhoda Walker, 1849
= Longicaudus van der Goot, 1913

HEMIPODAPHIS David, Narayanan & Rajasingh, 1971
Orient. Insects 5: 559
Type species Hemipodaphis monstrosa David, Narayanan & Rajasingh, 1971

H. monstrosa David, Narayanan & Rajasingh, 1971 HRL

HEMITRAMA Mordvilko, 1921
Izv. Sev. Oblast. Sta. Zashch. Rast. Vredit. 3: 63
Type species Hemitrama bykovi Mordvilko, 1921
= Paracletus von Heyden, 1837

H. bykovi Mordvilko, 1921 to Paracletus von Heyden, 1837
bykovi subsp. uzbekistanica Kan, 1970

HENNINGSIA Heie, 1967
Spolia zool. Mus. haun. 26: 86, described as subg. of Germaraphis Heie, 1967
Type species Germaraphis (Henningsia) ungulata Heie, 1967
Fossil, baltic amber

HETEROCALLIS Quednau, 1966
Can. Ent. 98: 422
Type species Heterocallis daviaulti Quednau, 1966
= Trichocallis Börner, 1930

H. daviaulti Quednau, 1966 to Thripsaphis (Trichocallis)

HETEROGENAPHIS Ivanovskaja, 1966
In Cherepanow, New Species of Fauna of Siberia and adjoining Region, Acad. Sci. U.S.S.R. Siberian branch, Novosibirsk, p. 18
Type species Heterogenaphis kunashyri Ivanovskaja, 1966
= Tuberocephalus Shinji, 1929

H. kunashyri Ivanovskaja, 1966 = Tuberocephalus sasakii (Matsumura, 1917)

HETERONEURA Davis, 1919 nec Fallen, 1810
Can. Ent. 51: 228
Type species Aphis setariae (Thomas, 1878) = Siphonophora setariae Thomas, 1878
= Hysteroneura Davis, 1919

HIBERAPHIS Börner, 1949
Beitr. tax. Zool. 1: 52
Type species Hiberaphis iberica Börner, 1949 = Saltusaphis scirpus Theobald, 1915
= Saltusaphis Theobald, 1915

H. hirtae Quednau ex Ossiannilsson, 1953 nomen nudum
iberica Börner, 1949 = Saltusaphis scirpus Theobald, 1915
lasiocarpae Ossiannilsson, 1953 to Saltusaphis Theobald, 1915

HILLERISLAMBERSIA A.N. Basu, 1968 (1967)
Bull. ent. Soc. India 8: 147
Type species Hillerislambersia darjeelingi A.N. Basu, 1968

H. darjeelingi A.N. Basu, 1968 HRL

HIMALAYAPHIS A.K. Ghosh & Verma, 1973
Orient. Insects 7: 271
Type species Himalayaphis anemones A.K. Ghosh & Verma, 1973

H. anemones A.K. Ghosh & Verma, 1973 HRL

HOLCAPHIS Hille Ris Lambers, 1939
Zool. Meded., Leiden 22: 97
Type species Aphis holci Hardy of Hille Ris Lambers, 1939 nec Hardy = Holcaphis holci Hille Ris Lambers, 1959
Here used as subg. of Diuraphis Aizenberg, 1935

H. agrostidis Muddathir, 1965 to Diuraphis (Holcaphis)
bromicola Hille Ris Lambers, 1959 to Diuraphis (Holcaphis)
calamagrostis Ossiannilsson, 1959 to Diuraphis (Holcaphis)
holci Hille Ris Lambers, 1959 (1939 as holci Hardy) to Diuraphis (Holcaphis)

HOLMANIA Szelegiewicz, 1964
Bull. Acad. pol. Sci. Cl. II Sér. biol. 12: 211
Type species Holmania chaetosiphon Szelegiewicz, 1964

H. chaetosiphon Szelegiewicz, 1964 BM HRL

HOLOTRICHOSIPHON RayChaudhuri, 1956
Zool. Verh., Leiden 31: 75
Type species Holotrichosiphon heterotrichus RayChaudhuri, 1956
= Eutrichosiphum Essig & Kuwana, 1918

H. heterotrichus RayChaudhuri, 1956 to Eutrichosiphum Essig & Kuwana, 1918
russellae A.K. Ghosh, M.R. Ghosh & RayChaudhuri, 1971 to Eutrichosiphum Essig & Kuwana, 1918

HOLZNERIA Lichtenstein, 1875
Bull. Soc. ent. Fr. (5) 5: lxxvi
Type species Pemphigus poschingeri Holzner, 1874 = Aphis bumeliae Schrank, 1801
= Prociphilus Koch, 1857

HOPLOCALLIS Pintera, 1952
Zool. ent. Listy 3: 151
Type species Hoplocallis ruperti Pintera, 1952
*Chaetocallis* Borner, 1952

H. komareki Pašek, 1953 to Myzocallis Passerini, 1860
pictus (Ferrari, 1872) (Pterocallis) BM HRL
*bodenheimeri* Hille Ris Lambers, 1948 (*Myzocallis*)
*boudyi* Mimeur, 1934 (*Myzocallis*)
*grodsinskyi* E.E. Blanchard, 1939 (*Tuberculoides*)
*suberis* Tavares, 1900 (*Aphis*)
ruperti Pintera, 1952 BM HRL

HOPLOCHAETAPHIS Aizenberg, 1959
Zool. Zh. 38: 1674
Type species Hoplochaitophorus zachvatkini Aizenberg & Moravskaja, 1950

H. parvula Hille Ris Lambers & van den Bosch, 1966 BM HRL
zachvatkini (Aizenberg & Moravskaja, 1950) (Hoplochaitophorus) HRL

HOPLOCHAITOPHORUS Granovsky, 1933
Proc. ent. Soc. Wash. 35: 29
Type species Chaitophorus quercicola Monell, 1879

H. heterotrichus Quednau, 1971 paratype BM
quercicola (Monell, 1879) (Chaitophorus) BM HRL
*quercifolii* Thomas, 1879 (*Callipterus*)
*spinosus* Oestlund, 1886 (*Chaitophorus*)
zachvatkini Aizenberg & Moravskaja, 1950 type of Hoplochaetaphis Aizenberg, 1959 q.v.

HORMAPHIDULA Börner, 1952
Mitt. thüring. bot. Ges. 4 (3): 183
Type species Hormaphis betulae (Mordvilko, 1901) = Cerataphis betulae Mordvilko, 1901
= Hormaphis Osten-Sacken, 1861

HORMAPHIS Osten-Sacken, 1861
Stettin. ent. Ztg 22: 422
Type species Hormaphis hamamelidis Osten-Sacken, 1861 = Byrsocrypta hamamelidis Fitch, 1851
*Hormaphidula* Börner, 1952

H. betulae (Mordvilko, 1901) (Cerataphis) type of Hormaphidula Börner, 1952
BM HRL
*? betulifoliae* Shinji, 1924 (*Thoracaphis*)
*? crassicornis* Dahl, 1912 (*Phylloxera*)
gallifoliae (Monzen, 1929) (Mansakia)
gibberi (Monzen, 1954) (Hamamelistes)
gibberi biological race grossae (Monzen, 1954) (Hamamelistes)
hamamelidis (Fitch, 1851) (Byrsocrypta) BM HRL
*cornus* Shimer, 1867 (*Hamamelistes*)
*hamamelidis* Osten-Sacken, 1861
hamamelidis Osten-Sacken, 1861 = hamamelidis (Fitch, 1851)
papyraceae Oestlund, 1887 = Hamamelistes spinosus Shimer, 1867
populi (Maskell, 1898) (Sphaerococcus) to Doraphis Matsumura & Hori ex Hori, 1929, but placed in Hormaphis by Takahashi, 1920: 14 and sometimes referred to as H. populi Takahashi.
shulliana Börner, 1952 nomen nudum

HOTTESINA Börner, 1950
Neue europäische Blattlausarten, privately published, p. 12
Type species Hottesina superba Börner, 1950 = Acyrthosiphon nigripes Hille Ris Lambers, 1935
= Acyrthosiphon Mordvilko, 1914

H. peucedani Bozhko, 1959 = Acyrthosiphon nigripes subsp. peucedani Bozhko, 1959
superba Börner, 1950 = Acyrthosiphon nigripes Hille Ris Lambers, 1935

HYADAPHIS Kirkaldy, 1904
Entomologist 37: 279
Type species Aphis xylostei Schrank, 1801 nec De Geer, 1773 = Siphocoryne foeniculi Passerini, 1860
*Miraphis* Nevsky, 1928
*Neohayhurstia* Aizenberg, (1954) 1956
*Siphocoryne* Passerini, 1863 nec 1860

H. agabiformis (Nevsky, 1928) (Miraphis)
aizenbergi Narzikulov, 1957 (Hayhurstia)
albus (Monzen, 1929) (Hyalopterus) HRL
apii Hall, 1932 = ? foeniculi (Passerini, 1860)
bicincta Börner, 1942 HRL
bupleuri Börner, 1939 BM HRL
*bupleuri* Cairaschi & Arnoux, 1943 (*Rhopalosiphum*)

H. coerulescens (Narzikulov, 1965) (Rhopalomyzus) HRL
conica Börner, (ex Bodenheimer, 1930: 430 nomen nudum) 1932 = coriandri (Das, 1918)
coniellum Theobald, 1925 = foeniculi (Passerini, 1860)
coriandri (Das, 1918) (Brevicoryne) BM HRL
*carii* Theobald, 1929 (*Hyalopterus*)
*conica* Börner, 1932
*obscurus* Theobald, 1922 (*Hyalopterus*) types BM
*peucedani* Hall, 1932 (*Hyalopterus*) types BM
eonica Börner ex Bodenheimer, 1930 errore pro conica
foeniculi (Passerini, 1860) (Siphocoryne) BM HRL
? *apii* Hall, 1932
*coniellum* Theobald, 1925
*conii* Davidson, 1909 (*Siphocoryne*)
*hyadaphis* Kirkaldy, 1905
*lonicerae* Boyer de Fonscolombe, 1841 (*Aphis*) nec Siebold, 1839
*lonicerae* Mosley, 1841 (*Aphis*)nec Siebold, 1839
*mellifera* Hottes, 1930
*schranki* Hille Ris Lambers, 1931
*umbellulariae* Davidson, 1911
*xylostei* Schrank, 1801 (*Aphis*) nec DeGeer, 1773
foeniculi subsp. hirsuta Börner, 1952
*mellifera* subsp. *hirsuta* Börner, 1952
galaganiae Nevsky, 1951
? hissarica (Narzikulov, 1965) (Rhopalomyzus)
hyadaphis Kirkaldy, 1905 = foeniculi (Passerini, 1860)
lonicerae Börner, 1939 = passerinii (del Guercio, 1911)
mellifera Hottes, 1930 = foeniculi (Passerini, 1860)
mellifera subsp. hirsuta Börner, 1952 = foeniculi subsp. hirsuta Börner, 1952
passerinii (del Guercio, 1911) (Siphocoryne) HRL
*lonicerae* Börner, 1939
polonica Szelegiewicz, 1959 HRL
schranki Hille Ris Lambers, 1931 = foeniculi (Passerini, 1860)
sparganii Theobald, 1925 = Rhopalosiphum nymphaeae (Linnaeus, 1761)
sphondylii (Koch, 1854) (Hyalopterus) HRL
tataricae (Aizenberg, 1935) (Hayhurstia) BM HRL
umbellulariae Davidson, 1911 = foeniculi (Passerini, 1860)
veratri Shinji, 1942

HYALOMYZUS Richards, 1958
Fla Ent. 41: 169
Type species Myzus eriobotryae Tissot, 1935

H. collinsoniae (Pepper, 1950) (Micromyzus) BM HRL
eriobotryae (Tissot, 1935) (Myzus) BM HRL
jussiaeae Smith, 1960 paratypes BM HRL
monardae (Davis, 1912) (Rhopalosiphum) BM HRL
raoi Hille Ris Lambers, 1973 (nomen nudum in A.K. Ghosh, M.R. Ghosh & RayChaudhuri, 1971) HRL
scabripes David & Narayanan, 1968 HRL
sensoriatus (Mason, 1940) (Myzus) BM HRL

HYALOPTEROIDES Theobald, 1916
Entomologist 39: 51
Type species Hyalopteroides pallida Theobald, 1916 = Aphis humilis Walker, 1852
*Hayhurstia* Mordvilko, 1921 nec del Guercio, 1917

H. blattnyi Pintera, 1959 to Chondrillobium Bozhko, 1961
humilis (Walker, 1852) (Aphis) BM HRL
*dactylis* Hayhurst, 1909 (*Hyalopterus*)
*pallida* Theobald, 1916 BM
? *slavae* Mordvilko, 1921 (*Brachycolus*)
pallida Theobald, 1916 = humilis (Walker, 1852)
palmerae Hille Ris Lambers, 1949 to Metopolophium Mordvilko, 1914
sinensis Tao, 1963 (type of Longicaudinus Hille Ris Lambers, 1965) = Longicaudinus corydalisicola (Tao, 1962)

HYALOPTERUS Koch, 1854
Die Pflanzenläuse Aphiden, Nürnberg, p. 16
Type species Aphis pruni Fabricius, 1775 = Aphis pruni Geoffroy, 1762

H. abietinus Matsumura, 1971 nomen dubium
abrotani Koch, 1854 (type of Lidaja Börner, 1952) to Coloradoa Wilson, 1910
albus Monzen, 1929 to Hyadaphis Kirkaldy, 1904
amygdali (Blanchard, 1840) (Aphis) BM HRL
*amygdali persicae* Mosley, 1841 (*Aphis*)
*arundiformis* Ghulamullah, 1942
*mimulus* Börner, 1950
*persicariae* Hartig, 1841 (*Aphis*) partim
aquilegiae Koch, 1854 = Longicaudus trirhodus (Walker, 1849)

H. arundiformis Ghullamullah, 1942 = amygdali (Blanchard, 1840)
carii Theobald, 1929 = Hyadaphis coriandri (Das,1918)
dactylidis Hayhurst, 1909 (type of Hayhurstia Mordvilko, 1921 nec del Guercio, 1917) = Hyalopteroides humilis (Walker, 1852)
dilineatus Buckton, 1879 = Longicaudus trirhodus (Walker, 1849)
eriophori Haliday ex Buckton, 1877 = Ceruraphis eriophori (Walker, 1848)
flavus Schouteden, 1906 = Longicaudus trirhodus (Walker, 1849)
insignis Theobald, 1918 = Melanaphis donacis (Passerini, 1862)
melanocephalus Buckton, 1879 = Brachycolus cucubali (Passerini, 1863)
mimulus Börner, 1950 = amygdali (Blanchard, 1840)
obscurus Theobald, 1922 = Hyadaphis coriandri (Das, 1918)
peucedani Hall, 1932 = Hyadaphis coriandri (Das, 1918)
pruni (Geoffroy, 1762) (Aphis) BM HRL
*arundinis* auctt. ? nec Fabricius, 1775 (*Aphis*)
*gracilis* Walker, 1852 (*Aphis*)
*phragmitidicola* Oestlund, 1886 (*Aphis*)
*pruni* Fabricius, 1775 (*Aphis*)
*spinarum* Hartig, 8141 (*Aphis*)
skorkini Mordvilko, 1929 type of Mordvilkoiella Shaposhnikov, 1964 q.v.
sphondylii Koch, 1854 to Hyadaphis Kirkaldy, 1904
suaedus Paik, 1965 = Xerophilaphis suaedae (Mimeur, 1934)

HYDAPHIAS Börner, 1930
Arch. klassif. phylogen. Ent. 1: 136
Type species Aphis bicolor Koch, 1855 nec Haldeman, 1844 = Hydaphias helvetica Hille Ris Lambers, 1947

H. helvetica Hille Ris Lambers, 1947 BM HRL
*bicolor* Koch, 1855 (Aphis) nec Haldeman, 1844
hofmanni Börner, 1950 BM HRL
molluginis Börner, 1939 BM HRL
mosana Hille Ris Lambers, 1956 BM HRL
necopinata Börner, 1939 type of Staegeriella Hille Ris Lambers, 1947 q.v.
tricolor Koch of Wahlgren, 1938 (errore pro bicolor Koch) nomen nudum

HYDRONAPHIS Shinji, 1922
Zool. Mag., Tokyo 34: 790
Type species Hydronaphis impatiens Shinji, 1922

H. impatiens Shinji, 1922 BM HRL
laporteae Miyazaki, 1971 BM
oenanthi Shinji, 1922 to Cavariella del Guercio, 1911
oenauthi Shinji, 1922 lapsus pro oenanthi Shinji, 1922

HYPEROMYZELLA Hille Ris Lambers, 1949
Temminkia 8: 279, 298, described as subg. of Hyperomyzus Börner, 1933
Type species Nectarosiphon rhinanthi Schouteden, 1903
Here used as subg. of Hyperomyzus Börner, 1933

HYPEROMYZUS Börner, 1933
Kleine Mitteilungen über Blattläuse, privately published, p. 2
Type species Aphis lactucae Linnaeus, 1758

H. börneri Prevost, 1959 = Hyperomyzus (Neonasonovia) zirnitsi Hille Ris Lambers, 1952
boerneri subsp. thorsteinni Stroyan, 1960 = Hyperomyzus (Neonasonovia) thorsteinni Stroyan, 1960
carduellinus (Theobald, 1915) (Rhopalosiphum) types BM HRL
*oleraceae* van der Goot, 1917 (*Rhopalosiphum*)
*sonchifoliae* Takahashi, 1923 (*Amphorophora*)
crepidis Heinze, 1961 = Hyperomyzus (Neonasonovia) picridis (Börner & Blunck, 1916)
davazhamci Holman & Szelegiewicz, 1972 nomen nudum
euphrasiae Walker of Börner, 1952 nec Walker, 1849 = Hyperomyzus (Neonasonovia) zirnitsi Hille Ris Lambers, 1952
franzi Börner, 1942 = Amphorophora gei (Börner, 1939)
lactucae (Linnaeus, 1758) (Aphis) BM HRL
*cosmopolitana* Mason, 1925 (*Amphorophora*)
*erraticum* Koch, 1854 (*Rhopalosiphum*)
*ribijaponica* Shinji, 1924 (*Rhopalosiphum*)
*sonchi* Oestlund, 1886 (*Rhopalosiphum*)
*sonchicola* Shinji, 1939 (*Amphorophora*)
*triticum* Theobald, 1923 (*Amphorophora*)
lampsanae (Börner, 1932) (Rhopalosiphoninus) BM HRL
pallidus Hille Ris Lambers, 1935 BM HRL
sobrinus F.P. Müller, 1966 = Hyperomyzus (Neonasonovia) zirnitsi Hille Ris Lambers, 1952
theobaldi Börner, 1950 = Rhopalosiphoninus (Myzosiphon) staphyleae (Koch, 1854)

Subgen. Hyperomyzella Hille Ris Lambers, 1949

H. (Hyperomyzella) rhinanthi (Schouteden, 1903) (Nectarosiphon) BM HRL
*affine* Börner, 1921 (*Rhopalosiphum*)
*britteni* Theobald, 1913 (*Rhopalosiphum*)
*tuberculatus* Theobald, 1929 (*Rhopalosiphoninus*) BM

Subgen. Neonosonovia Hille Ris Lambers, 1949

H. (Neonasonovia) ? accidentalis (Knowlton, 1929) (Amphorophora)
(Neonosanovia) hieracii (Börner, 1939) (Rhopalosiphoninus) BM HRL
(Neonasonovia) inflatus (Richards, 1962) (Nasonovia (Hyperomyzus)) HRL
(Neonasonovia) nabali (Oestlund, 1886) (Rhopalosiphum) BM HRL
*? braggi* Mason, 1925 (*Amphorophora*)
*hayhursti* Mason, 1925 (*Amphorophora*)
*mitchelli* Mason, 1925 (*Amphorophora*)
*pergandei* Mason, 1925 (*Amphorophora*)
(Neonasonovia) nabali subsp. fronki (Knowlton, 1945) (Amphorophora) HRL
(Neonasonovia) niger (J.M. Baker, 1934) (Amphorophora) HRL
(Neonasonovia) nigricornis (Knowlton, 1927) (Amphorophora)
(Neonasonovia) petiolaris (Knowlton & Allen, 1945) (Amphorophora)
(Neonasonovia) picridis (Börner & Blunck, 1916) (Rhopalosiphum) BM HRL
*crepidis* Heinze, 1961 (*Hyperomyzus*)
*hieracioides* Theobald, 1926 (*Amphorophora*) types BM HRL
(Neonasonovia) ribiellus (Davis, 1919) (Macrosiphum) BM HRL
*osborni* Knowlton, 1942 (*Amphorophora*)
(Neonasonovia) sandilandicus (Robinson, 1974) (Nasonovia (Hyperomyzus))
(Neonasonovia) thorsteinni Stroyan, 1960 (Hyperomyzus) BM HRL
*boerneri* subsp. *thorsteinni* Stroyan, 1960 (*Hyperomyzus*) BM HRL
(Neonasonovia) zirnitsi Hille Ris Lambers, 1952 BM HRL
*boerneri* Prevost, 1959 (*Hyperomyzus*) BM HRL
*euphrasiae* Walker of Börner, 1952 nec Walker, 1849 (*Hyperomyzus*)
*sobrinus* F.P. Müller, 1966 (*Hyperomyzus*) BM HRL

HYSTERONEURA Davis, 1919
Can. Ent. 51: 263
Type species Siphonophora setariae Thomas, 1878
*Heteroneura* Davis, 1919 nec Fallen, 1810

H. gregoryi Eastop, 1954 = setariae (Thomas, 1878)
ogloblini E.E. Blanchard, 1939 = setariae (Thomas, 1878)

H. setariae (Thomas, 1878) (Siphonophora) BM HRL
*bituberculata* Wilson, 1914 (*Aphis*)
*gregoryi* Eastop, 1954 type BM
*ogloblini* E.E. Blanchard, 1939
*panicola* Thomas, 1878 (*Siphonophora*)
*prunicoleus* Ashmead, 1881 (*Aphis*)
*scotti* Sanderson, 1905 (*Aphis*)

HYSTRICHIELLA Börner, 1909
Zool. Anz. 33: 609, described as subg. of Phylloxera Boyer de Fonscolombe, 1834
Type species Phylloxera spinulosa Targioni-Tozzetti, 1875 = P. quercina Ferrari, 1872
= Phylloxera Boyer de Fonscolombe, 1834
Phylloxeridae

IDIOPTERUS Davis, 1909
Ann. ent. Soc. Am. 2: 198
Type species Idiopterus nephrelepidis Davis, 1909
*Fullawayella* del Guercio, 1911

I. brasiliensis Moreira, 1925 to Microparsus (Picturaphis)
nephrelepidis Davis, 1909 BM HRL
*kirkaldyi* Fullaway, 1910 (*Macrosiphum*)

IDIOVATUS Börner, 1944
In Brohmer, Fauna von Deutschland, ed. 5, Leipzig, p. 217
Type species Ovatus titschaki Börner, 1942 (combination in 1952)
= Nectarosiphon Schouteden, 1901

ILLINOIA Wilson, 1910
Ann. ent. Soc. Am. 3: 318
Type species Siphonophora liriodendri Monell, 1879
*Ericobium* MacGillivray, 1958

I. alni (Mason, 1925) (Amphorophora) BM HRL
andromedae (MacGillivray, 1958) (Masonaphis (Ericobium)) HRL
azaleae (Mason, 1925) (Amphorophora) HRL
*vaccinii* Mason, 1925 (*Amphorophora*)
azaleae subsp. kalmiaflora (Tissot & Pepper, 1944) (Amphorophora) BM HRL
*kalmiaflora* Tissot & Pepper, 1944 (*Amphorophora*)
azaleae subsp. rhododendronia (Mason, 1925) (Amphorophora) HRL
*rhododendronia* Mason, 1925 (*Amphorophora*)

I. borealis (Mason, 1925) (Amphorophora) BM HRL
brevitarsis (Gillette & Palmer, 1933) (Amphorophora) BM HRL
canadensis (MacGillivray, 1958) (Masonaphis (Ericobium)) BM HRL
ceanothi (Bartholomew, 1932) (Amphorophora) BM HRL
corylina (Davidson, 1914) (Rhopalosiphum) HRL
dzhibladzeae (Shaposhnikov, 1964) (Masonaphis) BM HRL
finni (MacGillivray, 1958) (Masonaphis (Ericobium)) HRL
goldamaryae (Knowlton, 1938) (Amphorophora) BM HRL
*patchiae* Essig, 1942 (*Amphorophora*)
gracilicornis (MacGillivray, 1958) (Masonaphis (Ericobium)) HRL
grindeliae (Williams, 1911) (Siphonophora) BM HRL
*aridus* Knowlton, 1929 (*Amphorophora*)
grindeliae subsp. palmerae (MacGillivray, 1958) (Masonaphis (Ericobium)) BM HRL
kerriae Shinji, 1930 to Aulacorthum Mordvilko, 1914
liriodendri (Monell, 1879) (Siphonophora) BM HRL
macgillivrayae (Hille Ris Lambers, 1966) (Masonaphis (Ericobium)) BM HRL
macrosiphum Wilson, 1912 to Acyrthosiphon Mordvilko, 1914
masoni (Knowlton, 1928) (Neomyzus) BM HRL
morrisoni (Swain, 1918) (Nectarosiphon) BM HRL
*pseudomorrisoni* MacGillivray, 1958 (*Masonaphis (Ericobium)*) HRL
osmaroniae Wilson, 1912 to Macrosiphum Passerini, 1860
pallida (Mason, 1925) (Amphorophora) HRL
pepperi (MacGillivray, 1958) (Masonaphis (Ericobium)) BM HRL
phacelia (Essig, 1942) (Amphorophora) HRL
reticulata (Mason, 1925) (Amphorophora)
rhododendri (Wilson, 1918) (Macrosiphum) (type of Masonaphis Hille Ris Lambers, 1939) to Illinoia (Masonaphis)
richardsi (MacGillivray, 1958) (Masonaphis (Ericobium)) BM HRL
richardsi subsp. pacifica (Hille Ris Lambers, 1966) (Masonaphis (Ericobium)) BM HRL
simpsoni (MacGillivray, 1958) (Masonaphis (Ericobium)) HRL
spiraeae (MacGillivray, 1958) (Masonaphis (Ericobium)) HRL
spiraecola (Patch, 1914) (Macrosiphum) BM HRL
thalictri (MacGillivray, 1958) (Masonaphis (Ericobium)) HRL
wahnaga (Hottes, 1952) (Amphorophora) paratype BM HRL

Subgen. Amphorinophora MacGillivray, 1958

I. (Amphorinophora) crystleae (Smith & Knowlton, 1939) (Amphorophora) BM HRL

I. (Amphorinophora) crystleae subsp. bartholomewi (Essig, 1942) (Amphorophora) BM HRL
*bartholomewi* Essig, 1942 (*Amphorophora*)

Subgen. Masonaphis Hille Ris Lambers, 1939

I. (Masonaphis) lambersi (MacGillivray, 1960) (Masonaphis) paratypes BM HRL
(Masonaphis) magna (Hille Ris Lambers, 1974) (Masonaphis) HRL
(Masonaphis) menziesiae (Robinson, 1969) (Masonaphis) paratypes BM HRL
(Masonaphis) paqueti (MacGillivray, 1958) (Masonaphis) BM HRL
(Masonaphis) patriciae (Robinson, 1969) (Masonaphis) paratypes BM HRL
(Masonaphis) rhododendri (Wilson, 1918) (Macrosiphum) BMHRL
(Masonaphis) rhokalaza (Tissot & Pepper, 1944) (Amphorophora) paratype BM HRL
(Masonaphis) wilhelminae (Hille Ris Lambers, 1962) (Masonaphis) HRL

Subgen. Oestlundia Hille Ris Lambers, 1949

I. (Oestlundia) davidsoni (Mason, 1925) (Amphorophora) BM HRL
*arnicae* Glendenning, 1926 (*Amphorophora*) BM HRL
*arnicae* subsp. *thatcheri* Knowlton & Allen, 1937 (*Amphorophora*)
*utahensis* Knowlton & Allen, 1945 (*Amphorophora*)
(Oestlundia) maxima (Mason, 1925) (Amphorophora) BM HRL
(Oestlundia) rubicola (Oestlund, 1886) (Macrosiphum Oestlund, 1886 nec Passerini, 1860) BM HRL

IMPATIENTINUM Mordvilko, 1914
Faune Russie, Ins. Hém. 1: 72
Type species Impatientinum fuscum Mordvilko, 1928 = Aphis balsamines Kaltenbach, 1862
*Tuberosiphum* Shinji, 1922

I. asiaticum Nevsky, 1929 BM HRL
*balsamines* subsp. *asiaticum* Nevsky, 1929
asiaticum subsp. dalhousiensis Verma, 1969 HRL
*impatiensae* subsp. *dalhousiensis* Verma, 1969 HRL
balsamines (Kaltenbach, 1862) (Aphis) BM HRL
*fuscum* Mordvilko, 1928
balsamines subsp. asiaticum Nevsky, 1929 = asiaticum Nevsky, 1929
fuscum Mordvilko, 1928 = balsamines (Kaltenbach, 1862)
impatiens (Shinji, 1922) (Tuberosiphum) BM HRL
*impatiensae* Shinji, 1935 (*Macrosiphum*)
? *smilaceti* Takahashi, 1924 (*Macrosiphum*)

I. impatiensae subsp. dalhousiensis Verma, 1969 = asiaticum subsp. dalhousiensis Verma, 1969
pilosellae Börner, 1933 to Nasonovia Mordvilko, 1914

INDIAPHIS A.N. Basu, 1969
Orient. Insects 3: 175
Type species Indiaphis crassicornis A.N. Basu, 1969

I. crassicornis A.N. Basu, 1969 HRL
rostrata A.K. Ghosh & RayChaudhuri, 1972 HRL
setosa (Hille Ris Lambers & A.N. Basu, 1966) (Ericolophium (Neoacyrthosiphon)) HRL

INDIOCHAITOPHORUS Verma, 1970 (1969)
Bull. Ent., ent. soc. India 10: 134
Type species Indiochaitophorus furcatus Verma, 1970 (1969)

I. furcatus Verma, 1970 (1969) HRL

INDOCINARA A.K. Ghosh, R.C. Basu & RayChaudhuri, 1969
Orient. Insects 3: 249
Type species Indocinara hottesis A.K. Ghosh, R.C. Basu & RayChaudhuri, 1969

I. hottesis A.K. Ghosh, R.C. Basu & RayChaudhuri, 1969 BM

INDOIDIOPTERUS Chakrabarti, A.K. Ghosh & RayChaudhuri, 1972
Orient. Insects 6: 391
Type species Capitophorus geranii Chowdhuri, R.C. Basu, Chakrabarti & RayChaudhuri, 1969

I. geranii (Chowdhuri, R.C. Basu, Chakrabarti & RayChaudhuri, 1969) (Capitophorus)

INDOMASONAPHIS Verma, 1972 (1971)
Bull. Ent., ent. Soc. India 12: 97
Type species Indomasonaphis indicum Verma, 1972

I. indica Verma, 1972

INDOMEGOURA Hille Ris Lambers, 1958
In MacGillivray, Temminckia 10: 25
Type species Rhopalosiphum indicum van der Goot, 1916
*Omeimegoura* Tao, 1963

I. indica (van der Goot, 1916) (Rhopalosiphum) BM HRL
*essigwanai* Mason, 1925 (*Amphorophora*)
*hemerocallidis* Matsumura, 1918 (*Rhopalosiphum*)
*lilicola* Shinji, 1933 (*Amphorophora*)
*miniatum* Matsumura, 1918 (*Rhopalosiphum*)
*mitsubautsugii* Shinji, 1923 (*Nectarosiphum*)
nigrotibiae (Tao, 1963) (Omeimegoura) BM HRL

INDOMYZUS A.K. Ghosh, M.R. Ghosh & RayChaudhuri, 1971
Orient. Insects 5: 328
Type species Indomyzus sensoriatus A.K. Ghosh, M.R. Ghosh & RayChaudhuri, 1971

I. sensoriatus A.K. Ghosh, M.R. Ghosh & RayChaudhuri, 1971

INDONIPPONAPHIS A.K. Ghosh & RayChaudhuri, 1973
Kontyû 41: 158
Type species Indonipponaphis tuberculata A.K. Ghosh & RayChaudhuri, 1973

I. tuberculata A.K. Ghosh & RayChaudhuri, 1973

IOWANA Hottes, 1954
Proc. biol. Soc. Wash. 67: 99
Type species Iowana frisoni Hottes, 1954
Here used as subg. of Aphis Linnaeus, 1758

I. frisoni Hottes, 1954 to Aphis (Iowana)

IPUKA van Harten & Ilharco, 1976
Agronomia lusit. 37: 16
Type species Aulacorthum dispersum van der Goot, 1917

I. dispersum (van der Goot, 1917) (Aulacorthum) BM HRL
*eastopi* Carver, 1965 (*Micromyzus*) BM HRL
melantherae van Harten & Ilharco, 1976 BM HRL

IRANAPHIS Remaudière & Davatchi, 1959
Rev. Path. vég. Ent. agric. Fr. 38: 142
Type species Iranaphis dehbani Remaudière & Davatchi, 1959

I. dehbani Remaudière & Davatchi, 1959 BM HRL

ISRAELAPHIS Essig, 1953
Pan-Pacif. Ent. 29: 127
Type species Israelaphis carmini Essig, 1953

I. carmini Essig, 1953 HRL
lambersi Ilharco, 1962 BM HRL
tavaresi Ilharco, 1961 BM HRL

IZIPHYA Nevsky, 1929
Trudȳ uzbekist. opȳt. Sta. Zashch. Rast. 16: 314
Type species Iziphya maculata Nevsky, 1929
*Caricaphis* Börner, 1930
*Juncobia* Quednau, 1954

I. albipes Richards, 1958
americana (Baker, 1917) (Saltusaphis) HRL
austriaca Börner, 1950 HRL
*? piluliferae* Quednau, 1954 HRL
*suecica* Hille Ris Lambers, 1952 HRL
borealis Aizenberg, 1954
brevipes Richards, 1968 paratype BM HRL
bufo (Walker, 1848) (Aphis) type BM HRL
*bufo* subsp. *ericetorum* Börner, 1952
*oettingenii* Quednau, 1954 HRL
bufo subsp. ericetorum Börner, 1952 = bufo (Walker, 1848)
familiaris (Walker, 1848) (Aphis) = Brachycaudus helichrysi (Kaltenbach, 1843) but the insect referred to as I. familiaris (Walker, 1848) is I. bufo (Walker, 1848)
flabella (Sanborn, 1904) (Chaitophorus) BM HRL
grandipes Richards, 1970 paratype BM
ingegardae Hille Ris Lambers, 1952 BM HRL
insessa (Walker, 1849) (Aphis) = Brachycaudus helichrysi (Kaltenbach, 1843) but the insect mentioned as I. insessa (Walker, 1849) is I. leegei Börner, 1940
leegei Börner, 1940 BM HRL
mackaueri Quednau, 1971
maculata Nevsky, 1929 HRL
*mongolica* Quednau, 1968 HRL
memorialis Börner, 1950 BM HRL
mongolica Quednau, 1968 = maculata Nevsky, 1929
montana Börner, 1950

I. nigriceps Richards, 1958 HRL
oettingenii Quednau, 1954 = fundatrix of bufo (Walker, 1848)
piluliferae Quednau, 1954 (= ? austriaca Börner, 1950) HRL
punctata Hille Ris Lambers, 1960 cotypes BM BRL
*punctatella* Richards, 1961
punctatella Richards, 1961 = punctata Hille Ris Lambers, 1960
spenceri Richards, 1958 BM HRL
suecica Hille Ris Lambers, 1952 = austriaca Börner, 1950
umbella Richards, 1968 HRL
variabilis Quednau, 1968/9 HRL
vittata Richards, 1958 BM HRL

IZIPHYOPSIS Börner, 1944
In Brohmer, Fauna von Deutschland, ed. 5, Leipzig, p. 214
Without species

JACEIFEX Amyot, 1847
Annls Soc. Ent. Fr. (2) 5: 481
Invalid

JACKSONIA Theobald, 1923
Scott. Nat. 1923: 19
Type species Jacksonia papillata Theobald, 1923

J. papillata Theobald, 1923
*morrisoni* Laing, 1928 (*Myzus*)

JARIA Quednau ex A.K. Ghosh & RayChaudhuri, 1972 lapsus pro Taoia Quednau, 1973

JAXARTAPHIS Mordvilko, 1914
Faune Russie, Ins. Hém. 1: 65
In key without species

JEZABURA lapsus pro Yezabura

JUDENKOA Hille Ris Lambers, 1946 & 1949
In Alta, H. & Docters van Leeuwen, W.M., Gallenboek, p. 158 (1946); Temminckia 9: 35, 37, 39, 46, (1949) described as subg. of Rhopalomyzus Mordvilko, 1921
Type species Aphis lonicerae Siebold, 1839
Here used as subg. of Rhopalomyzus Mordvilko, 1921

JUGLANDIFEX Amyot, 1847
Annls Soc. ent. Fr. (2) 5: 481
Invalid

JUNCOBIA Quednau, 1954
Mitt. biol. Reichsanst. Ld- u. Forstw. 78: 40
Type species cited as Caricaphis leegei Borner, 1940 = Iziphya leegei Börner, 1940
= Iziphya Nevsky, 1929

JUNCOMYZUS Hille Ris Lambers, 1965
Tijdschr. Ent. 7: 193
Type species Juncomyzus obscurus Hille Ris Lambers, 1965

J. floris Miyazaki, 1971 BM HRL
hillerislambersi Calilung, 1972
niger Miyazaki, 1971 HRL
obscurus Hille Ris Lambers, 1965 BM HRL
rhois (Takahashi, 1924) (Myzus) BM HRL
*japonicus* Takahashi, 1963 (*Sitomyzus*)

KABURAGIA Takagi, 1937
Bull. Forest Exp. Stn Seoul 26: 20, 22
Type species Kaburagia rhusicola Takagi, 1937
*Macrorhinarium* Tsai & Tang, 1945

K. ailanthi Chowdhuri, R.C. Basu, Chakrabarti & RayChaudhuri, 1969 paratype BM
rhusicola Takagi, 1937
*ensigallis* Tsai & Tang, 1945 (*Macrorhinarium*)
*lingi* Tao, 1943 (*Pemphigella*)
*ovogallis* Tsai & Tang, 1945 (*Macrorhinarium*)

KAKIMIA Hottes & Frison, 1931
Bull. Ill. St. nat. Hist. Surv. 19: 344, described as subg. of Myzus Passerini, 1860
Type species Myzus (Kakimia) thomasi Hottes & Frison, 1931

K. acyrthosiphon Richards, 1963
alpina (Gillette & Palmer, 1928) (Myzus) BM HRL
altaensis Stenseth, 1969 (Nasonovia) HRL
aquilegiae (Essig, 1917) (Myzus) BM HRL
*essigi* Gillette & Palmer, 1929 (*Myzus*)
canadensis Robinson, 1968 to Delphiniobium Mordvilko, 1914

K. castilleiae Sampson, 1939 BM HRL
cefsmithi (Knowlton, 1940) (Capitophorus) HRL
cerei Gillette & Palmer, 1933 BM HRL
collomiae Palmer, 1936 HRL
crenicorna (Smith & Knowlton, 1939) (Macrosiphum) BM HRL
cynosbati (Oestlund, 1887) (Nectarophora) BM HRL
heucherae (Thomas, 1879) (Siphonophora)
houghtonensis (Troop, 1906) (Aphis) BM HRL
jammuensis (Verma, 1970) (Nasonovia (Kakimia)) BM HRL
michaelseni (Schouteden, 1904) (Myzus)
mimulicola Drews & Sampson, 1937 to Ovatus van der Goot, 1913
muesebecki Knowlton & Allen, 1939
polemonii (Gillette & Palmer, 1929) (Myzus) BM HRL
potentillae (Williams, 1911) (Myzus) (preoccupied in Myzus) BM HRL
purpurascens (Oestlund, 1887) (Nectarophora) BM HRL
*thalictri* (Williams, 1911) (*Myzus*)
ribe-utahensis Knowlton, 1935
ribifolii (Davidson, 1917) (Myzus) BM HRL
*suguri* Shinji, 1927 (*Myzus*)
robinsoni Richards, 1958 = wahinkae (Hottes, 1933)
salviae (Nevsky, 1929) (Acyrthosiphon) HRL
takala (Hottes, 1933) (Dactynotus (Kakimia)) BM HRL
thomasi (Hottes & Frison, 1931) (Myzus (Kakimia))
utahensis Knowlton, 1943 HRL
vockerothi Richards, 1963
wahinkae (Hottes, 1933) (Dactynotus) BM HRL
*robinsoni* Richards, 1958 BM

Subgen. Neokakimia Doncaster & Stroyan, 1952

K. (Neokakimia) dasyphylli Stroyan, 1954 to Nasonovia (Neokakimia)
(Neokakimia) saxifragae Doncaster & Stroyan, 1952 (type of Neokakimia Doncaster & Stroyan, 1952)) to Nasonovia (Neokakimia)

KALLISTAPHIS Kirkaldy, 1905
Can. Ent. 37: 417
Type species Aphis betulicola Kaltenbach, 1843
= Calaphis Walsh, 1863

K. basalis Stroyan, 1957 = Calaphis flava Mordvilko, 1928

KALTANAPHIS Becker-Migdisova, 1959
Materiali K "Osnovam paleontologii" 3: 107
Type species Kaltanaphis permiensis Becker-Migdisova, 1959
Permaphidopseidae, Fossil, possibly Psylloidea, not Aphidoidea

KALTENBACHIELLA Schouteden, 1906
Mém. Soc. r. ent. Belg. 12: 194
Type species Kaltenbachiella menthae Schouteden, 1906 = Byrsocrypta pallida Haliday, 1838
*Gobaisha* Matsumura, 1917

K. elsholtriae (Shinji, 1935) (Eriosoma) BM HRL
japonica (Matsumura, 1917) (Gobaisha) HRL
menthae Schouteden, 1906 = pallida (Haliday, 1838)
pallida (Haliday, 1838) (Byrsocrypta) BM HRL
*alba* Ratzeburg, 1844 (*Aphis*)
*menthae* Passerini, 1860 (*Rhizobius*)
*menthae* Schouteden, 1906
*ulmi* Lichtenstein, 1879 (*Pemphigus*)
ulmifusa (Walsh & Riley, 1869) (Pemphigus) BM HRL
*? walshii* Williams, 1911 (*Pemphigus*)

KAOCHIAOJA Tao, 1963
Pl. Prot. Bull., Taiwan 5: 169
Type species Myzus arthraxonis Takahashi, 1921

K. arthraxonis (Takahashi, 1921) (Myzus) HRL
*pollinae* Shinji, 1924 (*Macrosiphum*)

KERMAPHIS Maskell, 1885
Trans. N. Z. Inst. 17: 19
Type species Anisophleba pini Koch, 1857 = Kermes pini Macquart, 1819 = Pineus Shimer, 1869
Adelgidae

KESSLERIA Lichtenstein, 1885
Monographie des Pucerons du peuplier, Montpellier, p. 16, described as subg. of

Pemphigus Hartig, 1839
Type species Pemphigus spirothecae Passerini, 1860 = Pemphigus spyrothecae Passerini, 1856
= Pemphigus Hartig, 1839

KUGEGANIA Eastop, 1955
Entomologist's mon. Mag. 91: 204, described as subg. of Micromyzus van der Goot, 1917
Type species Micromyzus (Kugegania) ageni Eastop, 1955
Here used as subg. of Micromyzella Eastop 1955

KURISAKIA Takahashi, 1924
Philipp. J. Sci. 24: 715
Type species Kurisakia juglandicola Takahashi, 1924 = Anoecia onigurumii Shinji, 1923
*Tuberocorpus* Shinji, 1929 (nec 1932)

K. ailanthi Takahashi, 1960
indica A.N. Basu, 1968 HRL
juglandicola Takahashi, 1924 = onigurumi (Shinji, 1923)
onigurumii (Shinji, 1923) (Anoecia) BM HRL
*juglandicola* Takahashi, 1924
*onigurumii* Shinji, 1924 (*Glyphina*)
*pterocaryae* Monzen, 1927 (*Glyphina*)
onigurumi subsp. querciphila Takahashi, 1960 HRL

LACHNAPHIS Amyot, 1847
Annls Soc. ent. Fr. (2) 5: 533
Invalid

LACHNAPHIS Shinji, 1922
Zool. Mag., Tokyo 34: 729
Type species Lachnaphis yomogi Shinji, 1922 = Sappaphis piri Matsumura, 1918
= Sappaphis Matsumura, 1918

L. yomogi Shinji, 1922 = Sappaphis piri Matsumura, 1918

LACHNIELLA del Guercio, 1909
Redia 5: 286
Type species Lachnus fasciatus Burmeister, 1835 of del Guercio = Aphis costata Zetterstedt, 1828
= Cinara Curtis, 1835

L. arizonica Wilson, 1919 to Cinara Curtis, 1835
atlantica Wilson, 1919 to Cinara Curtis, 1835
burrilli Wilson, 1919 to Cinara Curtis, 1835
caudelli Wilson, 1919 to Cinara Curtis, 1835
cilicica del Guercio, 1909 = Cinara confinis (Koch, 1856)
cilicica var. cecconii del Guercio, 1909 (type of Dinolachnus Börner, 1940) = Cinara confinis (Koch, 1856)
comata Doncaster, 1956 to Cinara Curtis, 1835
edulis Wilson, 1919 to Cinara Curtis, 1835
gracilis Wilson, 1919 (type of Dilachnus Baker, 1919; and of Panimerus Laing, 1926 nec Eaton, 1913; and of Wilsonia Baker, 1919 nec Bonaparte, 1838; and of Neochmosis Laing ex Theobald, 1929) to Cinara Curtis, 1835
inoptis Wilson, 1919 = Cinara pinea (Mordvilko, 1895)
juniperi var. signata del Guercio, 1909 = Cinara costata (Zetterstedt, 1828)
laricis subsp. cuneomaculata del Guercio, 1909 = Cinara cuneomaculata (del Guercio, 1909)
montana Wilson, 1919 = Lachnus allegheniensis McCook, 1877
montanensis Wilson, 1919 to Cinara Curtis, 1835
nigra Wilson, 1919 to Cinara Curtis, 1835
nigrotuberculata del Guercio, 1909 = Cinara laricis (Hartig, 1839)
oblonga del Guercio, 1909 to Cinara Curtis, 1835
pacifica Wilson, 1919 to Cinara Curtis, 1835
pergandei Wilson, 1919 to Cinara Curtis, 1835
picta del Guercio, 1909 to Cinara Curtis, 1835
pinivora Wilson, 1919 to Cinara Curtis, 1835
schwarzii Wilson, 1919 to Cinara Curtis, 1835
thujafolia Theobald, 1914 = Cinara tujafilina (del Guercio, 1909)
thunbergii Wilson, 1919 to Eulachnus del Guercio, 1909
tujae del Guercio, 1909 = Cinara cupressi (Buckton, 1881)
tujafilina del Guercio, 1909 to Cinara Curtis, 1835
vandykei Wilson, 1919 to Cinara Curtis, 1835

LACHNOCHAITOPHORUS Granovsky, 1933
Proc. ent. Soc. Wash. 35: 33
Type species Lachnochaitophorus querceus Granovsky, 1933

bisselli Granovsky, 1933 = obscurus (Tissot, 1932)
obscurus (Tissot, 1932) (Patchia) paratype BM
*bisselli* Granovsky, 1933
querceus Granovsky, 1933

LACHNUS Burmeister, 1835
Handbuch der Entomologie, Berlin, 2: 91 (genus attributed to Illiger)
Type species Lachnus fasciatus Burmeister, 1835 = Aphis roboris Linnaeus, 1758
*Dryaphis* Kirkaldy, 1904
*Dryobius* Koch, 1855 nec Le Conte, 1850
*Pterochlorus* Passerini, 1860 emendation pro Pteroclorus Rondani, 1848
*Pteroclorus* Rondani, 1848
*Schizodryobius* van der Goot, 1913
*Sublachnobius* Heinze, 1962

L. abieticola Cholodkovsky, 1899 = Cinara confinis (Koch, 1856)
abietis Fitch, 1851 to Cinara Curtis, 1835
acutihirsutus Kumar & Burkhardt, 1970 BM
agilis Kaltenbach, 1843 type of Eulachnus del Guercio, 1909 q.v.
allegheniensis McCook, 1877 BM HRL
*montana* Wilson, 1919 (*Lachniella*)
*quercicolens* Ashmead, 1881
alnifoliae Fitch, 1851 to Pterocallis Passerini, 1860
apini Gillette & Palmer, 1924 to Cinara Curtis, 1835
ater Gillette & Palmer, 1924 to Cinara Curtis, 1835
aucupariae Dahlbom, 1851 nomen dubium
australi Ashmead, 1881 nomen dubium
bignoniae Macchiati, 1884 nomen dubium
biotae van der Goot, 1917 = Cinara tujafilina (del Guercio, 1909)
bogdanowi Mordvilko, 1895 = Cinara pruinosa (Hartig, 1841)
bonneti Heer, 1852 fossil not lachnid, vide Heie, 1967: 222
braggi Gillette, 1917 to Cinara Curtis, 1835
brevispinosus Gillette & Palmer, 1924 to Cinara Curtis, 1835
californicus Essig, 1909 type of Essigella del Guercio, 1909 q.v.
carificex Cook, 1879 errore pro laricifex Fitch, 1858
castaneae Hille Ris Lambers, 1967 = roboris (Linnaeus, 1758)
cembrae Seitner, 1936 to Cinara Curtis, 1835

L. chosoni Szelegiewicz, 1975 (Lachnus (Schizodryobius))
cimicoides Germar & Berendt, 1856 type of Berendtaphis Heie, 1971 Fossil q.v.
coloradensis Gillette, 1917 to Cinara Curtis, 1835
confinis Koch, 1856 to Cinara Curtis, 1835
costatus Hartig, 1839 = Cinara costata (Zetterstedt, 1828)
crassicornis Hille Ris Lambers, 1948 BM HRL
cupressi Buckton, 1881 to Cinara Curtis, 1835
curvipes Patch, 1912 to Cinara Curtis, 1835
dentatus LeBaron, 1872 = Tuberolachnus salignus (Gmelin, 1790)
distinguendus Dahlbom, 1851 nomen dubium
dryoides Germar & Berendt, 1856 type of Germaraphis Heie, 1967 q.v. Fossil
engelmanniensis Gillette & Palmer, 1924 to Cinara Curtis, 1835
exsiccator Altum, 1882 = pallipes (Hartig, 1841)
farinosus Cholodkovsky, 1891 = Cinara costata (Zetterstedt, 1828)
fasciatus Burmeister, 1835 (nominal type of Lachniella del Guercio, 1909)= roboris (Linnaeus, 1758), but Kaltenbach, 1843 and many later authors applied the name to Cinara costata (Zetterstedt, 1828)
fasciatus Shinji, 1922 nec Burmeister, 1835 = Cinara shinji Inouye, 1939
ferrisi Swain, 1918 to Cinara Curtis, 1835
fici Takahashi, 1934
flavus Mordvilko, 1895 = Cinara pilicornis (Hartig, 1841)
flexilis Gillette & Palmer, 1924 to Cinara Curtis, 1835
flocculosa Williams, 1911 to Schizolachnus Mordvilko, 1909 (1908)
formicophilus Buckton, 1901 nomen dubium
fuliginosus Buckton, 1891 = Tuberolachnus salignus (Gmelin, 1790)
fuscus Passerini, 1863 nomen dubium
glabra Gillette & Palmer, 1924 to Cinara Curtis, 1835
glandulosus Menge, 1856 type of Mengeaphis Heie, 1967, q.v. fossil
glehnus Essig, 1915 to Cinara Curtis, 1835
greeni Schouteden, 1905 to Cinara Curtis, 1835
grossus Kaltenbach, 1846 = Cinara piceae (Panzer, 1801)
himalayensis van der Goot, 1918 to Nippolachnus Matsumura, 1917
hottesi Gillette & Palmer, 1924 to Cinara Curtis, 1835
hyalinus Koch, 1856 = Cinara pilicornis (Hartig, 1841)
hyperophilus Koch, 1855 to Cinara Curtis 1835
iliciphilus (del Guercio, 1909) (Dryaphis) BM HRL
    *minor* del Guercio, 1909 (*Dryaphis*) HRL
    *roboris* var. *longirostris* Mordvilko, 1909 (*Pterochlorus*)
incertus Schouteden, 1906 = Maculolachnus submacula (Walker, 1848)
inflatus Shinji, 1922

L. japonicus (Matsumura, 1917) (Pterochlorus) HRL
juglandicola Kaltenbach, 1843 type of Chromaphis Walker, 1870 q.v.
juniperensis Gillette & Palmer, 1925 to Cinara Curtis, 1835
juniperinus Mordvilko, 1895 = Cinara cupressi (Buckton, 1881)
juniperivora Wilson, 1919 to Cinara Curtis, 1835
laricicolus Matsumura, 1917 to Cinara Curtis, 1835
laricifex Fitch, 1858 to Cinara Curtis, 1835
laricifoliae Wilson, 1915 to Cinara Curtis, 1835
laricis Hartig, 1839 to Cinara Curtis, 1835
laricis Koch, 1856 = Cinara laricis (Hartig, 1839)
lasiocarpae Gillette & Palmer, 1930 = Cinara confinis (Koch, 1856)
lepineyi Mimeur, 1934 = roboris (Linnaeus, 1758)
longicorpi Shinji, 1922 (longicarpi in caption) = Eulachnus thunbergii (Wilson, 1919)
longipennis Matsumura, 1917 to Cinara Curtis, 1835
longistigma Monell, 1878 = Longistigma caryae (Harris, 1841)
longitarsus Ferrari, 1872 to Protrama Baker, 1920
longulus Germar & Berendt, 1856 to Palaeophyllaphis Heie, 1967 fossil
macrocephalus Buckton, 1881 = Cinara pilicornis (Hartig, 1841)
maculatus Lichtenstein, 1884 nomen nudum
maculosus Cholodkovsky, 1899 = Cinara laricis (Hartig, 1839)
medispinosus Gillette & Palmer, 1929 to Cinara Curtis, 1835
muravensis Arnhardt, 1927 = Cinara laricis (Hartig, 1839)
murrayanae Gillette & Palmer, 1924 to Cinara Curtis, 1835
neubergi Arnhardt, 1930 (used as type of Pseudocinara Pašek, 1951 m.s. ex Pintera, 1966) to Cinara Curtis, 1835
nigripes Takahashi, 1932 = Tuberolachnus salignus (Gmelin, 1790)
niitakayamensis Takahashi, 1925 type of Sinolachnus Hille Ris Lambers, 1956 q.v.
occidentalis Davidson, 1909 to Cinara Curtis, 1835
oregonensis Wilson, 1915 to Cinara Curtis, 1835
padi Hartig, 1841 nomen dubium
pallipes (Hartig, 1841) (Aphis) BM HRL
*exsiccator* Altum, 1882
palmerae Gillette, 1917 = Cinara pruinosa (Hartig, 1841)
parvus Wilson, 1915 (as Unilachnus parvus type of Unilachnus Wilson, 1919) to Schizolachnus Mordvilko, 1909 (1908)
pectorosus Heer, 1853 nomen dubium, fossil
persicae Cholodkovsky, 1899 type of Pterochloroides Mordvilko, 1914 q.v. and of Tuberodryobius Das, 1918

L. petrorum Scudder, 1877 type of Geranchon Scudder, 1890 fossil
piceicola Cholodkevsky, 1898 = Cinara pilicornis (Hartig, 1841)
piceicola var. viridescens Cholodkovsky, 1898 (type of Cinarina Börner, 1939) = Cinara pilicornis (Hartig, 1841)
pichtae Mordvilko, 1895 = Cinara pectinatae (Nördlinger, 1880)
pineti Hartig, 1839 = Cinara pini (Linnaeus, 1758)
pineus Mordvilko, 1895 (type of Cinarella Hille Ris Lambers, 1948) to Cinara Curtis, 1835
pineus var. curtipilosa Mordvilko, 1895 = Cinara piniphila (Ratzeburg, 1844)
pini var. cembrae Cholodkovsky, 1891 = Cinara cembrae (Seitner, 1936)
pinicola Kaltenbach, 1843 (type of Cinaropsis Börner, 1939) to Cinara Curtis, 1835
pinidensiflorae Essig & Kuwana, 1918 to Cinara Curtis, 1835
pinihabitans Mordvilko, 1895 (type of Neodimosis Toth, 1935) to Cinara Curtis 1835
piniradiatae Davidson, 1909 to Schizolachnus Mordvilko, 1909 (1908)
platani Kaltenbach, 1843 to Tinocallis Matsumura, 1919
platanicola Riley, 1883 = Longistigma caryae (Harris, 1841)
ponderosae Williams, 1911 to Cinara Curtis, 1835
pruinosa Hartig, 1841 (type of Pityaria Börner, 1949) to Cinara Curtis, 1835
pseudosabinae Nevsky, 1929 = Cinara tujafilina (del Guercio, 1909)
pseudotsugae Wilson, 1912 to Cinara Curtis, 1835
pulverulens Gillette & Palmer, 1924 to Cinara Curtis, 1835
punctatus Burmeister, 1835 = Tuberolachnus salignus (Gmelin, 1790)
pyri Buckton, 1899 type of Pyrolachnus A.N. Basu & Hille Ris Lambers, 1968 q.v.
quercicolens Ashmead, 1881 = allegheniensis McCook, 1877
quercifoliae Fitch, 1851 ? to Chaitophorus Koch, 1854
quesneli Scudder, 1878 type of Sbenaphis Scudder, 1890, fossil
rileyi Williams, 1911 to Eulachnus del Guercio, 1909
roboris (Linnaeus, 1758) (Aphis) BM HRL
*boerneri* Pašek, 1953 (*Lachnus (Schizodryobius)*)
*castaneae* Hille Ris Lambers, 1967
*cerricola* del Guercio, 1909 (*Dryaphis*)
*croaticus* Koch, 1855 (*Dryobius*) ♂
*fasciatus* Burmeister, 1835
*hyalinatus* Fabricius, 1794 (*Lygaeus*)
*ilicicola* Boisduval, 1867 (*Aphis*)
*ilicina* del Guercio, 1909 (*Dryaphis*)
*lepineyi* Mimeur, 1934

*longipes* Dufour, 1833 (*Aphis*)
? *roboris nigra* del Guercio, 1909 (*Dryaphis*)
? *roboris* subsp. *georgica* Abashidze, 1951
*sachtlebeni* Börner, 1952
*sessilis* Börner, 1940

L. roboris subsp. georgica Abashidze, 1951 = ? roboris (Linnaeus, 1758)
rosae Cholodkovsky, 1899 (type of Maculolachnus Gaumont, 1920; and of Neolachnus Mordvilko, 1929) = Maculolachnus submacula (Walker, 1848)
rosarum van der Goot, 1912 = Maculolachnus submacula (Walker, 1848)
rubicundus Wilson, 1915 to Cinara Curtis, 1835
sabinae Gillette & Palmer, 1924 to Cinara Curtis, 1835
sabinae Nevsky, 1929 nec Gillette & Palmer, 1924 = Cinara tujafilina (del Guercio, 1909)
sabinianus Swain, 1919 = Cinara arizonica (Wilson, 1919)
sachtlebeni Börner, 1952 = roboris (Linnaeus, 1758)
salicellus Fitch, 1851
salicicola Uhler, 1862 to Pterocomma Buckton, 1879
sessilis Börner, 1940 = roboris (Linnaeus, 1758)
sibiricae Gillette & Palmer, 1924 to Cinara Curtis, 1835
similis Gillette & Palmer, 1924 nec van der Goot, 1917 = Cinara medispinosa (Gillette & Palmer, 1929)
similis van der Goot, 1917 to Cinara Curtis, 1835
smilacis Williams, 1911 = Neoprociphilus aceris (Monell, 1882)
solitarius Gillette & Palmer, 1924 to Cinara Curtis, 1835
splendens Gillette & Palmer, 1924 to Cinara Curtis, 1835
subnudus Börner ex Bodenheimer, 1930 & 1937 nomen nudum
subterraneus del Guercio, 1900 = Maculolachnus submacula (Walker, 1848)
subterraneus Hartig, date ?, vide Schouteden, 1906: 203, footnote
swirskii Hille Ris Lambers, 1954 BM HRL
swirskii subsp. persica Davatchi, Hille Ris Lambers & Remaudière, 1957 paratype BM HRL
taeniatoides Mordvilko, 1901 to Cinara Curtis, 1835
taeniatus Koch, 1856 to Cinara Curtis, 1835
tatakaensis Takahashi, 1937 = ? Tuberolachnus salignus (Gmelin, 1790)
taxifolia Swain, 1918 to Cinara Curtis, 1835
taxifoliae Swain, errore pro taxifolia Swain. 1918
terminalis Gillette & Palmer, 1932 to Cinara Curtis, 1835
titabarensis RayChaudhuri & A.K. Ghosh, 1964 to Aiceona Takahashi, 1921
tropicalis (van der Goot, 1916) (Pterochlorus) BM HRL
*bogoriensis* Franssen, 1932 (*Pterochlorus*)

L. vanduzei Swain, 1919 = Cinara confinis (Koch, 1856)
wichmanni Hille Ris Lambers, 1956 type of Sublachnobius Heinze, 1962 BM HRL

Subgen. Schizodryobius van der Goot, 1913

L. (Schizodryobius) boerneri Pašek, 1953 = Lachnus roboris (Linnaeus, 1758)
(Schizodryobius) chosoni Szelegiewicz, (1974) 1975 to Lachnus Burmeister, 1835

LACTUCOBIUM Hille Ris Lambers, 1947
Temminckia 7: 255, described as subg. of Acyrthosiphon Mordvilko, 1914
Type species Acyrthosiphon scariolae Nevsky, 1929 = Siphonophora lactucae Passerini, 1860
= Acyrthosiphon Mordvilko, 1914

LAINGIA Theobald, 1922
Bull. ent. Res. 12: 429
Type species Laingia psammae Theobald, 1922
*Anochetium* Wood-Baker, 1943

L. psammae Theobald, 1922
*nondescriptum* Wood-Baker, 1943 (*Anochetium*)
? *pilosa* Dahl, 1912 (*Glyphina*) nec Buckton, 1883

LAMBERSAPHIS Narzikulov, 1961
Trudȳ Inst. Zool. Parazit. tadzhik. 20: 85
Type species Neothomasia populicola subsp. pruinosa Narzikulov, 1954

L. pruinosa (Narzikulov, 1954) (Neothomasia) BM HRL
*populicola* subsp. *pruinosa* Narzikulov, 1954 (*Neothomasia*)

LAMBERSIUS Olive, 1965
Ann. ent. Soc. Am. 58: 284, described as subg. of Dactynotus Rafinesque, 1818
Type species Siphonophora erigeronensis Thomas, 1878
Here used as subg. of Uroleucon Mordvilko, 1914

LANDISAPHIS Knowlton & Ma, 1949
J. Kans. ent. Soc. 22: 147
Type species Landisaphis davisi Knowlton & Ma, 1949

L. davisi Knowlton & Ma, 1949 BM

LARICARIA Börner, 1949
Beitr. tax. Zool. 1: 59, described as subg. of Cinaria Börner, 1939
Type species Cinaria kochiana Börner, 1939
= Cinara Curtis, 1835

LARICETHUS Amyot, 1847
Annls Soc. ent. Fr. (2) 5: 485
Invalid

LARSSONAPHIS Heie, 1967
Spol. zool. Mus. haun. 26: 169
Type species Larssonaphis obnubila Heie, 1967
Fossil

LARVAPHIS Ossiannilsson, 1953
Opusc. ent. 18: 237, described as subg. of Thripsaphis Gillette, 1917
Type species Thripsaphis (Larvaphis) brevicornis Ossiannilsson, 1953
Here used as subg. of Thripsaphis Gillette, 1917

LAUFFERELLA Lindinger, 1933
Ent. Rdsch. 50: 32
Type species Pseudochermes populi Bonfigli, 1909 = Phylloxera populi del Guercio, 1900
= Phylloxerina Börner, 1908
Phylloxeridae

LEPTOPTERYX Zetterstedt, 1837 nec Horsfield, 1821
Isis (von Oken) 1837: 39
Type species Leptopteryx nivalis Zetterstedt, 1837 = Aphis punctipennis Zetterstedt, 1828
= Euceraphis Walker, 1870
L. nivalis Zetterstedt, 1837 = Euceraphis punctipennis (Zetterstedt, 1828)

LEUCOSIPHON Börner, 1952
Mitt. thüring. bot. Ges. 4 (3): 78, 245, described as subg. of Medoralis Börner, 1952
Type species Aphis farinosa Gmelin, 1790
= Aphis Linnaeus, 1758

LIDAJA Börner, 1952
Mitt. thüring. bot. Ges. 4 (3): 117, 252
Type species Hyalopterus abrotani Koch, 1854
= Coloradoa Wilson, 1910

L. heinzei Börner, 1952 to Coloradoa Wilson, 1910
taurica Mamontova, 1963 to Coloradoa Wilson, 1910

LINEOMYZOCALLIS Richards, 1965
Mem. ent. Soc. Can. 44: 29, described as subg. of Myzocallis Passerini, 1860
Type species Aphis bella Walsh, 1863
Here used as subg. of Myzocallis Passerini, 1860

L. ephemerata Richards, 1965 = Myzocallis (Lineomyzocallis) walshii (Monell, 1879)
longirostris Richards, 1965 to Myzocallis (Lineomyzocallis)
occultus Richards, 1965 to Myzocallis (Lineomyzocallis)

LINOSIPHON Börner, (1944) 1950
In Brohmer, P., Fauna von Deutschland, ed. 5, Leipzig, p. 217 in key, without species (1944)
Neue europäische Blattlausarten, privately published, p. 16 (1950)
Type species Macrosiphum galiophagum Theobald [sic] = Wimshurst, 1923

L. asperulophagum Holman, 1961 paratypes BM HRL
galii (Mamontova, 1961) (Anthracosiphon) paratypes BM HRL
galiophagum (Wimshurst, 1923) (Macrosiphum) presumed type BM HRL
sanguinarium (Hottes & Frison, 1931) (Macrosiphum) BM HRL

LIOSOMAPHIS Walker, 1868
Zoologist (2) 3: 1119
Type species Aphis berberidis Kaltenbach, 1843

L. atra Hille Ris Lambers, 1966 BM HRL
? *neoempetri* A.K. Ghosh, R.C. Basu & RayChaudhuri, 1971 (*Wahlgreniella*)
berberidis (Kaltenbach, 1843) (Aphis) BM HRL
? *berberidis* Fitch, 1851 (*Aphis*)
? *berberidis* Narzikulov, 1957 (*Rhopalomyzus*)
evadens Rusanova, 1951 = ? Myzus (Nectarosiphon) ligustri (Mosley, 1851)
himalayensis A.N. Basu, 1964 BM HRL
laricis Rupais, 1974 to Elatobium Mordvilko, 1914
lydiae Narzikulov, 1957 type of Berberidaphis Narzikulov, 1960 q.v.
ornata Miyazaki, 1971 HRL
turanicus Narzikulov, 1960

LIPAMYZODES Heinze, 1960
Beitr. Ent. Berlin 10: 817
Type species Lipaphis matthiolae Doncaster, 1954

L. matthiolae (Doncaster, 1954) (Lipaphis) types BM HRL

LIPAPHIDIELLA Doncaster, 1954
Proc. R. ent. Soc. Lond. 23: 83, described as subg. of Lipaphis Mordvilko, 1928
Type species Myzaphis lepidii Nevsky, 1929
Here used as subg. of Lipaphis Mordvilko, 1928

LIPAPHIDOIDES Börner, 1939
Arb. physiol. angew. Ent. Berl. 6: 50, described as subg. of Lipaphis Mordvilko, 1928
Type species Lipaphis rossi Borner, 1939
= Lipaphis Mordvilko, 1928

LIPAPHIS Mordvilko, 1928
In Philipjev, Keys for the identification of Russian Insects, Moscow, p. 200
Type species Aphis erysimi Kaltenbach, 1843
*Lipaphidoides* Börner, 1939

L. alliariae F.P. Müller, 1952 BM HRL
cochleariae Jacob, 1956 type BM HRL
erysimi (Kaltenbach, 1843) (Aphis) BM HRL
*contermina* Walker, 1849 (*Aphis*) lectotype BM
*indobrassicae* Das, 1918 (*Siphocoryne*)
*mathiolella* Theobald, 1918 (*Aphis*)
*papaveri* Takahashi, 1921 (*Rhopalosiphum*)
*pseudobrassicae* Davis, 1914 (*Aphis*)
*? sisymbrii* del Guercio, 1913 (*Rhopalosiphum*)
fritzmuelleri Börner, 1950 BM HRL
hedickei Börner, 1952 nomen nudum
matthiolae Doncaster, 1954 type of Lipamyzodes Heinze, 1960 q.v.
rossi Börner, 1939 type of Lipaphidoides Börner, 1939 BM HRL
rossi subsp. conringiae Bozhko, 1961 HRL
ruderalis Börner, 1939 to Lipaphis (Lipaphidiella)
turritella (Wahlgren, 1838) (Brachycolus) BM HRL

Subgen. Lipaphidiella Doncaster, 1954

L. (Lipaphidiella) lepidii (Nevsky, 1929) (Myzaphis) BM HRL
(Lipaphidiella) lepidii subsp. lepidiicardariae (Knechtel & Manolache, 1944) (Myzaphis) BM HRL
(Lipaphidiella) ruderalis (Börner, 1939) (Lipaphis) BM HRL

Subgen. Volutaphis Börner, 1939

L. (Volutaphis) centaureae Börner, 1939 type of Volutaphis Börner, 1939 q.v.

LIPORRHINUS Börner, 1939
Arb. physiol. angew. Ent. Berl. 6: 82
Type species Aphis chelidonii Kaltenbach, 1843
Here used as subg. of Acyrthosiphon Mordvilko, 1914

LITHAPHIS Scudder, 1890
U. S. Geol. Survey Territories, 13: 257
Type species Lithaphis diruta Scudder, 1890
Fossil

LITHOAPHIS Takahashi, 1959
Bull. Univ. Osaka Prefect. (Ser. B) 9: 1
Type species Lithoaphis shiiae Takahashi, 1959

L. lithocarpi (Takahashi, 1929) (Astegopteryx)
shiiae Takahashi, 1959 BM HRL

LIZERIUS E.E. Blanchard, 1923
Physis, B. Aires 7: 120
Type species Lizerius ocoteae E.E. Blanchard, 1923
*Neolizerius* E.E. Blanchard, 1939

L. acunai (Holman, 1974) (Neolizerius) BM HRL
melanocallis (Quednau, 1974) (Paoliella) BM
ocoteae E.E. Blanchard, 1923 BM HRL
reticulosiphon (Quednau, 1974) (Paoliella) BM
tuberculatus (E.E. Blanchard, 1939) (Neolizerius) BM HRL

Subgen. Paralizerius Quednau, 1974

L. (Paralizerius) brasiliensis Quednau, 1974 BM
(Paralizerius) cermelii Quednau, 1974 BM HRL
(Paralizerius) costai Quednau, 1974 BM
(Paralizerius) intermedius Quednau, 1974 BM

LOEWIA Lichtenstein, 1886 nec Egger, 1856
Monographie des Pucerons du peuplier, Montpellier, p. 37
Type species Schizoneura passerinii Signoret, 1875
= Phloeomyzus Horvath, 1896

LONGICAUDINUS Hille Ris Lambers, 1965
Tijdschr. Ent. 108: 197
Type species Hyalopteroides sinensis Tao, 1963 = Pergandeida corydalisicola Tao, 1962

L. corydalisicola (Tao, 1962) (Pergandeida) BM HRL
*sinensis* Tao, 1963 (*Hyalopteroides*) HRL

LONGICAUDUS van der Goot, 1913
Tijdschr. Ent. 56: 105
Type species Aphis trirhoda Walker, 1849
*Hemiaphis* Börner, 1926
*Yezosiphum* Matsumura, 1919

L. dunlopi Hille Ris Lambers, 1965 HRL
hameliae Theobald, 1929 = Toxoptera odinae (van der Goot, 1917)
hamelii David, 1956 mis-spelling of hameliae
himalayensis Hille Ris Lambers, 1965 HRL
suaedae Mimeur, 1934 to Xerophilaphis Nevsky, 1928
trirhodus (Walker, 1849) (Aphis) lectotype BM HRL
*aquilegiae* Koch, 1854 (*Hyalopterus*)
*dilineatus* Buckton, 1879 (*Hyalopterus*)
*flavus* Schouteden, 1906 (*Hyalopterus*)
*microrosae* Shinji, 1930 (*Pergandeida*)
*thalictri* Essig & Kuwana, 1918 (*Aphis*) nec Koch, 1854
*thalictri* Matsumura, 1919 (*Yezosiphum*)
*trirhodus* subsp. *japonicus* Hille Ris Lambers, 1965
trirhodus subsp. japonicus Hille Ris Lambers, 1965 = trirhodus (Walker, 1849)

LONGIROSTRINA Kumar & Burkhardt, 1971
Jl. Kans. ent. Soc. 44: 149
Type species Longirostris raji Kumar & Burkhardt, 1970
= Aphis Linnaeus, 1758

LONGIROSTRIS Kumar & Burkhardt, 1970 nec S.W.D., 1836
J. Kans. ent. Soc. 43: 458
Type species Longirostris raji Kumar & Burkhardt, 1970
= Aphis Linnaeus, 1758

L. raji Kumar & Burkhardt, 1970 (type of Longirostrina Kumar & Burkhardt, 1971), to Aphis Linnaeus, 1758

LONGISTIGMA Wilson, 1909
Can. Ent. 41: 385
Type species Aphis caryae Harris, 1841
*Davisia* del Guercio, 1909

L. caryae (Harris, 1841) (Aphis) BM HRL
*chantali* Quednau, 1971 HRL
*longistigma* Monell, 1878 (*Lachnus*)
*platanicola* Riley, 1883 (*Lachnus*)
chantali Quednau, 1971 = caryae (Harris, 1841)
liquidambarus (Takahashi, 1925) (Dilachnus)

LONGITARSUS Shinji, 1930 nec Berthold ex Latreille, 1827; Latreille ex Cuvier, 1829
Rep. Jap. Ass. advmt Sci. 5: 189
Type species Trama taraxaci Shinji, 1929
= Neotrama Baker, 1920

LONGIUNGUIS van der Goot, 1917
Contrib. Faune Indes Néerl. 1 (3): 112
Type species Aphis sacchari Zehntner, 1897
= Melanaphis van der Goot, 1917

L. daisenensis Sorin, 1970 to Melanaphis van der Goot, 1917
dassi David, 1956 = Melanaphis sacchari forma dassi (David, 1956)
elizabethae Ossiannilsson, 1967 to Melanaphis van der Goot, 1917
indosacchari David, 1956 = Melanaphis sacchari forma indosacchari (David, 1956)
jamatonica Sorin, 1970 to Melanaphis van der Goot, 1917
koreanus Sorin, 1972 to Melanaphis van der Goot, 1917
luzulella Hille Ris Lambers, 1939 to Melanaphis van der Goot, 1917
montana Sorin, 1970 to Melanaphis van der Goot, 1917
odinae van der Goot, 1917 to Toxoptera Koch, 1856
spathodeae van der Goot, 1918 = Toxoptera odinae (van der Goot, 1917)
tateyamaensis Sorin, 1970 to Melanaphis van der Goot, 1917
yasumatsui Sorin, 1970 to Melanaphis van der Goot, 1917

LONICERAPHIS Narzikulov, 1962
Dokl. Akad. Nauk. Tadzhik. SSR. 5: 41
Type species Loniceraphis paradoxa Narzikulov, 1962

L. paradoxa Narzikulov, 1962

LÖWIA vide LOEWIA

LOXERATES Rafinesque, 1817
Am. mon. Mag. Crit. Rev. 1: 361 described as subg. of Aphis Linnaeus, 1758
Type species Aphis diervilla-lutea Rafinesque, 1817 invalid
Invalid

LULAPHIS lapsus pro Lutaphis

LUTAPHIS Shinji, 1924
Zool. Mag., Tokyo 36: 346
Type species Lutaphis nirecola Shinji, 1924
= Tinocallis Matsumura, 1919

L. alnifoliae Shinji, 1924 nomen dubium
nirecola Shinji, 1924 to Tinocallis Matsumura, 1919

LYGAEUS (Lygaeidae)

L. hyalinatus Fabricius, 1794 = Lachnus roboris (Linnaeus, 1758)

MACCHIATIELLA del Guercio, 1909 nec 1917
Riv. Patol. Veg., Padova (n.s.) 4: 5
Type species Aphis rhamni Boyer de Fonscolombe, 1841
*Neanuraphis* Nevsky, 1926
*Neolachnaphis* Shinji, 1924

M. itadori (Shinji, 1924) (Neolachnaphis) BM HRL
*etadorii* Shinji, 1924 (*Acauda*)
*jozankeanus* Hori & Matsumura ex Hori, 1927 (*Acaudus*)
*rhamni* Hori, 1927 (*Acaudus*)
*sanguisorbae* Shinji, 1924 (*Acauda*) teste Shinji, 1941: 461
rhamni (Boyer de Fonscolombe, 1841) (Aphis) BM HRL
rhamni subsp. tarani (Nevsky, 1928) (Neanuraphis) HRL
*catharticae* Nevsky, 1929 (*Neanuraphis*)
*tarani* Nevsky, 1928 (*Neanuraphis*)

MACCHIATIELLA del Guercio, 1917 nec 1909
Redia 12: 210
Type species Macchiatiella trifolii del Guercio, 1917 = Aphis pisum Harris, 1776
= Acyrthosiphon Mordvilko, 1914

M. trifolii del Guercio, 1917 = Acyrthosiphon pisum (Harris, 1776)

MACHILAPHIS Takahashi, 1960
Kontyû 28: 11
Type species Phyllaphis machili Takahashi, 1928

M. machili (Takahashi, 1928) (Phyllaphis) BM HRL

MACRHYNCHUS Haupt, 1913
Mitt. ent. Ges. Halle 5/7: 45
Type species Macrhynchus pini Haupt, 1913 = Aphis quercus Linnaeus, 1758
= Stomaphis Walker, 1870

M. pini Haupt, 1913 = Stomaphis quercus (Linnaeus, 1758)

MACROCAUDUS Shinji, 1930
Lansania 2: 72, 78
Type species Macrocaudus phaseoli Shinji, 1930
= ? Acyrthosiphon Mordvilko, 1914

M. azukii Shinji, 1932
clematii Shinji, 1941 = Macrosiphum clematifoliae Shinji, 1924
phaseoli Shinji, 1930 to ? Acyrthosiphon Mordvilko, 1914

MACROMYZUS Takahashi, 1960
Kontyû 28: 225
Type species Myzus woodwardiae Takahashi, 1921

M. indicus David & Narayanan, 1968 HRL
polypodicola (Takahashi, 1921) (Myzus) BMHRL
woodwardiae (Takahashi, 1921) (Myzus) BM HRL
*woodwardiae* subsp. *hinoi* Moritsu, 1951 (*Myzus*)

MACROPODAPHIS Remaudière & Davatchi, 1958
Revue Path. vég. Ent. agric. 37: 241
Type species Macropodaphis rechingeri Remaudiere & Davatchi, 1958

M. alexandri Ivanovskaja, 1965
kuraijensis Ivanovskaja, 1965 HRL
paradoxa Zachvatkin & Aizenberg, 1960
primigenius Ivanovskaja, 1965 HRL
rechingeri Remaudière & Davatchi, 1958 paratypes BM HRL

MACRORHINARIUM Tsai & Tang, 1945 & 1946
Agric. Assoc. China, Circular 50: 33 (1945); Trans. R. ent. Soc. Lond. 97: 412 (1946)

Type species Macrorhinarium ovogallis Tsai & Tang, 1945 = Kaburagia rhusicola Takagi, 1937
= Kaburagia Takagi, 1937

M. ensigallis Tsai & Tang, 1945 = Kaburagia rhusicola Takagi, 1937
ovogallis Tsai & Tang, 1945 = Kaburagia rhusicola Takagi, 1937

MACROSIPHON del Guercio, 1913 vide Macrosiphum Passerini, 1860
Redia 9: 188
Without species

MACROSIPHONIELLA del Guercio, 1911
Redia 7: 331
Type species Siphonophora atra Ferrari, 1872
*Dielcysmura* Mordvilko, 1914
*Phalangomyzus* Börner, 1939
*Pyrethromyzus* Börner, (1944) 1950

M. abrotani (Walker, 1852) (Aphis) BM HRL
absinthii (Linnaeus, 1758) (Aphis) BM HRL
*fasciata* del Guercio, 1913
*lineata* del Guercio, 1913
aetnensis Barbagallo, 1969 HRL
affinis Hille Ris Lambers, 1938 = atra (Ferrari, 1872)
affinis subsp. turanica Narzikulov & Umarov, 1969
alatavica (Nevsky, 1928) (Macrosiphum) HRL
*lambersi* Verma, 1971 HRL
albicola Shinji, 1932
altaica Ivanovskaja, 1971
arenariae Bozhko, 1954
artemisiae (Boyer de Fonscolombe, 1841) (Aphis) BM HRL
*basilicum* Hottes, 1930 (*Macrosiphum (Macrosiphoniella)*)
*inflatus* Theobald, 1928 (*Neoacaudus*) type BM
artemisiae subsp. meridionalis Barbagallo, 1969 HRL
astericola Okamoto & Takahashi, 1927 = Macrosiphoniella (Asterobium) yomenae (Shinji, 1922)
asteris subsp. aktaschica Nevsky, 1929 = Macrosiphoniella (Asterobium) aktaschica Nevsky, 1929
asteris var. dilatosiphon Rusanova, 1943 nomen nudum
atra (Ferrari, 1872) (Siphonophora) BM HRL
*affinis* Hille Ris Lambers, 1938 HRL

M. atrata Umarov, 1964
aurantiaca del Guercio, 1913 = Sitobion luteum (Buckton, 1876)
austriacae Bozhko, (1957) 1961
bedfordi Theobald, 1914 = sanborni (Gillette, 1908)
beretica E.E. Blanchard, 1922 to Uroleucon (Lambersius)
? berkemiae (Shinji, 1941) (Acyrthosiphon) identity and position uncertain
cayratiae Tseng & Tao, 1936
cegmidi Szelegiewicz, 1963
chaetosiphon Takahashi & Moritsu, 1963 to Obtusicauda Soliman, 1927
chamaemelifoliae Remaudière & Leclant, 1972 paratypes BM HRL
chamomillae Hille Ris Lambers, 1947 = tapuskae (Hottes & Frison, 1931)
chrysanthemi del Guercio, 1913 = sanborni (Gillette, 1908)
chrysanthemi var. brevicauda del Guercio, 1913 = sanborni (Gillette, 1908)
cinerescens Hille Ris Lambers, 1966 HRL
citricola van der Goot, 1917 to Sinomegoura Takahashi, 1960
crassipes Shinji, 1924 (legend to figure)
crassitubia Narzikulov & Umarov, 1969 to Obtusicauda Soliman, 1927
dimidiata Börner, 1942 BM HRL
dracunculi Umarov, 1964
? dubia (Ferrari, 1872) (Siphonophora)
erigeronis Nevsky, 1928 to Uroleucon (Uromelan)
fasciata del Guercio, 1913 = absinthii (Linnaeus, 1758)
formosartemisiae Takahashi, 1921 BM HRL
    *japonica* Shinji, 1942
frigidae Ivanovskaja, 1971
frigidicola Gillette & Palmer, 1928 BM HRL
frigidivora Szelegiewicz, 1974 HRL
fulvicola Shinji, 1933 = yomogicola (Matsumura, 1917)
glabra (Gillette & Palmer, 1928) (Macrosiphum) HRL
grandicauda Takahashi & Moritsu, 1963 BM HRL
    *macrocaudus* Tao, 1964 (*Dactynotus*)
haerelli Heinze, 1960 = subaequalis Börner, 1942
heinzei Börner, 1950 = staegeri Hille Ris Lambers, 1947
helichrysi Remaudiere, 1952 BM HRL
hidaensis Takahashi & Moritsu, 1963 HRL
hikosanensis Moritsu, 1949 HRL
hikosanensis subsp. matsumurana A.K. Ghosh, R.C. Basu & RayChaudhuri, 1973 HRL
    *matsumurana* A.K. Ghosh, R.C. Basu & RayChaudhuri, 1973

M. hokkaidensis Miyazaki, 1971 HRL
janckei Borner, 1939 (type of Ramitrichophorus Hille Ris Lambers, 1947) to Macrosiphoniella (Ramitrichophorus)
japonica Shinji, 1942 = formosartemisiae Takahashi, 1921
kaufmanni Börner, 1940 BM HRL
kikungshana Takahashi, 1937 HRL
kirgisica Umarov, 1964
kuwakusae Uye, 1924 nomen dubium
kuwayamai Takahashi, 1941 (type of Sinosiphoniella Tao, 1963) to Metopeurum Mordvilko, 1914
lambersi Verma, 1971 = alatavica (Nevsky, 1928)
leucanthemi (Ferrari, 1872) (Siphonophora) BM HRL
lidiae Umarov, 1966
lineata del Guercio, 1913 = absinthii (Linnaeus, 1758)
lithospermi Bozhko, 1959
longirostrata Holman & Szelegiewicz, 1974
lopatini (Umarov, 1964) (Paczoskia)
ludovicianae (Oestlund, 1886) (Siphonophora) BM HRL
*artemisivulgaris* Knowlton & Allen, 1938 (*Macrosiphum*)
macrura Hille Ris Lambers, 1948 to Macrosiphoniella (Papillomyzus)
maculata Nevsky, 1937/8
matsumurana A.K. Ghosh, R.C. Basu & RayChaudhuri, 1973 = hikosanensis subsp. matsumurana A.K. Ghosh, R.C. Basu & RayChaudhuri, 1973
miestingeri (Börner, 1950) (Paczoskia) HRL
millefolii (DeGeer, 1773) (Aphis) BM HRL
moldavica Bozhko, (1957) 1961 HRL
mutellinae Börner, 1950 HRL
nidensis Szelegiewicz, 1960 = subaequalis Börner, 1942
nigropilosa Nevsky, 1929
nigropilosa subsp. pamirica Umarov, 1964
nitida Börner, 1950 BM HRL
oblonga (Mordvilko, 1901) (Siphonophora) BM HRL
*lineatum* van der Goot, 1912 (*Macrosiphum*)
obtecta (Börner, 1950) (Paczoskia) BM HRL
olgae Nevsky, 1929
papillata Holman, (1961) 1962 to Macrosiphoniella (Papillomyzus)
parthenii El Kady, 1960 = tapuskae (Hottes & Frison, 1931)
parthinii Habib & El Kady, 1961 = tapuskae (Hottes & Frison, 1931)
pennsylvanica (Pepper, 1950) (Macrosiphum) BM HRL

M. persequens (Walker, 1852) (Aphis) lectotype BM HRL
*pseudolineata* Hille Ris Lambers, 1931 HRL
physaliae Shinji, 1924 nomen dubium
procerae Bozhko, date ?
prunifoltae Shinji, 1924 to Ryoichitakahashia Hille Ris Lambers, 1965
pseudoartemisiae Shinji, 1933 BM HRL
pseudolineata Hille Ris Lambers, 1931 = persequens (Walker, 1852)
ptarmicae Hille Ris Lambers, 1956 BM HRL
pulvera (Walker, 1848) (Aphis) lectotype BM HRL
*amica* Walker, 1848 (*Aphis*) BM
*atomaria* Walker, 1849 (*Aphis*) type BM
*collega* Walker, 1848 (*Aphis*)
*reducta* Walker, 1848 (*Aphis*) BM
sainshandi Szelegiewicz, 1973
sanborni (Gillette, 1908) (Macrosiphum) BM HRL
*bedfordi* Theobald, 1914 type BM
*chrysanthemi* del Guercio, 1913
*chrysanthemi* var. *brevicauda* del Guercio, 1913
*chrysanthemicolens* Williams, 1911 (*Siphonophora*)
*nishigaharae* Essig & Kuwana, 1918 (*Macrosiphum*)
scopariae Bozhko, 1959
sejuncta (Walker, 1848) (Aphis) type BM HRL
*formicarium* Theobald, 1913 (*Macrosiphum*) type BM
sibirica Ivanovskaja, 1971
sieversianae Holman & Szelegiewicz, 1974
silvestrii Roberti, 1954 BM HRL
sojaki Holman & Szelegiewicz, 1974 HRL
soongarica Szelegiewicz, 1963 HRL
spinipes A.N. Basu, 1968 HRL
spinipes subsp. rhododendri A.K. Ghosh, R.C. Basu & RayChaudhuri, 1969
staegeri Hille Ris Lambers, 1947 cotypes BM HRL
*heinzei* Börner, 1950
staegeri subsp. ucrainica Bozhko, 1959
staticis Theobald, 1923 to Staticobium Mordvilko, 1914
subaequalis Börner, 1942 BM HRL
*haerelli* Heinze, 1960 HRL
*nidensis* Szelegiewicz, 1960 paratypes BM HRL
subterranea (Koch, 1855) (Siphonophora) BM HRL
*trimaculata* Hille Ris Lambers, 1938 BM HRL

M. sudhakaris H. Banerjee, A.K. Ghosh & RayChaudhuri, 1969
tadshikana Narzikulov, 1958
tanacetaria (Kaltenbach, 1843) (Aphis) BM HRL
*lilacina* Ferrari, 1872 (*Siphonophora*)
tanaretaria var. bonariensis E.E. Blanchard, 1922 = tanacetaria subsp. bonariensis E.E. Blanchard, 1932
tanacetaria subsp. bonariensis E.E. Blanchard, 1932 BM HRL
*tanacetaria* var. *bonariensis* E.E. Blanchard, 1922
? *tanacetaria* subsp. *italica* Hille Ris Lambers, 1966 BM HRL
tanacetaria subsp. italica Hille Ris Lambers, 1966 = ? tanacetaria subsp. bonariensis E.E. Blanchard, 1932
tapuskae (Hottes & Frison, 1931) (Macrosiphum) paratypes BM HRL
*ceratus* Börner, 1940 (*Phalangomyzus*)
*chamomillae* Hille Ris Lambers, 1947 BM HRL
*gulbenkiani* Ilharco, 1968 (*Uroleucon*) paratype BM HRL
*parthenii* El kady, 1960 paratype BM
*parthinii* Habib & El Kady, 1961
teriolana Hille Ris Lambers, 1931 BM HRL
triglochiniella Theobald, 1928 = Sitobion avenae (Fabricius, 1775)
trimaculata Hille Ris Lambers, 1938 = subterranea (Koch, 1855)
tuberculatumartemisicola Bozhko, (1957) 1961
umarovi (Narzikulov ex Narzikulov & Umarov, 1972 nomen nudum) Narzikulov, 1972
usquertensis Hille Ris Lambers, 1935 BM HRL
xeranthemi Bozhko, 1959 BM HRL
yangi Takahashi, 1937 to Macrosiphoniella (Asterobium)
yomogicola (Matsumura, 1917) (Macrosiphum) BM HRL
*fulvicola* Shinji, 1933
*parvum* Shinji, 1922 (*Macrosiphum*)
yomogifoliae (Shinji, 1924) (Macrosiphum) BM HRL
zerogutierreziae (Smith & Knowlton, 1937) (Macrosiphum) BM HRL

Subgen. Asterobium Hille Ris Lambers, 1938

M. (Asterobium) aktaschica Nevsky, 1929 (Macrosiphoniella)
*asteris* subsp. *aktaschica* Nevsky, 1929 (*Macrosiphoniella*)
(Asterobium) asteris (Walker, 1849) (Aphis) types BM HRL
(Asterobium) crepidis Holman & Szelegiewicz, 1974
(Asterobium) davazhamci Holman & Szelegiewicz, 1974 HRL
(Asterobium) linariae (Koch, 1855) (Siphonophora) BM HRL
*ohatensis* Szelegiewicz, 1966 paratypes BM HRL

M. (Asterobium) ohatensis Szelegiewicz, 1966 = linariae (Koch, 1855)
(Asterobium) soosi Szelegiewicz, 1966 paratypes BM HRL
(Asterobium) yangi Takahashi, 1937 (Macrosiphoniella) BM HRL
(Asterobium) yomenae (Shinji, 1922) (Macrosiphum) BM HRL
*astericola* Okamoto & Takahashi, 1927 (*Macrosiphoniella*)
*moriokae* Shinji, 1924 (*Macrosiphum*)
*yomenafoliae* Shinji, 1922 (*Macrosiphum*)

Subgen. Papillomyzus Szelegiewicz, 1963

M. (Papillomyzus) macrura Hille Ris Lambers, 1948 (Macrosiphoniella) BM HRL
(Papillomyzus) papillata Holman, (1961) 1962 (Macrosiphoniella) BM HRL
(Papillomyzus) riedeli Szelegiewicz, 1963 type of Papillomyzus Szelegiewicz, 1963 HRL
(Papillomyzus) tuberculata (Nevsky, 1928) (Macrosiphum)

Subgen. Ramitrichophorus Hille Ris Lambers, 1947

M. (Ramitrichophorus) hillerislambersi Ossiannilsson, 1954 BM HRL
(Ramitrichophorus) janckei Börner, 1939 (Macrosiphoniella) BM HRL
(Ramitrichophorus) medvedevi (Bozhko, 1957) (Ramitrichophorus) BM HRL
*nasti* Szelegiewicz, 1958 BM HRL
(Ramitrichophorus) nasti Szelegiewicz, 1958 = medvedevi Bozhko, 1957
(Ramitrichophorus) paradoxa (Bozhko, 1957) (Ramitrichophorus)

MACROSIPHUM del Guercio, 1900 nec Passerini, 1860
Nuove Relaz. R. Staz. Ent. agr. 1: 159
Type species Aphis convolvuli Kaltenbach, 1843 = Aphis persicae Sulzer, 1776
= Nectarosiphon Schouteden, 1901

MACROSIPHUM Oestlund, 1886 nec Passerini, 1860
A. Rep. Minn. St. Geol. 14: 27
Type species Macrosiphum rubicola Oestlund, 1886
= Oestlundia Hille Ris Lambers, 1949

M. rubicola Oestlund, 1886 to Illinoia (Oestlundia)

MACROSIPHUM Passerini, 1860
Gli Afidi, Parma, p. 27, footnote
Type species Aphis rosae Linnaeus, 1758
*Nectarophora* Oestlund, 1887

*Passerinia* Macchiati, 1882
*Siphonophora* Koch, 1855 nec Fischer, 1823

M. aaroni Knowlton, 1949 to Uroleucon Mordvilko, 1914
adenocaulonae Essig, 1936 to Uroleucon Mordvilko, 1914
adenophorae Matsumura, 1918 to Uroleucon (Uromelan)
aetheocornum Smith & Knowlton, 1939 BM HRL
agrimoniellum (Cockerell, 1903) (Nectarophora) HRL
agrostemnium Theobald, 1913 = Nasonovia ribisnigri (Mosley, 1941)
akebiae Shinji, 1935 & 1941 to Sitobion Mordvilko, 1914
aktashicum Nevsky, 1928 (spelled aktaschicum in 1929)
alatavicum Nevsky, 1928 to Macrosiphoniella del Guercio, 1911
albertinae Hille Ris Lambers, 1966 HRL
albifrons Essig, 1911 BM HRL
*puyallupsi* Knowlton, 1943
allii Jackson, 1918 = Sitobion avenae (Fabricius, 1775)
alopeculi mis-spelling of alopecuri
alopecuri Takahashi, 1921 to Sitobion Mordvilko, 1914
alpinum Meier, 1967 BM HRL
amamiamum Takahashi, 1930 to Uroleucon (Uromelan)
amelanchiericolens Patch, 1919
americanum Mordvilko, 1919
amurense Mordvilko, 1919: key
amygdaloides Theobald, 1925 BM HRL
anomalae Hottes & Frison, 1931 to Uroleucon (Lambersius)
anomellus Knowlton & Allen, 1938 to Obtusicauda Soliman, 1927
aquilegiae Theobald, 1913 = Aulacorthum solani (Kaltenbach, 1843)
artemisiae var. citrinum Schouteden, 1901
artemisiphilum Knowlton & Allen, 1938 to Obtusicauda Soliman, 1927
artemisivulgaris Knowlton & Allen, 1938 = Macrosiphoniella ludovicianae (Oestlund, 1886)
arundinis Theobald, 1913 = Metopolophium dirhodum (Walker, 1849)
asterifoliae Strom, 1934 to Pleotrichophorus Börner, 1950
asterophagum Nevsky, 1928
atripes Gillette & Palmer, 1933 to Uroleucon Mordvilko, 1914
aucubae Bartholomew, 1932 = Aulacorthum solani (Kaltenbach, 1843)
audeni MacDougall, 1926 HRL
aulacorthoides David, Narayan & Rajasingh, 1971 HRL
avenivorum Kirkaldy, 1905 = Sitobion fragariae (Walker, 1848)
barri Essig, 1949 = Acyrthosiphon lactucae (Passerini, 1860)

M. begoniae Schouteden, 1901 = Aulacorthum solani (Kaltenbach, 1843)
berchemiae Takahashi, 1938 to Sitobion Mordvilko, 1914
bicolor Nevsky, 1929 to Uroleucon Mordvilko, 1914
bisensoriatum MacDougall, 1926 HRL
bonitum Hottes, 1950 to Uroleucon (Uromelan)
? bonnevilla (Knowlton & Allen, 1937) (Amphorophora)
bosqi E.E. Blanchard, 1932 = Acyrthosiphon malvae (Mosley, 1841)
brachytarsus Hille Ris Lambers, 1933 = Sitobion nigronectarium (Theobald, 1915)
breviscriptum Palmer, 1936 to Uroleucon (Lambersius)
calendulae Nevsky, 1951
californicum (Clarke, 1903) (Nectarophora) BM HRL
*laevigatae* Essig, 1911 cotype BM HRL
capitophoroides E.E. Blanchard, 1944
caricis Glendenning, 1926 to Sitobion Mordvilko, 1914
carpinicolens Patch, 1919 (type of Neocorylobium MacGillivray, 1968) to Macrosiphum (Neocorylobium)
cefsmithi Knowlton & Allen, 1938 to Obtusicauda Soliman, 1927
centranthi Theobald, 1915 types BM HRL
cercidiphylli Matsumura, 1918 to Aulacorthum Mordvilko, 1914
cerinthiacum Börner, 1950 BM HRL
chilensis Essig, 1953 to ? Uroleucon (Lambersius) or ? Uroleucon Mordvilko, 1914
cholodkovskyi (Mordvilko, 1909) (Siphonophora) BM HRL
*? portschinskyi* Mordvilko, 1909 (*Siphonophora*)
*? rushkovskii* Mordvilko, 1919
chondrillae Nevsky, 1929 to Uroleucon Mordvilko, 1914
circicola Takahashi, 1923 mis-spelling of cirsicola Takahashi, 1923
cirsicola Takahashi, 1923 to Aulacorthum Mordvilko, 1914
cissi Theobald, 1920 to Sitobion Mordvilko, 1914
clematidis Takahashi, 1929 = clematifoliae Shinji, 1924
clematifoliae Shinji, 1924 BM HRL
*clematidis* Takahashi, 1929
*clematii* Shinji, 1941 (*Macrocaudus*)
cockerelli Hottes, 1950 BM HRL
cocoensis E.E. Blanchard, 1932 to Uroleucon (Lambersius)
colylicola errore pro corylicola Shinji, 1930
compositae Theobald, 1915 to Uroleucon (Uromelan)
constrictum Patch, 1923

M. corallinum Theobald, 1925 type BM
corallorhizae (Cockerell, 1903) (Nectarophora) BM
cordobensis E.E. Blanchard, 1932 to Uroleucon (Lambersius)
coriariae Shinji, 1922 nomen dubium
cornelli Patch, 1926 = Acyrthosiphon malvae (Mosley, 1841)
cornifoliae Shinji, 1924
coryli Davis, 1914 to Macrosiphum (Neocorylobium)
corylicola Shinji, 1930 to Macrosiphum (Neocorylobium)
creelii Davis, 1914 BM
crenicornum Smith & Knowlton, 1939 to Kakimia Hottes & Frison, 1931
cuscutae Holman, 1974 to Sitobion Mordvilko, 1914
cyparissiae var. cucurbitae del Guercio, 1913 = euphorbiae (Thomas, 1878)
cystopteridis Robinson, 1966 to Sitobion Mordvilko, 1914
dahliafolii Theobald, 1918 = Uroleucon (Uromelan) compositae (Theobald, 1915)
daphnes Ossiannilsson, 1959 = daphnidis Börner, 1950
daphnidis Börner, 1950 BM HRL
*daphnes* Ossiannilsson, 1959
debilis Takahashi, 1923 to Uroleucon Mordvilko, 1914
diervillae Patch, 1919 BM HRL
dirhodum (Walker, 1849) (Aphis) (type of Goidanichiellum Martelli, 1950) to Metopolophium Mordvilko, 1914
doronicicola Leclant, 1972 HRL
duffieldii Theobald, 1913 = Aulacorthum solani (Kaltenbach, 1843)
echinocysti Bartholomew, 1932 HRL
edrossi Essig, 1953 BM
elegans Bozhko, 1959 nomen nudum
eleusinae errore pro eleusines Theobald, 1929
eleusines Theobald, 1929 = Sitobion miscanthi Takahashi, 1921
eoessigi Knowlton, 1947 to Uroleucon (Uromelan)
epilobiellum Theobald, 1923 = ? tinctum (Walker, 1849)
epilobii Theobald, 1919 = tinctum (Walker, 1849)
erigeronella Soliman, 1927 = Uroleucon (Lambersius) erigeronensis (Thomas, 1878)
esakii Takahashi, 1924 to Aulacorthum Mordvilko, 1914
escalantii Knowlton, 1928 to Uroleucon (Lambersius)
eupatoricolens Patch, 1919 to Uroleucon Mordvilko, 1914
eupatorii (Williams, 1911) (Siphonophora)
euphorbiae (Thomas, 1878) (Siphonophora) BM HRL
*asclepiadifolii* Thomas, 1878 (*Siphonophora*)
*asclepiadis* Cowen ex Gillette & Baker, 1895 (*Nectarophora*)

*citrifolii* Ashmead, 1880 (*Siphonophora*)
*cucurbitae* Middleton ex Thomas, 1878 (*Siphonophora*)
*cyparissiae* var. *cucurbitae* del Guercio, 1913
*euphorbicola* Thomas, 1878 (*Siphonophora*)
*? euphorbiellum* Theobald, 1917 BM HRL ?
*heleniella* Cockerell, 1903 (*Nectarophora*)
*koehleri* Börner, 1937
*lycopersici* Clarke, 1903 (*Nectarophora*)
*rosaeollae* Theobald, 1915
*solanifolii* Ashmead, 1882 (*Siphonophora*)
*tabaci* Pergande, 1898 (*Nectarophora*)
*tulipae* Monell, 1879 (*Siphonophora*)

M. euphorbiellum Theobald, 1917 = ? euphorbiae (Thomas, 1878)
euryae Takahashi, 1937 to Metopolophium Mordvilko, 1914
evodiae Takahashi, 1929 to Acyrthosiphon Mordvilko, 1914
fagopyri A.K. Ghosh & RayChaudhuri, 1972 HRL
fallacis Nevsky, 1929 to Uroleucon Mordvilko, 1914
filifoliae Gillette & Palmer, 1928 to Obtusicauda Soliman, 1927
flavum Tao, 1963
floridae (Ashmead, 1882) (Siphonophora) (described as variety of Siphonophora rosae but first referred to as Siphonophora floridae)
formicarium Theobald, 1913 = Macrosiphoniella sejuncta (Walker, 1848)
formosanum Takahashi, 1921 to Uroleucon Mordvilko, 1914
fuchuensis Shinji, 1942 to Uroleucon Mordvilko, 1914
funestum (Macchiati, 1885) (Siphonophora) BM HRL
*funestum* subsp. *shelkovnikovi* Mordvilko, 1919: 506
*rubi* var. *rufa* Buckton, 1883 (*Siphonophora*)
*rubifolium* Theobald, 1917
*shelkovnikovi* Mordvilko, 1919: 361

funestum subsp. shelkovnikovi Mordvilko, 1919 = funestum (Macchiati, 1885)
fuscicornis MacDougall, 1926
galiophagum Wimshurst, 1923 type of Linosiphon Börner, 1950 q.v.
gaurae (Williams, 1911) (Siphonophora) BM HRL
*gaurina* Williams, 1911 (*Siphonophora*)
*onagrae* Patch, 1919

gei (Koch, 1855) (Siphonophora) BM HRL
gei of virologists = euphorbiae (Thomas, 1878)
gentneri Mason, 1947 to Fimbriaphis Richards, 1959
geranii Chowdhuri, R.C. Basu, Chakrabarti & RayChaudhuri, 1969 = pseudogeranii Chakrabarti & RayChaudhuri, 1974

M. geranii (Oestlund, 1887) (Nectarophora) BM HRL
giganteum Matsumura, 1918 to Uroleucon (Uromelan)
githago Theobald, 1913 nomen dubium
glabrum Gillette & Palmer, 1928 to Macrosiphoniella del Guercio, 1911
gobonis Matsumura, 1917 to Uroleucon (Uromelan)
graminum Theobald, 1913 = Metopolophium dirhodum (Walker, 1849)
gravelii van der Goot, 1917 to Sitobion Mordvilko, 1914
gravicornis Patch, 1919 to Uroleucon (Lambersius)
griersoni E.E. Blanchard, 1932 to Uroleucon (Uromelan)
hagi Essig & Kuwana, 1918 = Aulacorthum solani (Kaltenbach, 1843)
hagicola Matsumura, 1917 = Aulacorthum solani (Kaltenbach, 1843)
hamiltoni Robinson, 1968 paratypes BM HRL
harpagorubus Knowlton, 1935 = Sitobion fragariae (Walker, 1848)
hartigi Hille Ris Lambers, 1947 BM HRL
*montanum* Hille Ris Lambers, 1931 nec Nevsky ex Mordvilko, 1929
hederae Theobald, 1915 = Aulacorthum solani (Kaltenbach, 1843)
?helianthi (Tao, 1963) (Dactynotus)
hellebori Theobald & Walton, 1923 BM HRL
holmani Leclant, 1972 HRL
holsti Takahashi, 1935 (type of Pseudoacyrthosiphon A.K. Ghosh & Ray-Chaudhuri, 1969) to Neoacyrthosiphon (Pseudoacyrthosiphon)
huantana Essig, 1953 to Uroleucon (Lambersius)
hyptis E.E. Blanchard, 1944 = Sitobion salviae (Bartholomew, 1932)
ibarae Matsumura, 1917 to Sitobion Mordvilko, 1914
ibotum Essig & Kuwana, 1918 to Aulacorthum Mordvilko, 1914
idahoensis Miller, 1933 to Uroleucon (Lambersius)
illini Hottes & Frison, 1931 to Uroleucon (Uromelan)
illini var. crudae Hottes & Frison, 1931 to Uroleucon (Uromelan)
illini var. sangamonensis Hottes & Frison, 1931 to Uroleucon (Uromelan)
impatiensae Shinji, 1935 = Impatientinum impatiens (Shinji, 1922)
impatiensicolens Patch, 1919 to Uroleucon Mordvilko, 1914
incertum Mordvilko, 1919 = ? stellariae Theobald, 1913
inexspectatum Leclant, 1974 BM HRL
inulifoliae Mimeur, 1934 = Uroleucon (Belochilum) inulae (Ferrari, 1872)
itoe Takahashi, 1925 type of Ericolophium Tao, 1963 q.v.
jasmini (Clarke, 1903) (Nectarophora)
jeanae Robinson, 1972 BM HRL
jonesi Gillette & Palmer, 1928 to Obtusicauda Soliman, 1927
kaltenbachi Schouteden, 1906 = Nasonovia ribisnigri (Mosley, 1841)
kickapoo Hottes & Frison, 1931 to Catamergus Oestlund, 1922

M. kikioensis Shinji, 1942 to Uroleucon Mordvilko, 1914
kiowanepus (Hottes, 1933) (Adactynus) paratypes BM HRL
kirkaldyi Fullaway, 1910 (type of Fullawayella del Guercio, 1911) = Idiopterus nephrelepidis Davis, 1909
knautiae Holman, 1972 HRL
koehleri Börner, 1937 = euphorbiae (Thomas, 1878)
kuricola Matsumura, 1917 nomen dubium
lactucae (Passerini, 1860) (Siphonophora) (type of Tlja Mordvilko, 1914) to Acyrthosiphon Mordvilko, 1914
lactucarius Börner, 1931 = Acyrthosiphon lactucae (Passerini, 1860)
lactucicola Strand, 1928 to Uroleucon (Uromelan)
laevigatae Essig, 1911 = californicum (Clarke, 1903)
lamii Theobald, 1915 = Aulacorthum solani (Kaltenbach, 1843)
lanceolatum Patch, 1919 to Uroleucon Mordvilko, 1914
lapponicum Shaposhnikov, 1964
laseri Holman, 1962 paratypes BM HRL
ligustrumae Shinji, 1927 = Aulacorthum ibotum (Essig & Kuwana, 1918)
lilii (Monell, 1879) (Siphonophora) BM HRL
linderae Shinji, 1922 to Aulacorthum Mordvilko, 1914
lineatum van der Goot, 1912 = Macrosiphoniella oblonga (Mordvilko, 1901)
lisae Heie, 1965 HRL
littoralis Bertels, 1973 lapsus pro littoralis Blanchard, 1939
littoralis E.E. Blanchard, 1939 to Uroleucon (Lambersius)
lizerianum E.E. Blanchard, 1922 to Uroleucon Mordvilko, 1914
longipes Gillette & Palmer, 1928 to Pleotrichophorus Börner, 1930
longirostris Gillette & Palmer, 1933 to Uroleucon (Lambersius)
lonicerae Uye, 1923 to Trichosiphonaphis Takahashi, 1922
lophospermum Theobald, 1914 = Myzus (Nectarosiphon) persicae (Sulzer, 1776)
loti Theobald, 1912 to Acyrthosiphon Mordvilko, 1914
lycopersicella Theobald, 1914 = Myzus (Nectarosiphon) persicae (Sulzer, 1776)
macolai E.E. Blanchard, 1932
malvicola Matsumura, 1917 nomen dubium
martini (Cockerell, 1903) (Nectarophora) BM
martorelli Smith, 1960 to Sitobion Mordvilko, 1914
matsumuraeanum Hori, 1926 = Aulacorthum solani (Kaltenbach, 1843)
meixneri Börner, 1950 HRL
melampyri Mordvilko, 1919 BM HRL
mentzeliae Wilson, 1915 BM HRL
mertensiae Gillette & Palmer, 1933 HRL
mesosphaeri Tissot, 1934 = Sitobion salviae (Bartholomew, 1932)

M. minutum van der Goot, 1916 to Uroleucon (Uromelan)
miscanthi Takahashi, 1921 to Sitobion Mordvilko, 1914
montanum Hille Ris Lambers, 1931 = hartigi Hille Ris Lambers, 1947
montanum Nevsky ex Mordvilko, 1929 nomen nudum = ? nevskyanum MacGillivray, 1958
monticolum Takahashi, 1935 to Uroleucon Mordvilko, 1914
mordvilkoi Miyazaki, 1968 BM HRL
*rosae* subsp. *orientale* Mordvilko, 1919 nec Macrosiphum orientale van der Goot, 1912
moriokae Shinji, 1924 = Macrosiphoniella (Asterobium) yomenae (Shinji, 1922)
muermosa Essig, 1953 to Uroleucon Mordvilko, 1914
mulgedifolii Tashev, 1967
mulgedii Nevsky, 1928
mulgedii subsp. ucrainica Bozhko, 1959
multipilosum Nevsky, 1951
mumecola Matsumura, 1917 to Myzus Passerini, 1860
muradachi Shinji, 1928 to Aulacorthum Mordvilko, 1914
myrmecophilum Theobald, 1916 to Metopolophium Mordvilko, 1914
naazamiae Shinji, 1935 nomen dubium
nasonovi Mordvilko, 1918 HRL
neavi Theobald, 1915 nomen dubium
neoartemisiae Takahashi, 1921 to Titanosiphon Nevsky, 1928
nevskyanum MacGillivray, 1958
*? montanum* Nevsky ex Mordvilko, 1929 nomen nudum
nickeli Essig, 1956 to Uroleucon Mordvilko, 1914
nigrinectaria Theobald, 1915 to Sitobion Mordvilko, 1914
nigrocampanulae Theobald, 1928 to Uroleucon (Uromelan)
nigromaculosum MacDougall, 1926 type of Eomacrosiphon Hille Ris Lambers, 1958 q.v.
nipponicum Essig & Kuwana, 1918 to Aulacorthum Mordvilko, 1914
nipponicum Shinji, 1942 = Uroleucon (Uromelan) lactucicola (Strand, 1928)
nishigaharae Essig & Kuwana, 1918 = Macrosiphoniella sanborni (Gillette, 1908)
nishikigi Shinji, 1928 (type of Pseudomegoura Shinji, 1929) = Aulacorthum magnoliae (Essig & Kuwana, 1918)
nuble Essig, 1953 to Uroleucon (Lambersius)
oljatae Hottes, 1950 = Sitobion avenae (Fabricius, 1775)
olmsteadi Robinson, 1965 paratypes BM HRL
onagrae Patch, 1919 = gaurae (Williams, 1911)
orchidacearum Franssen & Tiggelovend, 1935 to Sitobion Mordvilko, 1914
oredonensis Börner, 1952 = oredonense Remaudière, 1952

M. oredonense Remaudière, 1952 BM HRL
*oredonensis* Börner, 1952
orientale van der Goot, 1912 to Uroleucon (Uromelan)
orthocarpus Davidson, 1909
osmaliae Shinji, 1924
osmaroniae (Wilson, 1912) (Illinoia) BM HRL
*occidentale* Essig, 1942 (*Amphorophora*) BM HRL
pachysiphon Hille Ris Lambers, 1966 BM HRL
packi Knowlton, 1928 to Pleotrichophorus Börner, 1930
paczoskii subsp. turanicum Nevsky, 1929 = Paczoskia turanica (Nevsky, 1929)
paederiae Takahashi, 1921 = Aulacorthum nipponicum (Essig & Kuwana, 1918)
pallens Hottes & Frison, 1931 paratype BM
pallidum (Oestlund, 1887) (Nectarophora) BM HRL
*pseudorosae* Patch, 1919
parvifolii Richards, 1967 HRL
parvum Shinji, 1922 = Macrosiphoniella yomogicola (Matsumura, 1917)
pechumani MacGillivray, 1966 BM HRL
pelargonii Kaltenbach of van der Goot, 1915 (type of Neomacrosiphum van der Goot, 1915) = Aulacorthum solani (Kaltenbach, 1843)
pennsylvanicum Pepper, 1950 to Macrosiphoniella del Guercio, 1911
perillae Shinji, 1924 (type of Perillaphis Takahashi, 1965) to Aulacorthum (Perillaphis)
perillae Takahashi, 1924 = Aulacorthum (Perillaphis) perillae (Shinji, 1924)
petasitis Matsumura, 1917 nomen dubium
phillipsii Theobald, 1925 = Uroleucon cichorii (Koch, 1855)
phyllanthi Takahashi, 1939 = Sitobion takahashi (Eastop, 1959)
piceaellum Theobald, 1916 = Aulacorthum solani (Kaltenbach, 1843)
polanense Pašek, 1955
pollinae Shinji, 1924 = Kaochiaoja arthraxonis (Takahashi, 1921)
polygoni-japonica Shinji, 1941 = Trichosiphonaphis (Xenomyzus) tade (Shinji, 1927)
polystachyae Franssen & Tiggelovend, 1935 = Sitobion luteum (Buckton, 1876)
potentillae (Oestlund, 1887) (Nectarophora)
potentillicaulis Miller, 1933
prenanthidis Börner, 1940 BM HRL
primulae Theobald, 1913 to Microlophium Mordvilko, 1914
primulana Matsumura, 1917 = Aulacorthum (Neomyzus) circumflexum (Buckton, 1876)
pseudocoryli Patch, 1919 BM HRL

M. pseudodirhodum Patch, 1919 to Acyrthosiphon Mordvilko, 1914
pseudogeranii Chakrabarti & RayChaudhuri, 1974 HRL
*geranii* Chowdhuri, R.C. Basu, Chakrabarti & RayChaudhuri, 1969 nec Oestlund, 1887
pseudohieracii Theobald, 1912 = Nasonovia ribisnigri (Mosley, 1841)
pseudorosae Patch, 1919 = pallidum (Oestlund, 1887)
ptericolens Patch, 1919 to Sitobion Mordvilko, 1914
pteridis Wilson, 1915 to Sitobion Mordvilko, 1914
pulcherimum Nevsky, 1928 misprint for pulcherrimum Nevsky, 1928
pulcherrimum Nevsky, 1928
purshiae Palmer, 1938 to Acyrthosiphon Mordvilko, 1914
puyallupsi Knowlton, 1943 = ? albifrons Essig, 1911
pyrifoliae MacDougall, 1926
ranunculi Pašek, 1955
raysmithi Hille Ris Lambers, 1966 HRL
rhododendri Wilson, 1918 to Illinoia (Masonaphis)
ribiellum Davis, 1919 to Hyperomyzus (Neonasonovia)
rogersii Theobald, 1913 = Acyrthosiphon malvae subsp. rogersii (Theobald, 1913)
rosae (Linnaeus, 1758) (Aphis) BM HRL
*dipsaci* Schrank, 1801 (*Aphis*)
*fragariae* Koch, 1855 (*Siphonophora*)
*rosae* Macchiati, 1882 (*Passerinia*)
*rosae* subsp. *fragaricola* Hille Ris Lambers, 1939 HRL
*rosae* var. *glauca* Buckton, 1876 (*Siphonophora*)
*rosaecola* Passerini, 1871 (*Siphonophora*)
*scabiosae* Scopoli, 1763 (*Aphis*)
rosae var. azerbaidshanica Rusanova, 1947
rosae var. azerbajdshanica Rusanova, 1943 nomen nudum
rosae var. scabiosae Rusanova, 1943 nomen nudum
rosae subsp. fragaricola Hille Ris Lambers, 1939 = rosae (Linnaeus, 1758)
rosae subsp. orientale Mordvilko, 1919 = mordvilkoi Miyazaki, 1968
rosae subsp. vasiljevi Mordvilko, 1919
rosaefolium Theobald, 1915 (type of Rhodobium Hille Ris Lambers 1947) = Rhodobium porosum (Sanderson, 1900)
rosaeibarae Matsumura, misnomer of ibarae Matsumura, 1917
rosaeiformis Das, 1918 to Sitobion Mordvilko, 1914
rosaeollae Theobald, 1915 = euphorbiae (Thomas, 1878)
rubiarctici Heikinheimo, 1946 BM HRL
rubicola (Oestlund, 1886) (Macrosiphum Oestlund, 1886 nec Passerini, 1860) (type of Oestlundia Hille Ris Lambers, 1949) to Illinoia (Oestlundia)

M. rubiellum Theobald, 1913 = Sitobion fragariae (Walker, 1848)
rubifoliae Shinji, 1922 nomen dubium
rubifolium Theobald, 1917 = funestum (Macchiati, 1885)
rubiformosanum Takahashi, 1927 to Microlophium Mordvilko, 1914. Takahashi, 1927 refers to 1923 where a Macrosiphum illustrated by Tao, 1968: 36 is described.
rudbeckiae var. madia Swain, 1919 = Uroleucon (Lambersius) madia Swain, 1919
rudbeckiarum (Cockerell, 1903) (Nectarophora) BM HRL
ruralis Hottes & Frison, 1931 to Uroleucon (Uromelan)
rushkovskii Mordvilko, 1919 = ? cholodkovskyi (Mordvilko, 1909)
salviae Bartholomew, 1932 to Sitobion Mordvilko, 1914
sanborni Gillette, 1908 (type of Pyrethromyzus Börner, (1944) 1950) to Macrosiphoniella del Guercio, 1911
sanguinarium Hottes & Frison, 1931 to Linosiphon Börner, 1950
schranki Theobald, 1927 = Microlophium carnosum (Buckton, 1876)
scoliopi Essig, 1936 to Fimbriaphis Richards, 1959
seneciocolum Paik, 1965 to Uroleucon Mordvilko, 1914
senecionis Matsumura, 1917 = Aulacorthum solani (Kaltenbach, 1843)
shelkovnikovi Mordvilko, 1919 = funestum (Macchiati, 1885)
sileneum Theobald, 1913 types BM HRL
sleesmani Pepper, 1950 type of Papulaphis Robinson, 1966 q.v.
smilaceti Takahashi, 1924 = ? Impatientinum impatiens (Shinji, 1922) or to ? Sitobion Mordvilko, 1914
smilacicola Takahashi, 1924 to Sitobion Mordvilko, 1914
smilacifoliae Takahashi, 1921 to Sitobion Mordvilko, 1914
sobae Shinji, 1922 = ? Aulacorthum solani (Kaltenbach, 1843)
solanifolii (Ashmead, 1882) (Siphonophora) = euphorbiae (Thomas, 1878)
solanum Bertels, 1973 lapsus pro solutum E.E. Blanchard, 1939
solutum E.E. Blanchard, 1939 incertae sedis
sonchi var. flavomarginata del Guercio, 1917 = Uroleucon (Uromelan) compositae (Theobald, 1915)
sonchicola Matsumura, 1917 = Uroleucon sonchi (Linnaeus, 1767)
sorbi Matsumura, 1918 (type of Unisitobion Takahashi, 1961) to Macrosiphum (Unisitobion)
spinotibium M.R. Ghosh, A.K. Ghosh & RayChaudhuri, 1971 = Sitobion gravelii (van der Goot, 1917)
spiraecola Patch, 1914 to Illinoia Wilson, 1910
sporadicum Knowlton, 1935 to Pleotrichophorus Börner, 1930
squarrosa Sanborn, 1904 to Uroleucon (Uromelan)
stanleyi Wilson, 1915 BM HRL

M. stellariae Theobald, 1913 types BM HRL
*? incertum* Mordvilko, 1919
subviride MacDougall, 1926
suguri Shinji, 1924 nomen dubium
sumatraensis Mason, 1927 = Sitobion ibarae (Matsumura, 1917)
sumomocola Monzen, 1929 = Myzus varians Davidson, 1912
syringae Matsumura, 1918 to Aulacorthum Mordvilko, 1914
tadecola Shinji, 1935 nomen dubium
taiheisanum Takahashi, 1935 to Neoacyrthosiphon Tao, 1963
taiwanum Takahashi, 1923 (type of Sumoia Tao, 1963) to Aulacorthum (Neomyzus)
tapuskae Hottes & Frison, 1931 to Macrosiphoniella del Guercio, 1911
tardae Hottes & Frison, 1931 to Uroleucon (Uromelan)
tenuicauda Bartholomew, 1932 BM HRL
tenuitarsus Gillette & Palmer, 1933 to Uroleucon (Lambersius)
theobaldii Davis, 1915 = Acyrthosiphon pisum (Harris, 1776)
thermopsaphis Knowlton, 1938 BM
tiliae (Monell, 1879) (Siphonophora) BM HRL
timpanogos Knowlton, 1942
tinctum (Walker, 1849) (Aphis) BM HRL
*epilobii* Theobald, 1919 nec Nectarophora epilobii Pergande, 1900
*? epilobiellum* Theobald, 1923 BM
tissoti Boudreaux, 1949 (1948) to Uroleucon (Lambersius)
tolmiea (Essig, 1942) (Amphorophora) paratypes BM
tricholobicola Strand, 1929 = Aulacorthum cirsicola (Takahashi, 1923)
trifolii Pergande, 1904 = Acyrthosiphon pisum (Harris, 1776)
trifolii Theobald, 1913 = Acyrthosiphon pisum (Harris, 1776)
trollii Börner, 1915 BM HRL
tsutae Shinji, 1935
tuberculaceps (Essig, 1942) (Amphorophora) HRL ?
tuberculatum Nevsky, 1928 to Macrosiphoniella (Papillomyzus)
tucumani Essig, 1953 to Uroleucon Mordvilko, 1914
valerianae (Clarke, 1903) (Nectarophora) BM HRL
venaefuscae Davis, 1914 BM
verbenae (Thomas, 1878) (Siphonophora)
verbesinae Boudreaux, 1949 (1948) to Uroleucon (Uromelan)
vereshtshagini Mordvilko, 1919
vernoniae van der Goot, 1915 = Uroleucon (Uromelan) compositae (Theobald, 1915)
veronicae Theobald, 1913 = Aulacorthum solani (Kaltenbach, 1843)

M. weberi Börner, 1933 BM HRL
williamsi Gillette & Palmer, 1929 = Uroleucon sonchellum (Monell, 1879)
yagasogae Hottes, 1948 = Sitobion insularis subsp. yagasogae (Hottes, 1948)
yasumatsui Moritsu, 1958 to Sitobion Mordvilko, 1914
yomenae Shinji, 1922 to Macrosiphoniella (Asterobium)
yomenafoliae Shinji, 1922 = Macrosiphoniella (Asterobium) yomenae (Shinji, 1922)
yomogicola Matsumura, 1917 to Macrosiphoniella del Guercio, 1911
yomogifoliae Shinji, 1924 to Macrosiphoniella del Guercio, 1911
zerogutierreziae Smith & Knowlton, 1937 to Macrosiphoniella del Guercio, 1911
zerothermum Knowlton & Allen, 1938 to Obtusicauda Soliman, 1927
zerozalphum Knowlton, 1935 = Acyrthosiphon malvae (Mosley, 1841)
zinzalae Hottes & Frison, 1931 to Uroleucon Mordvilko, 1914
zionense Knowlton, 1935 BM HRL
zoorosarum Knowlton & Smith, 1936 = Rhodobium porosum (Sanderson, 1900)
zymozionensis Knowlton, 1946 to Uroleucon (Lambersius)

Subgen. Macrosiphoniella del Guercio, 1911

M. (Macrosiphoniella) basilicum Hottes, 1930 nom. nov. pro Aphis artemisiae Ph. F. Gmelin, 1758 (invalid !) nec Boyer de Fonscolombe, 1841 = Macrosiphoniella artemisiae (Boyer de Fonscolombe, 1841)

Subgen. Neocorylobium MacGillivray, 1968

M. (Neocorylobium) carpinicolens Patch, 1919 (Macrosiphum) BM HRL
(Neocorylobium) coryli Davis, 1914 (Macrosiphum) BM HRL
(Neocorylobium) corylicola Shinji, 1930 (Macrosiphum) BM HRL
(Neocorylobium) vandenboschi (Hille Ris Lambers, 1966) (Corylobium) HRL

Subgen. Sitobion Mordvilko, 1914

M. (Sitobion) adgnatum F.P. Müller, 1959 to Sitobion Mordvilko, 1914
(Sitobion) africanum Hille Ris Lambers, 1954 to Sitobion Mordvilko, 1914
(Sitobion) africanum subsp. matatum Eastop, 1958 to Sitobion Mordvilko, 1914
(Sitobion) africanum subsp. neusi Eastop, 1958 = Sitobion neusi (Eastop, 1958)
(Sitobion) bamendae Eastop, 1959 to Sitobion Mordvilko, 1914
(Sitobion) chanikiwiti Eastop, 1959 to Sitobion Mordvilko, 1914
(Sitobion) colei Eastop, 1959 to Sitobion Mordvilko, 1914
(Sitobion) congolensis Hille Ris Lambers & Doncaster, 1956 to Sitobion Mordvilko, 1914

M. (Sitobion) gathaka Eastop, 1955 = Sitobion graminis (Takahashi, 1950)
(Sitobion) halli Eastop, 1959 to Sitobion Mordvilko, 1914
(Sitobion) hirsutirostris Eastop, 1959 to Sitobion Mordvilko, 1914
(Sitobion) howlandae Eastop, 1959 to Sitobion Mordvilko, 1914
(Sitobion) krahi Eastop, 1959 to Sitobion Mordvilko, 1914
(Sitobion) leelamaniae David, 1958 to Sitobion Mordvilko, 1914
(Sitobion) manitobensis Robinson, 1965 to Sitobion Mordvilko, 1914
(Sitobion) microspinulosum David, Narayanan & Rajasingh, 1972 to Sitobion Mordvilko, 1914
(Sitobion) mucatha Eastop, 1955 to Sitobion Mordvilko, 1914
(Sitobion) nigeriensis Eastop, 1959 to Sitobion Mordvilko, 1914
(Sitobion) ochnearum Eastop, 1959 to Sitobion Mordvilko, 1914
(Sitobion) plectranthi M.R. Ghosh, A.K. Ghosh & RayChaudhuri, 1971 to Sitobion Mordvilko, 1914
(Sitobion) rubiphila Takahashi, 1964 = ? Sitobion kamtshaticum (Mordvilko 1919)
(Sitobion) salicicornii Richards, 1963 to Sitobion Mordvilko, 1914
(Sitobion) schoelli F.P. Müller, 1959 = Sitobion anselliae (Hall, 1932)
(Sitobion) scoticum Stroyan, 1969 to Sitobion Mordvilko, 1914
(Sitobion) sijui Eastop, 1953 type of Pseudaphis Hille Ris Lambers, 1956 q.v.
(Sitobion) smilacicola subsp. sikkimensis A.K. Ghosh & RayChaudhuri, 1968 to Sitobion Mordvilko, 1914
(Sitobion) takahashii Eastop, 1959 to Sitobion Mordvilko, 1949
(Sitobion) yakini Eastop, 1959 to Sitobion Mordvilko, 1914
(Sitobion) yongyooti Robinson, 1972 to Sitobion Mordvilko, 1914

Subgen. Unisitobion Takahashi, 1961

M. (Unisitobion) sorbi Matsumura, 1918 (Macrosiphum) BM HRL

MACROTRICHAPHIS Miyazaki, 1971
Insecta matsum. 34: 28
Type species Macrotrichaphis yatsugatakensis Miyazaki, 1971

M. yatsugatakensis Miyazaki, 1971 BM HRL

MACULODRYAPHIS Gaumont, 1923
Annls Épiphyt. 9: 340
Lapsus pro Maculolachnus Gaumont, 1920

MACULOLACHNUS Gaumont, 1920
Bull. Soc. ent. Fr. 1920: 26-31
Type species Lachnus rosae Cholodkovsky, 1899 = Aphis submacula Walker, 1848
*Maculodryaphis* Gaumont, 1923
*Neolachnus* Mordvilko, 1929

M. jachonthovi Juchnevitch & Kan ex Narzikulov, Juchnevitch & Kan, 1971
rubi A.K. Ghosh & RayChaudhuri, 1972 HRL
sijpkensi Hille Ris Lambers, 1962 types BM HRL
submacula (Walker, 1848) (Aphis) type BM HRL
*incertus* Schouteden, 1906 (*Lachnus*)
*ogasawarae* Matsumura, 1917 (*Pterochlorus*)
*rosae* Cholodkovsky, 1899 (*Lachnus*)
*rosarum* van der Goot, 1912 (*Lachnus*)
*subterraneus* del Guercio, 1900 (*Lachnus*)

MALAPHIS Lichtenstein, 1885 misprint for Melaphis

MALAPHIS Shaposhnikov, 1951
Ént. Obozr. 31: 517
Type species Malaphis magna Shaposhnikov, 1951 = Neanuraphis quaestionis Börner, 1942
= Allocotaphis Börner, 1950

M. magna Shaposhnikov, 1951 = Allocotaphis quaestionis (Börner, 1942)

MAMONTOVA Shaposhnikov, 1964
In Bei-Bienko, G.Y., Keys to the Insects of the European part of the USSR 1: 589
Type species Mamontova vera Shaposhnikov, 1964
= ? Nearctaphis Shaposhnikov, 1950

M. vera Shaposhnikov, 1964 to ? Nearctaphis Shaposhnikov, 1950

MANSAKIA Matsumura, 1917
In Nagano, K., A collection of Essays for Mr. Yasushi Nawa, Gifu, 3: 59
Type species Mansakia miyabei Matsumura, 1917
= Hamamelistes Shimer, 1867

M. gallifoliae Monzen, 1929 to Hormaphis Osten-Sacken, 1861
kagamii Monzen, 1929 = ? Hamamelistes betulinus subsp. miyabei (Matsumura, 1917)
miyabei Matsumura, 1917 = Hamamelistes betulinus subsp. miyabei (Matsumura, 1917)

MARIAELLA Szelegiewicz, 1961
Bull. Acad. pol. Sci. Cl. II. Sér. Sci. biol. 9: 191
Type species Mariaella lambersi Szelegiewicz, 1961

M. lambersi Szelegiewicz, 1961 paratypes BM HRL

MASONAPHIS Hille Ris Lambers, 1939
Temminckia 4: 122
Type species Illinoia rhododendri (Wilson, 1918) = Macrosiphum rhododendri Wilson, 1918
Here used as subg. of Illinoia Wilson, 1910

M. anaphalidis A.N. Basu, 1964 type of Neomasonaphis A.K. Ghosh & RayChaudhuri, 1972 q.v.
dzhibladzeae Shaposhnikov, 1964 to Illinoia Wilson, 1910
lambersi MacGillivray, 1960 to Illinoia (Masonaphis)
magna Hille Ris Lambers, 1974 to Illinoia (Masonaphis)
menziesiae Robinson, 1969 to Illinioa (Masonaphis)
paqueti MacGillivray, 1958 to Illinoia (Masonaphis)
patriciae Robinson, 1969 to Illinoia (Masonaphis)
wilhelminae Hille Ris Lambers, 1962 to Illinoia (Masonaphis)

Subgen. Ericobium MacGillivray, 1958

M. (Ericobium) andromedae MacGillivray, 1958 to Illinoia Wilson, 1910
(Ericobium) azaleae subsp. caucasicum Bozhko, 1959 nomen nudum
(Ericobium) canadensis MacGillivray, 1958 to Illinoia Wilson, 1910
(Ericobium) finni MacGillivray, 1958 to Illinoia Wilson, 1910
(Ericobium) gracilicorne MacGillivray, 1958 to Illinoia Wilson, 1910
(Ericobium) grindeliae subsp. palmerae MacGillivray, 1958 to Illinoia Wilson, 1910
(Ericobium) macgillivrayae Hille Ris Lambers, 1966 to Illinoia Wilson, 1910
(Ericobium) pepperi MacGillivray, 1958 to Illinoia Wilson, 1910
(Ericobium) pseudomorrisoni MacGillivray, 1958 = Illinoia morrisoni (Swain, 1918)
(Ericobium) richardsi MacGillivray, 1958 to Illinoia Wilson, 1910
(Ericobium) richardsi subsp. pacifica Hille Ris Lambers, 1966 to Illinoia Wilson, 1910
(Ericobium) simpsoni MacGillivray, 1958 to Illinoia Wilson, 1910
(Ericobium) spiraeae MacGillivray, 1958 to Illinoia Wilson, 1910
(Ericobium) thalictri MacGillivray, 1958 to Illinoia Wilson, 1910

Subgen. Neomasonaphis A.K. Ghosh & RayChaudhuri, 1972

M. (Neomasonaphis) inulae A.K. Ghosh & RayChaudhuri, 1972 to Neomasonaphis A.K. Ghosh & RayChaudhuri, 1972
(Neomasonaphis) rumicis Chakrabarti & RayChaudhuri, 1975

MASRAPHIS Soliman, 1938
Bull. Ministr. Agric. Egypt tech. scient. Serv. 208: 4
Type species Masraphis phyllostachia Soliman, 1938 = Aphis bambusae Fullaway, 1910
= Melanaphis van der Goot, 1917

M. phyllostachia Soliman, 1938 = Melanaphis bambusae (Fullaway, 1910)

MASTOPODA Oestlund, 1886
Rep. Minn. geol. Surv. 14: 52
Type species Mastopoda pteridis Oestlund, 1886

M. pteridis Oestlund, 1886 BM HRL

MATSUMURAJA Schumacher, 1921
Zool. Anz. 53: 186
Type species Acanthaphis rubi Matsumura, 1918
*Acanthaphis* Matsumura, 1918 nec del Guercio, 1908
*Neophorodon* Takahashi, 1922

M. calorai Calilung, 1972
capitophoroides Hille Ris Lambers, 1966 paratype BM HRL
formosana Takahashi, 1925
hirakurensis Sorin, 1971 BM
indica A.K. Ghosh, M.R. Ghosh & RayChaudhuri, 1971 = Tuberoaphis hydrangeae subsp. digitata Hille Ris Lambers & A.N. Basu, 1966
nuditerga Hille Ris Lambers, 1965 paratypes BM HRL
ribi Shinji, 1941 lapsus pro rubi Matsumura, 1918
rubea Sorin, 1965
rubi (Matsumura, 1918) (Acanthaphis) HRL
rubicola Takahashi, 1927 BM HRL
rubifoliae Takahashi, 1931 BM HRL
*rubi* Takahashi, 1922 (*Neophorodon*) nec Acanthaphis rubi Matsumura 1918
*zeni* Tseng & Tao, 1938 (*Acanthaphis*) teste Tao, 1963
rubiphila Takahashi, 1965 HRL
sorini Takahashi, 1965 HRL
taisetsusana Miyazaki, 1971 HRL
urticae A.K. Ghosh, M.R. Ghosh & RayChaudhuri, 1971

MECINARIA Börner, 1949
Beitr. tax. Zool. 1: 59, described as subg. of Cinaria Börner, 1939
Type species Aphis piceae Panzer, 1801
= Cinara Curtis, 1835

MECONAPHIS Amyot, 1847
Annls Soc. ent. Fr. (2) 5: 478
Invalid

MECYNAPHIS Amyot, 1847
Annls. Soc. ent. Fr. (2) 5: 380
Invalid

MEDORALIS Börner, 1952
Mitt. thüring. bot. Ges. 4 (3): 78
Type species Aphis pomi DeGeer, 1773
= Aphis Linnaeus, 1758

MEGALOCALLIS Takahashi, 1963
Kontyû 31: 160
Type species Megalocallis takagii Takahashi, 1963
= Yamatocallis Matsumura, 1917

M. takagii Takahashi, 1963 to Yamatocallis Matsumura, 1917

MEGALOPHYLLAPHIS M.R. Ghosh, A.K. Ghosh & RayChaudhuri, 1971 (1970)
Orient. Insects 4: 383
Type species Megalophyllaphis obscura M.R. Ghosh, A.K. Ghosh & RayChaudhuri, 1971 (1970)
Here used as subg. of Yamatocallis Matsumura, 1917

M. obscura M.R. Ghosh, A.K. Ghosh & RayChaudhuri, 1971 (1970) to Yamatocallis (Megalophyllaphis)

MEGALOSIPHUM Mordvilko, 1919
Faune Russie, Ins. Hém. 1 (2): 357
Type species Aphis sonchi Linnaeus, 1767
= Uroleucon Mordvilko, 1914

M. oxyacantha MB ex Rusanova, 1947: 172
sonchi var. nigrosiphon Rusanova, 1943 nomen nudum

MEGANTENNAPHIS Heie, 1967
Spolia zool. Mus. haun. 26: 142
Type species Megantennaphis hauniensis Heie, 1967
Fossil

MEGAPODAPHIS Heie, 1967
Spolia zool. Mus. haun. 26: 155
Type species Megapodaphis monstrabilis Heie, 1967
Fossil

MEGOURA Buckton, 1876
Monograph of the British Aphides, London, 1: 188
Type species Megoura viciae Buckton, 1876
*Drepaniella* del Guercio, 1913

M. abnormis L.K. Ghosh, 1970 = lespedezae (Essig & Kuwana, 1918)
aconiti subsp. sylvanae Knechtel & Manolache, 1943 = Delphiniobium junackianum subsp. sylvanae (Knechtel & Manolache, 1943)
bibula Hottes, 1930 = viciae Buckton, 1876
brevipilosa Miyazaki, 1971 HRL
cajanae M.R. Ghosh, A.K. Ghosh & RayChaudhuri, 1971 = lespedezae (Essig & Kuwana, 1918)
crassicauda Mordvilko, 1919 BM HRL
*japonica* Okamoto & Takahashi, 1927 (*Rhopalosiphum*)
*lathyri* Shinji, 1924 (*Amphorophora*)
*moriokae* Shinji, 1923 (*Nectarosiphon*)
*viciae* subsp. *coreana* Moritsu, 1948
*viciae* subsp. *crassicauda* Mordvilko, 1919
*viciae* subsp. *ussuriensis* Mordvilko, 1919
*viciae* var. *japonica* Matsumura, 1918 (*Rhopalosiphum*)
*vicicola* Shinji, 1941 (*Amphorophora*)
dryopteridis Matsumura, 1918 = Amphorophora ampullata Buckton, 1876
jacobsoni Mason, 1927 = Sinomegoura citricola (van der Goot, 1917)
kaltenbachi Hille Ris Lambers, 1938 = viciae Buckton, 1876
lespedezae (Essig & Kuwana, 1918) (Rhopalosiphum) BM HRL
*abnormis* L.K. Ghosh, 1970 RHL
*cajanae* M.R. Ghosh, A.K. Ghosh & RayChaudhuri, 1971
litoralis F.P. Müller, 1952 BM HRL
pallipes A.N. Basu, 1969 HRL

M. solani Thomas, 1879 to Rhopalosiphoninus (Myzosiphon)
viciae Buckton, 1876 types BM HRL
*bibula* Hottes, 1930
*kaltenbachi* Hille Ris Lambers, 1938
*papilionacearum* Lindlinger, 1932 (*Rhopalosiphum*)
*viciae* Kaltenbach, 1843 (*Aphis*) nec Fabricius, 1781
viciae subsp. abchasica Mordvilko, 1919
viciae subsp. coreana Moritsu, 1948 = crassicauda Mordvilko, 1919
viciae subsp. crassicauda Mordvilko, 1919 = crassicauda Mordvilko, 1919
viciae subsp. nigricauda Mordvilko ex Rusanova, 1943 nomen nudum
viciae subsp. ussuriensis Mordvilko, 1919 = crassicauda Mordvilko, 1919

MEGOURELLA Hille Ris Lambers, 1949
Temminckia 8: 268
Type species Aphis tribulis Walker, 1849

M. purpurea Hille Ris Lambers, 1949 BM HRL
tribulis (Walker, 1849) (Aphis) type BM HRL

MEGOURINA Hille Ris Lambers, 1974
Tijdschr. Ent. 117: 128
Type species Megourina lagacei Hille Ris Lambers, 1974

M. lagacei Hille Ris Lambers, 1974 HRL

MEGOUROLEUCON Miyazaki, 1971
Insecta matsum. 34: 37
Type species Megouroleucon codonopsicola Miyazaki, 1971

M. codonopsicola Miyazaki, 1971 BM HRL

MEGOUROPARSUS Smith & Heie, 1963
Ann. ent. Soc. Am. 56: 401
Type species Neoamphorophora tephrosiae Smith, 1948
= Microparsus Patch, 1919

M. dooarsis A.K. Ghosh & RayChaudhuri, 1969 to ? Microparsus Patch, 1919
kislankoi Smith & Heie, 1963 to Microparsus Patch, 1919

MEITANAPHIS Tsai & Tang, 1946
Trans. R. ent. Soc. Lond. 97: 410
Type species Meitanaphis elongallis Tsai & Tang, 1946
= Schlechtendalia Lichtenstein, 1883

M. elongallis Tsai & Tang, 1946 to Schlechtendalia Lichtenstein, 1893

MELANAPHIS van der Goot, 1917
Contrib. Faune Indes Néerl. 1 (3): 61
Type species Aphis bambusae Kirkaldy lapsus pro Fullaway, 1910
*Geoktapia* Mordvilko, 1921
*Longiunguis* van der Goot, 1917
*Masraphis* Soliman, 1938
*Nevskia* or *Nevskya* Mordvilko, 1932
*Piraphis* Börner, 1932
*Pyraphis* Börner, 1931
*Schizaphidiella* Hille Ris Lambers, 1939
*Yezabura* Matsumura, 1917

M. arundinariae (Takahashi, 1937) (Aphis) BM HRL
bambusae (Fullaway, 1910) (Aphis) BM HRL
? *photiniae* Matsumura, 1918 (*Yezabura*)
*phyllostachia* Soliman, 1938 (*Masraphis*)
*sasae* Matsumura, 1917 (*Yezabura*)
daisenensis (Sorin, 1970) (Longiunguis)
donacis (Passerini, 1862) (Aphis) BM HRL
*insignis* Theobald, 1918 (*Hyalopterus*)
elizabethae (Ossiannilsson, 1967) (Longiunguis) BM HRL
formosana (Takahashi, 1921) (Aphis) HRL
*miscanthi* Takahashi, 1921 (*Aphis*)
jamatonica (Sorin, 1970) (Longiunguis) BM HRL
japonica (Takahashi, 1919) (Brachysiphum) BM HRL
koreana (Sorin, 1972) (Longiunguis) HRL
luzulella (Hille Ris Lambers, 1939) (Longiunguis) BM HRL
meghalayensis RayChaudhuri & C. Banerjee, 1974
meghalayensis subsp. bengalensis RayChaudhuri & C. Banerjee, 1974
montana (Sorin, 1970) (Longiunguis) BM HRL
pahanensis (Takahashi, 1950) (Aphis) types BM HRL
? *pyrivora* Tao, 1962 (*Sappaphis*)
pyraria (Passerini, 1861) (Myzus) BM HRL
*areshensis* Mordvilko, 1921 (*Geoktapia*)

*pyrastri* Boisduval, 1867 (*Aphis*)
*pyrinus* Ferrari, 1872 (*Myzus*)
*quinquarticulata* Hille Ris Lambers, 1939) (*Schizaphidiella*) HRL
*streili* Börner, 1931 (*Pyraphis*)

M. sacchari (Zehntner, 1897) (Aphis) BM HRL
*pheidolei* Theobald, 1916 (*Aphis*)
*sorghella* Schouteden, 1906 (*Aphis*)
*sorghi* Theobald, 1904 (*Aphis*)
sacchari forma dassi (David, 1956) (Longiunguis) HRL
*dassi* David, 1956 (*Longiunguis*)
sacchari forma indosacchari (David, 1956) (Longiunguis) HRL
*indosacchari* David, 1956 (*Longiunguis*)
siphonella (Essig & Kuwana, 1918) (Aphis) HRL
tateyamaensis (Sorin, 1970) (Longiunguis) HRL
vandergooti RayChaudhuri & C. Banerjee, 1974 = Brachysiphoniella montana (van der Goot, 1917)
yasumatsui (Sorin, 1970) (Longiunguis) BM HRL

MELANOCALLIS Oestlund, 1922
Rep. Minn. St. Ent. 19: 136
Type species Callipterus caryaefoliae Davis, 1910 = Aphis fumipennella Fitch, 1855

M. fumipennellus (Fitch, 1855) (Aphis)
*caryaefoliae* Davis, 1910 (*Callipterus*)

MELANOSIPHON Börner, 1944
In Brohmer, P., Fauna von Deutschland, ed. 5, Leipzig, p. 217
Type species Aulacorthum speyeri Börner, 1939
= Aulacorthum Mordvilko, 1914

MELANOSIPHUM Shinji, 1942
Insect World 46: 228
Type species Melanosiphum lonicericola Shinji, 1942
= Amphicercidus Oestlund, 1922

M. lonicericola Shinji, 1942 = Amphicercidus japonicus (Hori, 1927)

MELANOXANTERIUM Schouteden, 1901
Annls Soc. ent. Belg. 45: 113
Type species Aphis salicis Linnaeus, 1758
= Pterocomma Buckton, 1879

M. antennatum Patch, 1913 type of Paducia Hottes & Frison, 1931 q.v.
coreanum Okamoto & Takahashi, 1927 to Plocamaphis Oestlund, 1922
yezoense Hori, 1929 to Pterocomma Buckton, 1879

MELANOXANTHUS Buckton, 1879 nec Eschscholtz, 1836
Monograph of the British Aphides, London, 2: 21
Type species Aphis salicis Linnaeus, 1758
= Pterocomma Buckton, 1879

M. bicolor Oestlund, 1887 to Pterocomma Buckton, 1879
flocculosus Weed, 1891 type of Plocamaphis Oestlund, 1922 q.v.
salijaponica Shinji, 1924 to Plocamaphis Oestlund, 1922
vignae Matsumura, 1917 to ? Pterocomma Buckton, 1879

MELAPHIS Walsh, 1867
Proc. ent. Soc. Philad. 6: 281
Type species Byrsocrypta rhois Fitch, 1866
*Truncaphis* Theobald, 1918

M. minutus Baker, 1919 = rhois (Fitch, 1866)
peitan Tsai & Tang, 1946 = Schlechtendalia chinensis (Bell, 1851)
rhois (Fitch, 1866) (Byrsocrypta) BM HRL
*minutus* Baker, 1919
*newsteadi* Theobald, 1918 (*Truncaphis*)

MELIARHIZOPHAGUS Smith, 1974
Tech. Bull. N. Carol. Agric. Exp. St. 226: 17
Type species Pemphigus fraxinifolii Riley, 1879
= Prociphilus Koch, 1857

MENGEAPHIS Heie, 1967
Spolia zool. Mus. haun. 26: 113
Type species Lachnus glandulosus Menge, 1856
Fossil

MERINGOSIPHON Carver, 1959
Proc. R. ent. Soc. Lond. (B) 28: 19
Type species Meringosiphon paradisiacus Carver, 1959

M. paradisiacum Carver, 1959 paratypes BM HRL

MESOCALLIS Matsumura, 1919
Trans. Sapporo nat. Hist. Soc. 7: 103
Type species Myzocallis sawashibae Matsumura, 1917
*Neocallis* Matsumura, 1919
*Nippochaitophorus* Takahashi, 1961
Here used as subg. of Pterocallis Passerini, 1860

M. alnicola A.K. Ghosh, 1974 to Pterocallis (Mesocallis)
fagicola Matsumura, 1919 to Pterocallis (Mesocallis)
obtusirostris A.K. Ghosh, 1974 to Pterocallis (Mesocallis)
pteleae Matsumura, 1919 to Pterocallis (Mesocallis)

MESOTRICHOSIPHUM Calilung, 1967
Philipp. Agric. 51: 89
Type species Mesotrichosiphum uichancoi Calilung, 1967

M. uichancoi Caillung, 1967 paratypes BM HRL

METANIPPONAPHIS Takahashi, 1959
Bull. Univ. Osaka Prefect. (Ser. B.) 9: 5
Type species Metanipponaphis rotunda Takahashi, 1959

M. assamensis A.K. Ghosh & RayChaudhuri, 1973
cuspidatae (Essig & Kuwana, 1918) (Nipponaphis) BM HRL
*gigantea* Takahashi, 1921 (*Astegopteryx*)
echinata A.K. Ghosh, 1974 HRL
lithocarpicola (Takahashi, 1933) (Thoracaphis)
rotunda Takahashi, 1959 BM HRL
shiicola Takahashi, 1959 HRL
? silvestrii (Takahashi, 1935) (Thoracaphis)

METAPHIS Matsumura, 1918
Trans. Sapporo nat. Hist. Soc. 7: 1
Type species Metaphis angelicae Matsumura, 1918
= Cavariella del Guercio, 1911

M. angelicae Matsumura, 1918 to Cavariella del Guercio, 1911

METAPHORODON Takahashi, 1961
Bull. Univ. Osaka Prefect. (Ser. B) 11: 1
Type species Phorodon ishimikawae Shinji, 1941
= Xenomyzus Aizenberg, 1935

M. isodonis Takahashi, 1965 to Myzus Passerini, 1860

METATRICHOSIPHON RayChaudhuri, 1956
Zoöl. Verh., Leiden 31: 79
Type species Trichosiphum nigrofasciatum Maki, 1917
Here used as subg. of Mollitrichosiphum Suenaga, 1934

METOPEURUM Mordvilko, 1914
Faune Russie, Ins. Hém. 1: 50, 56, 67
Type species Aphis tanaceti Linnaeus of Mordvilko, 1914 nec Linnaeus, 1758
= Metopeurum fuscoviride Stroyan, 1950
*Pharalis* auctt. nec Leach, 1826 invalid
*Sinosiphoniella* Tao, 1963

M. achilleae Bozhko, 1959
borysthenicum Bozhko, 1959 HRL
capillatum (Börner, 1950) (Pharalis)
enslini (Börner, 1933) (Microsiphon) HRL
fuscoviride Stroyan, 1950 BM HRL
*tanaceti* auctt. nec Linnaeus, 1758 (*Aphis*)
kuwayamai (Takahashi, 1941) (Macrosiphoniella) HRL
matricariae Bozhko, 1959
paeke Umarov, 1964

METOPOLOPHINUM L.K. Ghosh, 1970
Bull. Ent., ent. Soc. India 11 (2): 116, described as subg. of Metopolophium Mordvilko, 1914
Type species Metopolophium (Metopolophinum) darjeelingensis L.K. Ghosh, 1970
Here used as subg. of Metopolophium Mordvilko, 1914

METOPOLOPHIUM Mordvilko, 1914
Faune Russie, Ins. Hém. 1: 270, described as subg. of Acyrthosiphon Mordvilko, 1914
Type species Aphis dirhoda Walker, 1849
*Goidanichiellum* Martelli, 1950

M. albidum Hille Ris Lambers, 1947 BM HRL
alpinum Hille Ris Lambers, 1966 BM HRL
arctogenicolens (Richards, 1964) (Acyrthosiphon (Metopolophium))
beiquei Hille Ris Lambers, 1960 HRL
caudatum (Pergande, 1900) (Nectarophora)
chandrani (David & Narayanan, 1968) (Acyrthosiphon (Metopolophium) HRL
dirhodum (Walker, 1849) (Aphis) lectotype BM HRL

*arundinis* Theobald, 1913 (*Macrosiphum*) type BM
*gracilis* Buckton, 1876 (*Myzus*) type BM
*graminum* Theobald, 1913 (*Macrosiphum*) type BM
*haywardi* Knowlton, 1942 (*Myzus*)
*longipennis* Buckton, 1876 (*Siphonophora*) types BM

M. euryae Takahashi, 1937 (Macrosiphum)
festucae (Theobald, 1917) (Myzus) types BM HRL
frisicum Hille Ris Lambers, 1947 BM HRL
gracilipes Börner, 1950 = Acyrthosiphon loti (Theobald, 1913)
knechteli Börner, 1950 to Acyrthosiphon Mordvilko, 1914
longicaudatum (David & Hameed, 1975) (Acyrthosiphon (Metopolophium)) HRL
montanum Hille Ris Lambers, 1966 HRL
myrmecophilum (Theobald, 1916) (Macrosiphum)
palmerae (Hille Ris Lambers, 1949) (Hyalopteroides) BM HRL
? pedicularis (Richards, 1972) (Acyrthosiphon)
potha Börner ex Franz, 1943 = Acyrthosiphon malvae subsp. potha (Börner ex Franz, 1943)
simlaense (Chakrabarti & RayChaudhuri ex Chakrabarti, A.K. Ghosh & Ray-Chaudhuri, 1974) (Acyrthosiphon (Metopolophium)) HRL
tenerum Hille Ris Lambers, 1947 BM HRL

Subgen. Metopolophinum L.K. Ghosh, 1970

M. (Metopolophinum) darjeelingense L.K. Ghosh, 1970 type of Metopolophinum L.K. Ghosh, 1970 BM HRL

MICRANTHAPHIS Grassi, 1912
In Grassi, B. & Foa, A., Contribute alla Conoscenza delle Fillosserine, .. p. 48
described as subg. of Phylloxera Boyer de Fonscolombe, 1834
Type species Micranthaphis foae (Börner, 1909) = Phylloxera foae Börner, 1909
= Phylloxera Boyer de Fonscolombe, 1834
Phylloxeridae

MICRAPHIS Takahashi, 1931
Rep. Dep. Agric. Govt. res. Inst. Formosa 53: 53
Type species Anuraphis artemisiae Takahashi, 1923

M. artemisiae (Takahashi, 1923) (Anuraphis) BM
takahashii Tseng & Tao, 1936 = ? Coloradoa viridis (Nevsky, 1929)

MICRELLA Essig, 1912
Pomona Coll. J. Ent. 4: 716
Type species Micrella monelli Essig, 1912
= Chaitophorus Koch, 1854

M. monelli Essig, 1912 to Chaitophorus Koch, 1854

MICROLOPHIUM Mordvilko, 1914
Faune Russie, Ins. Hém. 1 (1): 198, described as subg. of Acyrthosiphon Mordvilko, 1914
Type species Aphis urticae Schrank, 1801 nec Linnaeus, 1758 = Siphonophora carnosa Buckton, 1876

M. carnosum (Buckton, 1876) (Siphonophora) BM HRL
*evansi* Theobald, 1923 (*Amphorophora*) types BM
*schranki* Theobald, 1927 (*Macrosiphum*) BM
*urticae* Schrank, 1801 (*Aphis*) nec Linnaeus, 1758
*urticae* subsp. *meridionale* Mordvilko, 1914 (*Acyrthosiphon (Microlophium)*)
carnosum subsp. tenuicauda Hille Ris Lambers, 1949 = sibiricum subsp. tenuicauda Hille Ris Lambers, 1949
primulae (Theobald, 1913) (Macrosiphum) BM HRL
rjabushinskyi (Mordvilko, 1914) (Acyrthosiphon) keyed as subsp. of Acyrthosiphon (Microlophium) sibiricum on p. 81; as A. (M.) rjabushinskyi in Addenda on p. 5; and described as A. rjabushinskii Mordvilko, 1919: 244.
rubiformosanum (Takahashi, 1927) (Macrosiphum) BM HRL
sibiricum (Mordvilko, 1914) (Acyrthosiphon (Microlophium)) BM HRL
*sibiricum* subsp. *kirgiz* Mordvilko, 1914 (*Acyrthosiphon (Microlophium)*)
sibiricum subsp. tenuicauda Hille Ris Lambers, 1949 HRL
*carnosum* subsp. *tenuicauda* Hille Ris Lambers, 1949 HRL

MICROMYZELLA Eastop, 1955
Entomologist's mon. Mag. 91: 203, described as subg. of Micromyzus van der Goot, 1917
Type species Myzus pterisoides Theobald, 1918

M. mitegoni (Eastop, 1955) (Micromyzus (Micromyzella)) types BM HRL
pterisoides (Theobald, 1918) (Myzus) types BM HRL

Subgen. Kugegania Eastop, 1955

M. (Kugegania) ageni (Eastop, 1955) (Micromyzus (Kugegania)) types BM HRL

MICROMYZODIUM David, 1958
Indian J. Ent. 20: 175
Type species Micromyzodium filicium David, 1958
*Eomyzus* Takahashi, 1960

M. dasi Verma, 1969
filicium David, 1958 BM HRL
nipponicum (Moritsu, 1949) (Myzus) BM HRL
polypodii Takahashi, 1963 HRL
spinulosum Miyazaki, 1971 BM
strobilanthi L.K. Ghosh, 1971 HRL

MICROMYZUS van der Goot, 1917
Contrib. Faune Indes Néerl. 1 (3): 52
Type species Micromyzus nigrum van der Goot, 1917

M. alliumcepa Essig, 1936 = Neotoxoptera formosana (Takahashi, 1921)
collinsoniae Pepper, 1950 to Hyalomyzus Richards, 1958
eastopi Carver, 1965 = Ipuka dispersum (van der Goot, 1917)
fuscus Richards, 1956 = Neotoxoptera formosana (Takahashi, 1921)
granotiae A.K. Ghosh, M.R. Ghosh & RayChaudhuri, 1970
jos Eastop, 1958 (Micromyzus (Picturaphis)) type BM
judenkoi Carver, 1965 paratypes BM HRL
kalimpongensis A.N. Basu, 1968 HRL
katoi (Takahashi, 1925) (Amphorophora) BM HRL
mawphlangensis A.K. Ghosh, 1974
niger van der Goot, 1917 BM HRL
nikkoensis Miyazaki, 1971 BM HRL
oliveri Essig, 1935 to Neotoxoptera Theobald, 1915
osmundae Takahashi, 1965 BM HRL
sleonensis Eastop, 1958 (Micromyzus (Picturaphis)) types BM HRL
varicolor van der Goot, 1918 nomen nudum
weigelae Takahashi, 1965 = Myzopsis diervillae Matsumura, 1918

Subgen. Kugegania Eastop, 1955

M. (Kugegania) ageni Eastop, 1955 (type of Kugegania Eastop, 1955) to Micromyzella (Kugegania)

Subgen. Micromyzella Eastop, 1955

M. (Micromyzella) mitegoni Eastop, 1955 to Micromyzella Eastop, 1955

Subgen. Picturaphis E.E. Blanchard, 1922

M. (Picturaphis) jos Eastop, 1958 to Micromyzus van der Goot, 1917
(Picturaphis) sleonensis Eastop, 1958 to Micromyzus van der Goot, 1917

MICROPARSUS Patch, 1919
Ent. News 20: 337
Type species Microparsus variabilis Patch, 1919 = Siphonophora desmodii Williams, 1911
*Megouroparsus* Smith & Heie, 1963

M. desmodii (Williams, 1911) (Siphonophora) BM HRL
*variabilis* Patch, 1919
desmodiorum Smith & Tuatay, 1960 paratypes BM HRL
? dooarsis (A.K. Ghosh & RayChaudhuri, 1969) (Megouroparsus)
kislankoi (Smith & Heie, 1963) (Megouroparsus) paratypes BM HRL
olivei Smith & Tuatay, 1960 paratype BM HRL
tephrosiae (Smith, 1948) (Neoamphorophora) BM
variabilis Patch, 1919 = desmodii (Williams, 1911)

Subgen. Picturaphis E.E. Blanchard, 1922

M. (Picturaphis) brasiliensis (Moreira, 1925) (Idiopterus) BM HRL
(Picturaphis) puertoricensis (Smith, 1970) (Picturaphis) HRL
(Picturaphis) vignaphilus (E.E. Blanchard, 1922) (Picturaphis) BM HRL

MICROSIPHON Börner, mis-spelling of Microsiphum Cholodkovsky, 1902

MICROSIPHON del Guercio, 1907
Redia 4: 191, 192
Type species Aphis tormentillae Passerini, 1879
= Aphis Linnaeus, 1758

MICROSIPHONIELLA Hille Ris Lambers, 1947
Temminckia 7: 186
Type species Chaitophorus artemisiae Gillette, 1911

M. acophorum (Smith & Knowlton, 1938) (Microsiphum) BM HRL
artemisiae (Gillette, 1911) (Chaitophorus) BM HRL
canadensis (Williams, 1911) (Cryptosiphum)
oregonensis (Wilson, 1915) (Microsiphum) BM

MICROSIPHUM Cholodkovsky, 1902
Izvest. Lesn. Inst. 8 (2): 5
Type species Microsiphum ptarmicae Cholodkovsky, 1902

M. acophorum Smith & Knowlton, 1938 to Microsiphoniella Hille Ris Lambers, 1947
enslini Börner, 1933 to Metopeurum Mordvilko, 1914
giganteum, Nevsky 1928 BM HRL
jazykovi Nevsky, 1928 BM HRL
millefolii Wahlgren, 1940 BM HRL
nudum Holman, 1961 BM HRL
oregonensis Wilson, 1915 to Microsiphoniella Hille Ris Lambers, 1947
ptarmicae Cholodkovsky, 1902
wahlgreni Hille Ris Lambers, 1947 BM HRL
woronieckae Judenko, 1931 BM HRL

MICROTARSUS Shinji, 1929 nec Eyton, 1839
Lansania 1 (3): 43
Rept. Jap. Assoc. f. Adv. Sci. 5: 188 (1930)
Type species Microtarsus pterydifoliae Shinji, 1929 = Atarsos orientalis Mordvilko, 1929
= Shinjia Takahashi, 1938

M. pterydifoliae Shinji, 1929 = Shinjia orientalis (Mordvilko, 1929)

MICROTHORACAPHIS Takahashi, 1958
Bull. Univ. Osaka Prefect. (B) 8: 4
Type species Microthoracaphis elongata Takahashi, 1958
= Neothoracaphis Takahashi, 1958

M. elongata Takahashi, 1958 to Neothoracaphis Takahashi, 1958
glaucae Takahashi, 1958 to Neothoracaphis Takahashi, 1958
querciphaga Takahashi, 1958 to Neothoracaphis Takahashi, 1958

MICROUNGUIS Tao, 1966
Q. Jl Taiwan Mus. 19: 175
Type species Thoracaphis depressus Takahashi, 1933
= Neothoracaphis Takahashi, 1958

MIMAPHIDUS Rondani, 1848
Nuovi Ann. Sci. nat. Bologna 9: 35
Type species Aphis ulmi Fabricius, 1775 nec Linnaeus, 1758 = Schizoneura lanuginosa Hartig, 1841
= Schizoneura Hartig, 1839

MIMEURIA Börner, 1952
Mitt. thüring. bot. Ges. 4 (3): 197, 259
Type species Neorhizobius ulmiphilus del Guercio, 1917
*Neorhizobius* del Guercio, 1917 nec Crotch, 1874

M. ulmiphila (del Guercio, 1917) (Neorhizobius) BM HRL
*ucrainensis* Mamontova, 1955 (*Paraprociphilus*) HRL

MIMOCALLIS Matsumura, 1919
Trans. Sapporo nat. Hist. Soc. 7: 109
Type species Mimocallis betuli-japonicae Matsumura, 1919
= Euceraphis Walker, 1870

M. betuli-japonicae Matsumura, 1919 to Euceraphis Walker, 1870

MINDARUS Koch, 1857
Die Pflanzenläuse Aphiden, Nürnberg, p. 277
Type species Mindarus abietinus Koch, 1857
*Pterostigma* Buckton, 1883 fossil
*Schizoneuroides* Buckton, 1883 fossil
*Sychnobrochus* Scudder, 1890 fossil

M. abietinus Koch, 1857 BM HRL
*pinicola* Thomas, 1879 (*Schizoneura*)
japonicus Takahashi, 1931 BM HRL
magnus Baker, 1922 fossil = ? transparens (Germar & Berendt, 1856)
obliquus (Cholodkovsky, 1896) (Schizoneura) BM HRL
parvus Heie, 1967 fossil
podocarpi Shinji, 1922 = Neophyllaphis podocarpi Takahashi, 1920
scudderi (Buckton, 1883) (Schizoneuroides) fossil
transparens (Germar & Berendt, 1856) (Aphis) fossil
*? magnus* Baker, 1922 fossil
victoria Essig, 1939 paratypes BM HRL

MINUTICORNIS Knowlton, 1928
Fla Ent. 12: 60
Type species Minuticornis gravidus Knowlton, 1928
= Siphonatrophia Swain, 1918

gravidus Knowlton, 1928 to Siphonatrophia Swain, 1918

MIRAPHIS Nevsky, 1928
Acta Univ. Asiae mediae (Ser. 8a, Zoologia) 3: 25
Type species Miraphis agabiformis Nevsky, 1928
= Hyadaphis Kirkaldy, 1904

M. agabiformis Nevsky, 1928 to Hyadaphis Kirkaldy, 1904
asparagi Börner ex Bodenheimer, 1930 & 1937 nomen nudum

MIRAPHOIDES Rusanova, 1943 (1942)
Trudy Azerb. Gos. Univ. 3(1): 36
Type species Miraphoides dilucidus Rusanova, 1943 nomen nudum
Without described species

M. dilucidus Rusanova, 1943 nomen nudum

MIROTARSUS Börner, 1939
Arb. physiol. angew. Ent. Berl. 6: 83, described as subg. of Acyrthosiphon Mordvilko, 1914
Type species Siphonophora cyparissiae Koch, 1855
= Acyrthosiphon Mordvilko, 1914

MISTURAPHIS Robinson, 1967
Can. Ent. 99: 565
Type species Misturaphis shiloensis Robinson, 1967

M. shiloensis Robinson, 1967 paratypes BM HRL

MIZELLA errore pro Myzella Börner, 1930

MIZOCALLIS mis-spelling of Myzocallis Passerini, 1860

MOLLITRICHOSIPHUM Suenaga, 1934
Bull. Kagoshima imp. Coll. Agric. For. 1: 798
Type species Greenidea tenuicorpus (Okajima, 1908) = Trichosiphum tenuicorpus Okajima, 1908
*Neotrichosiphon* RayChaudhuri, 1956

M. shinjii RayChaudhuri, M.R. Ghosh, M. Banerjee & A.K. Ghosh, 1973
tenuicorpus (Okajima, 1908) (Trichosiphum) BM HRL
yamabiwae Suenaga, 1934 to Mollitrichosiphum (Metatrichosiphon)

Subgen. Metatrichosiphon RayChaudhuri, 1956

M. (Metatrichosiphon) alni A.K. Ghosh, M.R. Ghosh & RayChaudhuri, 1970 = ? montanum (van der Goot, 1918)
(Metatrichosiphon) buddleiae A.K. Ghosh, H. Banerjee & RayChaudhuri, 1971
(Metatrichosiphon) glaucae Takahashi, 1962 HRL
*nigrofasciatum* subsp. *glaucae* Takahashi, 1962
(Metatrichosiphon) kazirangi A.K. Ghosh, 1974
(Metatrichosiphon) lithocarpi (Takahashi, 1931) (Paratrichosiphum)
(Metatrichosiphon) loochuanum Takahashi, 1962 = luchuanum (Takahashi, 1930)
(Metatrichosiphon) luchuanum (Takahashi, 1930) (Greenidea) HRL
*loochuanum* Takahashi, 1962
(Metatrichosiphon) montanum (van der Goot, 1918) (Trichosiphum) BM HRL
*? alni* A.K. Ghosh, M.R. Ghosh & RayChaudhuri, 1970
(Metatrichosiphon) nandii A.N. Basu, 1964 HRL
(Metatrichosiphon) nigrofasciatum (Maki, 1917) (Trichosiphum) BM HRL
(Metatrichosiphon) nigrofasciatum subsp. glaucae Takahashi, 1962 = glaucae Takahashi, 1962
(Metatrichosiphon) niitakaense (Takahasi, 1937) (Paratrichosiphum) HRL
(Metatrichosiphon) rhusae A.K. Ghosh, 1974
(Metatrichosiphon) taiwanum (Takahashi, 1921) (Greenidea) HRL
(Metatrichosiphon) yamabiwae Suenaga, 1934 (Mollitrichosiphum) HRL

MONAPHIS Walker, 1870
Zoologist 3: 2001
Type species Aphis antennata Kaltenbach, 1843
*Bradyaphis* Mordvilko, 1894

M. antennata (Kaltenbach, 1843) (Aphis) BM HRL

MONELLIA Oestlund, 1887
Bull. Minn. geol. nat. Hist Surv. 4: 44
Type species Aphis caryella Fitch, 1855
*Monelliopsis* Richards, 1960
*Protopterocallis* Richards, 1965

M. bisetosa (Richards, 1966) (Monelliopsis)
californica Essig, 1912 to Chromaphis Walker, 1870
canadensis (Richards, 1965) (Protopterocallis) BM HRL
caryae (Monell, 1879) (Callipterus) BM HRL
caryella (Fitch, 1855) (Aphis) BM HRL

M. ? costalis (Fitch, 1855) (Aphis) BM HRL
*caryella* var. *costalis* Fitch, 1855 (*Aphis*)
*marginella* Fitch, 1855 (*Aphis*)
formosana (Takahashi, 1924) (Myzocallis) (type of Cranaphis Takahashi, 1939)
= Cranaphis formosana (Takahashi, 1924)
hispida Quednau, 1971
lagerstroemiae Takahashi, 1920 = Sarucallis kahawaluokalani (Kirkaldy, 1907)
maculella (Fitch, 1855) (Aphis)
microsetosa Richards, 1960 BM HRL
nigropunctata Granovsky, 1931 cotype BM HRL
*pleurialis* Richards, 1965 (*Monelliopsis*)
tuberculata (Richards, 1966) (Monelliopsis) HRL

MONELLIOPSIS Richards, 1960
Mem. ent. Soc. Can. 44: 89
Type species Callipterus caryae Monell, 1879
= Monellia Oestlund, 1887

M. bisetosa Richards, 1966 to Monellia Oestlund, 1887
pleurialis Richards, 1965 = Monellia nigropunctata Granovsky, 1931
tuberculata Richards, 1966 to Monellia Oestlund, 1887

MONZENIA Takahashi, 1962
Bull. Univ. Osaka Prefect. (B) 13: 7
Type species Nipponaphis globuli Monzen, 1934

M. globuli (Monzen, 1934) (Nipponaphis) BM HRL

MORDVILKOIELLA Shaposhnikov, 1964
In Bei-Bienko, G.Y., Keys to the Insects of the European part of the USSR 1: 569
Type species Hyalopterus skorkini Mordvilko, 1929

M. skorkini (Mordvilko, 1929) (Hyalopterus)

MORDVILKOMEMOR Shaposhnikov, 1950
Ént. Obozr. 31: 9
Type species Dentatus pilosus Mordvilko ex Nevsky, 1929
Here used as subg. of Brachycaudus van der Goot, 1913

M. macrotuberculatum Narzikulov, 1950 = Brachycaudus (Mordvilkomemor) pilosus (Mordvilko ex Nevsky, 1929)

MORDWILKOJA del Guercio, 1909
Riv. Patol. veg. Padova (N.S.) 4: 11
Type species Byrsocrypta vagabunda Walsh, 1863

M. vagabunda (Walsh, 1863) (Byrsocrypta) BM
*oestlundi* Cockerell, 1906 (*Pemphigus*)

MORITZIELLA Börner, 1908
Zool. Anz. 33: 608
Type species Phylloxera corticalis Kaltenbach, 1867
*Acanthaphis* del Guercio, 1908
Phylloxeridae

MUCROTRICHAPHIS Knowlton & Allen, 1940
Ohio J. Sci. 40: 31
Type species Mucrotrichaphis toti Knowlton & Allen, 1940 = Macrosiphum zerothermum Knowlton & Allen, 1938
= Obtusicauda Soliman, 1927

M. albicornis Knowlton & Allen, 1940 to Obtusicauda Soliman, 1927
flavila Knowlton & Allen, 1940 to Obtusicauda Soliman, 1927
toti Knowlton & Allen, 1940 = Obtusicauda zerotherma (Knowlton & Allen, 1938)
zerohypsi Knowlton & Allen, 1940 to Obtusicauda Soliman, 1927

MUSCAPHIS Börner, 1933
Kleine Mitteilungen über Blattläuse, privately published, p. 4
Type species Muscaphis musci Börner, 1933

M. musci Börner, 1933 BM HRL
*stammeri* Börner, 1952
stammeri Börner, 1952 = musci Börner, 1933

MYZAKKAIA A.N. Basu, 1969
Orient. Insects 3: 177
Type species Myzakkaia himalayensis A.N. Basu, 1969 = Vesiculaphis verbasci Chowdhuri, R.C. Basu, Chakrabarti & RayChaudhuri, 1969

M. himalayensis A.N. Basu, 1969 = verbasci (Chowdhuri, R.C. Basu, Chakrabarti & RayChaudhuri, 1969)
niitakaensis (Takahashi, 1937) (Myzus)
polygonicola A.N. Basu, 1969 HRL
verbasci (Chowdhuri, R.C. Basu, Chakrabarti & RayChaudhuri, 1969) (Vesiculaphis) BM HRL
*himalayensis* A.N. Basu, 1969 HRL

MYZAPHIS van der Goot, 1913
Tijdschr. Ent. 56: 96
Type species Aphis rosarum Kaltenbach, 1843
*Francoa* del Guercio, 1916

M. amygdalina Nevsky, 1928 to Myzus Passerini, 1860
avariolosa David, Rajasingh & Narayanan, 1971 HRL
beybienkoi Narzikulov, 1957 to Myzus (Nectarosiphon)
bucktoni Jacob, 1946 types BM HRL
canadensis Richards, 1963
ericae Börner, 1933 type of Ericaphis Börner, 1939 q.v.
komatsubarae Shinji, 1922 incertae sedis
lepidii Nevsky, 1929 (type of Lipaphidiella Doncaster, 1954) to Lipaphis (Lipaphidiella)
lepidii subsp. lepidiicardariae Knechtel & Manolache, 1944 to Lipaphis (Lipaphidiella)
lepidii var. cubensis Rusanova, 1943 nomen nudum
pujoli Gomez-Menor, 1950 to Aphidura Hille Ris Lambers, 1956
rosarum (Kaltenbach, 1843) (Aphis) BM HRL
*elegans* del Guercio, 1916 (*Francoa*)
*rhodolestes* Wood-Baker, 1943 (*Trilobaphis*)
rosarum subsp. turanicum Nevsky, 1929 = turanica Nevsky, 1929
turanica Nevsky, 1929 HRL
*rosarum* subsp. *turanica* Nevsky, 1929

MYZELLA Börner, 1930
Arch. klassif. phylogen. Ent. 1: 139
Type species Aphis galeopsidis Kaltenbach, 1843
= Cryptomyzus Oestlund, 1922

M. ulmeri Börner, 1952 = Cryptomyzus alboapicalis (Theobald, 1916)

MYZOCALLIDIUM Börner, 1949
Beitr. tax. zool. 1: 49
Type species Myzocallidium riehmi Börner, 1949
= Therioaphis Walker, 1870

M. astragali Dzhibladze, 1959 to Therioaphis Walker, 1870
dorycnii Pintera, 1956 to Therioaphis Walker, 1870

M. riehmi Börner, 1949 to Therioaphis Walker, 1870
tenerum Aizenberg, 1956 to Therioaphis Walker, 1870

MYZOCALLIS Passerini, 1860
Gli Afidi, Parma, p. 28
Type species Aphis coryli Goeze, 1778
*Agrioaphis* Walker, 1870
*Dryomyzus* Hille Ris Lambers, 1948
*Pasekia* Aizenberg, 1959
*Neomyzocallis* Richards, 1965

M. agrifolicola Richards, 1966 HRL
*dicksoni* Hille Ris Lambers, 1966
alhambra Davidson, 1920 = punctatus (Monell, 1879)
alnicola Shinji, 1924 = ? Pterocallis (Recticallis) alnijaponicae (Matsumura, 1919)
arundinariae Essig, 1917 to Takecallis Matsumura, 1917
assimilis Börner, 1940 = castanicola Baker, 1917
bambusicola Takahashi, 1921 type of Chucallis Tao, 1964 q.v.
bambusifoliae Takahashi, 1921 = Takecallis arundinariae (Essig, 1917)
bodenheimeri Hille Ris Lambers, 1948 = Hoplocallis pictus (Ferrari, 1872)
boerneri Stroyan, 1957 BM HRL
*? schreiberi* Hille Ris Lambers & Stroyan, 1959
boudyi Mimeur, 1934 = Hoplocallis pictus (Ferrari, 1872)
californicus Baker, 1917 to Tuberculatus (Pacificallis)
californicus var. pallida Davidson, 1919 = Tuberculatus (Pacificallis) pallidus (Davidson, 1919)
capitatus Essig & Kuwana, 1918 to Tuberculatus (Orientuberculoides)
carpini (Koch, 1855) (Callipterus) BM HRL
castanicola Baker, 1917
*assimilis* Börner, 1940
*castaneae* Buckton, 1881 (*Callipterus*) nec Fitch, 1856
*davidsoni* Swain, 1918
*petraeae* Bozhko, 1957 (*Tuberculoides*)
coryli (Goeze, 1778) (Aphis) BM HRL
*avellanae* Blanchard, 1840 (*Aphis*) nec Schrank, 1801
*coryli* Koch, 1855 (*Callipterus*)
cyperis Macchiati, 1884 incertae sedis
davidsoni Swain, 1918 = castanicola Baker, 1917
dicksoni Hille Ris Lambers, 1966 = agrifolicola Richards, 1966
discolor (Monell, 1879) (Callipterus) BM HRL

M. edwardsi Laing, 1927 type of Neuquenaphis E.E. Blanchard, 1939 q.v.
elliotti Boudreaux & Tissot, 1962 to Myzocallis (Lineomyzocallis)
essigi Shinji, 1917 = Tuberculatus (Tuberculoides) annulatus (Hartig, 1841)
exultans Boudreaux & Tissot, 1962 to Myzocallis (Lineomyzocallis)
formosana Takahashi, 1924 to Cranaphis Takahashi, 1939
frisoni Boudreaux & Tissot, 1962 to Myzocallis (Lineomyzocallis)
glandulosus Hille Ris Lambers, 1948 (Myzocallis (Dryomyzus)) BM HRL
granovskyi Boudreaux & Tissot, 1962 to Myzocallis (Lineomyzocallis)
hypeici Thomas, 1879 errore pro hyperici
hyperici Thomas, 1879 = Brachysiphum hyperici (Monell, 1879)
insularis Takahashi, 1927 to Tinocallis Matsumura, 1919
kahawaluokalani Kirkaldy, 1907 to Sarucallis Shinji, 1922
kashiwae Matsumura, 1917 to Tuberculatus (Orientuberculoides)
kiowanicus Hottes, 1933 to Tuberculatus (Pacificallis)
komareki (Pašek, 1953) (Hoplocallis) type of Pasekia Aizenberg, 1959 HRL
kunugi Shinji, 1924 to ? Tuberculatus (Orientuberculoides)
kuricola var. cantonensis Takahashi, 1936
longiunguis Boudreaux & Tissot, 1962 to Myzocallis (Lineomyzocallis)
macrotuberculata Essig & Kuwana, 1918 = Tuberculatus (Acanthocallis) quercicola (Matsumura, 1917)
maureri Swain, 1918 to Tuberculatus (Pacificallis)
melanocera Boudreaux & Tissot, 1962 to Myzocallis (Lineomyzocallis)
meridionalis Granovsky, 1939 to Myzocallis (Lineomyzocallis)
mimicus Richards, 1965 (Myzocallis (Neomyzocallis))
montana Higuchi, 1972 to Pterocallis Passerini, 1860
multisetis Boudreaux & Tissot, 1962 to Myzocallis (Lineomyzocallis)
mushensis Takahashi, 1925 to Tinocallis Matsumura, 1919
myricae (Kaltenbach, 1843) (Aphis) BM HRL
naracola Matsumura, 1918 = Tuberculatus (Orientuberculoides) kashiwae (Matsumura, 1917)
nigra Okamoto & Takahashi, 1927 = Tuberculatus (Acanthocallis) stigmatus (Matsumura, 1917)
occidentalis Remaudière & Nieto Nafria, 1974 BM HRL
pasaniae Davidson, 1915 to Tuberculatus (Pacificallis)
pepperi Boudreaux & Tissot, 1962 to Myzocallis (Lineomyzocallis)
pilosus Takahashi,1929 to Tuberculatus (Acanthocallis)
polychaetus David, 1969 (Dryomyzus) BM HRL
*aptera* Richards & Kumar, 1971 (*Myzocallis (Agrioaphis)*)

M. pseudoalni Takahashi, 1921 to Pterocallis (Recticallis)
pulchellus Glendenning, 1929 = Tinocallis platani (Kaltenbach, 1843)
punctatus (Monell, 1879) (Callipterus) BM HRL
*alhambra* Davidson, 1920
*asclepiadis* Monell, 1879 (*Callipterus*)
*hyalinus* Monell, 1879 (*Callipterus*)
quercifolii Davidson, 1919 to Tuberculatus (Pacificallis)
querciformosanus Takahashi, 1921 to Tuberculatus (Orientuberculoides)
querciplatensis E.E. Blanchard, 1926 = Tuberculatus (Tuberculoides) annulatus (Hartig, 1841)
quercus var. insignis Ferrari, 1872 = Tuberculatus (Tuberculoides) eggleri Börner, 1950
rhombifoliae Granovsky, 1928 to Pterocallis Passerini, 1860
rostropunctatus Richards, 1965 (Myzocallis (Neomyzocallis)) paratype BM
saccharinus del Guercio, 1913 = Chaitophorus populialbae (Boyer de Fonscolombe, 1841)
sasae Matsumura, 1917 to Takecallis Matsumura, 1917
sawashibae Matsumura, 1917 (type of Mesocallis Matsumura, 1919) to Pterocallis (Mesocallis)
schreiberi Hille Ris Lambers & Stroyan, 1959 = ? boerneri Stroyan, 1957 BM HRL
spinosa Boudreaux & Tissot, 1962 to Myzocallis (Lineomyzocallis)
subulmifolii Miller, 1936 = ? Tinocallis ulmifolii (Monell, 1879)
taiwanus Takahashi, 1926 to Takecallis Matsumura, 1917
tonkawa Hottes, 1949 = Tuberculatus (Pacificallis) kiowanicus (Hottes, 1933)
tuberculatus (de Geer) Mordvilko, 1935 lapsus = Pterocallis albida Börner, 1940
ulmifoliae Shinji, 1924 (legend to figure)
viridis Takahashi, 1929 to Tinocallis Matsumura, 1919
woodworthi Shinji, 1917 nomen dubium
yokoyamai Takahashi, 1923 to Tuberculatus (Orientuberculoides)
youngii Knowlton & Hall, (1949) 1950 = Tuberculatus (Pacificallis) kiowanicus (Hottes, 1933)
zelkowae Takahashi, 1919 = Tinocallis ulmiparvifoliae Matsumura, 1918

Subgen. Agrioaphis Walker, 1870

M. (Agrioaphis) aptera Richards & Kumar, 1971 = Myzocallis polychaetus (David, 1969)

Subgen. Dryomyzus Hille Ris Lambers, 1948

M. (Dryomyzus) glandulosus Hille Ris Lambers, 1948 (type of Dryomyzus Hille Ris Lambers, 1948) to Myzocallis Passerini, 1860

Subgen. Lineomyzocallis Richards, 1965

M. (Lineomyzocallis) bellus (Walsh, 1863) (Aphis) BM HRL
(Lineomyzocallis) elliotti Boudreaux & Tissot, 1962 (Myzocallis) cotype BM
(Lineomyzocallis) ephemerata Richards, 1965 = walshii (Monell, 1879)
(Lineomyzocallis) exultans Boudreaux & Tissot, 1962 (Myzocallis) cotype BM HRL
(Lineomyzocallis) frisoni Boudreaux & Tissot, 1962 (Myzocallis) cotype BM HRL
(Lineomyzocallis) granovskyi Boudreaux & Tissot, 1962 (Myzocallis) cotype BM HRL
(Lineomyzocallis) longirostris Richards, 1965 paratype BM
(Lineomyzocallis) longiunguis Boudreaux & Tissot, 1962 (Myzocallis) cotype BM
(Lineomyzocallis) melanocera Boudreaux & Tissot, 1962 (Myzocallis) cotype BM HRL
(Lineomyzocallis) meridionalis Granovsky, 1939 (Myzocallis) BM HRL
(Lineomyzocallis) multisetis Boudreaux & Tissot, 1962 (Myzocallis) cotype BM HRL
(Lineomyzocallis) occultus Richards, 1965
(Lineomyzocallis) pepperi Boudreaux & Tissot, 1962 (Myzocallis) cotype BM
(Lineomyzocallis) spinosus Boudreaux & Tissot, 1962 (Myzocallis) cotype BM HRL
(Lineomyzocallis) walshii (Monell, 1879) (Callipterus) BM HRL
*ephemerata* Richards, 1965 paratype BM

Subgen. Neomyzocallis Richards, 1965

M. (Neomyzocallis) mimica Richards, 1965 to Myzocallis Passerini, 1860
(Neomyzocallis) rostropunctata Richards, 1965 to Myzocallis Passerini, 1860
(Neomyzocallis) tuberculatus Richards, 1965 to Tuberculatus Mordvilko, 1894

Subgen. Nippocallis Matsumura, 1917

M. (Nippocallis) kuricola (Matsumura, 1917) (Nippocallis) BM HRL

MYZODES Mordvilko, 1914
Faune Russie, Ins. Hém. 1: 52, 69
Type species Myzodes tabaci Mordvilko, 1914 = Aphis persicae Sulzer, 1776
= Nectarosiphon Schouteden, 1901

M. certus subsp. pseudopersicae Börner, 1952 = Myzus (Nectarosiphon) certus (Walker, 1849)
myosotidis Börner, 1950 to Myzus (Nectarosiphon)
osmundae Mordvilko, 1929 nomen nudum
prokovskajae Mordvilko, 1914 to ? Utamphorophora Knowlton, 1947
rubioides Bozhko, 1959 nomen nudum
tabaci Mordvilko, 1914 = Myzus (Nectarosiphon) persicae (Sulzer, 1776)

Subgen. Myzodium Börner, 1949

M. (Myzodium) brevirostris Börner, 1950 to Dysaphis (Pomaphis)
(Myzodium) rabeleri Börner, 1950 (type of Myzodium Börner, 1949) = Myzodium modestum (Hottes, 1926)

MYZODIUM Börner, 1949
In Janetschek, H., Ber. natur.-med. Ver. Innsbruck 4849: 153; also Schlern-Schriften 67: 53 as genus; also in Börner, Neue europäische Blattlausarten, privately published (1950), p. 11, described as subg. of Myzodes Mordvilko, 1914
Type species Myzodes (Myzodium) rabeleri Börner, (1949 nomen nudum) 1950 = Carolinaia modesta Hottes, 1926

M. knowltoni Smith & Robinson, 1975 BM
lagarriguei Remaudière ex Börner, 1952 = Ericaphis latifrons (Börner, 1942)
modestum (Hottes, 1926) (Carolinaia) BM HRL
*rabeleri* Börner, (1949 nomen nudum) 1950 (Myzodes (Myzodium))

MYZOIDES van der Goot, 1913
Tijdschr. Ent. 56: 84
Type species Aphis cerasi Fabricius, 1775
= Myzus Passerini, 1860

MYZOPSIS Matsumura, 1918
Trans. Sapporo nat. Hist. Soc. 7: 19
Type species Myzopsis diervillae Matsumura, 1918

M. diervillae Matsumura, 1918 HRL
    *? plectranthi* Shinji, 1939 (*Myzus*)
    *weigelae* Takahashi, 1965 (*Micromyzus*)
  plectranthi Shinji, 1939 (Myzus) BM HRL (= ? diervillae Matsumura, 1918)
    *siegesbeckiae* Takahashi, 1965 (*Myzus*)

MYZOPSIS Pašek, 1955 nec Matsumura, 1918
  Pr. výzk. Úst. lesn. 8: 127
  Type species Myzopsis plantaginis Pašek, 1955
  = Pomaphis Börner, 1939

M. plantaginis Pašek, 1955 to Dysaphis (Pomaphis)

MYZOSIPHON Hille Ris Lambers, 1946
  In Alta, H. & Docters van Leeuwen, W.M., Gallenboek, p. 244
  Type species Rhopalosiphum staphyleae Koch, 1854
    *Arthromyzus* Börner, 1950
  Here used as subg. of Rhopalosiphoninus Baker, 1920

MYZOSIPHUM Tao, 1964
  Q. Jl Taiwan Mus. 17: 229
  Type species Myzosiphum ryukyuensis Tao, 1964

M. ryukyuense Tao, 1964

MYZOTOXOPTERA Theobald, 1927
  Entomologist 60: 31
  Type species Myzotoxoptera wimshurstae Theobald, 1927

M. heikinheimoi Börner, 1952 (type of Submegoura Hille Ris Lambers, 1953) to
    Rhopalosiphoninus (Submegoura)
  wimshurstae Theobald, 1927 types BM HRL

MYZOXILUS Rondani, 1848 errore pro Myzoxylus

MYZOXYLUS Blot, 1831
  Mém. Soc. r. Agric. Comm. Caen 3: 332
  Type species Myzoxylus mali Blot, 1831 = Aphis lanigera Hausmann, 1802
  = Eriosoma Leach, 1818

M. mali Blot, 1831 = Eriosoma lanigerum (Hausmann, 1802)

MYZUS Passerini, 1860
Gli Afidi, Parma, p. 27
Type species Aphis cerasi Fabricius, 1775
*Myzoides* van der Goot, 1913
*Prunomyzus* Hille Ris Lambers & Rogerson, 1946
*Spinaspidaphis* Heinze, 1961

M. ajugae Schouteden, 1903 to Myzus (Nectarosiphon)
alectorolophi Heinze, 1961 = cerasi (Fabricius, 1775)
alpinus Gillette & Palmer, 1928 to Kakimia Hottes & Frison, 1931
amygdalinus (Nevsky, 1928) (Myzaphis)
? *persicarum* Boisduval, 1867 (*Aphis*)
anselliae Hall, 1932 to Sitobion Mordvilko, 1914
aquilegiae Essig, 1917 to Kakimia Hottes & Frison, 1931
arthraxonis Takahashi, 1921 type of Kaochiaoja Tao, 1963 q.v.
asamensis Takahashi, 1965 BM
asclepiadis Passerini, 1863 = Aphis nerii Boyer de Fonscolombe, 1841
asiaticus (Szelegiewicz, 1969) (Prunomyzus) HRL
asteriae Shinji, 1941 to Myzus (Nectarosiphon)
berchemiae Takahashi, 1938
biennis Sanborn, 1904
boehmeriae Takahashi, 1922 BM HRL
borealis Ossiannilsson, 1959 BM HRL
braggii Gillette, 1908 = Capitophorus elaeagni (del Guercio, 1894)
brevipilosus Baerg, 1922 = Chaetosiphon (Pentatrichophus) minus (Forbes, 1884)
brevisiphon A.N. Basu, 1969 = formosanus Takahashi, 1923
callange Essig, 1954 = cerasi (Fabricius, 1775)
carthusianus Haviland, 1918 = Capitophorus elaeagni (del Guercio, 1894)
cerasi (Fabricius, 1775) (Aphis) BM HRL
*alectorolophi* Heinze, 1961
*aparines* Kaltenbach, 1843 (*Aphis*) nec Fabricius, 1775
*asperulae* Walker, 1848 (*Aphis*)
*callange* Essig, 1954
*cerasi* O.F. Müller, 1776 (*Aphis*)
*euphrasiae* Walker, 1849 (*Aphis*)
*galiifolium* Theobald, 1919
*langei* Essig, 1936 nec Myzus (Galiobium) langei (Börner, 1933)
*molluginis* Koch, 1854 (*Aphis*)
*pruniavium* Börner, 1926

*quasipyrinus* Theobald, 1929
*veronicae* Walker, 1848 (*Aphis*)
M. cerasi subsp. umefoliae (Shinji, 1924) (Akkaia) BM HRL
chaenomelis Dzhibladze, 1951 = Ovatus malisuctus (Matsumura, 1918)
citricidus Kirkaldy, 1907 to Toxoptera Koch, 1856
clavatus Paik, 1965 = Neotoxoptera oliveri (Essig, 1935)
clematifoliae Shinji, 1924 = varians Davidson, 1912
compositae (Takahashi, 1965) (Ovatus) HRL
cydonii Abashidze, 1951 nomen nudum
dispar Patch, 1914 = Cryptomyzus galeopsidis (Kaltenbach, 1843)
distinctus Nevsky, 1929
dycei Carver, 1961 BM HRL
*urticae* Tao, 1963
elaeagni del Guercio, 1894 to Capitophorus van der Goot, 1913
erigerionella Theobald, 1926 index, lapsus pro erigeroniella
erigeroniella Theobald, 1926 = Acyrthosiphon malvae (Mosley, 1841)
eriobotryae Tissot, 1935 type of Hyalomyzus Richards, 1958 q.v.
essigi Gillette & Palmer, 1929 = Kakimia aquilegiae (Essig, 1917)
fataunae Shinji, 1924 BM HRL
fatouae Shinji, miscorrection for fataunae Shinji, 1924
ferulaginis Macchiati of Lichtenstein, 1885: 133 nomen nudum?
festucae Theobald, 1917 to Metopolophium Mordvilko, 1914
filaginis Schouteden, 1906 to Pleotrichophorus Börner, 1930
filicis A.N. Basu, 1969 HRL
formosanus Takahashi, 1923 BM HRL
*brevisiphon* A.N. Basu, 1969 HRL
*kawatabiensis* Miyazaki, 1971 HRL
formosartemisiae Takahashi, 1921 to Capitophorus van der Goot, 1913
fragaefolii Cockerell, 1901 to Chaetosiphon (Pentatrichopus)
fragariae Theobald, 1912 = Chaetosiphon (Pentatrichops) fragaefolii (Cockerell, 1901)
fukis Monzen, 1927 = Aphis fukii Shinji, 1922
galiifolium Theobald, 1919 = cerasi (Fabricius, 1775)
gei Theobald, 1919 = Aulacorthum solani (Kaltenbach, 1843)
geranicola Shinji, 1935 to Cryptaphis Hille Ris Lambers, 1947
glaucii Theobald, 1923 = Aulacorthum solani (Kaltenbach, 1843)
godetiae Shinji, 1917
gracilis Buckton, 1876 = Metopolophium dirhodum (Walker, 1849)
gumi Shinji, 1922 nomen dubium
harmstoni Knowlton, 1943 to ? Fimbriaphis Richards, 1959
haywardi Knowlton, 1942 = Metopolophium dirhodum (Walker, 1849)

M. hemerocallis Takahashi, 1921 BM HRL
higansakurae Monzen, 1927 to Tuberocephalus Shinji, 1929
humboldti Essig, 1941 to Utamphorophora Knowlton, 1947
hydrocotylei Theobald, 1925 = Aulacorthum solani (Kaltenbach, 1843)
impatiensae Monzen, 1929 = Eumyzus impatiensae (Shinji, 1924)
inuzakurae Shinji, 1930 nomen dubium
isodonis (Takahashi, 1965) (Metaphorodon)
japonensis Miyazaki, 1968 HRL
japonicus Monzen, 1929 = Ovatus malisuctus (Matsumura, 1918)
junackianus Karsch, 1887 to Dephiniobium Mordvilko, 1914
kawatabiensis Miyazaki, 1971 = formosanus Takahashi, 1923
kokusaki Shinji, 1935
komaumii Shinji, 1943
*lonicerae* Shinji, 1944
kusaki Shinji, 1941 = ? Aulacorthum solani (Kaltenbach, 1843)
lactucicola Takahashi, 1934 to Myzus (Nectarosiphon)
lamii van der Goot, 1912 = Cryptomyzus galeopsidis (Kaltenbach, 1843)
langei Essig, 1936 = cerasi (Fabricius, 1775)
laricellus Theobald, 1927 = Sitobion fragariae (Walker, 1848)
leptotrichus David, Rajasingh & Narayanan, 1972 HRL
leucocrini Gillette & Palmer, 1929 to Ovatus van der Goot, 1913
lilii Mason, 1940 to Fimbriaphis Richards, 1959
lonicerae Shinji, 1944 = komaumii Shinji, 1943
lythri (Schrank, 1801) (Aphis) BM HRL
*droserae* Heinze, 1961 (*Spinaspidaphis*) HRL
*mahaleb* Koch, 1854 (*Aphis*)
*pruni* Ferrari, 1872 (*Phorodon*)
*pruni mahaleb* Boyer de Fonscolombe, 1841 (*Aphis*)
magnolifoliae Shinji, 1941 = Aulacorthum magnoliae (Essig & Kuwana, 1918)
mali Ferrari, 1872 = Dysaphis (Pomaphis) plantaginea (Passerini, 1860)
malicolens Hori, 1929 = Ovatus crataegarius (Walker, 1850)
malisuctus Matsumura, 1918 to Ovatus van der Goot, 1913
malvae Oestlund, 1886 = Myzus (Nectarosiphon) persicae (Sulzer, 1776)
matricariae Macchiati, 1882 nomen dubium
mercurialis Theobald, 1919 = Aulacorthum solani (Kaltenbach, 1843)
mespiliella Theobald, 1920 = ? Ovatus crataegarius (Walker, 1850)
michaelseni Schouteden, 1904 to Kakimia Hottes & Frison, 1931
molluginellus Theobald, 1924 = Sitobion fragariae (Walker, 1848)
momonis Matsumura, 1917 to Tuberocephalus Shinji, 1929

M. montanus Takahashi, 1925 (type of Taiwanomyzus Tao, 1963) to Utamphorophora Knowlton, 1947
moriokae Shinji, 1941 nomen dubium
morrisoni Laing, 1928 = Jacksonia papillata Theobald, 1923
mumecola (Matsumura, 1917) (Macrosiphum) BM HRL
*umecola* Shinji, 1924 (mis-spelled ume in Shinji, 1941, p. 946)
mushaensis Takahashi, 1931 BM HRL
negifoliae Shinji, 1932
neogei Theobald, 1926 = Aulacorthum solani (Kaltenbach, 1843)
neomexicanus W.P. Cockerell & T.D.A. Cockerell, 1901 to Aphis Linnaeus, 1758
neorosarum Theobald, 1915 = Chaetosiphon (Pentatrichopus) tetrarhodum (Walker, 1849)
niitakaensis Takahashi, 1937 to Myzakkaia A.N. Basu, 1969
nipponicus Moritsu, 1949 to Micromyzodium David, 1958
obtusirostris David, Narayanan & Rajasingh, 1971 BM HRL
oenotherae Williams, 1911
ornatus Laing, 1932 types BM HRL
padellus Hille Ris Lambers & Rogerson, 1946 (Myzus (Prunomyzus)) types BM HRL
parthenocissi Takahashi, 1965 to Aulacorthum (Neomyzus)
pergandei Sanderson, 1901 = Myzus (Nectarosiphon) persicae (Sulzer, 1776)
persicae Passerini, 1860 to Brachycaudus (Acaudus)
persicae (Sulzer, 1776) (Aphis) to Myzus (Nectarosiphon)
persicae var. cerastii Theobald, 1926 = Myzus (Nectarosiphon) certus (Walker, 1849)
persicae var. portulacella Theobald, 1926 = Myzus (Nectarosiphon) persicae (Sulzer, 1776)
persicae var. sanguisorbiella Theobald, 1926 = Myzus (Nectarosiphon) persicae (Sulzer, 1776)
persicae subsp. dyslycialis F.P. Müller, 1955 = Myzus (Nectarosiphon) persicae (Sulzer, 1776)
phenax Cockerell, 1903 = Aphis helianthi Monell, 1879
philadelphi Takahashi, 1965 BM HRL
physocarpi Pepper, 1950 = Utamphorophora humboldti (Essig, 1941)
pileae Takahashi, 1965 BM HRL
pilosus van der Goot, 1912 = Pleotrichophorus glandulosus (Kaltenbach, 1846)
plantagicola Takahashi, 1931 = ? Dysaphis (Pomaphis) plantaginea (Passerini, 1860)

M. plantagifoliae Shinji, 1924 = ? Dysaphis (Pomaphis) plantaginea (Passerini, 1860)
plantagineus Passerini, 1860 to Dysaphis (Pomaphis)
plectranthi Shinji, 1939 to Myzopsis Matsumura, 1918
polemonii Gillette & Palmer, 1929 to Kakimia Hottes & Frison, 1931
polyanthi Theobald, 1926 = Aulacorthum solani (Kaltenbach, 1843)
polygonifoliae Shinji, 1944 (type of Aphorodon Takahashi, 1961) to Tricho-siphonaphis (Xenomyzus)
polygoniformosanus Takahashi, 1921 type of Trichosiphonaphis Takahashi, 1922 q.v.
polygoniyonai Shinji, 1938 nomen dubium
polypodiacicola Takahashi, 1921 corrected to polypodicola
polypodicola Takahashi, 1921 to Macromyzus Takahashi, 1960
porosus Sanderson, 1900 to Rhodobium Hille Ris Lambers, 1947
portulacae Macchiati, 1883 nomen dubium
potentillae Oestlund, 1886 = Chaetosiphon (Pentatrichopus) thomasi (Hille Ris Lambers, 1953)
potentillae Williams, 1911 to Kakimia Hottes & Frison, 1931
pruniavium Börner, 1926 = cerasi (Fabricius, 1775)
pseudolamii Theobald, 1926 = Aulacorthum solani (Kaltenbach, 1843)
pseudosolani Theobald, 1922 = Aulacorthum solani (Kaltenbach, 1843)
pterisoides Theobald, 1918 type of Micromyzella Eastop, 1955 q.v.
pyrarius Passerini, 1861 (type of Nevskia Mordvilko, 1932) to Melanaphis van der Goot, 1917
pyrinus Ferrari, 1872 = Melanaphis pyraria (Passerini, 1861)
quasipyrinus Theobald, 1929 = cerasi (Fabricius, 1775)
ranunculi del Guercio, 1900 = Tubaphis ranunculina (Walker, 1852)
rarus Monzen, 1927 = Tuberocephalus sakurae (Matsumura, 1917)
rhois Takahashi, 1924 to Juncomyzus Hille Ris Lambers, 1965
ribifolii Davidson, 1917 to Kakimia Hottes & Frison, 1931
ribis Shinji, 1922 = ? Nasonovia ribisnigri (Mosley, 1841)
ribis subsp. bucktonii W.P. Cockerell & T.D.A. Cockerell, 1903
ribis var. bucktoni del Guercio, 1894 = Nasonovia ribisnigri (Mosley, 1841)
ribis subsp. trifasciata W.P. Cockerell & T.D.A. Cockerell, 1903
ribis var. trifasciata del Guercio, 1894
roseus Macchiati, 1881 = Aphis fabae Scopoli, 1763
rubra Macchiati, 1884 = Aphis fabae Scopoli, 1763
rubrum del Guercio, 1900 = Aphis fabae Scopoli, 1763
sakurae Matsumura, 1917 to Tuberocephalus Shinji, 1929
sasakii Matsumura, 1917 to Tuberocephalus Shinji, 1929
scammelli Mason, 1942 to Ericaphis Börner, 1939

M. scrophulariae (Thomas, 1879) (Phorodon)
sensoriatus Mason, 1940 to Hyalomyzus Richards, 1958
siegesbeckiae Takahashi, 1965 = Myzopsis plectranthi Shinji, 1939
siegesbeckicola Strand, 1929 HRL
suguri Shinji, 1927 = Kakimia ribifolii (Davidson, 1917)
takahashii Strand, 1929 = Ovatus malisuctus (Matsumura, 1918)
targionii del Guercio, 1894 nomen dubium
thalictri Williams, 1911 = Kakimia purpurascens (Oestlund, 1887)
tropicalis Takahashi, 1923 = varians Davidson, 1912
tsengi Tao, 1963 = Tuberocephalus sasakii (Matsumura, 1917)
ume Shinji, 1941 mis-spelling of umecola Shinji, 1924
umecola Shinji, 1924 = mumecola (Matsumura, 1917)
urticae Tao, 1963 = dycei Carver, 1961
vaccinii Theobald, 1924 to Wahlgreniella Hille Ris Lambers, 1949
valuliae Robinson, 1974 to Ovatus van der Goot, 1913
varians Davidson, 1912 BM HRL
*clematifoliae* Shinji, 1924
*sumomocola* Monzen, 1929 (*Macrosiphum*)
*tropicalis* Takahashi, 1923
veronicae del Guercio, 1900 = Aulacorthum solani (Kaltenbach, 1843)
veronicellus Theobald, 1926 = Aulacorthum solani (Kaltenbach, 1843)
vincae Gillette, 1908 = Aulacorthum (Neomyzus) circumflexum (Buckton, 1876)
whitei Theobald, 1912 = Cryptomyzus galeopsidis (Kaltenbach, 1843)
woodwardiae Takahashi, 1921 type of Macromyzus Takahashi, 1960 q.v.
woodwardiae subsp. hinoi Moritsu, 1951 = Macromyzus woodwardiae (Takahashi, 1921)
xanthomelii Shinji, 1941 nomen dubium
yamatonis Miyazaki, 1971 HRL
yangi Takahashi, 1938

Subgen. Cerasomyzus Narzikulov, 1958

M. (Cerasomyzus) bozhkoae Narzikulov, 1958 (type of Cerasomyzus Narzikulov, 1958) to Aphidura Hille Ris Lambers, 1956

Subgen. Galiobium Börner, 1933

M. (Galiobium) galinarium (Narzikulov ex Juchnevitch, 1968) (Galiobium)
(Galiobium) langei (Börner, 1933) (Trilobaphis (Galiobium)) BM HRL

Subgen. Kakimia Hottes & Frison, 1931

M. (Kakimia) thomasi Hottes & Frison, 1931 type of Kakimia Hottes & Frison, 1931 q.v.

Subgen. Nectarosiphon Schouteden, 1901

M. (Nectarosiphon) ajugae Schouteden, 1903 (Myzus) BM HRL
(Nectarosiphon) ascalonicus Doncaster, 1946 types BM HRL
(Nectarosiphon) asteriae Shinji, 1941 (Myzus) BM HRL
(Nectarosiphon) beybienkoi (Narzikulov, 1957) (Myzaphis) HRL
(Nectarosiphon) caryophyllacearum Hille Ris Lambers, 1946 = certus (Walker, 1849)
(Nectarosiphon) certus (Walker, 1849) (Aphis) BM HRL
*caryophyllacearum* Hille Ris Lambers, 1946 HRL
*certus* subsp. *pseudopersicae* Börner, 1952 (*Myzodes*)
*persicae* var. *cerastii* Theobald, 1926 (*Myzus*)
(Nectarosiphon) dianthicola Hille Ris Lambers, 1966 BM HRL
(Nectarosiphon) lactucicola Takahashi, 1934 (Myzus) BM HRL
(Nectarosiphon) ligustri (Mosley, 1841) (Aphis) BM HRL
*ligustri* Kaltenbach, 1843 (*Aphis*)
? *evadens* Rusanova, 1951 (*Liosomaphis*)
(Nectarosiphon) linariae Holman, 1965 paratypes BM HRL
(Nectarosiphon) myosotidis (Börner, 1950) (Myzodes) BM HRL
(Nectarosiphon) persicae (Sulzer, 1776) (Aphis) BM HRL
*achyrantes* Monell, 1879 (*Siphonophora*)
*antirrhinii* Macchiati, 1883 (*Siphonophora*)
*betae* Theobald, 1913 (*Rhopalosiphum*)
*callae* Koch, 1854 in litteris synonym of Rhopalosiphum dianthi Schrank
? *calendulella* Monell, 1879 (*Siphonophora*)
*consors* Walker, 1848 (*Aphis*) type BM
*convolvuli* Kaltenbach, 1843 (*Aphis*)
*cymbalariae* Schouteden, 1900 (*Aphis*)
*cynoglossi* Williams, 1911 (*Phorodon*)
*deposita* Walker, 1848 (*Aphis*) lectotype BM
*derelicta* Walker, 1849 (*Aphis*)
*dianthi* Schrank, 1801 (*Aphis*)
*dubia* Curtis, 1842 (*Aphis*)
*dyslycialis* F.P. Müller, 1955 BM HRL
*egressa* Walker, 1849 (Aphis)
*galeactitis* Macchiati, 1883 (*Rhopalosiphum*)
*lactucellum* Theobald, 1914 (*Rhopalosiphum*)
*lophospermum* Theobald, 1914 (*Macrosiphum*)

*lycopersicella* Theobald, 1914 (*Macrosiphum*)
*malvae* Oestlund, 1886 (*Myzus*)
*nasturtii* Koch, 1855 (*Siphonophora*)
*particeps* Walker, 1845 (*Aphis*)
*pergandei* Sanderson, 1901 (*Myzus*)
*persicae* Morren, 1836 (*Aphis*)
*persicae* subsp. *dyslycialis* F.P. Müller
*persicae* var. *portulacella* Theobald, 1926 (*Myzus*)
*persicae* var. *sanguisorbiella* Theobald, 1926 (*Myzus*)
*persicophila* Rndn. in Scheda, Passerini, 1860 (*Aphis*)
*? persola* Walker, 1848 (*Aphis*)
*rapae* Curtis, 1842 (*Aphis*) partim
*redundans* Walker, 1849 (*Aphis*) type BM
*tabaci* Mordvilko, 1914 (*Myzodes*)
*trilineatum* del Guercio, 1920 (*Rhopalosiphum*)
*tulipae* Thomas, 1879 (*Rhopalosiphum*)
*vastator* Smee, 1846 (*Aphis*)
*vulgaris* Kyber, 1815 (*Aphis*)

M. (Nectarosiphon) polaris Hille Ris Lambers, 1952 BM HRL
(Nectarosiphon) stellariae Strand, 1929 (Amphorophora) BM
(Nectarosiphon) titschaki (Börner, 1942) (Ovatus) HRL

Subgen. Prunomyzus Hille Ris Lambers & Rogerson, 1946

M. (Prunomyzus) padellus Hille Ris Lambers & Rogerson, 1946 (type of Prunomyzus Hille Ris Lambers & Rogerson, 1946) to Myzus Passerini, 1860

Subgen. Sciamyzus Stroyan, 1954

M. (Sciamyzus) cymbalariae Stroyan, 1954 (type of Sciamyzus Stroyan, 1954) BM HRL
*cymbalariellus* Stroyan, 1967
(Sciamyzus) cymbalariellus Stroyan, 1967 = cymbalariae Stroyan, 1954

Subgen. Tubaphis Hille Ris Lambers, 1947

M. (Tubaphis) clematophilus Takahashi, 1965 to Tubaphis Hille Ris Lambers, 1947

NARZICULOVIA Umarov, 1964
Dokl. Akad. Nauk. Tadzh. S.S.R. 7: 25
Type species Narziculovia dolychosiphon Umarov, 1964
= Obtusicauda Soliman, 1927

N. dolychosiphon Umarov, 1964 to Obtusicauda Soliman, 1927

NARZIKULOVIA mis-spelling of Narziculovia Umarov, 1964

NASONOVIA Mordvilko, 1914
Faune Russie, Ins. Hém. 1: 72
Type species Aphis ribicola Kaltembach, 1843 = A. ribis-nigri Mosley, 1841
*Submacrosiphum* Hille Ris Lambers, 1931

N. altaensis Stenseth, 1969 to Kakimia Hottes & Frison, 1931
compositellae (Theobald, 1924) (Aphis) types BM HRL
nigra (Hille Ris Lambers, 1931) (Submacrosiphum) BM HRL
*hieracii* Kaltenbach, 1843 (*Aphis*) nec Schrank, 1801
*hieracii* subsp. *nigrum* Hille Ris Lambers, 1931 (*Submacrosiphum*) HRL
nivalis (Börner, 1950) (Submacrosiphum) BM HRL
pilosellae (Börner, 1933) (Impatientinum) BM HRL
ribisnigri (Mosley, 1841) (Aphis) BM HRL
*agrostemnium* Theobald, 1913 (*Macrosiphum*)
*alliariae* Koch, 1855 p. 177 nec p. 160 (*Siphonophora*)
*hieracii* subsp. *teriolanum* Hille Ris Lambers, 1931 (*Submacrosiphum*) HRL
*kaltenbachi* Schouteden, 1906 (*Macrosiphum*)
*polygoni* Buckton, 1876 (*Siphonophora*)
*pseudohieracii* Theobald, 1912 (*Macrosiphum*)
*ribicola* Kaltenbach, 1843 (*Aphis*)
*? ribis* Shinji, 1922 (*Myzus*)
*ribis* var. *bucktoni* del Guercio, 1894 (*Myzus*)
vannesii Stenseth, 1968 to Nasonovia (Neokakimia)

Subgen. Hyperomyzus Börner, 1933

N. (Hyperomyzus) inflata Richards, 1962 to Hyperomyzus (Neonasonovia)
(Hyperomyzus) lactucae (Linnaeus 1758) to Hyperomyzus Börner, 1933
(Hyperomyzus) sandilandica Robinson, 1974 to Hyperomyzus (Neonasonovia)

Subgen. Kakimia Hottes & Frison, 1931

N. (Kakimia) jammuensis Verma, 1970 to Kakimia Hottes & Frison, 1931
(Kakimia) rostrata David & Hameed, 1974 HRL

Subgen. Neokakimia Doncaster & Stroyan, 1952

N. (Neokakimia) brachycyclica Holman, 1972 HRL
(Neokakimia) brevipes (Börner, 1950) (Submacrosiphon) HRL ?
(Neokakimia) dasyphylli (Stroyan, 1954) (Kakimia (Neokakimia)) BM HRL

N. (Neokakimia) saxifragae (Doncaster & Stroyan, 1952) (Kakimia (Neokakimia))
type BM HRL
(Neokakimia) vannesii Stenseth, 1968 (Nasonovia) HRL

NEANOECIA Börner, 1950
Neue europäische Blattlausarten, privately published, p. 16
Type species Neanoecia krizusi Börner, 1950
= Anoecia Koch, 1857

N. krizusi Börner, 1950 to Anoecia Koch, 1857

NEANURAPHIS Nevsky, 1928
Ent. Mitt. 17: 192
Type species Neanuraphis tarani Nevsky, 1928
= Macchiatiella del Guercio, 1909 nec 1917

N. catharticae Nevsky, 1929 = Macchiatiella rhamni subsp. tarani (Nevsky, 1928)
newskii Börner, 1940 to Dysaphis Börner, 1931
quaestionis Börner, 1942 type of Allocotaphis Börner, 1950 q.v.
tarani Nevsky, 1928 = Macchiatiella rhamni subsp. tarani (Nevsky, 1928)

NEAPHIS Nevsky, 1929
Zool. Anz. 82: 206
Type species Neaphis viridis Nevsky, 1929
= Coloradoa Wilson, 1910

N. artemisiae Narzikulov, 1958 = Coloradoa tadzhikistanica Narzikulov, 1974
brevipilosa Ivanovskaja, 1958 to Coloradoa Wilson, 1910
ponticae Börner, 1940 to Coloradoa Wilson, 1910
procerae Bozhko, 1959 to Coloradoa Wilson, 1910
viridis Nevsky, 1929 to Coloradoa Wilson, 1910

NEAQUENAPHIS lapsus pro Neuquenaphis

NEARCTAPHIS Shaposhnikov, 1950
Ént. Obozr. 31: 223
Type species Aphis bakeri Cowen, 1895
*Amelanchieria* Shaposhnikov, 1950
*Fitchiella* Shaposhnikov, 1950
? *Mamontova* Shaposhnikov, 1964

N. argentinaeradicis (Gillette & Palmer, 1932) (Aphis) HRL
bakeri (Cowen, 1895) (Aphis) BM HRL
*cephalicola* Cowen, 1895 (Aphis)
californica Hille Ris Lambers, 1970 paratypes BM HRL
californica subsp. nigrescens Hille Ris Lambers, 1970 HRL
clydesmithi Hille Ris Lambers, 1970 HRL
crataegifoliae (Fitch, 1851) (Aphis) BM HRL
*brevis* Sanderson, 1902 (*Aphis*)
*malifoliae* Fitch, 1855 (*Aphis*)
crataegifoliae subsp. occidentalis Hille Ris Lambers, 1970 BM HRL
hottesi Hille Ris Lambers, 1970 HRL
kachena (Hottes, 1934) (Aphis) BM HRL
multisetis (Richards, 1969) (Roepkea)
punctata (Richards, 1969) (Roepkea)
sclerosa (Richards, 1969) (Roepkea)
sensoriata (Gillette & Bragg, 1918) (Aphis) BM HRL
*viridicaudus* Miller, 1933 (*Rhopalosiphum*)
? vera (Shaposhnikov, 1964) (Mamontova)
yohoensis Bradley, 1965 BM HRL
zababsis (Richards, 1969) (Roepkea)

NECTAROPHORA Oestlund, 1887
Bull. Minn. geol. Surv. 4: 78
Type species Aphis rosae Linnaeus, 1758
= Macrosiphum Passerini, 1860

N. agrimoniella Cockerell, 1903 to Macrosiphum Passernini, 1860
artemisiae Cowen ex Gillette & Baker, 1895 to Obtusicauda Soliman, 1927
asclepiadis Cowen ex Gillette & Baker, 1895 = Macrosiphum euphorbiae (Thomas, 1878)
baccharidis Clarke, 1903 to Uroleucon (Lambersius)
californica Clarke, 1903 to Macrosiphum Passerini, 1860
caudata Pergande, 1900 to Metopolophium Mordvilko, 1914
corallorhizae Cockerell, 1903 to Macrosiphum Passerini, 1860
coweni Hunter, 1901 = Obtusicauda artemisiae (Cowen, 1895)
cynosbati Oestlund, 1887 to Kakimia Hottes & Frison, 1931
destructor Johnson, 1900 = Acyrthosiphon pisum (Harris, 1776)
epilobii Pergande, 1900 to Uroleucon Mordvilko, 1914

N. fulvae Oestlund, 1887 type of Catamergus Oestlund, 1922 q.v.
geranii Oestlund, 1887 to Macrosiphum Passerini, 1860
heleniella Cockerell, 1903 = Macrosiphum euphorbiae (Thomas, 1878)
insularis Pergande, 1900 to Sitobion Mordvilko, 1914
jasmini Clarke, 1903 to Macrosiphum Passerini, 1860
lycopersici Clarke, 1903 = Macrosiphum euphorbiae (Thomas, 1878)
martini Cockerell, 1903 to Macrosiphum Passerini, 1860
pallida Oestlund, 1887 to Macrosiphum Passerini, 1860
potentillae Oestlund, 1887 to Macrosiphum Passerini, 1860
purpurascens Oestlund, 1887 to Kakimia Hottes & Frison, 1931
rhamni Clarke, 1903 to Sitobion Mordvilko, 1914
rudbeckiarum Cockerell, 1903 to Macrosiphum Passerini, 1860
tabaci Pergande, 1898 = Macrosiphum euphorbiae (Thomas, 1878)
valerianae Clarke, 1903 to Macrosiphum Passerini, 1860

NECTAROSIPHON Schouteden, 1901
Annls Soc. ent. Belg. 45: 112, nomen novum pro Macrosiphum del Guercio, 1900 nec Passerini, 1860, Oestlund, 1886
Type species Aphis convolvuli Kaltenbach, 1843 = Aphis persicae Sulzer, 1776
Here used as subg. of Myzus Passerini, 1860
*Amyzus* Hille Ris Lambers, 1946
*Idiovatus* Börner, 1944
*Macrosiphum* del Guercio, 1900 nec Passerini, 1860
*Myzodes* Mordvilko, 1914
*Rhopalosiphum* Passerini, 1860 nec Koch, 1854

N. franzi Börner ex Franz, 1943 nomen nudum
gei Börner, 1939 to Amphorophora Buckton, 1876
idaei Börner, 1939 = Amphorophora rubi subsp. idaei (Börner, 1939)
mitsubautsugii Shinji, 1923 (as Nectarosiphum) = Indomegoura indica (van der Goot, 1916)
moriokae Shinji, 1923 = Megoura crassicauda Mordvilko, 1919
morrisoni Swain, 1918 to Illinoia Wilson, 1910
obako Shinji, 1922 nomen dubium
rhinanthi Schouteden, 1903 (type of Hyperomyzella Hille Ris Lambers, 1949) to Hyperomyzus (Hyperomyzella)

NECTAROSIPHUM errore pro Nectarosiphon Schouteden, 1901

NEOACAUDUS Theobald, 1927
The Plant Lice or Aphididae of Great Britain 2: 326
Type species Acaudus bipapillatus Theobald, 1923 = Aphis helichrysi Kaltenbach, 1843
= Brachycaudus van der Goot, 1913

N. inflatus Theobald, 1928 = Macrosiphoniella artemisiae (Boyer de Fonscolombe, 1841)

NEOACYRTHOSIPHON Tao, 1963
Pl. Prot. Bull., Taiwan 5: 189
Type species Acyrthosiphon taiheisanum (Takahashi, 1935) = Macrosiphum taiheisanum Takahashi, 1935

N. rhododendri M.R. Ghosh, A.K. Ghosh & Raychaudhuri, 1971
taiheisanum (Takahashi, 1935) (Macrosiphum)
taiheisanum subsp. ovalifolii M.R. Ghosh, A.K.Ghosh & Raychaudhuri, 1971

Subgen. Pseudoacyrthosiphon A.K. Ghosh & RayChaudhuri, 1969

N. (Pseudoacyrthosiphon) holsti (Takahashi, 1935) (Macrosiphum) HRL
(Pseudoacyrthosiphon) nepalensis A.K. Ghosh, R.C. Basu & RayChaudhuri, 1973
(Pseudoacyrthosiphon) takahashii A.K. Ghosh, 1969

NEOAMPHOROPHORA Mason, 1924
Proc. ent. Soc. Wash. 26: 49
Type species Neoamphorophora kalmiae Mason, 1924
*Testataphis* Börner, 1952

N. kalmiae Mason, 1924 BM HRL
ledi (Wahlgren, 1938) (Amphorophora) BM HRL
tephrosiae Smith, 1948 (type of Megouroparsus Smith & Heie, 1963) to Microparsus Patch, 1919

NEOBACILLAPHIS Huculak, 1968
Annls zool. Warsz. 25: 425
Type species Neobacillaphis szelegiewiczi Huculak, 1968

N. striata (Bozhko, 1961) (Thripsaphis)
szelegiewiczi Huculak, 1968 HRL

NEOBETULAPHIS A.N. Basu, 1964
J. Linn. Soc. (Zool.) 45: 226
Type species Neobetulaphis pusilla A.N. Basu, 1964

N. alba Higuchi, 1972
pusilla A.N. Basu, 1964 HRL

NEOBRACHYCAUDUS Narzikulov, 1963
Ént. Obozr. 94: 614, described as subg. of Brachycaudus van der Goot, 1913
Type species Aphis cardui Linnaeus, 1758
= Acaudus van der Goot, 1913

NEOCALAPHIS Shinji, 1927
Bull. Morioka imp. Coll. For. Agric. 11: 28
Type species Calaphis magnoliae Essig & Kuwana, 1918

N. magnoliae (Essig & Kuwana, 1918) (Calaphis) HRL
*magnolifoliae* Shinji, 1923 (*Euceraphis*)
magnolicolens (Takahashi, 1921) (Calaphis) BM HRL

NEOCALLIPTERUS van der Goot, 1915
Beiträge zur Kenntnis der Holländischen Blattläuse, Haarlem – Berlin, p. 320
Type species Aphis betulicola Kaltenbach, 1843
= Calaphis Walsh, 1863

NEOCALLIS Matsumura, 1919
Trans. Sapporo nat. Hist. Soc. 7: 104
Type species Neocallis carpinicola Matsumura, 1919 = Myzocallis sawashibae Matsumura, 1917
= Mesocallis Matsumura, 1919

N. carpinicola Matsumura, 1919 = Pterocallis (Mesocallis) sawashibae (Matsumura, 1917)

NEOCAVARIELLA Shinji, 1932
Oyo-Dobuts.-Zasshi 4: 122
Type species Cavariella araliae Takahashi, 1921
= Cavariella del Guercio, 1911

NEOCERURAPHIS Shaposhnikov, 1956
Trudy zool. Inst., Leningr. 23: 285
Type species Ceruraphis viburnicola (Gillette, 1909) nec Börner, 1926 = Aphis viburnicola Gillette, 1909
= Ceruraphis Börner, 1926

NEOCHMOSIS Laing ex Theobald, 1929
In Theobald, F.V., The Plant Lice or Aphididae of Great Britain 3: 129 footnote
Type species Lachniella gracilis Wilson, 1919
= Cinara Curtis, 1835

NEOCHROMAPHIS Takahashi, 1921
Japanese Aphidae, privately published, p. 28
Type species Neochromaphis carpini Takahashi, 1921 = Chromaphis carpinicola Takahashi, 1921

N. carpini Takahashi, 1921 = carpinicola (Takahashi, 1921)
carpinicola (Takahashi, 1921) (Chromaphis) HRL
*carpini* Takahashi, 1921
coryli Takahashi, 1961
ostryae (Börner, 1949) (Pterocallis) BM HRL

NEOCINARIA Pašek ex Pintera, 1966
In Pintera, Acta ent. bohemoslev. 63: 281, described as subg. of Cinara Curtis, 1835
Type species Cinaria escherichi Börner, 1950
= Cinara Curtis, 1835

NEOCORYLOBIUM MacGillivray, 1968
Ann. ent. Soc. Am. 61: 341
Type species Macrosiphum carpinicolens Patch, 1919
Here used as subg. of Macrosiphum Passerini, 1860

NEODECOROSIPHON Heinze, 1960
Zool. Anz. 165: 193
Type species Neodecorosiphon muscicolens Heinze, 1960 = Aphis poae Hardy, 1850
= Cryptaphis Hille Ris Lambers, 1947

N. muscicolens Heinze, 1960 = Cryptaphis poae (Hardy, 1850)

NEODIMOSIS Toth, 1935
Z. Morph. Ökol. Tiere 30: 495 (errore pro Neochmosis Laing ex Theobald, 1929)
Type species Lachnus pinihabitans Mordvilko, 1895
= Cinara Curtis, 1835

NEODYSAPHIS Hille Ris Lambers, 1965 nec Narzikulov, 1961
Tijdschr. Ent. 108: 191
Type species Neodysaphis deutziae Hille Ris Lambers, 1965
= Nippodysaphis Hille Ris Lambers, 1965

N. deutziae Hille Ris Lambers, 1965 type of Nippodysaphis Hille Ris Lambers, 1965 q.v.

NEODYSAPHIS Narzikulov, 1961
Trudy Inst. Zool. Parazit., Dushanbe 20: 82, described as subg. of Dysaphis Börner, 1931
Type species Dysaphis pseudomolli Narzikulov, 1961
Here used as subgen. of Dysaphis Börner, 1931

NEOGREENIDEA RayChaudhuri, M.R. Ghosh, M. Banerjee & A.K. Ghosh, 1973
Kontyû 41: 62, described as subg. of Greenidea Schouteden, 1905
Type species Greenidea (Neogreenidea) ayyari RayChaudhuri, M.R. Ghosh, M. Banerjee & A.K. Ghosh, 1973
Here used as subg. of Greenidea Schouteden, 1905

NEOGREENIDEOIDA RayChaudhuri, 1956
Zool. Verh., Leiden 31: 73, described as subg. of Greenideoida van der Goot, 1917
Type species Greenideoida (Neogreenideoida) philippensis RayChaudhuri, 1956
Here used as subg. of Greenideoida van der Goot, 1917

NEOHAYHURSTIA Aizenberg, (1954) 1956
Diss. biol. Sci. Lenin Univ., Moscow, p. 9 (1954)
Trudȳ vses. ent. Obshch. 45: 154 (1956)
Type species Hayhurstia tataricae Aizenberg, 1935
= Hyadaphis Kirkaldy, 1904

NEOKAKIMIA Doncaster & Stroyan, 1952
Ann. Mag. nat. Hist. 5: 984, described as subg. of Kakimia Hottes & Frison, 1931
Type species Kakimia (Neokakimia) saxifragae Doncaster & Stroyan, 1952
Here used as subg. of Nasonovia Mordvilko, 1914

NEOLACHNAPHIS Shinji, 1924
Zool. Mag., Tokyo 36: 351
Type species Neolachnaphis itadori Shinji, 1924
= Macchiatiella del Guercio, 1909 nec 1917

N. itadori Shinji, 1924 to Macchiatiella del Guercio, 1909 nec 1917

NEOLACHNUS Mordvilko, 1929
Trudȳ prikl. Ent. 14: 55
Type species Neolachnus rosae (Cholodkovsky, 1899) = Aphis submacula Walker, 1848
= Maculolachnus Gaumont, 1920

NEOLACINAPHIS lapsus pro Neolachnaphis

NEOLIZERIUS E.E. Blanchard, 1939
Physis, B. Aires 17: 873
Type species Neolizerius tuberculatus E.E. Blanchard, 1939
= Lizerius E.E. Blanchard, 1923

N. acunai Holman, 1974 to Lizerius E.E. Blanchard, 1923
tuberculatus E.E. Blanchard, 1939 to Lizerius E.E. Blanchard, 1923

NEOMACROSIPHUM van der Goot, 1915
Beiträge zur Kenntnis der Holländischen Blattläuse, Haarlem-Berlin, p. vii
Type species Macrosiphum pelargonii (Kaltenbach) of van der Goot, 1915 nec Kaltenbach, 1843 = Aphis solani Kaltenbach, 1843
= Aulacorthum Mordvilko, 1914

NEOMASONAPHIS A.K. Ghosh & RayChaudhuri, 1972
Orient. Insects 6: 376, described as subg. of Masonaphis Hille Ris Lambers, 1939
Type species Masonaphis anaphilidis (mis-spelling of anaphalidis) A.N. Basu, 1964

N. anaphalidis (A.N. Basu, 1964) (Masonaphis) HRL
inulae (A.K. Ghosh & RayChaudhuri, 1972) (Masonaphis (Neomasonaphis)) HRL

NEOMEGOURA Shinji, 1930
Lansania 2: 72
Without species

NEOMYRUS errore pro Neomyzus

NEOMYZAPHIS Theobald, 1926
The Plant Lice or Aphididae of Great Britain 1: 262
Type species Aphis abietina Walker, 1849
= Elatobium Mordvilko, 1914

N. piceana Inouye, 1939 to Elatobium Mordvilko, 1914

NEOMYZOCALLIS Richards, 1965
Mem. ent. Soc. Can. 44: 29, described as subg. of Myzocallis Passerini, 1860
Type species Callipterus punctata Monell, 1879
= Myzocallis Passerini, 1860

NEOMYZODES Aizenberg, 1954
New data on the system and fauna of aphids, p. 10
Diss. biol. Sci. Lenin Univ. Moscow, p. 9
Type species Myzus ornatus minutus Aizenberg, 1954 nomen nudum
Without validly described species

N. ornatus minutus Aizenberg, 1954 nomen nudum

NEOMYZUS van der Goot, 1915 & 1917
Beiträge zur Kenntnis der Holländischen Blattläuse, Haarlem-Berlin, p. VII, as Neomyrus (1915)
Contrib. Faune Indes Néerl. 3: 50 (1917)
Type species Siphonophora circumflexa Buckton, 1876
Here used as subg. of Aulacorthum Mordvilko, 1914
*Sumoia* Tao, 1963

N. dendrobii A.N. Basu, 1969 to Aulacorthum (Neomyzus)
dicentrae A.N. Basu, 1968 to Aulacorthum (Neomyzus)
masoni Knowlton, 1928 to Illinoia Wilson, 1910
primulum A.K. Ghosh, H. Banerjee & RayChaudhuri, 1971 to Aulacorthum (Neomyzus)

NEONASONOVIA Hille Ris Lambers, 1949
Temminckia 8: 277, described as subg. of Hyperomyzus Börner, 1933
Type species Rhopalosiphum picridis Börner & Blunck, 1916
Here used as subg. of Hyperomyzus Börner, 1933

NEONIPPOLACHNUS Shinji, 1924
Zool. Mag., Tokyo 36: 343
Type species Neonippolachnus betulae Shinji, 1924

N. betulae Shinji, 1924

NEONIPPONAPHIS Takahashi, 1962
Bull. Univ. Osaka Pref. (B) 13: 9
Type species Neonipponaphis shiiae Takahashi, 1962

N. shiiae Takahashi, 1962 HRL

NEOPARACLETUS Strom, 1942
Ann. ent. Soc. Am. 35: 332
Type species Neoparacletus caricis Strom, 1942 = Pemphigus corrugatans Sirrine, 1894
Here used as subg. of Prociphilus Koch, 1857

N. caricis Strom, 1942 = Prociphilus (Neoparacletus) corrugatans (Sirrine, 1894)

NEOPARATRICHOSIPHUM A.K. Ghosh & RayChaudhuri, 1962
Sci. Cult. 29: 107, described as subg. of Paratrichosiphum Takahashi, 1931
Type species Paratrichosiphum (Neoparatrichosiphum) khasyanum A.K. Ghosh & RayChaudhuri, 1962
= Eutrichosiphum Essig & Kuwana, 1918

NEOPHORODON Takahashi, 1922
Proc. ent. Soc. Wash. 24: 204
Type species Neophorodon rubi Takahashi, 1922
= Matsumuraja Schumacher, 1921

N. rubi Takahashi, 1922 = Matsumuraja rubifoliae Takahashi, 1931

NEOPHYLLAPHIS Takahashi, 1920
Can. ent. 52: 20
Type species Neophyllaphis podocarpi Takahashi, 1920

N. araucariae Takahashi, 1937 BM HRL
brimblecombei Carver, 1971 paratypes BM HRL
fransseni Hille Ris Lambers, 1967 HRL
gingerensis Carver, 1959 paratypes BM HRL
grobleri Eastop, 1955 types BM HRL
podocarpi Takahashi, 1920 BM HRL
*podocarpi* Shinji, 1922 (*Mindarus*)
rappardi Hille Ris Lambers, 1967 HRL
totarae Cottier, 1953 paratypes BM HRL
viridis Ilharco, 1973 HRL

Subgen. Chileaphis Essig, 1953

N. (Chileaphis) michelbacheri Essig, 1953 (Chileaphis) paratypes BM HRL

NEOPROCIPHILUS Patch, 1912
Bull. Me agric. Exp. Stn 202: 174
Type species Pemphigus attenuatus Osborne & Sirrine, 1892 = Pemphigus aceris Monell, 1882

N. aceris (Monell, 1882) (Pemphigus) BM
*attenuatus* Osborne & Sirrine, 1892 (*Pemphigus*)
*smilacis* Williams, 1911 (*Lachnus*)

NEOPTEROCOMMA Hille Ris Lambers, 1935
Arb. morph. taxon. Ent. Berl. 2: 52
Type species Neopterocomma asiphum Hille Ris Lambers, 1935

N. asiphum Hille Ris Lambers, 1935 HRL
verhoeveni Hille Ris Lambers, 1956 BM HRL

NEORHIZOBIUS del Guercio, 1917 nec Crotch, 1874
Redia 12: 251
Type species Neorhizobius ulmiphilus del Guercio, 1917
= Mimeuria Börner, 1952

N. poae del Guercio, 1917 = Aploneura lentisci (Passerini, 1856)
stramineus del Guercio, 1917 = Aploneura lentisci (Passerini, 1856)
ulmiphilus del Guercio, 1917 type of Mimeuria Börner, 1952 q.v.

NEORHOPALOMYZUS Tao, 1963
Pl. Prot. Bull., Taiwan 5: 181
Type species Amphorophora lonicericola Takahashi, 1921

N. lonicericola (Takahashi, 1921) (Amphorophora) HRL

NEORHOPALOSIPHONINUS A.K. Ghosh & RayChaudhuri, 1968
Proc. zool. Soc. Calcutta 21: 187
Type species Neorhopalosiphoninus smilacifoliae A.K. Ghosh & RayChaudhuri, 1968

N. smilacifoliae A.K. Ghosh & RayChaudhuri, 1968

NEOSALTUSAPHIS Hille Ris Lambers, 1961
Bull. Res. Coun. Israel (B) Zool. 10B: 97
Type species Neosaltusaphis bodenheimeri Hille Ris Lambers, 1961

N. bodenheimeri Hille Ris Lambers, 1961 HRL

NEOSAPPAPHIS Hille Ris Lambers, 1959
Mitt. schweiz. ent. Ges. 32: 283
Type species Neosappaphis franzi Hille Ris Lambers, 1959
*Shaposhnikoviella* Mamontova, 1963

N. franzi Hille Ris Lambers, 1959 HRL
paradoxa (Mamontova, 1963) (Shaposhnikoviella)

NEOSCHOUTEDENIA Schumacher, 1923
Dt. ent. Z. 1923: 403
Type species Geoica cyperi Schouteden, 1902 = Pemphigus utricularius Passerini, 1856
= Geoica Hart, 1894

NEOSTOMAPHIS Takahashi, 1960
Bull. Univ. Osaka Pref. (B) 10: 2, described as subg. of Stomaphis Walker, 1870
Type species Stomaphis (Neostomaphis) fagi Takahashi, 1960
= Stomaphis Walker, 1870

NEOSYMYDOBIUS Baker, 1920
Bull. U.S. Dep. Agric. 826: 32
Type species Symydobius albasiphus Davis, 1914

N. agrifoliae (Essig, 1917) (Symydobius) BM HRL
albasiphus (Davis, 1914) (Symydobius) cotypes BM HRL
canadensis Richards, 1965
chrysolepis (Swain, 1918) (Symydobius) BM HRL
luteus Tissot, 1932 paratype BM
memorialis Hottes & Frison, 1931 paratype BM HRL
mimicus Hottes, 1926 BM ?
quercihabitus Miller, 1936 HRL

NEOTHOMASIA Baker, 1920
Bull. U.S. Dep. Agric. 826: 35
Type species Chaitophorus populicola Thomas, 1878
= Chaitophorus Koch, 1854

N. abditus Hottes, 1926 to Chaitophorus Koch, 1854
populicola subsp. pruinosa Narzikulov, 1954 (type of Lambersaphis Narzikulov, 1961) = Lambersaphis pruinosa (Narzikulov, 1954)
saliciniger Knowlton, 1927 to Chaitophorus Koch, 1854

N. utahensis Knowlton, 1928 to Chaitophorus Koch, 1854

NEOTHORACAPHIS Takahashi, 1958
Bull. Univ. Osaka Pref. (B) 8: 1
Type species Nipponaphis yanonis Matsumura, 1917
*Microthoracaphis* Takahashi, 1958
*Microunguis* Tao, 1966

N. depressa (Takahashi, 1933) (Thoracaphis)
elongata (Takahashi, 1958) (Microthoracaphis)
glaucae (Takahashi, 1958) (Microthoracaphis)
quercicola (Takahashi, 1921) (Astegopteryx)
*saitamaensis* Shinji, 1923 (*Glyphina*)
querciphaga (Takahashi, 1958) (Microthoracaphis) HRL
saramaoensis (Takahashi, 1935) (Thoracaphis)
sutepensis (Takahaish, 1941) (Thoracaphis)
tarakoensis (Takahashi, 1937) (Thoracaphis)
yanonis (Matsumura, 1917) (Nipponaphis) BM HRL
*distyfoliae* Takahashi, 1920 (*Nipponaphis*)
*distylifoliae* Takahashi, 1920 (*Nipponaphis*)
*distylii* Pergande of Essig & Kuwana, 1918 (*Nipponaphis*)

NEOTOXOPTERA Theobald, 1915
Bull. ent. Res. 6: 31
Type species Neotoxoptera violae Theobald, 1915 (nec Rhopalosiphum violae Pergande, 1900) = Micromyzus oliveri Essig, 1935

N. abeliae Takahashi, 1965 BM HRL
formosana (Takahashi, 1921) (Fullawayella) BM HRL
*alliumcepa* Essig, 1936 (*Micromyzus*) cotypes BM
*fuscus* Richards, 1956 (*Micromyzus*) paratype BM
oliveri (Essig, 1935) (Micromyzus) cotypes BM HRL
*clavatus* Paik, 1965 (*Myzus*) BM HRL
*violae* Theobald, 1915 nec Rhopalosiphum violae Pergande, 1900 BM
*violae* Theobald, 1920 (*Phorodon*) nec Rhopalosiphum violae Pergande, 1900
violae (Pergande, 1900) (Rhopalosiphum) BM HRL
*violae* Essig, 1909 (*Rhopalosiphum*)
violae Theobald, 1915 = oliveri (Essig, 1935)
yasumatsui Sorin, 1971 types BM HRL

NEOTRAMA Baker, 1920
Bull. U.S. Dep. Agric. 826: 20
Type species Neotrama delguercioi Baker, 1920 nomen novum pro Trama troglodytes von Heyden of Del Guercio, 1909 = Trama caudata del Guercio, 1909
*Longitarsus* Shinji, 1923/30 nec Berthold ex Latreille, 1827; Latreille ex Cuvier, 1829

N. caudata (del Guercio, 1909) (Trama) BM HRL
*delguercioi* Baker, 1920
*horvathi* del Guercio, 1909 (*Trama*)
caudata subsp. maritima Eastop, 1953 = maritima Eastop, 1953
delguercioi Baker, 1920 = caudata (del Guercio, 1909)
formicella Theobald, 1929 type BM
maritima Eastop, 1953 types BM HRL
*caudata* subsp. *maritima* Eastop, 1953 BM HRL
narzykulovi (Kan, 1962) (Protrama)
pamirica (Narzikulov, 1962) (Protrama)
taraxaci (Shinji, 1929) (Trama)

NEOTRICHOSIPHON RayChaudhuri, 1956
Zool. Verh., Leiden 31: 79, described as subg. of Metatrichosiphon RayChaudhuri, 1956
Type species Trichosiphum tenuicorpus Okajima, 1908
= Mollitrichosiphum Suenaga, 1934

NEOTRICHOSIPHUM RayChaudhuri, M.R. Ghosh, M. Banerjee & A.K. Ghosh, 1973
Kontyû 41: 59, described as subg. of Eutrichosiphum Essig & Kuwana, 1918
Type species Eutrichosiphum (Neotrichosiphum) subinoyi RayChaudhuri, M.R. Ghosh, M. Banerjee & A.K. Ghosh, 1973
Here used as subg. of Eutrichosiphum Essig & Kuwana, 1918

NEOTUBERCULATUS van der Goot, 1915
Beiträge zur Kenntnis der Holländischen Blattläuse, Haarlem-Berlin, p. 582, errore
Type species Aphis quercus Kaltenbach, 1843 = Aphis annulata Hartig, 1841
= Tuberculoides van der Goot, 1915

NEUQUENAPHIS E.E. Blanchard, 1939
Physis, B. Aires 17: 880
Type species Myzocallis edwardsi Laing, 1927
*Spicaphis* Essig, 1953

N. bulbicauda Hille Ris Lambers, 1968 = chilensis (Essig, 1953)
chilensis Essig, 1953 BM HRL
*bulbicauda* Hille Ris Lambers, 1968 paratypes BM HRL
edwardsi (Laing, 1927) (Myzocallis) types BM HRL
*flavipes* Hille Ris Lambers, 1968 paratypes BM HRL
*papillata* Richards, 1968 paratype BM
essigi Hille Ris Lambers, 1968 HRL
*michelbacheri* Essig, 1953 (*Spicaphis*) nec Neuquenaphis michelbacheri Essig, 1953
flavipes Hille Ris Lambers, 1968 = edwardsi (Laing, 1927)
michelbacheri Essig, 1953 paratypes BM HRL
palliceps Hille Ris Lamber, 1968 paratypes BM HRL
papillata Richards, 1968 = edwardsi (Laing, 1927)
schlingeri Hille Ris Lambers, 1968 HRL
sensoriata Hille Ris Lambers, 1968 paratypes BM HRL
similis Hille Ris Lambers, 1968 HRL

NEVADAPHIS Drews, 1941
Pan-Pacif. Ent. 17: 60
Type species Nevadaphis sampsoni Drews, 1941

N. sampsoni Drews, 1941 HRL

NEVSKIA Mordvilko, 1932
Trudȳ Zashch. Rast., Ent. 5: 236, not indicated as new
Type species Nevskia pyraria (Passerini) = Myzus pyrarius Passerini, 1861*
= Melanaphis van der Goot, 1917

N. armeniaca Mordvilko, 1932 nomen nudum

NEVSKYA Mordvilko correction of Nevskia Mordvilko, 1932 (see Mamontova, 1954, p. 9)

N. centralasiae Narzikulov & Umarov, 1969

NEVSKYA Ossiannilsson, 1953 nec Mordvilko, 1932 (and Mamontova, 1954)
Opusc. ent. 18: 233
Type species Nevskya fungifera Ossiannilsson, 1953
= Nevskyella Ossiannilsson, 1954

N. fungifera Ossiannilsson, 1953 type of Nevskyella Ossiannilsson, 1954 q.v.

NEVSKYAPHIS Shaposhnikov, 1950
Ént. Obozr. 31: 225
Type species Dentatus bicolor Nevsky, 1929
Here used as subg. of Brachycaudus van der Goot, 1931

NEVSKYELLA Ossiannilsson, 1954
Opusc. ent. 19: 54
Type species Nevskya fungifera Ossiannilsson, 1953
*Nevskya* Ossiannilsson, 1953

N. fungifera (Ossiannilsson, 1953) (Nevskya) BM HRL
meridionalis Hille Ris Lambers & van den Bosch, 1964 types BM HRL

NIPPOCALLIS Matsumura, 1917
J. Coll. Agric. Tohoku imp. Univ. 7: 365
Type species Nippocallis kuricola Matsumura, 1917
Here used as subg. of Myzocallis Passerini, 1860

N. kuricola Matsumura, 1917 to Myzocallis (Nippocallis)
lewi Tao, 1964 = Tuberculatus (Acanthocallis) pilosus (Takahashi, 1929)

* First described species included in the genus.

NIPPOCHAITOPHORUS Takahashi, 1961
Kontyû 29: 247
Type species Nippochaitophorus moriokaensis Takahashi, 1961 = Myzocallis sawashibae Matsumura, 1917
= Mesocallis Matsumura, 1919

N. moriokaensis Takahashi, 1961 = Pterocallis (Mesocallis) sawashibae (Matsumura, 1917)

NIPPODYSAPHIS Hille Ris Lambers, 1965
Tijdschr. Ent. 108: 389
Type species Neodysaphis deutziae Hille Ris Lambers, 1965
*Neodysaphis* Hille Ris Lambers, 1965 nec Narzikulov, 1961

N. deutziae (Hille Ris Lambers, 1965) (Neodysaphis) HRL

NIPPOLACHNUS Matsumura, 1917
J. Coll. Agric. Tohoku imp. Univ. 7: 382
Type species Nippolachnus piri Matsumura, 1917

N. abietinus Matsumura, 1917 to Cinara Curtis, 1835
abietis Matsumura, lapsus pro abietinus Matsumura, 1917
bengalensis A.N. Basu & Hille Ris Lambers, 1968 HRL
eriobotryae A.N. Basu & Hille Ris Lambers, 1968 = himalayensis (van der Goot, 1917)
himalayensis (van der Goot, 1917) (Lachnus) HRL
*eriobotryae* A.N. Basu & Hille Ris Lambers, 1968 HRL
micromeli Shinji, 1924 = piri Matsumura, 1917
piri Matsumura, 1917 BM HRL
*micromeli* Shinji, 1924
*piri* Essig & Kuwana, 1918 (*Anoecia*)

NIPPONAPHIS Pergande, 1906
Ent. News 17: 205
Type species Nipponaphis distychii Pergande, 1906

N. amamiana Takahashi, 1962
autumna Monzen, 1934 type of Dinipponaphis Takahashi, 1962 q.v.
*yanonis* var. *autumna* Monzen, 1934
autumnalis Monzen lapsus pro autumna Monzen, 1934
coreana (Paik, 1965) (Thoracaphis) BM HRL
cuspidatae Essig & Kuwana, 1918 to Metanipponaphis Takahashi, 1959

N. distychii Pergande, 1906 HRL
distyfoliae Takahashi, 1920 = Neothoracaphis yanonis (Matsumura, 1917)
distylifoliae Takahashi, 1920 = Neothoracaphis yanonis (Matsumura, 1917)
distyliicola Monzen, 1934 BM HRL
*gigantea* Takahashi, 1958 HRL
distylii Pergande irregular emendation of distychii Pergande, 1906 by Essig & Kuwana, 1918; they described Neothoracaphis yanonis (Matsumura, 1917)
ficicola Hille Ris Lambers & Takahashi, 1959 HRL
gigantea Takahashi, 1958 = distyliicola Monzen, 1934
globuli Monzen, 1934 type of Monzenia Takahashi, 1962 q.v.
holboelliae A.K. Ghosh & RayChaudhuri, 1974
karatanei (Sasaki, 1936) (Anoecia)
litseae Takahashi, 1959 HRL
machili (Takahashi, 1933) (Thoracaphis)
machilicola (Shinji, 1933) (Thoracaphis)
machiliphaga Takahashi, 1959 (type of Pseudonipponaphis A.K. Ghosh & RayChaudhuri, 1973) to Nipponaphis (Pseudonipponaphis)
manoji A.K. Ghosh & RayChaudhuri, 1974
monzeni Takahashi, 1958 to Sinonipponaphis Tao, 1966
yanonis Matsumura, 1917 type of Neothoracaphis Takahashi, 1958 q.v.
yanonis var. autumna Monzen, 1934 = Dinipponaphis autumna (Monzen, 1934)

Subgen. Pseudonipponaphis A.K. Ghosh & RayChaudhuri, 1973

(Pseudonipponaphis) himalayensis A.K. Ghosh & RayChaudhuri, 1973
(Pseudonipponaphis) machiliphaga Takahashi, 1959 (Nipponaphis)
(Pseudonipponaphis) querciphaga A.K. Ghosh & RayChaudhuri, 1973
(Pseudonipponaphis) siamensis (Takahashi, 1941) (Thoracaphis)

NIPPOSIPHUM Matsumura, 1917
J. Coll. Agric. Tohoku imp. Univ. 7: 410
Type species Nipposiphum salicicola Matsumura, 1917
= Cavariella del Guercio, 1911

N. salicicola Matsumura, 1917 to Cavariella del Guercio, 1911

NISHIYANA Matsumura, 1917
In Nagano, K., A Collection of Essays for Mr. Yasushi Nawa, Gifu, 3: 90
Type species Nishiyana aomoriensis Matsumura, 1917
= Prociphilus Koch, 1857

N. aomoriensis Matsumura, 1917 to Prociphilus Koch, 1857

NOTABILIA Mordvilko, 1909 (1908)
Biol. Zbl. 29: 159, footnote; and Ezheg. zool. Muz. 13: 362, 1909 (1908)
Type species Phylloxera notabilis Pergande, 1904
= Xerophylla Walsh, 1867
Phylloxeridae

NOVAPHIS Mordvilko, 1929
Trudȳ prikl. Ent. 14: 84 error pro Neaphis Nevsky, 1929

NURUDEA Matsumura, 1917
In Nagano, K., A Collection of Essays for Mr. Yasushi Nawa, Gifu, 3: 65
Type species Nurudea ibofushi Matsumura, 1917
*Floraphis* Tsai & Tang, 1946
*Fushia* Matsumura, 1917
*Nurudeopsis* Matsumura, 1917

N. ibofushi Matsumura, 1917
*sinica* Tsai & Tang, 1946
meitanensis Tsai & Tang, 1946 (Floraphis)
shiraii (Matsumura, 1917) (Nurudeopsis) HRL
sinica Tsai & Tang, 1946 = ibofushi Matsumura, 1917
yanoniella (Matsumura, 1917) (Nurudeopsis) HRL
*rosea* Matsumura, 1917 (*Fushia*)

NURUDEOPSIS Matsumura, 1917
In Nagano, K., A Collection of Essays for Mr. Yashushi Nawa, Gifu, 3: 67
Type species Nurudeopsis shiraii Matsumura, 1917
= Nurudea Matsumura, 1917

N. shiraii Matsumura, 1917 to Nurudea Matsumura, 1917
yanoniella Matsumura, 1917 to Nurudea Matsumura, 1917

NYMPHAEFEX Amyot, 1847
Annls Soc. ent. Fr. (2) 5: 479
Invalid

NYMPHAPHIS Takahashi, 1960
Kontyû 28: 14
Type species Nymphaphis quercus Takahashi, 1960
= Diphyllaphis Takahashi, 1960

N. quercus Takahashi, 1960 to Diphyllaphis Takahashi, 1960

OBTUSICAUDA Soliman, 1927
Univ. Calif. Publ. Ent. 4: 99
Type species Obtusicauda essigi Soliman, 1927
*Mucrotrichaphis* Knowlton & Allen, 1940
*Narziculovia* Umarov, 1964

O. albicornis (Knowlton & Allen, 1940) (Mucrotrichaphis) BM
anomella (Knowlton & Allen, 1938) (Macrosiphum)
artemisiae (Cowen ex Gillette & Baker, 1895) (Nectarophora) BM HRL
*coweni* Hunter, 1901 (*Nectarophora*) HRL
artemisiphila (Knowlton & Allen, 1938) (Macrosiphum) BM
cefsmithi (Knowlton & Allen, 1938) (Macrosiphum) BM
chaetosiphon Takahashi & Moritsu, 1963 (Macrosiphoniella)
crassitubia Narzikulov & Umarov, 1969 (Macrosiphoniella) HRL
crassitubia subsp. kazakii Narzikulov, 1974
dolychosiphon Umarov, 1964 (Narziculovia)
essigi Soliman, 1927 BM HRL
filifoliae (Gillette & Palmer, 1928) (Macrosiphum) BM HRL
flavila (Knowlton & Allen, 1940) (Mucrotrischaphis)
frigidae (Oestlund, 1886) (Siphonophora) BM HRL
jonesi (Gillette & Palmer, 1928) (Macrosiphum)
zerohypsi (Knowlton & Allen, 1940) (Mucrotrichaphis) BM
zerothermum (Knowlton & Allen, 1938) (Macrosiphum) BM
*toti* Knowlton & Allen, 1940 (*Mucrotrichaphis*)

Subgen. Artemisaphis Knowlton & Roberts, 1947

O. (Artemisaphis) artemisicola (Williams, 1911) (Aphis) BM HRL

OEDISIPHUM van der Goot, 1917
Contrib. Faune Indes Néerl. 1 (3): 122
Type species Oedisiphum compositarum van der Goot, 1917

O. compositarum van der Goot, 1917 BM HRL
indicum A.K. Ghosh, 1969 = soureni A.N. Basu, 1964
soureni A.N. Basu, 1964 HRL
*indicum* A.K. Ghosh, 1969

OESTLUNDIA Hille Ris Lambers, 1949
Temminckia 8: 225
Type species Macrosiphum rubicola Oestlund, 1886
*Macrosiphum* Oestlund, 1886 nec Passerini, 1860
Here used as subg. of Illinoia Wilson, 1910

OESTLUNDIELLA Granovsky, 1930
Proc. ent. Soc. Wash. 32: 63
Type species Euceraphis flavus Davidson, 1912

O. flava (Davidson, 1912) (Euceraphis) BM HRL

OKAJIMAIA Suenaga, 1933
Kontyû 7: 249
Type species Okajimaia japonica Suenaga, 1933 = Glyphinaphis bambusae van der Goot, 1917
= Glyphinaphis van der Goot, 1917

O. japonica Suenaga, 1933 = Glyphinaphis bambusae van der Goot, 1917

OLIGOCALLIS Heie, 1967
Spolia zool. Mus. haun. 26: 133
Type species Oligocallis larssoni Heie, 1967
Fossil, Baltic amber

OMEIMEGOURA Tao, 1963
Pl. Prot. Bull., Taiwan 5: 184
Type species Omeimegoura nigrotibiae Tao, 1963
= Indomegoura Hille Ris Lambers, 1958

O. nigrotibiae Tao, 1963 to Indomegroura Hille Ris Lambers, 1958

ONISCOMYZUS Börner, 1942
Veröff. dt. Kolon-u. Übersee-Mus. Bremen 3: 259
Type species Oniscomyzus bramstedti Börner, 1942 = Ctenocallis dobrovljanskyi Klodnitsky, 1924
= Ctenocallis Klodnitsky, 1924

O. bramstedti Börner, 1942 = Ctenocallis dobrovljanskyi Klodnitsky, 1924

OREGMA Buckton, 1893
Indian Mus. Notes 3:87
Type species Oregma bambusae Buckton, 1893
= Astegopteryx Karsch, 1890

O. aderuensis Takahashi, 1935 to Chaitoregma Hille Ris Lambers & A.N. Basu, 1966
alexanderi Takahashi, 1924 to Pseudoregma Doncaster, 1966
bambusae Buckton, 1893 to Astegopteryx Karsch, 1890
bambusae var. carolinensis Takahashi, 1941 to Pseudoregma Doncaster, 1966

O. bambusicola Takahashi, 1921 type of Pseudoregma Doncaster, 1966 q.v.
bambusifoliae Takahashi, 1921 (type of Trichoregma Takahashi, 1929) to Astegopteryx Karsch, 1890
basalis van der Goot, 1917 to Astegopteryx Karsch, 1890
cantonensis Takahashi, 1936 = ? Pseudoregma bambusicola (Takahashi, 1921)
dendrocalami Takahashi, 1935 to Pseudoregma Doncaster, 1966
formosana Takahashi, 1924 to Astegopteryx Karsch, 1890
gombakana Takahashi, 1950 to Pseudoregma Doncaster, 1966
graminum van der Goot, 1917 to Pseudoregma Doncaster, 1966
heitoensis Takahashi, 1927 = Pseudoregma koshuensis (Takahashi, 1924)
japonica Takahashi, 1924 to Ceratovacuna Zehntner, 1897
koshuensis Takahashi, 1924 to Pseudoregma Doncaster, 1966
longifila Takahashi, 1929 to Ceratovacuma Zehntner, 1897
loranthi van der Goot, 1912 to Tuberaphis Takahashi, 1933
lutescens van der Goot, 1917 = Astegopteryx bambusae (Buckton, 1893)
minuta van der Goot, 1917 to Astegopteryx Karsch, 1890
montana van der Goot, 1917 to Pseudoregma Doncaster, 1966
muiri van der Goot, 1918 to Astegopteryx Karsch, 1890
mysorensis David, 1956 = Astegopteryx bambusae (Buckton, 1893)
mysorensis Krishnamurti, 1930 = ? Astegopteryx insularis (van der Goot, 1912)
mysorensis Theobald, 1929 = ? Astegopteryx insularis (van der Goot, 1912)
nicolaiae Takahashi, 1935 to Pseudoregma Doncaster, 1966
nipae van der Goot, 1917 to Astegopteryx Karsch, 1890
nipae var. cocois van der Goot, 1918 nomen nudum
oplismeni Takahashi, 1924 = ? Ceratovacuna nekoashi (Sasaki, 1910)
orientalis Takahashi, 1923 = ? Ceratovacuna nekoashi (Sasaki, 1910)
pallida van der Goot, 1917 to Astegopteryx Karsch, 1890
panici van der Goot, 1917 to Ceratovacuna Zehntner, 1897
panicola Takahashi, 1921 to Pseudoregma Doncaster, 1966
pendleburyi Takahashi, 1950 to Pseudoregma Doncaster, 1966
pseudomontana Takahashi, 1924 = Pseudoregma koshuensis (Takahashi, 1924)
qanici Uye, 1924 = Ceratovacuna japonica (Takahashi, 1924)
rhapidis van der Goot, 1917 to Astegopteryx Karsch, 1890
salatigensis van der Goot, 1917 to Astegopteryx Karsch, 1890
sasae Monzen, 1927 = Ceratovacuna japonica (Takahashi, 1924)
silvestrii Takahashi, 1927 to Ceratovacuna Zehntner, 1897
similis van der Goot, 1917 to Astegopteryx Karsch, 1890
singaporensis van der Goot, 1918 to Astegopteryx Karsch, 1890
striata van der Goot, 1917 to Astegopteryx Karsch, 1890

O. subglandulosa Hille Ris Lambers & A.N. Basu, 1966 = Ceratovacuna silvestrii (Takahashi, 1927)
sumatrensis Takahashi, 1926 = Pseudoregma koshuensis (Takahashi, 1924)
sundanica van der Goot, 1917 to Pseudoregma Doncaster, 1966
tattakana Takahashi, 1925 type of Chaitoregma Hille Ris Lambers & A.N. Basu, 1966 q.v.
tattakana var. suishana Takahashi, 1929 to Chaitoregma Hille Ris Lambers & A.N. Basu, 1966
uscae Tao, 1964 errore pro uscare Tao, 1964
uscare Tao, 1964 to Ceratovacuna Zehntner, 1897

ORIENTUBERCULOIDES Hille Ris Lambers, 1974
Boll. Zool. agr. Bachi. (II) 11: 23, described as subg. of Tuberculatus Mordvilko, 1894
Type species Myzocallis yokoyamai Takahashi, 1923
Here used as subg. of Tuberculatus Mordvilko, 1894

OROBION Mordvilko, 1914
Faune Russie, Ins. Hém. 1: 67
Without species

ORYCTAPHIS Scudder, 1890
U.S. Geol. Survey Territories 13: 266
Type species Oryctaphis lesueurii Scudder, 1890
= ? Siphonophoroides Buckton, 1883
Fossil

OSSIANNILSSONIA Hille Ris Lambers, 1952
Ent. Tidskr. 73: 41
Type species Ossiannilssonia oelandica Hille Ris Lambers, 1952

O. oelandica Hille Ris Lambers, 1952 BM HRL

OVATOIDES Börner, 1939
Arb. physiol. angew. Ent. Berl. 6: 79
Type species Aphis inulae Walker, 1849
Here used as subg. of Ovatus van der Goot, 1913

OVATOMYZUS Hille Ris Lambers, 1947
Zoöl. Meded., Leiden 28: 306
Type species Ovatomyzus stachyos Hille Ris Lambers, 1947

O. boraginacearum Eastop, 1952 types BM HRL
calaminthae (Macchiati, 1885) (Phorodon) BM HRL
*minutus* Börner ex Eggler, 1951 (*Ovatus*)
*pusillus* Börner, 1950 (*Ovatus*)
chamaedrys (Passerini, 1879) (Phorodon) BM HRL
stachyos Hille Ris Lambers, 1947 BM HRL

OVATOPHORODON Aizenberg, (1954) 1966
Diss. biol. Sci. Lenin Univ., Moscow, p. 9, without species (1954)
Trudȳ vses. ent. Oboshch. 51: 136, 151 (1966)
Type species Siphonophora menthae Buckton of Shaposhnikov nec Buckton, 1876 = Phorodon mentharius van der Goot, 1913
= Ovatus van der Goot, 1913

OVATOPSIS Aizenberg, 1954
Diss. biol. Sci. Lenin Univ. Moscow p. 9
Type species Ovatopsis ranunculi Aizenberg, 1954 = Aphis ranunculina Walker, 1852
= Tubaphis Hille Ris Lambers, 1947

O. ranunculi Aizenberg, 1954 = Tubaphis ranunculina (Walker, 1852)

OVATUS van der Goot, 1913
Tijdschr. Ent. 56: 84
Type species Ovatus mespili van der Goot, 1913 = Aphis insita Walker, 1849
*Ovatophorodon* Aizenberg, (1954) 1966

O. compositae Takahashi, 1965 to Myzus Passerini, 1860
crataegarius (Walker, 1850) (Aphis) lectotype BM HRL
*crataegina* Walker m.s. ex Theobald, 1926: 281
*malicolens* Hori, 1929 (*Myzus*)
*melissae* Walker, 1852 (*Aphis*)
*menthae* Walker, 1852 (*Aphis*)
*? mespiliella* Theobald, 1920 (*Myzus*)
glechomae Hille Ris Lambers, 1947 BM HRL
insitus (Walker, 1849) (Aphis) BM HRL
*mespili* van der Goot, 1913
latifrons Börner, 1942 (as Ericaphis type of Boreamyzus Shaposhnikov, 1964) to Ericaphis Börner, 1939
leucocrini (Gillette & Palmer, 1929) (Myzus) paratypes BM HRL
lycopi (Nevsky, 1929) (Phorodon) HRL

O. malisuctus (Matsumura, 1918) (Myzus) BM HRL
*chaenomelis* Dzhibladze, 1951 (*Myzus*) HRL
*japonica* Essig & Kuwana, 1918 (*Aphis*)
*japonicus* Monzen, 1929 (*Myzus*)
*takahashii* Strand, 1929 (*Myzus*)
mentharius (van der Goot, 1913) (Phorodon) BM HRL
*menthastri* Hille Ris Lambers, 1947 HRL
menthastri Hille Ris Lambers, 1947 = mentharius (van der Goot, 1913)
mespili van der Goot, 1913 = insitus (Walker,1849)
mimulicola (Drews & Sampson, 1937) (Kakimia) HRL
minutus Börner ex Eggler, 1951 = Ovatomyzus calaminthae (Macchiati, 1885)
minutus (van der Goot, 1917) (Phorodon)
minutus subsp. nipponicus Takahashi, 1965 = nipponicus Takahashi, 1965
nipponicus Takahashi, 1965 BM HRL
*minutus* subsp. *nipponicus* Takahashi, 1965
phloxae (Sampson, 1939) (Phorodon) HRL
pusillus Börner, 1950 = Ovatomyzus calaminthae (Macchiati, 1885)
reticulatus Heie, 1972 HRL
titschaki Börner, 1942 (type of Idiovatus Börner, 1944) to Myzus (Nectarosiphon)
valuliae (Robinson, 1974) (Myzus) BM HRL

Subgen. Ovatoides Börner, 1939

O. (Ovatoides) inulae (Walker, 1849) (Aphis) BM HRL

PACHYPAPPA Koch, 1856
Die Pflanzenläuse Aphiden, Nürnberg, p. 269
Type species Pachypappa marsupialis Koch, 1856
*Pemphiglachnus* Knowlton, 1928

P. grandis Tullgren, 1925 = populi (Linnaeus, 1758)
lactea Tullgren, 1909 type of Pachypappa Tullgren, 1909 nec Koch, 1856; and of Pachypappella Baker, 1920 q.v.
marsupialis Koch, 1856 HRL
myrtilli Börner, 1950
populi (Linnaeus, 1758) (Aphis) BM HRL
*grandis* Tullgren, 1925 HRL
pseudobyrsa (Walsh, 1863) (Byrsocrypta) BM HRL
*populi* Gillette, 1908 (*Schizoneura*)
sacculi (Gillette, 1914) (Asiphum) BM HRL
*caudelli* MacDougall, 1926 (*Pachypapella*)
*kaibabensis* Knowlton, 1938 (*Pemphiglachnus*)

P. varsoviensis (Mordvilko, 1895) (Pemphigus) HRL
vesicalis Koch, 1856 BM HRL

PACHYPAPPA Tullgren, 1909 nec Koch, 1856
Ark. Zool. 5: 69
Type species Pachypappa lactea Tullgren, 1909
= Pachypappella Baker, 1920

P. lactea Tullgren, 1909 type of Pachypappella Baker, 1920 q.v.

PACHYPAPPELLA Baker, 1920
Bull. U.S. Dep. Agric. 826: 71
Type species Pachypappa lactea Tullgren, 1909
*Pachypappa* Tullgren, 1909 nec Koch, 1856

P. caudelli MacDougall, 1926 = Pachypappa sacculi (Gillette, 1914)
lactea (Tullgren, 1909) (Pachypappa) BM HRL
orientalis Mordvilko ex Shaposhnikov, 1955

PACIFICALLIS Richards, 1965
Mem. ent. Soc. Can. 44: 66, described as subg. of Tuberculatus Mordvilko, 1894
Type species Tuberculatus (Pacificallis) columbiae Richards, 1965
Here used as subg. of Tuberculatus Mordvilko, 1894

PACZOSKIA Mordvilko, 1914
Faune Russie, Ins. Hém. 1: 63, in key, without species
Type species Paczoskia paczoskii Mordvilko, 1919

P. brevipilosa Tashev, 1964 HRL
budhium H. Banerjee, A.K. Ghosh & RayChaudhuri, 1969 HRL
*tenuirostris* L.K. Ghosh, 1970 (*Uroleucon*) HRL
longipes Tashev, 1964 HRL
lopatini Umarov, 1964 to Macrosiphoniella del Guercio, 1911
major Börner, 1950 BM HRL
miestingeri Börner, 1950 to Macrosiphoniella del Guercio, 1911
obtecta Börner, 1950 to Macrosiphoniella del Guercio, 1911
paczoskii Mordvilko, 1919 BM HRL
turanica (Nevsky, 1929) (Macrosiphum)
*paczoskii* subsp. *turanicum* Nevsky, 1929 (*Macrosiphum*)
wagneri Remaudière & Tuatay, 1963 HRL
wintreberti Remaudière & Tuatay, 1963 HRL

PADUCIA Hottes & Frison, 1931
Bull. Ill. nat. Hist. Surv. 19: 167
Type species Melanoxantherium antennatum Patch, 1913

P. antennata (Patch, 1913) (Melanoxantherium) BM HRL
aterrima Hille Ris Lambers, 1952 HRL

PALAEOAPHIS Richards, 1966
Can. Ent. 98: 750
Type species Palaeoaphis archimedia Richards, 1966
Fossil, Cretaceous Canadian amber

PALAEOPHYLLAPHIS Heie, 1967
Spolia zool. Mus. haun. 26: 97
Type species Palaeophyllaphis longirostris Heie, 1967
Fossil, Baltic amber

PALAEOSIPHON Heie, 1967
Spolia zool. Mus. haun. 26: 119
Type species Aphis hirsuta Germar & Berendt, 1856
Fossil, Baltic amber

PALAEOTHELAXES Heie, 1967
Spolia zool. Mus. haun. 26: 42
Type species Palaeothelaxes setosa Heie, 1967
Fossil, Baltic amber

PANAPHIS Kirkaldy, 1904
Entomologist 37: 279
Type species Aphis juglandis Frisch, 1734 prebinominal = Aphis juglandis Goeze, 1778
= Callaphis Walker, 1870

PANIMERUS Laing, 1926 nec Eaton, 1913
Entomologist 59: 323 nom. nov. pro Dilachnus Baker, 1919 nec Fairmaire, 1896
Type species Lachniella gracilis Wilson, 1919
= Cinara Curtis, 1835

P. kiangsiensis Lou, 1935 = Cinara formosana (Takahashi, 1924)

PAOLIELLA Theobald, 1928
Bull. ent. Res. 19: 177
Type species Paoliella hystrix Theobald, 1928
*Unipterus* Hall, 1932

P. ayari (Eastop, 1955) (Unipterus) type BM HRL
browni Quednau, 1962 type BM
chiangae van Harten & Ilharco, 1972 BM HRL
commiphorae (Doncaster, 1954) (Unipterus) type BM HRL
commiphorae subsp. persimilis (Eastop, 1955) (Unipterus) types BM HRL
delottoi (Hille Ris Lambers, 1954) (Unipterus) BM HRL
eastopi Hille Ris Lambers, 1973 BM HRL
echinata Eastop, 1956 type BM HRL
harteni Ilharco, 1971 BM HRL
hystrix Theobald, 1928 type BM HRL
kenyensis (Eastop, 1956) (Unipterus) type BM HRL
*terminaliae* subsp. *kenyensis* Eastop, 1956 (*Unipterus*) BM
longirostris Quednau, 1974 type BM
melanocallis Quednau, 1974 to Lizerius E.E. Blanchard, 1923
monotuberculata van Harten & Ilharco, 1972 HRL
nachensis (Eastop, 1955) (Unipterus) type BM HRL
nirmalae (David, 1969) (Unipterus (Paoliella)) paratypes BM HRL
papillata (Hall, 1932) (Unipterus) types BM HRL
pteleopsidis Quednau, 1974 types BM HRL
reticulosiphon Quednau, 1974 to Lizerius E.E. Blanchard, 1923
sawus (Eastop, 1955) (Unipterus) type BM
terminaliae (Hall, 1932) (Unipterus) types BM HRL
ufuasi (Eastop, 1955) (Unipterus) types BM HRL
wettsteini Quednau, 1964 paratypes BM HRL

PAPILLAPHIS Börner, 1952
Mitt. thüring. bot. Ges. 4 (3): 459 nomen novum pro Tuberculaphis Börner, 1952 nec Das, 1918
Type species Doralina taraxacicola Börner, 1940
= Aphis Linnaeus, 1758

PAPILLOMYZUS Szelegiewicz, 1963
Annls zool., Warsz. 21: 54, described as subg. of Macrosiphoniella del Guercio, 1911
Type species Macrosiphoniella (Papillomyzus) riedeli Szelegiewicz, 1963
Here used as subg. of Macrosiphoniella del Guercio, 1911

PAPULAPHIS Robinson, 1966
Can. Ent. 98: 1256
Type species Macrosiphum sleesmani Pepper, 1950

P. sleesmani (Pepper, 1950) (Macrosiphum) BM HRL

PARABRACHYUNGUIS Remaudière & Davatchi, 1955
Rev. Path. vég. Ent. agric. Fr. 33: 241, described as subg. of Brachyunguis Das, 1918
Type species Brachyunguis (Parabrachyunguis) kaussarii Remaudière & Davatchi, 1955
= Brachyunguis Das, 1918

PARACALLIPTERUS RayChaudhuri & A.K. Ghosh, 1964
Zool. Meded., Leiden 39: 260
Type species Paracallipterus kalipadi RayChaudhuri & A.K. Ghosh, 1964

P. kalipadi RayChaudhuri & A.K. Ghosh, 1964 BM

PARACERATAPHIS Mordvilko, 1929
Trudȳ prikl. Ent. 14: 34
Type species Paracerataphis tremulae Mordvilko, 1929 = Sphaerococcus populi Maskell, 1898
= Doraphis Matsumura & Hori ex Hori, 1929

P. tremulae Mordvilko, 1929 = Doraphis populi (Maskell, 1898)

PARACHAITOPHORUS Takahashi, 1937
Konowia 16: 90
Type species Patchia spiraeae Takahashi, 1924

P. spiraeae (Takahashi, 1924) (Patchia) BM HRL

PARACLETIUS Amyot, 1847
Ann. Soc. ent. Fr. (2) 5: 487
Invalid

PARACLETUS von Heyden, 1837
Ent. Beitr. Mus. Senckenb. 2: 295
Type species Paracletus cimiciformis von Heyden, 1837
*Hemitrama* Mordvilko, 1921

P. bykovi (Mordvilko, 1921) (Hemitrama) BM HRL
cimiciformis von Heyden, 1837 BM HRL
*derbesi* Lichtenstein, 1880 (*Pemphigus*)
*harukawai* Tanaka, 1957 (*Forda*) BM HRL
*pallidoides* Lichtenstein, 1880 (*Pemphigus*)
*palidus* Derbes, 1869 (*Pemphigus*)
*portschinskyi* Mordvilko, 1921
donisthorpei Theobald, 1929 types BM HRL
portschinskyi Mordvilko, 1921 = camiciformis von Heyden, 1837
subnudus Hille Ris Lambers, 1954 HRL

PARACOLOPHA Hille Ris Lambers, 1966
Hilgardia 37: 600
Type species Dryopeia morrisoni Baker, 1919

P. morrisoni (Baker, 1919) (Dryopeia) BM HRL

PARAGREENIDEA RayChaudhuri, 1956
Zool. Verh., Leiden 31: 25, described as subg. of Greenidea Schouteden, 1905
Type species Greenidea viticola Takahashi, 1929
Here used as subg. of Greenidea Schouteden, 1905

PARALIZERIUS Quednau, 1974
Can. Ent. 106: 47, described as subg. of Lizerius E.E. Blanchard, 1923
Type species Lizerius (Paralizerius) cermelii Quednau, 1974
Here used as subg. of Lizerius E.E. Blanchard, 1923

PARAMORITZIELLA Grassi, 1912
Mentioned in Börner, 1930, p. 189 with type species caryaefoliae Fitch, 1856.
Grassi reference not found by us.
Phylloxeridae

PARAMYZUS Börner, 1933
Kleine Mitteilungen über Blattläuse, privately published, p. 2
Type species Paramyzus heraclei Börner, 1933

P. heraclei Börner, 1933 BM HRL
heraclei subsp. similis Takahashi, 1963 HRL
longirostris Miyazaki, 1971 BM HRL

PARANIPPONAPHIS Takahashi, 1959
Bull. Univ. Osaka Prefect. (B) 9: 3
Type species Paranipponaphis takaoensis Takahashi, 1959

P. takaoensis Takahashi, 1959 BM HRL

PARANOECIA Zwölfer, 1957
Z. angew. Ent. 40: 198
Type species Anoecia pskovica Mordvilko, 1916
Here used as subg. of Anoecia Koch, 1857

PARAPERGANDEA Börner, 1930
Arch. klassif. phylogen. Ent. 1: 160
Type species Pemphigus caryaevenae Fitch, 1856
= ? Phylloxera Boyer de Fonscolombe, 1834
Phylloxeridae

PARAPHIS Mordvilko m.s. name ex Shaposhnikov, 1955, not described
In Shaposhnikov, 1955, Forest Pests, Moscow & Leningrad, p. 795
Type species Paraphis prinsepiae Mordvilko ex Shaposhnikov, 1955, nomen nudum

P. prinsepiae Mordvilko ex Shaposhnikov, 1955 nomen nudum

PARAPHORODON Tseng & Tao, 1938
Jl W. China Border Res. Soc. 10: 205
Type species Paraphorodon omeishanensis Tseng & Tao, 1938 = Phorodon cannabis Passerini, 1860
*Diphorodon* Börner, 1939
Here used as subg. of Phorodon Passerini, 1860

P. omeishanensis Tseng & Tao, 1938 = Phorodon (Paraphorodon) cannabis Passerini, 1860

PARAPHYLLOXERA Grassi, 1909
Atti Accad. naz. Lincei Rc. 18 (2): 421, described as subg. of Phylloxera Boyer de Fonscolombe, 1834
Type species Paraphylloxera glabra (von Heyden) = Vacuna glabra von Heyden, 1837
= Phylloxera Boyer de Fonscolombe, 1834
Phylloxeridae

PARAPROCIPHILUS Mordvilko, 1923
Dokl. Acad. Nauk. SSR 1923: 44, described as subg. of Prociphilus Koch, 1857
Type species Pemphigus baicalensis Cholodkovsky, 1921
Here used as subg. of Prociphilus Koch, 1857

P. graminis E.E. Blanchard, 1944 = Asiphonella dactylonii Theobald, 1923
ucrainensis Mamontova, 1955 = Mimeuria ulmiphila (del Guercio, 1917)

PARASCHIZAPHIS Hille Ris Lambers, 1947
Zoöl. Meded., Leiden 28: 316, described as subg. of Schizaphis Bőrner, 1931
Type species Toxoptera typhae Laing, 1923 = Toxoptera scirpi Passerini, 1874
Here used as subg. of Schizaphis Börner, 1931

P. longisetosa Higuchi, 1970 to Schizaphis (Paraschizaphis)
rosazevedoi Ilharco, 1961 = Schizaphis (Paraschizaphis) scirpi (Passerini, 1874)
scirpi subsp. eriophori (errore eriophoris) F.P Müller, 1974 to Schizaphis (Paraschizaphis)
scirpicola Hille Ris Lambers, 1960 to Schizaphis (Paraschizaphis)

PARASTOMAPHIS Pašek, 1953
Véstn. csl. Spol. zool. 17: 163, 171, described as subg. of Stomaphis Walker, 1870
Type species Stomaphis graffii Cholodkovsky, 1894
= Stomaphis Walker, 1870

P. cupressi Pintera, 1965 to Stomaphis Walker, 1870

PARATHECABIUS Börner, 1950
Neue europäische Blattläuse, privately published, p. 18
Type species Thecabius lysimachiae Börner, 1916
Here used as subg. of Thecabius Koch, 1857

P. cerastii Borner, 1950 to Thecabius (Parathecabius)
saliciradicis Börner, 1950 to Pemphigus Hartig, 1839
stammeri Zwölfer, 1957 to Thecabius (Parathecabius)

PARATHORACAPHIS Takahashi, 1958
Insecta matsum. 22: 13
Type species Thoracaphis setigera Takahashi, 1932

P. cheni (Takahashi, 1936) (Thoracaphis) BM
elongata (Takahashi, 1941) (Thoracaphis)
gooti (Takahashi, 1950) (Thoracaphis) types BM HRL
kayashimai (Takahashi, 1950) (Thoracaphis) types BM
setigera (Takahashi, 1932) (Thoracaphis) BM HRL

PARATINOCALLIS Higuchi, 1972
Insecta matsum. 35: 30
Type species Paratinocallis corylicola Higuchi, 1972
Here used as subg. of Pterocallis Passerini, 1860

P. corylicola Higuchi, 1972 to Pterocallis (Paratinocallis)

PARATOXOPTERA E.E. Blanchard, 1944
Acta zool. lilloana 2: 19
Type species Paratoxoptera argentiniensis E.E. Blanchard, 1944 = Myzus citricidus Kirkaldy, 1907
= Toxoptera Koch, 1856

P. argentiniensis E.E. Blanchard, 1944 = Toxoptera citricidus (Kirkaldy, 1907)

PARATRICHOSIPHUM Takahashi, 1931
Rep. Govt Res. Inst. Dep. Agric. Formosa 53: 31
Type species Greenidea tattakana Takahashi, 1925
= Eutrichosiphum Essig & Kuwana, 1918

P. alnicola A.N. Basu, 1968 to Eutrichosiphum Essig & Kuwana, 1918
alnifoliae Tao, 1958 to Eutrichosiphum Essig & Kuwana, 1918
flavum Takahashi, 1941 to Eutrichosiphum Essig & Kuwana, 1918
javanicum RayChaudhuri, 1956 = Eutrichosiphum (Ditrichosiphon) elongatum Takahashi, 1940
kyushuensis Tao, 1960 = Allotrichosiphon kashicola (Kurisaki, 1920)
lithocarpi Takahashi, 1931 to Mollitrichosiphum (Metatrichosiphon)
niitakaensis Takahashi, 1937 to Mollitrichosiphum (Metatrichosiphon)
sensoriatus A.K. Ghosh, 1974 to Eutrichosiphum Essig & Kuwana, 1918
sikkimense RayChaudhuri, 1974 Ghosh, M. Banerjee & A.K. Ghosh, 1973 to Eutrichosiphum Essig & Kuwana, 1918
tattakanum subsp. assamensis A.K. Ghosh & RayChaudhuri, 1962 = Eutrichosiphum assamense (A.K. Ghosh & RayChaudhuri,1962)

Subgen. Neoparatrichosiphum A.K. Ghosh & RayChaudhuri, 1962

P. (Neoparatrichosiphum) khasyanum A.K. Ghosh & RayChaudhuri, 1962 (type of Neoparatrichosiphum A.K. Ghosh & RayChaudhuri, 1962) to Eutrichosiphum Essig & Kuwana, 1918
(Neoparatrichosiphum) raychaudhurii A.K. Ghosh, 1969 to Eutrichosiphum Essig & Kuwana, 1918

PARTHENOPHYLLOXERA Grassi & Foa, 1911
Atti Accad. naz. Lincei Rc. 20 (2): 611, described as subg. of Phylloxera Boyer de Fonscolombe, 1834
Type species Phylloxera (Parthenophylloxera) ilicis Grassi ex Grassi & Foa, 1911
= Phylloxera Boyer de Fonscolombe, 1834
Phylloxeridae

PASEKIA Aizenberg, 1959
Zool. Zh. 38: 1674, 1677, described as subg. of Myzocallis Passerini, 1860
Type species Myzocallis komareki Pašek, 1953 = Hoplocallis komareki Pašek, 1953
= Myzocallis Passerini, 1860

PASSERINIA Macchiati, 1882
Revista scientifico-industriale di G. Vimercati, Firenze 12: 356
Type species Passerinia rosae Macchiati, 1882 = Aphis rosae Linnaeus, 1758
= Macrosiphum Passerini, 1860

P. rosae Macchiati, 1882 = Macrosiphum rosae (Linnaeus, 1758)

PATCHIA Baker, 1920
Bull. U.S. Dept. Agric. 826: 34
Type species Patchia virginiana Baker, 1920

P. obscurus Tissot, 1932 to Lachnochaitophorus Granovsky, 1933
spiraeae Takahashi, 1924 type of Parachaitophorus Takahashi, 1937 q.v.
virginiana Baker, 1920 BM HRL
winforii Miller, 1933

PATCHIELLA Tullgren, 1925
Mddn CentAnst. Fors Vas. Jordbromrad. 280 (Ent. 44): 12
Type species Schizoneura reaumuri Kaltenbach, 1843

P. reaumuri (Kaltenbach, 1843) (Schizoneura) BM HRL
reaumuri subsp. orientalis Mordvilko ex Shaposhnikov, 1955

PAULIANAPHIS Essig, 1957
Naturaliste malgache 9: 208
Type species Paulianaphis madagascariensis Essig, 1957

P. madagascariensis Essig, 1957 BM HRL

PELTAPHIS Frison & Ross, 1933
Can. Ent. 65: 152, described as subg. of Thripsaphis Gillette, 1917
Type species Thripsaphis (Peltaphis) hottesi Frison & Ross, 1933
*Zabapaphis* Richards, 1971
Here used as subg. of Thripsaphis Gillette, 1917

PEMPHIGELLA Tullgren, 1909
Ark. Zool. 5: 171, footnote
= Meddn CentAnst. ForsVas. Jordbromrad. 14 (Ent. 5): 171, footnote
Type species Tetraneura cornicularia Passerini, 1856 errore pro Pemphigus cornicularius Passerini, 1856 = Aphis pistaciae (Linnaeus, 1767)
= Baizongia Rondani, 1848

P. foliodentata Tao, 1947 type of Chaetogeoica Remaudière & Tao, 1957 q.v.
lingi Tao, 1943 = Kaburagia rhusicola Takagi, 1937
marginata Tao, 1947 = Forda sichangensis Remaudière & Tao, 1957
mimeuri Gaumont, 1930 to Geoica Hart, 1894
paglianoi Gaumont, 1930 to Baizongia Rondani, 1848

PEMPHIGETUM Mordvilko, 1928
Bull. Soc. zool. Fr. 53: 359
Type species Pemphigetum muticae Mordvilko, 1928
= Geoica Hart, 1894

P. muticae Mordvilko, 1928 to Geoica Hart, 1894

PEMPHIGINUS Börner, 1930
Arch. klassif. phylogen. Ent. 1: 153
Type species Pemphigus populi Courchet, 1879
= Pemphigus Hartig, 1839

P. kelloggi Takahashi, 1939 to Thecabius Koch, 1857

PEMPHIGLACHNUS Knowlton, 1928
Ann. ent. Soc. Am. 21: 264
Type species Pemphiglachnus kaibabensis Knowlton, 1928 = Asiphum sacculi Gillette, 1914
= Pachypappa Koch, 1856

P. kaibabensis Knowlton, 1928 = Pachypappa sacculi (Gillette, 1914)

PEMPHIGUS Hartig, 1839
Jber. Fortschr. Forstwiss. u. forstl.-naturk. im Jahre 1836 u. 1837 1(4): 645; and also Z. Ent. (Germar) 3: 366 (1841)
Type species Aphis bursaria Linnaeus, 1758 (designated by Fitch, 1855, p. 73 footnote)
*Aphioides* Rondani, 1848
*Baizongiella* E.E. Blanchard, 1944
*Hamadryaphis* Kirkaldy, 1904
*Kessleria* Lichtenstein, 1885
*Pemphiginus* Börner, 1930
*Rhizobius* Burmeister, 1835
*Rhizobius* Passerini, 1860
*Rhizophthiridium* van der Hoeven, 1840
*Rhyzobius* Ferrari, 1872 nec Stephens, 1829
*Rhyzoicus* Passerini, 1860

P. acerifolii Riley, 1879 = Prociphilus (Paraprociphilus) tesselatus (Fitch, 1851)
aceris Monell, 1882 to Neoprociphilus Patch, 1912
aedificator Buckton, 1893 (type of Dasia van der Goot ex Das, 1918 nec Gray, 1839) = Baizongia pistaciae (Linnaeus, 1767)
affinis Kaltenbach, 1843 (type of Bucktonia Lichtenstein, 1896) to Thecabius Koch, 1857
alni Provancher, 1890 = Prociphilus (Paraprociphilus) tesselatus (Fitch, 1851)
alnifoliae Williams, 1911 to Prociphilus Koch, 1857
americanus Walker, 1852 to Prociphilus Koch, 1857
andropogiae Shinji, 1924
argrimoniae Shinji, 1924 = Tetraneura (Tetraneurella) nigriabdominalis (Sasaki, 1899)
asteris Lichtenstein, 1884 nomen nudum
attenuatus Osborne & Sirrine, 1892 (type of Neoprociphilus Patch, 1912) = Neoprociphilus aceris (Monell, 1882)
baicalensis Cholodkovsky, 1921 (type of Paraprociphilus Mordvilko, 1923) to Prociphilus (Paraprociphilus)
balsamiferae Williams, 1911 = betae Doane, 1900
betae Doane, 1900 BM HRL
*balsamiferae* Williams, 1911
biramatus Ivanovskaja, 1973
borealis Tullgren, 1909 HRL
boyeri Passerini, 1856 = Tetraneura ulmi (Linnaeus, 1758)
boyeri var. saccharata del Guercio, 1900 = Tetraneura ulmi (Linnaeus, 1758)
brevicornis (Hart, 1894) (Tychea) HRL

P. burrowi Sanborn, 1904
bursarius (Linnaeus, 1758) (Aphis) BM HRL
*glandiformis* Rudow, 1875
*lactucae* Fitch, 1871 (*Rhizobius*)
*lactucae* Mosley, 1841 (*Eriosoma*)
*lactucae* Westwood, 1849
*lactucarius* Passerini, 1856
*pilosellae* Burmeister, 1835 (*Rhizobius*)
*populi* Mosley, 1841 (*Eriosoma*)
*pyriformis* Lichtenstein, 1885
*sonchi* Passerini, 1860 (*Rhyzobius*)
bursifex Heer, 1853, only gall described, fossil
caerulescens Passerini, 1856 to Tetraneura Hartig, 1841
californicus Davidson, 1911 = Thecabius populiconduplifolius (Cowen, 1895)
canadensis del Guercio, 1913 BM HRL (= ? populitransversus Riley, 1879)
*solanophila* E.E. Blanchard, 1944 (*Baizongiella*)
caryaecaulis Fitch, 1855 to Phylloxera Boyer de Fonscolombe, 1834
caraevenae Fitch, 1856 (type of Parapergandea Borner, 1930) to Phylloxera Boyer de Fonscolombe, 1834
cinchonae Buckton, 1889 nomen nudum
clematicola Shinji, 1922 to Colopha Monell, 1877
clematis Shinji, 1922 to Eriosoma (Colophina)
coccus Buckton, 1889 nomen dubium
coluteae Passerini, 1863 HRL
cornicularius Passerini, 1856 (as Tetraneura type of Pemphigella Tullgren, 1909) = Baizongia pistaciae (Linnaeus, 1767)
corniculoides Lichtenstein, 1880 = Baizongia pistaciae (Linnaeus, 1767)
corrugatans Sirrine, 1893 to Prociphilus (Neoparacletus)
coweni Cockerell, 1905 type of Tamalia Baker, 1920 q.v.
cynodonti Das, 1918 to Asiphonella Theobald, 1923
degeeri Kaltenbach, 1843 = Prociphilus (Stagona) xylostei (de Geer, 1773)
derbesi Lichtenstein, 1880 = Paracletus cimiciformis von Heyden, 1837
diani Ferrari, 1872
dorocola Matsumura, 1917 BM HRL
edificator Buckton, 1893 lapsus pro aedificator Buckton, 1893
ephemeratus Hottes & Frison, 1931
euphraticus Börner, 1930 nomen nudum
fagifoliae del Guercio, 1913 nomen dubium
fataunae Shinji, 1924
follicularius Passerini, 1861 = Forda marginata Koch, 1857

P. folliculoides Lichtenstein, 1880 = Forda marginata Koch, 1857
formicarius Walsh, 1863
formicetorum Walsh, 1863
fraxini Hartig, 1841 = Prociphilus fraxini (Fabricius, 1777)
fraxini-dipetalae Essig, 1911 to Prociphilus Koch, 1857
fraxinifolii Riley, 1879 (type of Meliarhizophagus Smith, 1974) to Prociphilus Koch, 1857
fraxinifolii Thomas, 1879 = Prociphilus fraxinifolii (Riley, 1879)
fuscicornis (Koch, 1857) (Amycla) BM HRL
fuscifrons var. saccharata del Guercio, 1895 = Tetraneura ulmi (Linnaeus, 1758)
gairi Stroyan, 1964 BM HRL
glandiformis Rudow, 1875 = bursarius (Linnaeus, 1758)
glebae Jackson, 1918 = ? Thecabius (Parathecabius) lysimachiae (Börner, 1916)
globosus Walker, 1852 = Smynthurodes betae Westwood, 1849
globulosus Theobald, 1915 = immunis Buckton, 1896
gnaphalii Kaltenbach, 1843 = populinigrae (Schrank, 1801)
gramineus Shinji, 1923 = Watabura nishiyae Matsumura, 1917
gravicornis Patch, 1913 to Thecabius (Parathecabius)
groenlandicus (Rübsaamen, 1898) (Tychea) BM HRL
groenlandicus subsp. crassicornis Hille Ris Lambers, 1953 HRL
hederae Horvath, 1894 = populi Courchet, 1879
hydrophilus Narzikulov, 1964
imaicus Cholodkovsky, 1912 type of Epipemphigus Hille Ris Lambers, 1966 q.v.
immunis Buckton, 1896 lectotype BM HRL
*globulosus* Theobald, 1915
*lichtensteini* Tullgren, 1909
*paghmanensis* Gullamullah, 1941/2 HRL
indicus Kieffer, 1908
inflatae del Guercio, 1911 = spyrothecae Passerini, 1856
iskanderkuli Narzikulov, 1957 HRL
junctisensoriatus Maxson, 1934 BM HRL
knowltoni Stroyan, 1970 HRL
lactucae Westwood, 1849 = bursarius (Linnaeus, 1758)
lactucarius Passerini, 1856 = bursarius (Linnaeus, 1758)
lasii (Cockerell, 1903) (Tychea)
laurifoliae Dolgova, 1973 HRL
laurifoliae Holman & Szelegiewicz, 1972 nomen nudum
lichtensteini Tullgren, 1909 = immunis Buckton, 1896
longicornus Maxson, 1923 BM HRL
longistigma Monell, 1878 to ? Prociphilus (Stagona)

P. lonicerae Hartig, 1841 = Prociphilus (Stagona) xylostei (de Geer, 1773)
luppovae Narzikulov, 1963 to Thecabius (Parathecabius)
matsumurai Monzen, 1929 HRL
microsetosus Aoki, 1975 HRL
minor Derbes, 1869
mongolicus Holman & Szelegiewicz, 1974 HRL
monophagus Maxson, 1934 BM HRL
montanus Narzikulov, 1957 BM HRL
mordvilkoi Cholodkovsky, 1912/3 BM HRL
nainitalensis Cholodkovsky, 1912
napaeus Buckton, 1896 types BM HRL
niisimae Matsumura, 1917 to Epipemphigus Hille Ris Lambers, 1966
nortonii Maxson, 1934 BM HRL
oenotherae Williams, 1891 nomen nudum
oestlundi Cockerell, 1906 = Mordwilkoja vagabunda (Walsh, 1863)
oplismeni Shinji, 1922 = Ceratovacuna nekoashi (Sasaki, 1910)
ovato-oblongus Kessler, 1881 = populinigrae (Schrank, 1801)
paghmanensis Ghulamullah, 1941/2 = immunis Buckton, 1896
pallidoides Lichtenstein, 1880 = Paracletus cimiciformis von Heyden, 1837
pallidus Derbes, 1869 = Paracletus cimiciformis von Heyden, 1837
passeki Börner, 1952 HRL
pedunculi Hartig, 1841 = ? Thelaxes dryophila (Schrank, 1801)
phenax Börner & Blunck, 1916 BM HRL
plicatus Dolgova, 1973 HRL
popularius Fitch, 1859
populi Courchet, 1879 type of Pemphiginus Borner, 1930 BM HRL
*hederae* Horvath, 1894
*populicourcheti* Lichtenstein, 1886
populicaulis Fitch, 1859 BM HRL
populiconduplifolius Cowen, 1895 to Thecabius Koch, 1857
populicourcheti Lichtenstein, 1886 = populi Courchet, 1879
populiglobuli Fitch, 1859 BM HRL
populi-monilis Riley, 1879 to Thecabius Koch, 1857
populinigrae (Schrank, 1801) (Aphis)
*filaginis* Boyer de Fonscolombe, 1841 (*Aphis*)
*gnaphalii* Kaltenbach, 1843
*ovatooblongus* Kessler, 1881
populiramulorum Riley, 1879 BM HRL
populitransversus Riley, 1879 BM HRL
*? canadensis* del Guercio, 1913
*? rubi* Thomas, 1879
*? solanophila* E.E. Blanchard, 1944 (*Baizongiella*)

P. populivenae Fitch, 1859 BM HRL
poschingeri Holzner, 1874 (type of Holzneria Lichtenstein, 1875) = Prociphilus bumeliae (Schrank, 1801)
protospirae Lichtenstein, (1884) 1885 BM HRL
pseudoauriculae Theobald, 1929 = Thecabius (Parathecabius) auriculae (Murray, 1877)
pyri Fitch, 1856 = ? Prociphilus americanus (Walker, 1852)
pyriformis Lichtenstein, 1885 = bursarius (Linnaeus, 1758)
quercus Hartig, 1839 nomen dubium
radicicola Essig, 1909 (type of Trifidaphis del Guercio, 1909) = Smynthurodes betae Westwood, 1849
ranunculi Davidson, 1910 = Thecabius populiconduplifolius (Cowen, 1895)
ranunculi Kaltenbach, 1843 = Thecabius affinis (Kaltenbach, 1843)
ranunculi Shinji, 1922 nec Kaltenbach, 1843 = Thecabius orientalis Mordvilko, 1935
retroflexus Courchet, 1879 = Forda marginata Koch, 1857
riccobonii Stefani, 1899 to Forda von Heyden, 1837
rileyi Stebbins, 1910
rubi Thomas, 1879 = ? populitransversus Riley, 1879
rubiradicis Theobald, 1929 BM
saccharata del Guercio, 1895 described as variety of P. fuscifrons in 1895 or boyeri in 1900 = Tetraneura ulmi (Linnaeus, 1758)
saccosus Mordvilko ex Shaposhnikov, 1955
salicicola Hille Ris Lambers, 1952 = saliciradicis (Börner, 1950)
saliciradicis (Börner, 1950) (Parathecabius) HRL
*salicicola* Hille Ris Lambers, 1952 HRL
salicis Lichtenstein, 1884 (type of Phylloxerina Börner, 1908), lapsus pro Phylloxera salicis Lichtenstein, 1884
semenovi Mordvilko, 1935
semilunarius Passerini, 1856 = Forda formicaria von Heyden, 1837
semilunoides Lichtenstein, 1880 = Forda formicaria von Heyden, 1837
similis Börner, 1949
sinensis Walker, 1852 = Schlechtendalia chinensis (Bell, 1851)
siphunculatus Hille Ris Lambers, 1973 HRL
spiriformis Lichtenstein lapsus pro pyriformis Lichtenstein, 1885
spirothecae Passerini, 1860 mis-spelling of spyrothecae Passerini, 1856 type of Kessleria Lichtenstein, 1885; and of Hamadryaphis Kirkaldy, 1904
spirothecaefacies Ghulamullah, 1941 = vesicarius Passerini, 1861

P. spyrothecae Passerini, 1856 BM HRL
*inflatae* del Guercio, 1911
*spirothecae* Passerini, 1860
*tortuosus* Rudow, 1875
spyrothecae var. infaustus Ferrari, 1872
take Shinji, 1922 to Prociphilus Koch, 1857
tartareus Hottes & Frison, 1931
tortuosus Rudow, 1875 = spyrothecae Passerini, 1856
trehernei Foster, 1975 type BM HRL
trifolii del Guercio, 1915 = Smynthurodes betae Westwood, 1849
ulmi Lichtenstein, 1879 = Kaltenbachiella pallida (Haliday, 1838)
ulmi-fusus Walsh & Riley, 1869 to Kaltenbachiella Schouteden, 1906
utricularia Passerini, 1856 to Geoica Hart, 1894
utriculoides Lichtenstein, 1880 = Geoica utricularia (Passerini, 1856)
varsoviensis Mordvilko, 1895 to Pachypappa Koch, 1856
venafuscus Patch, 1909 = Prociphilus americanus (Walker, 1852)
vesicalis Passerini errore pro vesicarius Passerini, 1861
vesicarius Passerini, 1861 BM HRL
*spirothecaefacies* Ghulamullah, 1941 HRL
vitifoliae Fitch, 1855 type of Viteus Shimer, 1867 q.v.
walshii Williams, 1911 to Kaltenbachiella Schouteden, 1906 or Tetraneura Hartig, 1841
watamushie Matsumura ex Nishiya, 1917 = Watabura nishiyae Matsumura, 1917
yanagi Shinji, 1928

PENTACERATINAPHIS Ivanovskaja-Shubina, 1965
Izd. sib. Otdel. Akad. Nauk SSR 1965: 59
Type species Pentaceratinaphis samagaltaica Ivanovskaja-Shubina, 1965
= Vesiculaphis del Guercio, 1911

P. samagaltaica Ivanovskaja-Shubina, 1965 to Vesiculaphis del Guercio, 1911

PENTALONIA Coquerel, 1859
Annls Soc. ent. Fr. 7: 259
Type species Pentalonia nigronervosa Coquerel, 1859

P. caladii van der Goot, 1917 = nigronervosa f. caladii van der Goot, 1917
gavarri Eastop, 1967 BM HRL
nigronervosa Coquerel, 1859 BM HRL
nigronervosa forma caladii van der Goot, 1917 BM HRL
*caladii* van der Goot, 1917

PENTAMYZUS Hille Ris Lambers, 1966
Hilgardia 37: 601
Type species Pentamyzus graminis Hille Ris Lambers, 1966

P. acaenae (Schouteden, 1904) (Rhopalosiphum) HRL
falklandicus Hille Ris Lambers, 1974 type BMHRL
graminis Hille Ris Lambers, 1966 BM HRL

PENTAPHIS Horvath, 1896
Wien. ent. Ztg. 15: 2
Type species Forda marginata Koch, 1857
A type citation of Tychea trivialis Passerini, 1860 led to the combination Pentaphis del Guercio, 1909
= Forda von Heyden, 1837

P. apuliae del Guercio, 1920 = Forda marginata Koch, 1857
pawlowae Mordvilko, 1901 = Forda marginata Koch, 1857
viridescens del Guercio, 1920 = Forda formicaria von Heyden, 1837

PENTATRICHOPUS Börner, 1930
Arch. klassif. phylogen. Ent. 1: 140
Type species Aphis tetrarhoda Walker, 1849
*Chaitomyzus* Takahashi, 1960
Here used as subg. of Chaetosiphon Mordvilko, 1914

P. alpinus Börner, 1950 to Chaetosiphon (Pentatrichopus)
coreanus Paik, 1965 to Chaetosiphon (Pentatrichopus)
janetscheki Borner, 1949 to Chaetosiphon Mordvilko, 1914
thomasi Hille Ris Lambers, 1953 to Chaetosiphon (Pentatrichopus)
thomasi subsp. jacobi Hille Ris Lambers, 1953 to Chaetosiphon (Pentatrichopus)

PENTATRICHOSIPHUM A.N. Basu, 1969
Orient. Insects 3: 182
Type species Pentatrichosiphum luteum A.N. Basu, 1969

P. luteum A.N. Basu, 1969 HRL

PERGANDEA Börner, 1909 nec Ashmead, 1905
Zool. Anz. 33: 610, described as subg. of Dactylosphaera Shimer, 1867
Type species Phylloxera conica Shimer, 1869
Phylloxeridae

PERGANDEIDA Schouteden, 1903
Zool. Anz. 26: 685
Type species Pergandeida ononidis Schouteden, 1903 = Aphis kaltenbachi Hille Ris Lambers, 1955
= Aphis Linnaeus, 1758

P. cahuille Dickson, 1940 = Xerophilaphis tetrapteralis (Cockerell, 1902)
corni Tissot, 1929 to Pseudasiphonaphis Robinson, 1965
corydalisicola Tao, 1962 to Longicaudinus Hille Ris Lambers, 1965
esulae Börner, 1940 to Aphis Linnaeus, 1758
kalopanacis Hori, 1927 to Aphis Linnaeus, 1758
mercurialis Balachowsky & Cairaschi, 1941 = Hayhurstia atriplicis (Linnaeus, 1761)
microrosae Shinji, 1930 = Longicaudus trirhodus (Walker, 1849)
nigra Wilson, 1911 to Aphis Linnaeus, 1758
ononidis Schouteden, 1903 = Aphis kaltenbachi Hille Ris Lambers, 1955
palustris Börner, 1940 = Aphis paludicola Hille Ris Lambers, 1955
pegani Mimeur, 1935 = Brachyunguis harmalae Das, 1918
polygonata Nevsky, 1929 to Aphis Linnaeus, 1758
polygonata var. hyperici Rusanova, 1943 nomen nudum
quilisi Fresca ex Quilis, 1931 nomen nudum
stanilandi Laing, 1923 = Aphis urticata Gmelin, 1790
tamaricifoliae Hall, 1926 = Brachyunguis tamaricis (Lichtenstein, 1885)
zweigelti Börner, 1940 to Aphis Linnaeus, 1758

Subgen. Doralida Börner, 1950

P. (Doralida) loti subsp. gollmicki Börner, 1952 = Aphis craccivora Koch, 1854

PERGANDEIDEA errore pro Pergandeida Schouteden, 1903

PERGANDEIDIA errore pro Pergandeida Schouteden, 1903

PERILLAPHIS Takahashi, 1965
Insecta matsum. 72: 101, described as subg. of Aulacorthum Mordvilko, 1914
Type species Macrosiphum perillae Shinji, 1924 (of Takahashi, 1965 ? nec Shinji, 1924)
Here used as subg. of Aulacorthum Mordvilko, 1914

PERIPHYLLUS van der Hoeven, 1863
Tijdschr. Ent. 6: 7
Type species Periphyllus testudo van der Hoeven, 1863 = Phyllophora testudinacea Fernie, 1852
*Chaetophorella* Börner, 1940
*Chaetophoria* Börner, 1940
*Chaitophorinella* van der Goot, 1913
*Chaitophorinus* Börner, 1930
*Chelymorpha* Clarke, 1858
*Phillophorus* Thornton, 1852 nec Grube, 1840

P. acericola (Walker, 1848) (Aphis) lectotype BM HRL
*horridus* Theobald, 1927 types BM
*perforata* Signoret, 1867 (*Aphis*)
acerifoliae (Takahashi, 1919) (Chaitophorinella) HRL
aceris (Linnaeus, 1761) (Aphis) BM HRL
*uhlmanni* Borner, 1950 (*Chaetophoria*)
*xanthomelas* Koch, 1854 (*Chaitophorus*)
aceris subsp. minutus Shaposhnikov, 1952 = minutus Shaposhnikov, 1952
aesculi Hille Ris Lambers, 1933 BM HRL
allogenes Szelegiewicz, 1974 nomen nudum HRL
americanus (Baker, 1917) (Chaitophorus) BM HRL
*palmerae* Knowlton, 1947
bengalensis A.K. Ghosh & RayChaudhuri, 1972 HRL
brevispinosus Gillette & Palmer, 1930 BM HRL
bulgaricus Taschev, 1964 HRL
californiensis (Shinji, 1917) (Thomasia) BM HRL
*japonicus* Baker, 1918 (*Chaitophorus*)
coleoptis (Shinji, 1932) (Chaitophorus)
coracinus (Koch, 1854) (Chaitophorus) BM HRL
*viridulus* Mamontova, 1955 HRL
formosanus Takahashi, 1921 HRL
ginnalae Paik ex Szelegiewicz, 1974 nomen nudum HRL
hirticornis (Walker, 1848) (Aphis) holotype BM HRL
*granulatus* Koch 1854 (*Chaitophorus*)
*lambersi* Börner, 1952
*templi* Hille Ris Lambers, 1935 cotypes BM HRL
horridus Theobald, 1927 = acericola (Walker, 1848)
koelreuteriae (Takahashi, 1919) (Chaitophorinella) HRL
kuwanaii (Takahashi, 1919) (Chaitophorinella) HRL
*japonicus* Essig & Kuwana, 1918 (*Chaitophorus*)

P. lambersi Börner, 1952 = hirticornis (Walker, 1848)
laricae Haliday ex Signoret, 1868 = testudinaceus (Fernie, 1852)
lichtensteini Hille Ris Lambers, 1947 = rhenanus (Börner, 1940)
lyropictus (Kessler, 1886) (Chaitophorus) BM HRL
*aceris* of Linnaeus nec Koch of Börner, 1940 (*Chaetophorella*)
*fusca* Börner, 1940 (*Chaetophorella*)
mamontovae Narzikulov, 1957 BM HRL
minutus Shaposhnikov, 1952 HRL
*aceris* subsp. *minutus* Shaposhnikov, 1952
negundinis (Thomas, 1877) (Chaitophorus) BM HRL
nevskii Mamontova, 1955 HRL
obscurus Mamontova, 1955 BM HRL
palmerae Knowlton, 1947 = americanus (Baker, 1917)
rhenanus (Börner, 1940) (Chaetophoria) BM HRL
*lichtensteini* Hille Ris Lambers, 1947 HRL
singeri (Börner, 1952) (Chaetophoria)
*helferi* Quednau, 1954 (*Chaetophorella*) HRL
steveni Mamontova, 1962 HRL
templi Hille Ris Lambers, 1935 = hirticornis (Walker, 1848)
testudinaceus (Fernie, 1852) (Phyllophora) BM HRL
*laricae* Haliday ex Signoret, 1868
*phyllophora* Clarke, 1858 (*Chelymorpha*)
*testudinatus* Thornton, 1852 (*Phillophorus*)
*testudo* van der Hoeven, 1863
? *villosus* Hartig, 1841 (*Aphis*)
testudo van der Hoeven, 1863 = testudinaceus (Fernie, 1852)
vandenboschi Hille Ris Lambers, 1966 HRL
venetianus Hille Ris Lambers, 1966 BM HRL
viridis (Matsumura, 1919) (Chaitophorus) HRL
viridulus Mamontova, 1955 = coracinus (Koch, 1854)

PERITYMBIA Westwood, 1869
Gardners' Chron. 30.i. 1869: 109
Type species Peritymbia vitisana Westwood, 1869 = Pemphigus vitifoliae Fitch, 1855
= Viteus Shimer,1867
Phylloxeridae

PHALANGOMYZUS Börner, 1939
Arb. physiol. angew. Ent. Berl. 6: 83
Type species Siphonophora oblonga Mordvilko, 1901
= Macrosiphoniella del Guercio, 1911

P. ceratus Börner, 1940 = Macrosiphoniella tapuskae (Hottes & Frison, 1931)

PHARALIS Leach, 1826
In Risso, Hist. Nat. princip. product. Eur. Meridi. 5: 217
Type species Aphis tanaceti Leach, 1826 nomen nudum
Invalid. When used applied to Metopeurum Mordvilko, 1914

P. capillata Börner, 1950 to Metopeurum Mordvilko, 1914

PHILLOPHORUS Thornton, 1852 nec Grube, 1840
Proc. ent. Soc. Lond. 2: 78
Type species Phillophorus testudinatus Thornton, 1852 = Phyllophora testudinacea Fernie, 1852
= Periphyllus van der Hoeven, 1863

P. testudinatus Thornton, 1852 (type of Chaitophorinella van der Goot, 1913) = Periphyllus testudinaceus (Fernie, 1852)

PHILLOXERA mis-spelling of Phylloxera

PHILYRIPTUS Amyot, 1847
Annls Soc. ent. Fr, (2) 5: 479
Invalid

PHLOEOMYZUS Horvath, 1896
Wien. ent. Ztg. 15: 5
Type species Schizoneura passerinii Signoret, 1875
*Loewia* Lichtenstein, 1886 nec Egger, 1856

P. dearborni Smith, 1974 = passerinii (Signoret, 1875)
dubius Borner, 1931/2 = passerinii (Signoret, 1875)
konarae Shinji, 1924 to Diphyllaphis Takahashi, 1960
passerinii (Signoret, 1875) (Schizoneura) BM HRL
*dearborni* Smith, 1974 paratype BM HRL
*dubius* Börner, 1931/2
*redelei* Hille Ris Lambers, 1931 HRL
redelei Hille Ris Lambers, 1931 = passerinii (Signoret, 1875)

PHLOEOPHTHIRIDIUM van der Hoeven, 1849
Handboek der Dierkunde, Ed. 2, Amsterdam 1: 509
Type species Chermes abietis Linnaeus, 1758
= Sacchiphantes Curtis, 1844
Adelgidae

PHORODON Passerini, 1860
Gli Afidi, Parma, p. 27
Type species Aphis humuli Schrank, 1801

P. abietifoliae Shinji, 1924 = Elatobium momii (Shinji, 1922)
asacola Matsumura, 1917 = Phorodon (Paraphorodon) cannabis Passerini, 1860
biburni Matsumura, 1918 lapsus pro viburni Matsumura, 1918
calaminthae Macchiati, 1885 to Ovatomyzus Hille Ris Lambers, 1947
cannabis Passerini, 1860 (type of Diphorodon Börner, 1939) to Phorodon (Paraphorodon)
chamaedrys Passerini, 1879 to Ovatomyzus Hille Ris Lambers, 1947
cynoglossi Williams, 1911 = Myzus (Nectarosiphon) persicae (Sulzer, 1776)
enkianthai Shinji, 1923 = Akkaia polygoni Takahashi, 1919
humuli (Schrank, 1801) (Aphis) BM HRL
*pruni* Scopoli, 1763 (*Aphis*) nec Geoffroy, 1762
? *secunda* Walker, 1849 (*Aphis*)
humuli subsp. japonensis Takahashi, 1965 BM HRL
humulifoliae Tseng & Tao, 1938
inulae Passerini, 1860 to Capitophorus van der Goot, 1913
ishimikawai Shinji, 1941 (type of Metaphorodon Takahashi, 1961) to Trichosiphonaphis (Xenomyzus)
lycopi Nevsky, 1929 to Ovatus van der Goot, 1913
mentharius van der Goot, 1913 to Ovatus van der Goot, 1913
minutum van der Goot, 1917 to Ovatus van der Goot, 1913
monardae Williams, 1911 nomen nudum
persifoliae Shinji, 1922
phloxae Sampson, 1939 to Ovatus van der Goot, 1913
polygoni van der Goot, 1917 to Trichosiphonaphis (Xenomyzus)
pruni Ferrari, 1872 = Myzus lythri (Schrank, 1801)
scrophulariae Thomas, 1979 to Myzus Passerini, 1860
viburni Matsumura, 1918
violae Theobald, 1920 = Neotoxoptera oliveri (Essig, 1935)

Subgen. Paraphorodon Tseng & Tao, 1938

P. (Paraphorodon) cannabis Passerini, 1860 BM HRL
*asacola* Matsumura, 1917 (*Phorodon*)
*cannabifoliae* Shinji, 1924 (*Capitophorus*)
*omeishanensis* Tseng & Tao, 1938 (*Paraphorodon*)

PHYLLAPHIS Koch, 1857
Die Pflanzenläuse Aphiden, Nürnberg, p. 248
Type species Aphis fagi Linnaeus, 1767

P. fagi (Linnaeus, 1767) (Aphis) BM HRL
*fagifoliae* Takahashi, 1919
*grandifoliae* Richards, 1973
fagifoliae Takahashi, 1919 = fagi (Linnaeus, 1767)
grandifoliae Richards, 1973 = fagi (Linnaeus, 1767)
konarae (Shinji, 1924) (Phloeomyzus) type of Diphyllaphis Takahashi, 1960 q.v.
machili Takahashi, 1928 type of Machilaphis Takahashi, 1960 q.v.
nigra Ashmead, 1881
quercicola Baker, 1916 (type of Stegophylla Oestlund, 1922) = Stegophylla querci (Fitch, 1859)
quercifoliae Gillette, 1914 to Stegophylla Oestlund, 1922

PHYLLAPHOIDES Takahshi, 1921
Spec. Rep. Formosa agric. Exp. Stn 20: 75
Type species Phyllaphoides bambucicola Takahashi, 1921

P. bambucicola Takahashi, 1921 rectified to bambusicola
bambusicola Takahashi, 1921 (rectification of bambucicola) HRL

PHYLLOPHORA Fernie, 1852 variant of Phillophorus Thornton, 1852

P. testudinacea Fernie, 1852 = Periphyllus testudinaceus (Fernie, 1852)

PHYLLOXERA Boyer de Fonscolombe, 1834
Annls Soc. ent. Fr. 3: 223
Type species Phylloxera quercus Boyer de Fonscolombe, 1834
*Acanthaphis* del Guercio, 1912 nec 1908
*? Dactylosphaera* Shimer, 1867 rectification of Daktulosphaira?
*? Daktulosphaira* Shimer, 1866 lapsus pro Dactylosphaera ?
*Euphylloxera* del Guercio, 1908
*Hystrichiella* Börner, 1909
*Micranthaphis* Grassi, 1912
*Parapergandea* Börner, 1930

*Paraphylloxera* Grassi, 1909
*Parthenophylloxera* Grassi & Foa, 1911
? *Pergandea* Börner, 1909 nec Ashmead, 1905
*Phylloxerella* Grassi, 1909
*Phylloxeroides* Grassi, 1909
*Psylloptera* Ferrari, 1872
*Rhanis* von Heyden, 1837 in litteris nec Dejean, 1835
*Vacuna* von Heyden, 1837
Phylloxeridae

P. crassicornis Dahl, 1912 = ? Hormaphis betulae Mordvilko, 1901
longirostris Boyer de Fonscolombe, 1841 = Stomaphis quercus (Linnaeus, 1758)

PHYLLOXERELLA Grassi, 1909
Atti Accad. naz. Lincei Rc. 18 (2): 420, described as subg. of Phylloxera Boyer de Fonscolombe, 1834
Type species Phylloxerella confusa Grassi, 1909
= Phylloxera Boyer de Fonscolombe, 1834
Phylloxeridae

PHYLLOXERINA Börner, 1908
Arb. biol. Bund. Anst. Land-u. Forstw. 6: 94
Type species Pemphigus salicis Lichtenstein, 1884, lapsus pro Phylloxera salicis Lichtenstein, 1884
*Guercioja* Mordvilko, 1909
*Laufferella* Lindinger, 1933
*Pseudochermes* Bonfigli, 1909 nec Nitsche, 1895
Phylloxeridae

PHYLLOXEROIDES Grassi, 1909
Atti Accad. naz. Lincei Rc. 18 (2): 421, described as subg. of Phylloxera Boyer de Fonscolombe, 1834
Type species Phylloxeroides italicum Grassi, 1909
= Phylloxera Boyer de Fonscolombe, 1834
Phylloxeridae

PHYMATOSIPHUM, Davis, 1909
Ann. ent. Soc. Am. 2: 196
Type species Phymatosiphum monelli Davis, 1909
= Drepanaphis del Guercio, 1909

P. monelli Davis, 1909 to Drepanaphis del Guercio, 1909

PICTURAPHIS E.E. Blanchard, 1922
Physis., B. Aires 6: 44
Type species Picturaphis vignaphilus E.E. Blanchard, 1922
Here used as subg. of Microparsus Patch, 1919

P. puertoricensis Smith & Knowlton, 1970 to Microparsus (Picturaphis)
vignaphilus E.E. Blanchard, 1922 to Microparsus (Picturaphis)

PILOBTUSAPHIS Rusanova, 1943
Trudȳ Azerb. Gos. Univ. biol. 3(1): 42
Type species Pilobtusaphis dsengei Rusanova, 1943 nomen nudum
Without described species

P. dsengei Rusanova, 1943 nomen nudum

PINEODES Börner, 1926
In Abderhalden E., Handbuch der biologischen Arbeitsmethoden, Berlin & Wien, Abt. IX, Teil 1, 2 Hälfte, Heft 2: 240
Type species Chermes pinifoliae Fitch, 1858
Adelgidae

PINETIFEX Amyot, 1847
Annls Soc. ent. Fr. (2) 5: 481
Invalid

PINEUS Shimer, 1869
Trans. Am. ent. Soc. 2: 383
Type species Coccus pinicorticis Fitch, 1855 = Coccus strobus Hartig, 1837
*Chermaphis* Maskell, 1884
*Eopineus* Steffan, 1968
*Kermaphis* Maskell, 1885
Adelgidae

PIRAPHIS Börner, 1932 errore pro Pyraphis Börner, 1931
In Sorauer, P., Handbuch der Pflanzenkrankheiten, Berlin, 5 (2): 597
= Melanaphis van der Goot, 1917

PITYAPHIS Amyot, 1847
Annls. Soc. ent. Fr. (2) 5: 481
Invalid

PITYARIA Börner, 1949
Beitr. tax. Zool. 1: 59, described as subg. of Cinaropsis Börner, 1939
Type species Lachnus pruinosus Hartig 1841
= Cinara Curtis, 1835

PLACOAPHIS Richards, 1961
Can. Ent. 93: 624
Type species Placoaphis siphunculata Richards, 1961

P. siphunculata Richards, 1961

PLATANAPHIS Amyot, 1847
Annls. Soc. ent. Fr. (2) 5: 475
Invalid

PLATYAPHIS Takahashi, 1957
Proc. R. ent. Soc. Lond. (B) 26: 109
Type species Platyaphis fagi Takahashi, 1957

P. fagi Takahashi, 1957 BM HRL

PLEOTRICHOPHORUS Börner, 1930
Arch. klassif. phylogen. Ent. 1: 138, described as subg. of Capitophorus van der Goot, 1913
Type species Aphis glandulosa Kaltenbach, 1846

P. acanthovillus (Knowlton & Smith, 1936) (Capitophorus) HRL
achilleae Holman, 1965 paratype BM HRL
afghanensis Narzikulov & Umarov, 1972
ambrosiae Hille Ris Lambers, 1969 HRL
amsinckii Richards, 1968 paratype BM HRL
antennarius Corpuz-Raros & Cook, 1974
artemisicola (Williams, 1911) (Siphonophora)
asterifoliae (Strom, 1934) (Macrosiphum) BM HRL
brevinectarius (Gillette & Palmer, 1933) (Capitophorus) paratypes BM HRL
chrysanthemi (Theobald, 1920) (Capitophorus) type BM HRL
? *formosanus* Takahashi, 1929 (*Capitophorus*)
decampus (Knowlton & Smith, 1936) (Capitophorus)
deviatus F.P. Müller, 1972 HRL
diutius Corpus-Raros & Cook, 1974
duponti Hille Ris Lambers, 1935 BM HRL

P. elongatus (Knowlton, 1929) (Capitophorus) cotypes BM HRL
    *chlorophainus* Knowlton & Smith, 1936 (*Capitophorus*)
    *feragaeus* Knowlton & Smith, 1936 (*Capitophorus*)
  filaginis (Schouteden, 1906) (Myzus) BM HRL
  filifoliae (Palmer, 1938) (Capitophorus) HRL
  glandulosus (Kaltenbach, 1846) (Aphis) BM HRL
    *pilosus* van der Goot, 1912 (*Myzus*)
  gnaphalodes (Palmer, 1938) (Capitophorus) BM HRL
  gregarius (Knowlton, 1929) (Capitophorus) BM HRL
  helichrysi Bozhko, 1963 HRL
  heterohirsutus (Gillette & Palmer, 1933) (Capitophorus) paratypes BM HRL
    *bitrichus* Knowlton & Smith, 1936 (*Capitophorus*) HRL
  hottesi Hille Ris Lambers, 1969 HRL
  infrequens (Knowlton & Smith, 1936) (Capitophorus) BM HRL
  intermedius Corpuz-Raros & Cook, 1974
  knowltoni Corpuz-Raros & Cook, 1974
  lagacei Hille Ris Lambers, 1969 HRL
  longinectarius (Gillette & Palmer, 1933) (Capitophorus)
  longipes (Gillette & Palmer, 1928) (Macrosiphum) BM HRL
  longirostris Hille Ris Lambers, 1969 HRL
  magnautensus (Knowlton & Smith, 1936) (Capitophorus) BM HRL
  narzikulovi Umarov, 1964
  neosporadicus Corpuz-Raros & Cook, 1974
  obscuratus Hille Ris Lambers, 1966 BM HRL
  oestlundi (Knowlton, 1927) (Capitophorus) BM HRL
  ohioensis (Smith, 1940) (Capitophorus)
  packi (Knowlton, 1928) (Macrosiphum) paratype BM HRL
  packi subsp. brevis Corpuz-Raros & Cook, 1974
  palmerae (Knowlton, 1935) (Capitophorus)
  parilis Corpuz-Raros & Cook, 1974
  patonkus (Hottes & Frison, 1931) (Capitophorus) paratypes BM HRL
    *patonkusellus* Corpuz-Raros & Cook, 1974
    *pseudopatonkus* Corpuz-Raros & Cook, 1974
  patonkusellus Corpuz-Raros & Cook, 1974 = patonkus (Hottes & Frison, 1931)
  persimilis Börner, 1950 BM HRL
  pseudoglandulosus (Palmer, 1952) (Capitophorus) BM HRL
  pseudopatonkus Corpuz-Raros & Cook, 1974 = patonkus (Hottes & Frison, 1931)
  pullus (Gillette & Palmer, 1933) (Capitophorus) paratypes BM HRL
  pycnorhysus (Knowlton & Smith, 1936) (Capitophorus) paratypes BM HRL

P. quadritrichus (Knowlton & Smith, 1936) (Capitophorus) paratype BM HRL
quadritrichus subsp. pallidus Corpuz-Raros & Cook, 1974
quadritrichus subsp. vulgaris Corpuz-Raros & Cook, 1974
remaudierei Leclant, 1968 BM HRL
rusticatus (Knowlton & Smith, 1937) (Capitophorus) HRL
spatulavillus (Knowlton & Smith, 1936) (Capitophorus) BM HRL
sporadicus (Knowlton, 1935) (Macrosiphum) BM HRL
stroudi (Knowlton, 1948) (Capitophorus)
tetradymiae Smith & Knowlton, 1972 paratypes BM HRL
triangulatus Corpuz-Raros & Cook, 1974
utensis (Pack & Knowlton, 1929) (Capitophorus) BM HRL
villosae Robinson, 1974 = ? wasatchii Knowlton, 1927
wasatchii (Knowlton, 1927) (Capitophorus) BM HRL
*? villosae* Robinson, 1974 BM HRL
xerozoous (Knowlton & Smith, 1934) (Capitophorus) HRL
zoomontanus (Knowlton & Smith, 1936) (Capitophorus) BM HRL

PLOCAMAPHIS Oestlund, 1922
Rep. Minn. St. Ent. 19: 122
Type species Melanoxanthus flocculosus Weed, 1899

P. amerinae (Hartig, 1841) (Aphis) BM HRL
*bituberculata* Theobald, 1912 (*Pterocomma*) types BM
amerinae subsp. borealis Ossiannilsson, 1959
bradleyi Richards, 1966 to Fullawaya Essig, 1912
braggii Gillette & Palmer, 1929 to Fullawaya Essig, 1912
bulbosa Richards, 1966 to Fullawaya Essig, 1912
coreana (Okamoto & Takahashi, 1927) (Melanoxanterium) HRL
flocculosa (Weed, 1891) (Melanoxanthus) BM HRL
flocculosa subsp. brachysiphon Ossiannilsson, 1959 BM HRL
flocculosa subsp. goernitzi Börner, 1940 BM HRL
*görnitzi* Börner, 1940
flocculosa subsp. macrosiphon Ossiannilsson, 1959 HRL
görnitzi Börner, 1940 = flocculosa subsp. goernitzi Börner, 1940
martini Richards, 1963 paratypes BM HRL
ontarioensis Richards, 1966 to Fullawaya Essig, 1912
salijaponica (Shinji, 1924) (Melanoxanthus)
terricola Hottes & Frison, 1931 to Fullawaya Essig, 1912

POLYOCELLARIA Sharp, 1901 lapsus pro Pollyocellaria Imhof, 1900 nomen nudum, probably Coccoidea

POLYTRICHAPHIS Miyazaki, 1971
Insecta matsum. 34: 166
Type species Polytrichaphis fragilis Miyazaki, 1971

P. fragilis Miyazaki, 1971 HRL

POMAPHIS Börner, 1939
Arb. physiol. angew. Ent. Berl. 6: 78
Type species Aphis pyri Boyer de Fonscolombe, 1841
*Dentatus* van der Goot, 1913 nec Gray, 1847
*Myzopsis* Pašek, 1955 nec Matsumura, 1918
Here used as subg. of Dysaphis Börner, 1931

PROCALAPHIS Quednau, 1954
Mitt. biol. Reichsanst. Ld- u. Forstw. 78: 23
Type species Aphis tuberculata von Heyden, 1837
= Callipterinella van der Goot, 1913

PROCIPHILUS Koch, 1857
Die Pflanzenläuse Aphiden, Nürnberg, p. 279
Type species Aphis bumeliae Schrank, 1801
*Anocaudus* A.K. Ghosh, Chakrabarti, Chowdhuri & RayChaudhuri, 1969
*Holzneria* Lichtenstein, 1875
*Meliarhizophagus* Smith, 1974
*Nishiyana* Matsumura, 1917
*Pulvius* Sanborn, 1906

P. alnifoliae (Williams, 1911) (Pemphigus) BM HRL
alnifoliae subsp. arbutifoliae Smith, 1969 = arbutifoliae Smith, 1969
alnifoliae subsp. fitchii Baker & Davidson, 1917 BM HRL
*fitchii* Baker & Davidson, 1917
*pyri* Fitch, 1851 (*Eriosoma*) nec Westwood, 1849
americanus (Walker, 1852) (Pemphigus) BM HRL
? *pyri* Fitch, 1856 (*Pemphigus*)
*venafuscus* Patch, 1909 (*Pemphigus*)
anemone Shinji, 1922 to Thecabius Koch, 1857
aomoriensis (Matsumura, 1917) (Nishiyana)
approximatus Patch, 1917 = probosceus (Sanborn, 1906)
arbutifoliae Smith, 1969 paratypes BM HRL
*alnifoliae* subsp. *arbutifoliae* Smith, 1969 paratypes BM HRL

P. bumeliae (Schrank, 1801) (Aphis) BM HRL
*erraticus* Koch, 1857
*ligustrinellum* Koch, 1856 (*Asiphum*)
*poschingeri* Holzner, 1874 (*Pemphigus*)
bumeliaeformis Mordvilko ex Shaposhnikov, 1955
carolinensis Smith, 1969 paratype BM HRL
cheni Tao, 1970
clerodendri Okamoto & Takahashi, 1927
crataegi Tullgren, 1909 = Prociphilus (Stagona) pini (Burmeister, 1835)
crataegicola Shinji, 1922 HRL
*? sasakii* Monzen, 1927
erigeronensis (Thomas, 1879) (Tychea) BM HRL
*flavula* Rohwer, 1908 (*Forda*)
*radicola* Oestlund, 1886 (*Tychea*)
erraticus Koch, 1857 = bumeliae (Schrank, 1801)
fitchii Baker & Davidson, 1917 = alnifoliae subsp. fitchii Baker & Davidson, 1917
formosanus Takahashi, 1935
fraxini (Fabricius, 1777) (Aphis) BM HRL
*fraxini* Hartig, 1841 (*Pemphigus*)
*nidificus* Low, 1882
fraxinidipetalae (Essig, 1911) (Pemphigus)
fraxinifolii (Riley, 1879) (Pemphigus) BM HRL
*fraxinifolii* Thomas, 1879 (*Pemphigus*)
ghanii Hille Ris Lambers, 1973 to Prociphilus (Neoparacletus)
konoi Hori, 1938 to Prociphilus (Stagona)
kuwanai Monzen, 1927 BM HRL
*? orientalis* Mordvilko, 1929 teste Shaposhnikov, 1955, p. 836
*takahashii* Maxson & Knowlton, 1937 paratypes BM HRL
laricis Shinji, 1941 HRL ?
ligustrifoliae (Tseng & Tao, 1938) (Thecabius) HRL ?
longianus Smith, 1974 HRL
lonicerae Shinji, 1943
micheliae Hille Ris Lambers, 1933 BM HRL
nidificus Löw, 1882 = fraxini (Fabricius, 1777)
oleae Koroneos, 1939 = oleae (Leach ex Risso, 1826)
oleae (Leach ex Risso, 1826) (Eriosoma) HRL
*oleae* Koroneos, 1939 HRL
oriens Mordvilko, 1935 BM HRL
orientalis Mordvilko, 1929 = kuwanai teste Shaposhnikov, 1955, p. 836
osmanthae Essig & Kuwana, 1918 BM HRL

P. pergandei Smith, 1974 HRL
pini Tao, 1970 (homonym of pini Burmeister, 1835 in subg. Stagona)
piniradicivorus Smith, 1969 BM HRL
populi Tao, 1970
pourthiacae Monzen, 1929 = ushikoroshi Shinji, 1924
probosceus (Sanborn, 1906) (Pulvius) HRL
*approximatus* Patch, 1917
sasakii Monzen, 1927 = ? crataegicola Shinji, 1922
takahashii Maxson & Knowlton, 1937 = kuwanai Monzen, 1927
take (Shinji, 1922) (Pemphigus)
taxus (A.K. Ghosh, Chakrabarti, Chowdhuri & RayChaudhuri, 1969) (Anocaudus) paratype BM HRL
umarovi Narzikulov, 1964
ushikoroshi Shinji, 1924 HRL
*pourthiacae* Monzen, 1929
xylostei subsp. orientalis Mordvilko, 1936 preoccupied by orientalis Mordvilko, 1929

Subgen. Neoparacletus Strom, 1942

P. (Neoparacletus) corrugatans (Sirrine, 1893) (Pemphigus) BM HRL
*caricis* Strom, 1942 (*Neoparacletus*)
(Neoparacletus) ghanii Hille Ris Lambers, 1973 HRL

Subgen. Paraprociphilus Mordvilko, 1923

P. (Paraprociphilus) baicalensis (Cholodkovsky, 1921) (Pemphigus) BM HRL
(Paraprociphilus) tesselatus (Fitch, 1851) (Eriosoma) BM HRL
*acerifolii* Riley, 1879 (*Pemphigus*)
*? alni* Kalm, 1770 (*Chermes*) nec Linnaeus, 1758
*alni* Provancher, 1890 (*Pemphigus*)
*stamineus* Haldeman, 1859 (*Aphis (Pemphigus)*)

Subgen. Stagona Koch, 1856

P. (Stagona) crataegistrobi (Smith, 1969) (Stagona) BM HRL
(Stagona) himalayaensis Chakrabarti, 1976 HRL
(Stagona) konoi Hori, 1938 (Prociphilus)
(Stagona) longistigma (Monell, 1878) (Pemphigus)
(Stagona) picearubensis (Smith, 1969) (Stagona) paratype BM HRL
(Stagona) pini (Burmeister, 1835) (Rhizobius) BM HRL
*crataegi* Tullgren, 1909 (*Prociphilus*)
*crataegi* Tullgren, 1919 (*Aphis*, lapsus)
*subterraneus* Kaltenbach, 1843 (*Rhizobius*)

P. (Stagona) vesicalis (Rudow, 1875) (Stagona)
(Stagona) xylostei (de Geer, 1773) (Aphis) BM HRL
*degeeri* Kaltenbach, 1843 (*Pemphigus*)
*lonicerae* Hartig, 1841 (*Pemphigus*)
*viridis* Theobald, 1914 (*Rhizobius*)

PROMICRELLA Börner, 1949
Beitr. tax. Zool. 1: 54
Type species Promicrella ramicola Börner, 1949
= Chaitophorus Koch, 1854

P. ramicola Börner, 1949 to Chaitophorus Koch, 1854
ramicola subsp. capreae F.P. Müller ex Börner, 1952 = Chaitophorus ramicola (Börner, 1949)

PROTAPHIS Börner, 1952
Mitt. thüring. bot. Ges. 4 (3): 93
Type species Brachyunguis anthemidis Börner, 1940
*Alhambra* Gomez-Menor, 1958
*Dasia* Gomez-Menor, 1951 nec van der Goot, 1918
Here used as subg. of Aphis Linnaeus, 1758

P. albus Remaudière ex Davatchi, 1959 to Aphis (Absinthaphis)
anthemiae (Rusanova, 1943 nomen nudum) (Xerophilaphis) Ivanovskaja, 1960 to Aphis (Protaphis)
funicularis F.P. Müller, 1968 to Aphis (Protaphis)
nevskyi Ivanovskaja, 1960 to Aphis (Protaphis)
striata Hille Ris Lambers, 1962 to Aphis (Protaphis)
terraealbae Ivanovskaja, 1959 (1958) to Aphis (Protaphis)

PROTOLACHNUS Theobald, 1915
Bull. ent. Res. 6: 145
Type species Protolachnus tuberculostemmata Theobald, 1915
= Eulachnus del Guercio, 1909

P. cretacea Mamontova, 1968 to Eulachnus del Guercio, 1909
martellii* Börner, 1952 = Cinara pectinatae (Nördlinger, 1880)
nigricola Pašek, 1953 to Eulachnus del Guercio, 1909
tuberculostemmata Theobald, 1915 to Eulachnus del Guercio, 1909

*Börner proposed the name Protolachnus martellii for specifically the adults of what del Guercio (1909) described as Eulachnus agilis (Kalt.). The alatae described by del Guercio are Cinara pectinatae Nördlinger!

PROTOPTEROCALLIS Richards, 1965
Mem. ent. Soc. Can. 44: 69
Type species Protopterocallis canadensis Richards, 1965
= Monellia Oestlund, 1887

P. canadensis Richards, 1965 to Monellia Oestlund, 1887

PROTRAMA Baker, 1920
Bull. U.S. Dep. Agric. 826: 19
Type species Trama radicis Kaltenbach, 1843

P. baronii Hille Ris Lambers, 1969 BM HRL
flavescens (Koch, 1856) (Trama) BM HRL
*donisthorpei* Theobald, 1913 (*Trama*) BM
longitarsus (Ferrari, 1872) (Lachnus) HRL
longitarsus subsp. sclerodensis Kumar, 1973
luppovae Narzikulov, 1962
narzykulovi Kan, 1962 to Neotrama Baker, 1920
orientalis Narzikulov, 1962
pamirica Narzikulov, 1962 to Neotrama Baker, 1920
penecaeca Stroyan, 1964 type BM HRL
radicis (Kaltenbach, 1843) (Trama) BM HRL
radicis subsp. asiatica (Narzikulov, 1973) (Trama)
ranunculi (del Guercio, 1909) (Trama) BM HRL
taraxaci Kan, 1962

PRUNAPHIS Shaposhnikov, 1964
In Bei-Bienko, G.Y., Keys to the Insects of the European part of the USSR, 1: 586, described as subg. of Brachycaudus van der Goot, 1913
Type species Brachycaudus cardui Linnaeus, 1758 = Aphis cardui Linnaeus, 1758
= Acaudus van der Goot, 1913

PRUNOMYZUS Hille Ris Lambers & Rogerson, 1946
Proc. R. ent. Soc. Lond. (B) 15: 105, described as subg. of Myzus Passerini, 1860
Type species Myzus (Prunomyzus) padellus Hille Ris Lambers & Rogerson, 1946
= Myzus Passerini, 1860

P. asiaticus Szelegiewicz, 1969 to Myzus Passerini, 1860

PSEUDACAUDELLA Börner, 1944
In Brohmer, P., Fauna von Deutschland, ed. 5, Leipzig, p. 216
Type species Acaudella rubida Börner, 1939
*Schizomyzus* Börner, 1950

P. rubida (Börner, 1939) (Acaudella) BM HRL
*lindneri* Börner, 1950 (*Schizomyzus*)

PSEUDAMBRIA Richards, 1966
Can. Ent. 98: 758
Type species Pseudambria longirostris Richards, 1966
Fossil, Cretaceous Canadian amber

PSEUDAMPHOROPHORA Heie, 1967
Spolia zool. Mus. haun. 26: 175
Type species Pseudamphorophora succini Heie, 1967
Fossil, baltic amber

PSEUDAPHIS Hille Ris Lambers, 1956
Boll. Lab. Zool. gen. agr. Portici 33: 172
Type species Macrosiphum (Sitobion) sijui Eastop, 1953

P. abyssinica Hille Ris Lambers, 1956 BM HRL
sijui (Eastop, 1953) (Macrosiphum (Sitobion)) types BM HRL

PSEUDASIPHONAPHIS Robinson, 1965
Can. Ent. 97: 1009
Type species Asiphonaphis anogis Hottes & Frison, 1931 = Pergandeida corni Tissot, 1929

P. corni (Tissot, 1929) (Pergandeida) paratypes BM HRL
*anogis* Hottes & Frison, 1931 (*Asiphonaphis*)
*carolinensis* Knowlton, 1937 (*Asiphonaphis*)

PSEUDESSIGELLA Hille Ris Lambers, 1966
Tijdschr. Ent. 109: 219
Type species Pseudessigella brachychaeta Hille Ris Lambers, 1966

P. brachychaeta Hille Ris Lambers, 1966 BM HRL

PSEUDOACYRTHOSIPHON A.K. Ghosh & RayChaudhuri, 1969
Orient. Insects 3: 94, described as subg. of Neoacyrthosiphon Tao, 1963
Type species Macrosiphum holsti Takahashi, 1935
Here used as subg. of Neoacyrthosiphon Tao, 1963

PSEUDOBREVICORYNE Heinze, 1960
Beitr. Ent. 10: 763
Type species Brevicoryne buhri Börner, 1952

P. buhri (Börner, 1952) (Brevicoryne) BM HRL
erysimi Holman, 1963 paratypes BM HRL

PSEUDOCERCIDIS Richards, 1961
Can. Ent. 93: 622
Type species Pseudocercidis rosae Richards, 1961

P. rosae Richards, 1961 paratypes BM HRL

PSEUDOCEROSIPHA Shinji, 1932
Isayoshi Ishii (ed.) Practical Horticulture, Tokyo, special number 5: 238; and again as new genus, 1941, Monograph of Japanese Aphids, Tokyo, p. 188
Type species Pseudocerosipha pruni Shinji, 1932 = Toxoptera rufiabdominalis Sasaki, 1899
= Rhopalosiphum Koch, 1854

P. pruni Shinji, 1932 = Rhopalosiphum rufiabdominalis (Sasaki, 1899)

PSEUDOCHERMES Bonfigli, 1909 nec Nitsche, 1895
Atti Accad. naz. Lincei Rc. 18 (2): 397
Type species Chermes populi Bonfigli, 1908
= Phylloxerina Börner, 1908
Phylloxeridae

PSEUDOCINARA Pašek ex Pintera, 1966
In Pintera, Acta ent. bohemoslov. 63: 282, described as subgen. of Cinara Curtis, 1835
Type species Cinara neubergi Arnhart, 1930 = Lachnus neubergi Arnhart, 1930
= Cinara Curtis, 1835

PSEUDOEPAMEIBAPHIS Gillette & Palmer, 1932
Ann. ent. Soc. Am. 25: 145
Type species Pseudoepameibaphis glauca Gillette & Palmer, 1932

P. essigi Knowlton & Smith, 1938 BM HRL
glauca Gillette & Palmer, 1932 BM HRL
tridentatae (Wilson, 1915) (Aphis) BM HRL
xenotrichis Knowlton & Smith, 1938 BM HRL
*zavillis* Knowlton & Smith, 1938
zavillis Knowlton & Smith, 1938 = xenotrichis Knowlton & Smith, 1938

PSEUDOLACHNUS Shinji, 1922
Zool. Mag., Tokyo 34: 730
Type species Pseudolachnus yomogi Shinji, 1922 = Cryptosiphum artemisiae Buckton, 1879
= Cryptosiphum Buckton, (1875) 1879

P. yomogi Shinji, 1922 = Cryptosiphum artemisiae Buckton, 1879

PSEUDOMEGOURA Shinji, 1929
Lansania 1: 112
Type species Macrosiphum nishikigi Shinji, 1928 = Rhopalosiphum magnoliae Essig & Kuwana, 1918
= Aulacorthum Mordvilko, 1914

PSEUDOMICRELLA Börner, 1949
Beitr. tax. Zool. 1: 55
Type species Aphis vitellinae Schrank, 1801
= Chaitophorus Koch, 1854

P. jacobi Börner, 1950 = Chaitophorus salijaponicus subsp. niger Mordvilko, 1929
reticulata Börner, 1950 to Chaitophorus Koch, 1854

PSEUDONIPPONAPHIS A.K. Ghosh & RayChaudhuri, 1973
Kontyû 41: 483, described as subg. of Nipponaphis Pergande, 1906
Type species Nipponaphis machiliphaga Takahashi, 1959
Here used as subg. of Nipponaphis Pergande, 1906

PSEUDOPTEROCOMMA MacGillivray, 1963
Can. Ent. 95: 941, described as subg. of Fullawaya Essig, 1912
Type species Fullawaya (Pseudopterocomma) hughi MacGillivray, 1963

P. canadensis Richards, 1966 BM HRL
hughi (MacGillivray, 1963) (Fullawaya (Pseudopterocomma)) HRL

PSEUDOREGMA Doncaster, 1966
Entomologist 99: 159
Type species Oregma bambusicola Takahashi, 1921

P. alexanderi (Takahashi, 1924) (Oregma) BM HRL
bambusicola (Takahashi, 1921) (Oregma) BM HRL
*? cantonensis* Takahashi, 1936 (*Oregma*)
bambusicola var. carolinensis (Takahashi, 1941) (Oregma)
*bambusae* var. *carolinensis* Takahashi, 1941 (*Oregma*)
dendrocalami (Takahashi, 1935) (Oregma)
gombakana (Takahashi, 1950) (Oregma) types BM
graminum (van der Goot, 1917) (Oregma)
koshuensis (Takahashi, 1924) (Oregma) BM HRL
*heitoensis* Takahashi, 1927 (*Oregma*)
*pseudomontana* Takahashi, 1924 (*Oregma*) HRL
*sumatrensis* Takahashi, 1926 (*Oregma*)
montana (van der Goot, 1917) (Oregma)
nicolaiae (Takahashi, 1935) (Oregma) BM HRL
panicola (Takahashi, 1921) (Oregma) BM HRL
pendleburyi (Takahashi, 1950) (Oregma) types BM HRL
sundanica (van der Goot, 1917) (Oregma) BM HRL

PSEUDORHOPALOSIPHONINUS Heinze, 1961
Beitr. Ent. 11: 63, 90
Type species Rhopalosiphum calthae Koch, 1854
Here used as subg. of Rhopalosiphoninus Baker, 1920
*Rhopalomyzus* Börner, 1950 nec Mordvilko, 1921

PSEUDOSIPHUM Shinji, lapsus pro Pseudolachnus

PSODORAPHIS Amyot, 1847
Annls Soc. ent. Fr. (2) 5: 478
Invalid

PSYLLAPHIS lapsus pro Phyllaphis

PSYLLOPTERA Ferrari, 1872
Annali Mus. civ. Stor. nat. Giacomo Doria 2: 65
Type species Psylloptera quercina Ferrari, 1872
= Phylloxera Boyer de Fonscolombe, 1834
Phylloxeridae

PTERASTHENIA Stroyan, 1952
Proc. R. ent. Soc. Lond. (B) 21: 149
Type species Pterasthenia shiraensis Stroyan, 1952

P. shiraensis Stroyan, 1952 type BM HRL

PTERIAPHIS Gaumont, 1923
Annls Epiphyt. 9: 342
Without species

PTEROCALLIDIUM Börner, 1949
Beitr. tax. Zool. 1: 49
Type species Chaitophorus maculatus Buckton, 1899 = Callipterus trifolii Monell, 1882
= Therioaphis Walker, 1870

P. lydiae Börner, 1949 = Therioaphis trifolii (Monell, 1882)
propinquum Börner, 1949 = Therioaphis trifolii (Monell, 1882)

PTEROCALLIS Passerini, 1860
Gli Afidi, Parma, p. 28
Type species Aphis alni Fabricius, 1781 = Aphis alni de Geer, 1779
*Subcallipterus* Mordvilko, 1894

P. albida Börner, 1940 BM HRL
*tuberculatus* (de Geer) Mordvilko, 1935 (*Myzocallis*) lapsus
alni (de Geer, 1773) (Aphis) BM HRL
*alni* Fabricius, 1781 (*Aphis*)
*betulae alni* Rossi, 1790 (*Aphis*)
? alnicola (Shinji, 1924) (Myzocallis)
alnifoliae (Fitch, 1851) (Lachnus) BM HRL
maculata (von Heyden, 1837) (Aphis) BM HRL
minimum van der Goot, 1912 = Betulaphis quadrituberculata (Kaltenbach, 1843)
montana (Higuchi, 1972) (Myzocallis)
ostryae Börner, 1949 to Neochromaphis Takahashi, 1921
pictus Ferrari, 1872 to Hoplocallis Pintera, 1952
rhombifoliae (Granovsky, 1928) (Myzocallis) BM HRL

Subgen. Mesocallis Matsumura, 1919

P. (Mesocallis) alnicola (A.K. Ghosh, 1974) (Mesocallis) (homonym of ? Pterocallis alnicola (Shinji, 1924))
(Mesocallis) fagicola (Matsumura, 1919) (Mesocallis)
(Mesocallis) obtusirostris (A.K. Ghosh, 1974) (Mesocallis) HRL
(Mesocallis) pteleae (Matsumura, 1919) (Mesocallis) HRL
*colyricola* Shinji, 1935 (*Agrioaphis*)
*hashibamii* Shinji, 1935 (*Agrioaphis)*
(Mesocallis) sawashibae (Matsumura, 1917) (Myzocallis)
*carpinicola* Matsumura, 1919 (*Neocallis*)
*moriokaensis* Takahashi, 1961 (*Nippochaitophorus*)

Subgen. Paratinocallis Higuchi, 1972

P. (Paratinocallis) corylicola (Higuchi, 1972) (Paratinocallis) HRL

Subgen. Recticallis Matsumura, 1919

P. (Recticallis) alnijaponicae (Matsumura, 1919) (Recticallis) BM HRL
*? alnicola* Shinji, 1924 (*Myzocallis*)
*alnifoliae* Shinji, 1941 (*Tuberculoides*)
*moriokae* Shinji, 1935 (*Agrioaphis*)
(Recticallis) nigrostriata (Shinji, 1941) (Tuberculoides) BM HRL
(Recticallis) pseudoalni (Takahashi, 1921) (Myzocallis) HRL

PTEROCHLOROIDES Mordvilko, 1914
Faune Russie, Ins. Hém. 1: 23, footnote
Type speceis Lachnus persicae Cholodkovsky, 1899
*Tuberodryobius* Das, 1918

P. persicae (Cholodkovsky, 1899) (Lachnus) BM HRL
*amygdali* van der Goot, 1912 (*Dryobius*)
*salicicola* Franssen, 1932 (*Pterochlorus*)

PTEROCHLORUS Passerini, 1860, emendation pro Pteroclorus Rondani, 1848
Gli Afidi, Parma, p. 33
Type species Aphis longipes Dufour, 1833, the only species originally included by Passerini, is a synonym of Aphis roboris Linnaeus, 1758, which had previously been selected as the type species of Pteroclorus Rondani, 1848.
= Lachnus Burmeister, 1835

P. bogoriensis Franssen, 1932 = Lachnus tropicalis (van der Goot, 1916)
japonicus Matsumura, 1917 to Lachnus Burmeister, 1835
ogasawarae Matsumura, 1917 = Maculolachnus submacula (Walker, 1848)
roboris var. brevirostris Mordvilko, 1909
roboris var. fagarum Rusanova, 1943 nomen nudum
roboris var. longirostris Mordvilko, 1909 = Lachnus iliciphilus (del Guercio, 1909)
salicicola Franssen, 1932 = Pterochloroides persicae (Cholodkovsky, 1899)
tropicalis van der Goot, 1916 to Lachnus Burmeister, 1835

PTEROCLORUS Rondani, 1848
'Familia Hemipterorum Aphidinae' in Nuovi Annali delle Scienze Naturali, p. 35
Type species Aphis roboris Linnaeus, 1758
= Lachnus Burmeister, 1835

PTEROCOMMA Buckton, 1879
Monograph of the British Aphides, London, 2: 142
Type species Pterocomma pilosa Buckton, 1879
Pterocomma is not preoccupied by Pterocoma Dejean, 1834; Agassiz, 1836, spelled with only one 'm'
*Aphioides* Passerini, 1860 nec Rondani, 1848
*Aristaphis* Kirkaldy, 1905
*Cladobius* Koch, 1856 nec Dejean, 1837
*Clavigerus* Szepligeti, 1883
*Melanoxanterium* Schouteden, 1901
*Melanoxanthus* Buckton, 1879 nec Eschscholtz, 1836
*Stauroceras* Börner, 1940

P. americanum Szelegiewicz, 1965 BM HRL
beulahense (Cockerell, 1904) (Cladobius) BM HRL
bicolor (Oestlund, 1887) (Melanoxanthus) BM HRL
*rufulus* Davidson, 1909 (*Cladobius*)
bituberculatum Theobald, 1912 = Plocamaphis amerinae (Hartig, 1841)
chaetosiphon (Börner, 1940) (Stauroceras) HRL
dubium Börner, 1950 = populeum (Kaltenbach, 1843)
fraxini Theobald, 1921 = rufipes (Hartig, 1841)
groenlandicum Hille Ris Lambers, 1952 cotypes BM HRL
jacksoni Theobald, 1921 BM HRL
*morio* Hille Ris Lambers, 1947 HRL
konoi Hori ex Takahashi, 1939 = pilosum subsp. konoi Hori ex Takahashi, 1939
ligustri Shinji, 1923 = Aphis crinosa Paik, 1969
medium Baker, 1917

P. morio Hille Ris Lambers, 1947 = jacksoni Theobald, 1921
pilosum Buckton, 1879 lectotype BM HRL
pilosum subsp. konoi Hori ex Takahashi, 1939 BM
*konoi* Hori ex Takahashi, 1939
pilosum subsp. sarmaticum Szelegiewicz, 1967 HRL
populeum (Kaltenbach, 1843) (Aphis) BM HRL
*dubium* Börner, 1950
*populeus* var. *longirostris* Mordvilko, 1909 (*Cladobius*)
*populeus* subsp. *turanicum* Nevsky, 1929
populeus subsp. turanicum Nevsky, 1929 = populeum (Kaltenbach, 1843)
populifoliae (Fitch, 1851) (Aphis) BM HRL
pseudopopuleum Palmer, 1952 BM HRL
ringdahli Wahlgren, 1940 BM HRL
rufipes (Hartig, 1841) (Aphis) BM HRL correct name for Cladobius steinheili according to Börner, 1952
*fraxini* Theobald, 1921 BM
*steinheili* Mordvilko, 1901 (*Cladobius*)
salicicola (Uhler, 1862) (Lachnus) nom. nov. pro Aphis salicti Harris, 1841 nec Schrank, 1801
*salicti* Harris, 1841 (*Aphis*) nec Schrank, 1801
salicis (Linnaeus, 1758) (Aphis) BM HRL
*nervosa* Zetterstedt, 1840 (*Aphis*)
*? pilosa* Haldeman, 1844 (*Aphis*) nec Zetterstedt, 1840
*salicis* O.F. Müller, 1776 (*Aphis*)
*viminalis* Hartig, 1841 (*Aphis*) nec Boyer de Fonscolombe, 1841
salicis subsp. rohdendorfi (Szelegiewicz, 1965 nomen nudum) Holman & Szelegiewicz, 1974
sanguiceps Richards, 1967
smithiae (Monell, 1879) (Chaitophorus) BM HRL
tremulae Börner, 1940 BM HRL
*tremulae* subsp. *stroyani* Szelegiewicz, 1967
tremulae subsp. stroyani Szelegiewicz, 1967 = tremulae Börner, 1940
? vignae (Matsumura, 1917) (Melanoxanthus)
yezoense (Hori, 1929) (Melanoxanterium) BM HRL

PTEROSTIGMA Buckton, 1883
Monograph of the British Aphides, London, 4: 178
Type species Pterostigma recurvum Buckton, 1883
= Mindarus Koch, 1857
Fossil

PTYCHODES Buckton, 1881
Monograph of the British Aphides, London, 3: 39
Type species Aphis juglandis Frisch, 1734 prebinominal = Aphis juglandis Goeze, 1778
= Callaphis Walker, 1870

P. quercicola Matsumura, 1919 = Tuberculatus (Acanthocallis) quercicola Matsumura, 1917

PULVIUS Sanborn, 1906
Bull. Univ. Kans. 3: 225
Type species Pulvius probosceus Sanborn, 1906
= Prociphilus Koch, 1857

P. probosceus Sanborn, 1906 to Prociphilus Koch, 1857

PYRAPHIS Börner, 1931
Anz. Schädlingsk. 7: 10
Type species Pyraphis streili Börner, 1931 = Myzus pyrarius Passerini, 1861
= Melanaphis van der Goot, 1917

P. streili Börner, 1931 = Melanaphis pyraria (Passerini, 1861)

PYRETHROMYZUS Börner, (1944) 1950
In Brohmer, P., Fauna von Deutschland, ed. 5, Leipzig, p. 217, in key, without named species
Neue europäische Blattlausarten, privately published, p. 15
Type species Macrosiphum sanborni Gillette, 1908
= Macrosiphoniella del Guercio, 1911

PYROLACHNUS A.N. Basu & Hille Ris Lambers, 1968
Ent. Ber., Amst. 28: 13
Type species Lachnus pyri Buckton, 1899

P. imbricatus David, Narayanan & Rajasingh, 1971 BM HRL
pyri (Buckton, 1899) (Lachnus) BM HRL
*krishni* George, 1927 (*Dilachnus*)

QUADRARTUS Monzen, 1954
Rep. Gakugei Fac. Iwate Univ. 7: 54
Type species Quadrartus yoshinomiyai Monzen, 1954

Q. yoshinomiyai Monzen, 1954 BM HRL

QUERNAPHIS Takahashi, 1958
Kontyû 26: 182
Type species Thoracaphis tuberculatus Takahashi, 1933 = Hamamelistes tuberculatus Takahashi, 1933

Q. tuberculata (Takahashi, 1933) (Hamamelistes) HRL

QUIPPELACHNUS Oestlund, 1922
Rep. Minn. St. Ent. 19: 134
Type species Euceraphis gillettei Davidson, 1915*
= Euceraphis Walker, 1870

RAMITRICHOPHORUS Hille Ris Lambers, 1947
Zool. Meded., Leiden 28: 291, described as subg. of Macrosiphoniella del Guercio, 1911
Type species Macrosiphoniella janckei Börner, 1939
Here used as subg. of Macrosiphoniella del Guercio, 1911

R. medvedevi Bozhko, 1957 to Macrosiphoniella (Ramitrichophorus)
paradoxus Bozhko, 1957 to Macrosiphoniella (Ramitrichophorus)

RECTICALLIS Matsumura, 1919
Trans. Sapporo nat. Hist. Soc. 7: 105
Type species Recticallis alni-japonicae Matsumura, 1919
Here used as subg. of Pterocallis Passerini, 1860

R. alni-japonicae Matsumura, 1919 to Pterocallis (Recticallis)

RECTINASUS Theobald, 1914
Entomologist 47: 28
Type species Rectinasus buxtoni Theobald, 1914

R. bucktoni lapsus pro buxtoni Theobald, 1914
buxtoni Theobald, 1914 types BM HRL
*caucasicus* Mordvilko, 1921
*cockerelli* Nevsky, 1929
buxtoni subsp. kasachstanica Juchnevitch, 1971
caucasicus Mordvilko, 1921 = buxtoni Theobald, 1914
cockerelli Nevsky, 1929 = buxtoni Theobald, 1914
shelkovnikovi Mordvilko, 1916

*It seems probable that Oestlund misidentified Oestlundiella flava Granovsky, 1930 with Euceraphis gillettei but it is in the interests of nomenclatural stability to accept the nominal type citation for Quippelachnus, as otherwise Quippelachnus would have to be used to replace the well established genus Oestlundiella Granovsky, 1930

RETICULAPHIS Takahashi, 1958
Insecta matsum. 22: 11
Type species Reticulaphis shiiae Takahashi, 1958

R. distylii (van der Goot, 1917) (Schizoneuraphis) HRL
*javensis* Takahashi, 1921 (*Astegopteryx*) in litteris synonym
distylii subsp. asymmetrica Hille Ris Lambers & Takahashi, 1959 HRL
distylii subsp. fici (Takahashi, 1923) (Astegopteryx) BM HRL
*fici* Takahashi, 1923 (*Astegopteryx*)
distylii subsp. foveolatae (Takahashi, 1935) (Thoracaphis) BM HRL
*fici* var. *foveolatae* Takahashi, 1935 (*Thoracaphis*)
distylii subsp. minutissima Hille Ris Lambers & Takahashi, 1959 HRL
distylii subsp. rotifera Hille Ris Lambers & Takahashi, 1959 HRL
distylii subsp. similis Hille Ris Lambers & Takahashi, 1959 HRL
mirabilis (Takahashi, 1939) (Thoracaphis)
*fici* var. *mirabilis* Takahashi, 1939 (*Thoracaphis*)
shiiae Takahashi, 1958 BM

RHANIS von Heyden, 1837 nec Dejean, 1835
Ent. Beitr. Mus. senckenb. 2: 289
Invalid in litteris synonym of Vacuna von Heyden, 1837
= Phylloxera Boyer de Fonscolombe, 1834
Phylloxeridae

RHIZAPHIS Planchon ex Bazille, Planchon & Sahut, 1868
In Bazille, Planchon & Sahut, C. r. hebd. Seanc. Acad. Sci., Paris 67: 336 (and footnote with description as Rizaphis)
Type species R(h)izaphis vastatrix Planchon, 1868 = Pemphigus vitifoliae Fitch, 1855
= Viteus Shimer, 1867
Phylloxeridae

RHIZOBERLESIA del Guercio, 1915
Redia 10: 246
Type species Rhizoberlesia trifolii del Guercio, 1915
Here used as subg. of Therioaphis Walker, 1870

R. trifolii del Guercio, 1915 = Therioaphis (Rhizoberlesia) brachytricha Hille Ris Lambers & van den Bosch, 1964

RHIZOBIUS Buckton, 1883. No such genus was described by Buckton. He (1883: 189) described a coccid in Rhizobius without indicating the author of the genus but Burmeister, 1835 is indicated as the author on p. 92 and in Buckton's (1899) description of the coccid.

RHIZOBIUS Burmeister, 1835*
Handbuch der Entomologie, Berlin, p. 87
Type species Rhizobius pilosellae Burmeister, 1835 = Aphis bursaria Linnaeus, 1758
= Pemphigus Hartig, 1839, nomen conservandum

R. eleusinis Thomas, 1878 = Anoecia cornicola (Walsh, 1863)
graminis Buckton, 1883 = Aploneura lentisci (Passerini, 1856)
helianthemi Westwood, 1843 to Trama von Heyden, 1837
jujubae Buckton, 1883 & 1899, a coccid
lactucae Fitch, 1871 = Pemphigus bursarius (Linnaeus, 1758)
menthae Passerini, 1860 = Kaltenbachiella pallida (Haliday, 1838)
pilosellae Burmeister, 1835 (type of Rhizophthiridium van der Hoeven, 1840) = Pemphigus bursarius (Linnaeus, 1758)
pini Burmeister, 1835 to Prociphilus (Stagona)
poae Buckton, 1883 = Aploneura lentisci (Passerini, 1856)
poae Thomas, 1879 = ? Aploneura lentisci (Passerini, 1856)
radicum Craveri, 1915 nomen dubium
sonchi Passerini, 1860 = Pemphigus bursarius (Linnaeus, 1758)
spicatus Hart, 1895 = ? Colopha ulmicola (Fitch, 1859)
subterraneus Kaltenbach, 1843 = Prociphilus (Stagona) pini (Burmeister, 1835)
viridis Theobald, 1915 = Prociphilus (Stagona) sylostei (de Geer, 1773)

RHIZOBIUS Passerini, 1860
Gli Afidi, Parma, p. 30 as Rhizobius Burmeister, 1835
Type species Rhizobius sonchi Passerini, 1860 = Aphis bursaria Linnaeus, 1758
= Pemphigus Hartig, 1839

R. sonchi Passerini, 1860 (type of Rhyzoicus Passerini, 1860) = Pemphigus bursarius (Linnaeus, 1758)

RHIZOCERA Kirkaldy, 1897
Leaflets New Zealand Dep. Agric. 20: 3

* Rhizobius Burmeister, 1835 is not preoccupied by Rhizobius Stephens as this spelling is an emendation by Agassiz, 1846 of a genus originally spelled Rhyzobius Stephens, 1829

Type species Rhizocera vastatrix (Planchon, 1868) = Pemphigus vitifoliae Fitch, 1855
= Viteus Shimer, 1867
Phylloxeridae

RHIZOCTONUS Mokrzecky, 1896
Trudy russk. ent. Obshch. 30: 439, 441 (names attributed to Dr. Horvath)
Type species Rhizoctonus ampelinus Mokrzecky, 1896
= ? Aploneura Passerini, 1863

R. ampelinus Mokrzecky, 1896 to ? Aploneura Passerini, 1863

RHIZOICUS del Guercio, 1919 nec 1906
Redia 12: 251 (as Rhizoicus Passerini)
Type species Rhizobius jujubae Buckton, 1883
Del Guercio, 1906 (Redia 3: 370) emended Rhyzoicus Passerini, 1860 to Rhizoicus and used the genus for the aphids Rh. graminis (poae), sonchi and menthae.
Coccidae

RHIZOMARIA Hartig, 1857 (1856)
Verhandl. d. Hils-Solling-Forstvereins 1856 (1857): 53
Type species Rhizomaria piceae Hartig, 1857 = Aphis tremulae Linnaeus, 1761
= Asiphum Koch, 1856

R. piceae Hartig, 1857 = Asiphum tremulae (Linnaeus, 1761)

RHIZONIARIA Wilson, 1910 lapsus pro Rhizomaria Hartig, 1857

RHIZOPHTHIRIDIUM van der Hoeven, 1840
Handboek der Dierkunde Ed. 2, Amsterdam 1: 508
Type species Rhizobius pilosellae Burmeister, 1835 = Aphis bursaria Linnaeus, 1758
= Pemphigus Hartig, 1839

RHIZOTERUS Hartig, 1841
Z. Ent. (Germar) 3: 363
Type species Rhizoterus vacca Hartig, 1841 = Forda formicaria von Heyden, 1837
= Forda von Heyden, 1837

R. vacca Hartig, 1841 = Forda formicaria von Heyden, 1837

RHODOBIUM Hille Ris Lambers, 1947
Temminckia 7: 300
Type species Macrosiphum rosaefolium Theobald, 1915 = Myzus porosus Sanderson, 1900

R. porosum (Sanderson, 1900) (Myzus) BM HRL
*pseudorosaefolium* E.E. Blanchard, 1922 (*Aulacorthum*)
*rosaefoliae* Takahashi, 1931 (*Acyrthosiphon*)
*rosaefolium* Theobald, 1915 (*Macrosiphum*) types BM
*viride* van der Goot, 1917 (*Aulacorthum*)
*zoorosarum* Knowlton & Smith, 1936 (*Macrosiphum*)

RHOPALOMYZUS Börner, 1950 nec Mordvilko, 1921
Neue europäische Blattlausarten, privately published, p. 12, described as subg. of Hyperomyzus Börner, 1933
Type species Rhopalosiphum calthae Koch, 1854 vide Börner, 1952: 139
= Pseudorhopalosiphoninus Heinze, 1961

RHOPALOMYZUS Mordvilko, 1921
Izv. Sev. Oblast. Sta. Zashch. Rast. Vredit. 3: 45, described as subg. of Myzus Passerini, 1860
Type species Rhopalosiphum poae Gillette, 1908

R. berberidis Narzikulov, 1957 = ? Liosomaphis berberidis (Kaltenbach, 1843)
codonopsidis Umarov, 1963 HRL
coerulescens Narzikulov, 1965 to Hyadaphis Kirkaldy, 1904
grabhami (Cockerell, 1903) (Rhopalosiphum) BM HRL
hissarica Narzikulov, 1965 to ? Hyadaphis Kirkaldy, 1904
poae (Gillette, 1908) (Rhopalosiphum) BM HRL
*alpigenae* Börner, 1914 (*Rhopalosiphum*)
*dianthae* var. *poae* Williams, 1911 (*Rhopalosiphum*)
smilacis (Matsumura, 1918) (Rhopalosiphum)
tianshanica Narzikulov, 1963

Subgen. Judenkoa Hille Ris Lambers, 1946

R. (Judenkoa) lonicerae (Siebold, 1839) (Aphis) BM HRL

RHOPALOSIPHON Scudder, 1882 emendation of Rhopalosiphum Koch, 1854

RHOPALOSIPHONINUS Baker, 1920
Bull. U.S. Dep. Agric. 826: 58

Type species Amphorophora latysiphon Davidson, 1912
*Amrhopalosiphunicus* Shinji, 1933
*Clavosiphum* Shinji, 1922

R. celtifoliae Shinji, 1924 HRL
chicotei Gomez-Menor, 1950 = Eucarazzia elegans (Ferrari, 1872)
deutzifoliae Shinji, 1924 BM HRL = ? latysiphon (Davidson, 1912)
hieracii Börner, 1939 to Hyperomyzus (Neonasonovia)
hydrangeae (Matsumura, 1918) (Rhopalosiphum) BM HRL
*sambucifoliae* Shinji, 1922 (*Clavosiphum*)
ichigo (Shinji, 1922) (Amphorophora)
lampsanae Börner, 1932 to Hyperomyzus Börner, 1933
latysiphon (Davidson, 1912) (Amphorophora) BM HRL
*? deutzifoliae* Shinji, 1924
longisetosus Chakrabarti & A.K. Ghosh ex Chakrabarti, A.K. Ghosh & RayChaudhuri, 1974
maianthemi Stroyan, 1965 paratype BM HRL
lupulinus Theobald, 1926 = Rhopalosiphoninus (Myzosiphon) staphyleae (Koch, 1854)
nobukii adenocauli Shinji, 1933 = tiliae (Matsumura, 1918)
persimilis Hille Ris Lambers, 1960 = Rhopalosiphoninus (Myzosiphon) solani (Thomas, 1879)
platicaudus Narzikulov, 1953 type of Amegosiphon Narzikulov, 1958 q.v.
polygoni Narzikulov, 1953 type of Tricaudatus Narzikulov, 1957 q.v.
pseudorumicis Theobald, 1926 = Rhopalosiphoninus (Myzosiphon) staphyleae subsp. tulipaellus (Theobald, 1916)
pterocaryae Aizenberg, 1954 nomen nudum
ribesinus (van der Goot, 1912) (Rhopalosiphum) BM HRL
salviae Hall, 1926 = Eucarazzia elegans (Ferrari, 1872)
subterrans (Wilson, 1915) (Amphorophora)
takahashii Tao, 1963
tiliae (Matsumura, 1918) (Rhopalosiphum) HRL
*adenocauli* Shinji, 1922 (*Clavosiphum*)
*nobukii adenocauli* Shinji, 1933
tuberculatus Theobald, 1929 = Hyperomyzus (Hyperomyzella) rhinanthi (Schouteden, 1903)
waltoni Theobald, 1923 = Rhopalosiphoninus (Myzosiphon) staphyleae (Koch, 1854)

Subgen. Eucarazzia del Guercio, 1921

R. (Eucarazzia) caucasicus (Aizenberg, 1956) to Eucarazzia del Guercio, 1921

Subgen. Myzosiphon Hille Ris Lambers, 1946

R. (Myzosiphon) solani (Thomas, 1879) (Megoura) HRL
*persimilis* Hille Ris Lambers, 1960 (*Rhopalosiphoninus*) HRL
(Myzosiphon) staphyleae (Koch, 1854) (Rhopalosiphum) BM HRL
*lupulinus* Theobald, 1926 (*Rhopalosiphoninus*) types BM
*theobaldi* Börner, 1950 (*Hyperomyzus*)
*waltoni* Theobald, 1923 (*Rhopalosiphoninus*) types BM
(Myzosiphon) staphyleae subsp. tulipaellus (Theobald, 1916) (Rhopalosiphum) types BM HRL
*pseudorumicis* Theobald, 1926 (*Rhopalosiphoninus*) type BM
*tulipaella* Theobald, 1916 (*Rhopalosiphum*) BM

Subgen. Pseudorhopalosiphoninus Heinze, 1961

R. (Pseudorhopalosiphoninus) calthae (Koch, 1854) (Rhopalosiphum) BM HRL

Subgen. Submegoura Hille Ris Lambers, 1953

R. (Submegoura) heikinheimoi (Börner, 1952) (Myzotoxoptera) BM HRL
(Submegoura) obscurata (Hille Ris Lambers m.s. ex) Stroyan, 1950 nomen nudum

RHOPALOSIPHUM Koch, 1854
Die Pflanzenläuse Aphiden, Nürnberg, p. 23
Type species Aphis nymphaeae Linnaeus, 1761
*Aresha* Mordvilko, 1921
*Pseudocerosipha* Shinji, 1932 and 1941
*Siphocoryne* Passerini, 1860
*Siphonaphis* van der Goot, 1915
*Stenaphis* del Guercio, 1913
*Yamataphis* Matsumura, 1917

R. absinthii Lichtenstein, 1885 to Coloradoa Wilson, 1910
acaenae Schouteden, 1904 to Pentamyzus Hille Ris Lambers, 1966
aconiti van der Goot, 1912 (type of Delphiniobium Mordvilko, 1914) = Delphiniobium junackianum (Karsch, 1887)
affine Börner, 1921 = Hyperomyzus (Hyperomyzella) rhinanthi (Schouteden, 1903)
alismae Koch, 1854 = nymphaeae (Linnaeus, 1761)
alpigenae Börner, 1914 = Rhopalomyzus poae (Gillette, 1908)

R. arbuti Davidson, 1910 (type of Wahlgreniella Hille Ris Lambers, 1949) = Wahlgreniella nervata subsp. arbuti (Davidson, 1910)
arundinariae (Tissot, 1933) (Anuraphis) BM
betae Theobald, 1913 = Myzus (Nectarosiphon) persicae (Sulzer, 1776)
bossekiae Gillette & Palmer, 1929 to Utamphorophora Knowlton, 1947
brachytarsus Takahashi, 1961 to Schizaphis (Paraschizaphis)
britteni Theobald, 1913 = Hyperomyzus (Hyperomyzella) rhinanthi (Schouteden, 1903)
bupleuri Cairaschi & Arnoux, 1943 = Hyadaphis bupleuri Börner, 1939
calaminthae Lichtenstein, 1884 nomen nudum
callae Koch, 1854 in litteris synonym of dianthi = Myzus (Nectarosiphon) persicae (Sulzer, 1776)
calthae Koch, 1854 (type of Rhopalomyzus Börner, 1950 nec Mordvilko, 1931; and of Pseudorhopalosiphoninus Heine, 1961) to Rhopalosiphoninus (Pseudorhopalosiphoninus)
carduellinum Theobald, 1915 to Hyperomyzus Börner, 1933
cerasifoliae (Fitch, 1855) (Aphis) BM HRL
*? furcata* Patch, 1914 (*Aphis*)
*tahasa* Hottes, 1950 (*Aphis*)
cicutae Koch, 1854 to Cavariella del Guercio, 1911
corylinum Davidson, 1914 to Illinoia Wilson, 1910
davisi Mordvilko, 1921 = Glabromyzus howardii (Wilson, 1911)
dianthi var. poae Williams, 1911 = Rhopalomyzus poae (Gillette, 1908)
elegans Ferrari, 1872 to Eucarazzia del Guercio, 1921
enigmae Hottes & Frison, 1931 paratypes BM HRL
enigmae var. parvae Hottes & Frison, 1931 = parvae Hottes & Frison, 1931
erraticum Koch, 1854 = Hyperomyzus lactucae (Linnaeus, 1758)
fitchii (Sanderson, 1902) (Aphis) = insertum (Walker, 1849)*
fucanoi Moritsu, 1947 = rufiabdominalis (Sasaki, 1899)
galeactitis Macchiati, 1883 = Myzus (Nectarosiphon) persicae (Sulzer, 1776)
gnaphalii Tissot, 1932 = rufiabdominalis (Sasaki, 1899)
grabhami Cockerell, 1903 to Rhopalomyzus Mordvilko, 1921
hemerocallidis Matsumura, 1918 = Indomegoura indica (van der Goot, 1916)
hippophaes Koch, 1854 = Capitophorus hippophaes (Walker, 1852)
hydrangeae Matsumura, 1918 to Rhopalosiphoninus Baker, 1920
indicum van der Goot, 1916 type of Indomegoura Hille Ris Lambers, 1958 q.v.

* Colonies on Gramineae on subterranean parts only; most North American records of fitchii from aerial parts of Graminueae apply to R. padi (Linnaeus, 1758)

R. insertum (Walker, 1849) (Aphis) BM HRL
*crataegella* Theobald, 1912 (*Aphis*) BM
*edentula* Buckton, 1879 (*Aphis*) lectotype BM
*fitchii* Sanderson, 1902 (*Aphis*) HRL
*mactata* Walker, 1849 (*Aphis*) lectotype BM
*sanguinarium* J.M. Baker, 1934 HRL
*viridis* Richards, 1960 paratype BM
kiku Hori, 1929 = Coloradoa rufomaculata (Wilson, 1908)
lactucellum Theobald, 1914 = Myzus (Nectarosiphon) persicae (Sulzer, 1776)
lespedezae Essig & Kuwana, 1918 to Megoura Buckton, 1876
luteum Mordvilko, 1929 nomen nudum
magnoliae Essig & Kuwana, 1918 to Aulacorthum Mordvilko, 1914
maidis (Fitch, 1856) (Aphis) BM HRL
*adusta* Zehntner, 1897 (*Aphis*)
*africana* Theobald, 1914 (*Aphis*) types BM
*cooki* Essig, 1911 (*Aphis*)
*monticellii* del Guercio, 1913 (*Stenaphis*)
? *obnoxia* Mordvilko, 1916 (*Aphis*)
*setariae* Rusanova, 1962 (*Schizaphis*) (*Toxoptera* (*Schizaphis*))
*vulpiae* del Guercio, 1913 (*Aphis*)
*zeae* Rusanova, 1960 (*Rhopalosiphon*)
miniatum Matsumura, 1918 = Indomegoura indica (van der Goot, 1916)
momo Shinji, 1922
monardae Davis, 1912 to Hyalomyzus Richards, 1958
musae (Schouteden, 1906) (Aphis) BM HRL
*scirpifolii* Gillette & Palmer, 1932 BM paratypes HRL
nabali Oestlund, 1886 to Hyperomyzus (Neonasonovia)
najadum Koch, 1854 = nymphaeae (Linnaeus, 1761)
nervatum Gillette, 1908 to Wahlgreniella Hille Ris Lambers, 1949
nigrum Richards, 1960 paratypes BM HRL
nymphaeae (Linnaeus, 1761) (Aphis) BM HRL
*alismae* Koch, 1854
*aquatica* Jackson, 1908 (*Aphis*)
*butomi* Schrank, 1801 (*Aphis*)
*donarium* Matsumura, 1918 (*Siphocoryne*)
*infuscata* Koch, 1854 (*Aphis*)
*najadum* Koch, 1854
*plantarum-aquaticum* Fabricius, 1794 (*Aphis*)
*prunaria* Walker, 1850 (*Aphis*)
*prunorum* Dobrovljansky, 1913 (*Aphis*)
*sparganii* Theobald, 1925 (*Hyadaphis*) types BM

R. oleraceae van der Goot, 1917 = Hyperomyzus carduellinus (Theobald, 1915)
padi (Linnaeus, 1758) (Aphis) BM HRL
*annuae* Oestlund, 1886 (*Aphis*)
*avenae-sativae* Schrank, 1801 (*Aphis*)
*holci* Ferrari, 1872 (*Aphis*)
*padi* subsp. *americanum* Mordvilko, 1921
*prunifoliae* (Fitch, 1855) (*Aphis*) partim
*pseudoavenae* Patch, 1917 (*Aphis*)
*tritici* Lawson, 1866 (*Aphis*)
padi subsp. americanum Mordvilko, 1921 = padi (Linnaeus, 1758)
padiformis Richards, 1962 BM
papaveri Takahashi, 1921 = Lipaphis erysimi (Kaltenbach, 1843)
papilionacearum Lindlinger, 1932 = Megoura viciae Buckton, 1876
parvae Hottes & Frison, 1931
*enigmae* var. *parvae* Hottes & Frison, 1931
phragmitidis Börner ex Bodenheimer, 1937 nomen nudum
picridis Börner & Blunck, 1916 (type of Neonasonovia Hille Ris Lambers, 1949) to Hyperomyzus (Neonasonovia)
pilipes Ossiannilsson, 1959 to Schizaphis Börner, 1931
poae Gillette, 1908 type species of Rhopalomyzus Mordvilko, 1921 q.v.
rhois Monell, 1879 type of Glabromyzus Richards, 1960 q.v.
ribesina van der Goot, 1912 to Rhopalosiphoninus Baker, 1920
ribijaponica Shinji, 1924 = Hyperomyzus lactucae (Linnaeus, 1758)
rufiabdominalis (Sasaki, 1899) (Toxoptera) BM HRL
*californica* Essig, 1944 (*Cerosipha*)
*fucanoi* Moritsu, 1947
*gnaphalii* Tissot, 1932
*oryzae* Matsumura, 1917 (*Yamataphis*)
*papaveri* Takahashi, 1921 (*Yamataphis*)
*pruni* Shinji 1932 & 1941 (*Pseudocerosipha*)
*setigera* E.E. Blanchard, 1939 (*Aresha*)
*shelkovnikovi* Mordvilko, 1921 (*Aresha*)
*splendens* Theobald, 1915 (*Siphocoryne*)
*subterraneum* Mason, 1937
rufulum Richards, 1960 BM HRL
rumicis Theobald, 1925 = Cavariella pastinacae (Linnaeus, 1758)
sambuci Matsumura, 1918 = Aulacorthum magnoliae (Essig & Kuwana, 1918)
sambucicola Takahashi, 1918 = Aulacorthum magnoliae (Essig & Kuwana, 1918)
sanguinarium J.M. Baker, 1934 = insertum (Walker, 1849)

R. scirpifolii Gillette & Palmer, 1932 = musae (Schouteden, 1906)
serotinae Oestlund, 1887 type of Cachryphora Oestlund, 1922 q.v.
sisymbrii del Guercio, 1913 = ? Lipaphis erysimi (Kaltenbach, 1843)
smilacis Matsumura, 1918 to Rhopalomyzus Mordvilko, 1921
sonchi Oestlund, 1886 = Hyperomyzus lactucae (Linnaeus, 1758)
staphyleae Koch, 1854 (type of Arthromyzus Börner, 1950; and of Myzosiphon Hille Ris Lambers, 1946) to Rhopalosiphoninus (Myzosiphon)
subterraneum Mason, 1937 = rufiabdominalis (Sasaki, 1899)
tiliae Matsumura, 1918 to Rhopalosiphoninus Baker, 1920
trilineatum del Guercio, 1920 = Myzus (Nectarosiphon) persicae (Sulzer, 1776)
tulipae Thomas, 1879 = Myzus (Nectarosiphon) persicae (Sulzer, 1776)
tulipaella Theobald, 1916 = Rhopalosiphoninus (Myzosiphon) staphyleae subsp. tulipaellus (Theobald, 1916)
uwamizusakurae (Monzen, 1929) (Aphis)
vagans van der Goot, 1917 to Sinomegoura Takahashi, 1960 or ? Amphorophora Buckton, 1876
viciae var. japonica Matsumura, 1918 = Megoura crassicauda Mordvilko, 1919
violae Essig, 1909 = Neotoxoptera violae (Pergande, 1900)
violae Pergande, 1900 to Neotoxoptera Theobald, 1915
viridicaudus Miller, 1933 = Nearctaphis sensoriata (Gillette & Bragg, 1918)
viridis Richards, 1960 = insertum (Walker, 1849)
wahlgreni Ossiannilsson, 1959 to Schizaphis Börner, 1931
yoksumi A.K. Ghosh, H. Banerjee & RayChaudhuri, 1971
zeae Rusanova (1943 nomen nudum) 1960 = maidis (Fitch, 1856)

RHOPALOSIPHUM Passerini, 1860
Gli Afidi, Parma, p. 27 as Rhopalosiphum Koch, 1854
Type species Aphis persicae Sulzer, 1776 but Gerstaecker, 1856: 272 had previously selected Aphis nymphaeae Linnaeus, 1767 as type species.
= Nectarosiphon Schouteden, 1901

RHOPALOSIPHUM van der Goot, 1913
Tijdschr. Ent. 56: 146 as Rhopalosiphum Koch, 1854
Type species Amphorophora ampullata Buckton, 1876, but this was not originally included in Rhopalosiphum by Koch; and Gerstaecker, 1856: 272 had previously selected Aphis nymphaeae Linnaeus, 1767 as type species.
= Amphorophora Buckton, 1876

RHOPALOSIPHUNICUS Shinji, 1924
Zool. Mag., Tokyo 36: 365 lapsus pro Rhopalosiphoninus Baker, 1920

RHYNCHOCLES Altum, 1882
Forstzoologie, Berlin 3: 365
Type species Rhynchocles longirostris Altum, 1882 = Aphis quercus Linnaeus, 1758
= Stomaphis Walker, 1870

R. longirostris Altum, 1882 = Stomaphis quercus (Linnaeus, 1758)

RHYZOBIUS Ferrari, 1872 nec Stephens, 1829
Annali Mus. civ. Stor. nat. Giacomo Doria 3: 84 emendation of Rhizobius Burmeister, 1835
= Pemphigus Hartig, 1839

RHYZOICUS Passerini, 1860
Gli Afidi, Parma, p. 30 footnote, said to be nomen novum pro Rhizobius Burmeister, 1935 nec Rhyzobius Stephens, 1829
Type species Rhizobius sonchi Passerini, 1860 = Aphis bursaria Linnaeus, 1758
= Pemphigus Hartig, 1839

RIZAPHIS Planchon, 1868 lapsus pro Rhizaphis in original description in footnote, correct spelling indicated in text by Bazille, Planchon & Sahut on same page.

ROEPKEA Hille Ris Lambers, 1935
Memorie Mus. Stor. nat. Venezia trident. 3: 6
Type species Yezabura marchali Borner, 1931

R. multisetis Richards, 1969 to Nearctaphis Shaposhnikov, 1950
phlomicola (Nevsky, 1929) (Anuraphis) HRL
phlomicola subsp. marchali (Börner, 1931) (Yezabura) BM HRL
*marchali* Börner, 1931 (*Yezabura)*
punctata Richards, 1969 to Nearctaphis Shaposhnikov, 1950
sclerosa Richards, 1969 to Nearctaphis Shaposhnikov, 1950
zababsis Richards, 1969 to Nearctaphis Shaposhnikov, 1950

RUMIFEX Amyot, 1847
Annls Soc. ent. Fr. (2) 5: 478
Invalid

RUNGSIA Mimeur, 1933
Bull. Soc. Sci. nat. Maroc 13: 104
Type species Rungsia graminis Mimeur, 1933 = Sipha maydis Passerini, 1860
*Siphonella* Börner, 1939 nec Macquart, 1835; Hagenow in Goinitz, 1850; Issel, 1869; Verril, 1879; Ledenfeld, 1887
Here used as subg. of Sipha Passerini, 1860

R. aegilopis Bozhko, 1961 to Sipha (Rungsia)
graminis Mimeur, 1933 = Sipha (Rungsia) maydis Passerini, 1860
nemaydis Narzikulov, 1963 = Sipha (Rungsia) elegans del Guercio, 1905
praecocis Bozhko, 1959 to Sipha (Rungsia)
taurica Mamontova, 1959 to Sipha (Rungsia)

RYOICHITAKAHASHIA Hille Ris Lambers, 1965
Tijdschr. Ent. 108: 190
Type species Ryoichitakahashia ilicis Hille Ris Lambers, 1965 = Macrosiphoniella prunifoltae Shinji, 1924

R. ilicis Hille Ris Lambers, 1965 = prunifoltae (Shinji, 1924)
prunifoltae (Shinji, 1924) (Macrosiphoniella) BM HRL
*ilexis* Moritsu, 1958 (*Anuraphis*)
*ilicis* Hille Ris Lambers, 1965 HRL

SACCHIPHANTES Curtis, 1844
Gardners' Chronicle 1844: 831
Type species Chermes abietis Linnaeus, 1758
*Phloeophthiridium* van der Hoeven, 1849
Adelgidae

SALICETIFEX Amyot, 1847
Annls Soc. Ent. Fr, (2) 5: 479
Invalid

SALTUSAPHIS Theobald, 1915
Bull. ent. Res. 6: 138
Type species Saltusaphis scirpus Theobald, 1915
*Hiberaphis* Börner, 1949
*Stenaphis* Quednau, 1954 nec del Guercio, 1913

africana Eastop, 1953 = scirpus Theobald, 1915
americana Baker, 1917 to Iziphya Nevsky, 1929

S. cinerea Börner, 1940 = Subsaltusaphis picta (Hille Ris Lambers, 1939)
elongata Baker, 1917 type of Stenaphis Quednau, 1954 nec del Guercio, 1913 BM HRL
flava Hille Ris Lambers, 1939 to Subsaltusaphis Quednau, 1953
granulata Börner, 1940 = Subsaltusaphis pallida (Hille Ris Lambers, 1939)
intermedia Hille Ris Lambers, 1939 type of Subsaltusaphis Quednau, 1953 q.v.
kienshuensis Shinji, 1942
lasiocarpae (Ossiannilsson, 1953) (Hiberaphis) BM HRL
ornata Theobald, 1927 (type of Bacillaphis Quednau, 1954) to Subsaltusaphis Quednau, 1953
pallida Hille Ris Lambers, 1939 to Subsaltusaphis Quednau, 1953
picta Hille Ris Lambers, 1939 to Subsaltusaphis Quednau, 1953
quadrilineata Börner, 1939 = Subsaltusaphis ornata (Theobald, 1927)
rossneri Börner, 1940 to Subsaltusaphis Quednau, 1953
scirpus Theobald, 1915 types BM HRL
*afghanica* Narzikulov & Umarov, 1970 (*Bacillaphis*)
*africana* Eastop, 1953 BM
*iberica* Börner, 1949 (*Hiberaphis*)
virginica Baker, 1917 to Subsaltusaphis Quednau, 1953
wanica Hottes & Frison, 1931 to Subsaltusaphis Quednau, 1953

SAMBUCIFEX Amyot, 1847
Annls Soc. ent. Fr. (2) 5: 477
Invalid

SANBORNIA Baker, 1920
Bull. U.S. Dep. Agric. 826: 50
Type species Sanbornia juniperi Pergande ex Baker, 1920
S. juniperi Pergande ex Baker, 1920 BM HRL

SAPPAPHIS Matsumura, 1918
Trans. Sapporo nat. Hist. Soc. 7: 18
Type species Sappaphis piri Matsumura, 1918
*Lachnaphis* Shinji, 1922
ariae Börner, 1950 to Dysaphis (Pomaphis)
brancoi subsp. rogersoni Stroyan, 1955 to Dysaphis Börner, 1931

S. euscaphis Paik, 1965 = Aphis horii Takahashi, 1923
gallica Hille Ris Lambers, 1955 to Dysaphis (Pomaphis)
leefmansi Hille Ris Lambers, 1954 to Dysaphis Börner, 1931
maritima Hille Ris Lambers, 1955 to Dysaphis (Pomaphis)
maritima subsp. glabra Hille Ris Lambers, 1955 to Dysaphis (Pomaphis)
palaestinensis Hille Ris Lambers, 1948 = Dysaphis tulipae (Boyer de Fonscolombe, 1841)
parasorbi Börner, 1952 to Dysaphis (Pomaphis)
piri Matsumura, 1918 BM HRL
*artemirhizus* Shinji, 1924 (*Anuraphis*)
*nashi* Shinji, 1944 (*Anuraphis*)
*piricola* Okamoto & Takahashi, 1927 (*Anuraphis*)
*yomogi* Shinji, 1922 (*Lachnaphis*)
pyrivora Tao, 1962 = ? Melanaphis pahanensis (Takahashi, 1950)
ranunculi Miyazaki, 1971 HRL
sphondylii Stroyan, 1952 = Dysaphis newskyi (Borner, 1950)

Subgen. Zinia Shaposhnikov, 1950

S. (Zinia) albocinerea Hille Ris Lambers, 1956 to Dysaphis Börner, 1931

SAPPOCALLIS Matsumura, 1919
Trans. Sapporo nat. Hist. Soc. 7: 107
Type species Sappocallis ulmicola Matsumura, 1919
*Telocallis* Shinji, 1922

S. ulmicola Matsumura, 1919 BM HRL
*alnifoliae* Shinji, 1922 (*Telocallis*)

SARUCALLIS Shinji, 1922
Zool. Mag., Tokyo 34: 730
Type species Sarucallis lythrae Shinji, 1922 = Myzocallis kahawaluokalani Kirkaldy, 1907

S. kahawaluokalani (Kirkaldy, 1907) (Myzocallis) BM HRL
*lagerstroemiae* Takahashi, 1920 (*Monellia*)
*lythrae* Shinji, 1922
lythrae Shinji, 1922 = kahawaluokalani (Kirkaldy, 1907)
nigropunctatus Tao, 1964

SATULA Olive, 1963
Ann. ent. Soc. Am. 56: 556
Type species Satula brachychaeta Olive, 1963
Here used as subg. of Uroleucon Mordvilko, 1914

S. brachychaeta Olive, 1963 to Uroleucon (Satula)

SBENAPHIS Scudder, 1890
U.S. Geol. Survey Territories 13: 250
Type species Lachnus quesneli Scudder, 1878
Fossil

SCHIZAPHIDIELLA Hille Ris Lambers, 1939
Zool. Meded., Leiden 22: 100
Type species Schizaphidiella quinquarticulata Hille Ris Lambers, 1939 = Myzus pyrarius Passerini, 1861
= Melanaphis van der Goot, 1917

S. quinquarticulata Hille Ris Lambers, 1939 = Melanaphis pyraria (Passerini, 1861)

SCHIZAPHIS Börner, 1931
Anz. Schädlingsk. 7: 10
Type species Aphis graminum Rondani, (1847) 1852

S. agrostis Hille Ris Lambers, 1947 BM HRL
arrhenatheri Pettersson, 1971 BM HRL?
borealis Tambs-Lyche, 1959 paratypes BM HRL
dubia Huculak, 1968 BM HRL
eastopi van Harten & Ilharco, 1971 to Schizaphis (Paraschizaphis)
fritzmuelleri Leclant, 1967 = pyri Shaposhnikov, 1952
geijskesi Hille Ris Lambers, 1939 = rufula (Walker, 1849)
geijskesi subsp. priori Stroyan, 1958 = priori Stroyan, 1958
germanica Börner, 1950 = holci Hille Ris Lambers, 1947
gracilis Richards, 1963
graminum (Rondani, (1847) 1852) (Aphis) BM HRL
graminum subsp. gigjai Stroyan, 1959 BM HRL
graminum subsp. phlei Orlob, 1961 BM HRL
hirticola Börner, 1950 = Schizaphis (Paraschizaphis) caricis (Schouteden, 1906)
holci Hille Ris Lambers, 1947 BM HRL
*germanica* Börner, 1950
hypersiphonata A.N. Basu, 1969 BM HRL
jaroslavi (Mordvilko, 1921) (Toxoptera) BM HRL

S. laingi Eastop, 1961 BM HRL
longicaudata Hille Ris Lambers, 1939 BM HRL
mehijiwae (Uye, 1924) (Toxoptera)
minuta (van der Goot, 1917) (Toxoptera) BM HRL
muhlenbergiae (Phillips & Davis, 1912) (Toxoptera) BM
nigerrima (Hille Ris Lambers, 1931) (Toxoptera) BM HRL
pilipes (Ossiannilsson, 1959) (Rhopalosiphum) BM HRL
piricola (Matsumura, 1917) (Toxoptera) BM HRL
priori Stroyan, 1958 BM HRL
*geyskesi* subsp. *priori* Stroyan, 1958
punjabipyri (Das, 1918) (Toxoptera) HRL
pyri Shaposhnikov, 1952 BM HRL
*fritzmuelleri* Leclant, 1967 HRL
rotundiventris (Signoret, 1860) (Schizoneura) BM HRL
*acori* Shinji, 1922 (*Toxoptera*)
*acori* Theobald, 1923 (*Aphis*) type BM
*calami* Theobald, 1923 (*Acaudus*) type BM
*cyperi* van der Goot, 1917 (*Toxoptera*)
rufula (Walker, 1849) (Aphis) lectotype BM HRL
*geijskesi* Hille Ris Lambers, 1939 HRL
setariae Rusanova, (1960) 1962 p. 87 = Rhopalosiphum maidis (Fitch, 1856)
variegata Ossiannilsson, 1959 BM HRL
viridirubra (Gillette & Palmer, 1932) (Toxoptera) metatypes BM HRL
wahlgreni (Ossiannilsson, 1959) (Rhopalosiphum) BM HRL
weingaertneriae Hille Ris Lambers, 1947 BM HRL
werderi Börner, 1950 HRL

Subgen. Euschizaphis Hille Ris Lambers, 1947

S. (Euschizaphis longicornis Richards, 1961 (Euschizaphis)
(Euschizaphis) palustris (Theobald, 1929) (Aphis) BM HRL

Subgen. Paraschizaphis Hille Ris Lambers, 1947

S. (Paraschizaphis) brachytarsus (Takahashi, 1961) (Rhopalosiphum) HRL
(Paraschizaphis) caricis (Schouteden, 1906) (Aphis) BM HRL (types)
*hirticola* Borner, 1950 (*Schizaphis*) HRL
(Paraschizaphis) eastopi van Harten & Ilharco, 1971 (Schizaphis) BM HRL
(Paraschizaphis) longisetosa (Higuchi, 1970) (Paraschizaphis) HRL
(Paraschizaphis) nigra (Baker, 1917) (Toxoptera) HRL
(Paraschizaphis) scirpi (Passerini, 1874) (Toxoptera) BM HRL
*rosazevedoi* Ilharco, 1961 (*Paraschizaphis*)
*typhae* Laing, 1923 (*Toxoptera*) type BM

S. (Paraschizaphis) scirpi subsp. eriophori (F.P. Müller, 1974) (Paraschizaphis) BM HRL
(Paraschizaphis) scirpicola (Hille Ris Lambers, 1960) (Paraschizaphis) BM HRL

SCHIZODRYOBIUS van der Goot, 1913
Tijdschr. Ent. 56: 130
Type species Lachnus exsiccator Altum, 1862 = Aphis pallipes Hartig, 1841
= Lachnus Burmeister, 1835

SCHIZOLACHNUS Mordvilko, 1909 (1908)
Ezheg. zool. Mus. 13: 375
Type speciesAphis tomentosa de Geer, 1773 = pineti Fabricius, 1781
*Unilachnus* Wilson, 1919

S. curvispinosus Hottes, Essig & Knowlton, 1954 HRL
flocculosus (Williams, 1911) (Lachnus) BM HRL
*tusoca* Hottes & Wehrle, 1950
*wahlea* Hottes, 1952 paratypes BM
lanosus Hottes, 1959
obscurus Börner, 1940 BM HRL
*pineti* subsp. *obscurus* Börner, 1940
orientalis (Takahashi, 1924) (Unilachnus) BM HRL
parvus (Wilson, 1915) (Lachnus) BM HRL
pineti (Fabricius, 1781) (Aphis) BM HRL
*fuliginosa* Buckton, 1881 (*Schizoneura*) lectotype BM
*pilosa* Buckton, 1883 (*Glyphina*) lectotype BM
*tomentosa* Villers, 1789 (*Aphis*)
*tomentosa pini* De Geer, 1773 (*Aphis*) invalid trinominal
pineti subsp. obscurus Börner, 1940 = obscurus Börner, 1940
piniradiatae (Davidson, 1909) (Lachnus) BM HRL
tusoca Hottes & Wehrle, 1950 = flocculosus (Williams, 1911)
wahlea Hottes, 1952 = flocculosus (Williams, 1911)

SCHIZOMYZUS Börner, 1950
Neue europäische Blattlausarten, privately published, p. 11
Type species Schizomyzus lindneri Börner, 1950 = Acaudella rubida Börner, 1939
= Pseudacaudella Börner, 1944

S. lindneri Börner, 1950 = Pseudacaudella rubida (Börner, 1939)

SCHIZONEURA Hartig, 1839
Jber. Fortschr. Forstwiss. u. forst.-naturk. im Jahre 1836 u. 1837 1 (4): 654
Type species Aphis ulmi (Linnaeus) of Passerini, 1860 = Chermes ulmi Linnaeus, 1758
*Afghanaphis* Takahashi, 1966
*Mimaphidus* Rondani, 1848
Here used as subg. of Eriosoma Leach, 1818

S. albipennis Walker, 1852 nomen dubium
americana Riley, 1879 to Eriosoma Leach, 1818
ampelorrhiza del Guercio, 1900 = Eriosoma (Schizoneura) ulmi (Linnaeus, 1758)
astilbensis Shinji, 1939 nomen dubium
caucasica Dzhibladze, 1960 to Colopha Monell, 1877
cerealium Szanislo, 1880/ 1 = Anoecia vagans (Koch, 1856)
compressa Koch, 1856 to Colopha Monell, 1877
corni Hartig, 1841 = Anoecia corni (Fabricius, 1775)
costata Hartig, 1841 = Cinara costata (Zetterstedt, 1828)
crataegi Oestlund, 1887 to Eriosoma Leach, 1818
fodiens Buckton, 1881 = Eriosoma (Schizoneura) ulmi (Linnaeus, 1758)
fuliginosa Buckton, 1881 = Schizolachnus pineti (Fabricius, 1781)
fulviabdominalis Sasaki, 1899 to Anoecia Koch, 1857
graminis del Guercio, 1895 = Anoecia corni (Fabricius, 1775)
grossulariae Taschenberg, 1887 = Eriosoma (Schizoneura) ulmi (Linnaeus, 1758)
japonica Matsumura, 1917 to Eriosoma (Schizoneura)
karschii Lichtenstein, 1885 = ? Smynthurodes betae Westwood, 1849
kochii Lichtenstein, 1885 = Anoecia corni (Fabricius, 1775)
lanuginosa Hartig, 1839 to Eriosoma (Schizoneura)
lusitanica Horvath, 1908 nomen dubium
meunieri Heie, 1969 = patchi Meunier, 1917
nigriabdominalis Sasaki, 1899 to Tetraneura (Tetraneurella)
obliquus Cholodkovaky, 1896 to Mindarus Koch, 1857
obscura Walker, 1852 = Anoecia corni (Fabricius, 1775)
panicola Thomas, 1879 = Anoecia cornicola (Walsh, 1862)
passerinii Signoret, 1875 type of Loewia Lichtenstein, 1886 nec Egger, 1856; and of Phloeomyzus Horvath, 1896 q.v.
patchae later spelling of patchiae Borner & Blunck, 1916
patchi Meunier, 1917. Fossil
*meunieri* Heie, 1969
patchiae Börner & Blunck, 1916 to Eriosoma (Schizoneura)
pinicola Thomas, 1879 = Mindarus abietinus Koch, 1857

S. piri Goethe, 1884 = Eriosoma (Schizoneura) lanuginosum (Hartig, 1841)
populi Gillette, 1908 = Pachypappa pseudobyrsa (Walsh, 1863)
radicicola Mokrzecky, 1903 = ? Anoecia vagans Koch, 1856
reaumuri Kaltenbach, 1843 type of Patchiella Tullgren, 1925 q.v.
rosaefoliae Shinji, 1939 nomen dubium
rotundiventris Signoret, 1860 to Schizaphis Börner, 1931
soror Börner & Blunck, 1916 = Eriosoma (Schizoneura) ulmi (Linnaeus, 1758)
spartanthi Boisduval, lapsus pro sparthani
sparthani Boisduval, 1866 nomen dubium
stigma Curtis, 1844 = Cinara costata (Zetterstedt, 1828)
turbida Walker, 1852 nomen dubium
ulmi (Linnaeus, 1758) (Chermes) to Eriosoma (Schizoneura)
ulmi subsp. orientalis Mordvilko ex Shaposhnikov, 1955 to Eriosoma (Schizoneura)
vagans Koch, 1856 (type of Subanoecia Börner, 1952) to Anoecia Koch, 1857
venusta Passerini, 1860 = Anoecia vagans (Koch, 1856)

SCHIZONEURAPHIS van der Goot, 1917
Contrib. Faune Indes Néerl. 1 (3): 245
Type species Schizoneuraphis gallarum van der Goot, 1917

S. distylii van der Goot, 1917 to Reticulaphis Takahashi, 1958
foliorum van der Goot, 1917 HRL
gallarum van der Goot, 1917 BM HRL
*hongkongensis* van der Goot, 1918 (*Thoracaphis*)
malayana (Takahashi, 1950) (Thoracaphis) type BM

SCHIZONEURATA Hille Ris Lambers, 1973
Fla Ent. 56: 295
Type species Schizoneurata tissoti Hille Ris Lambers, 1973

S. tissoti Hille Ris Lambers, 1973 HRL

SCHIZONEURELLA Hille Ris Lambers, 1973
Orient. Insects 7: 249
Type species Schizoneurella indica Hille Ris Lambers, 1973

S. indica Hille Ris Lambers, 1973 HRL

SCHIZONEURITES Cockerell, 1915
Proc. U.S. natn. Mus. 49: 487
Type species Schizoneurites brevirostris Cockerell, 1915
*Antiquaphis* Heie, 1967
*Elektraphis* Steffan, 1968
Fossil

SCHIZONEUROIDES Buckton, 1883
Monograph of the British Aphides, London, 4: 178
Type species Schizoneuroides scudderi Buckton, 1883
= Mindarus Koch, 1856
Fossil

SCHLECHTENDALIA Lichtenstein, 1883
Stettin. ent. Ztg 44: 230
Type species Aphis chinensis Bell, 1851
*Abamalekia* del Guercio, 1906
*Meitanaphis* Tsai & Tang, 1946

S. chinensis (Bell, 1851) (Aphis) BM HRL
*? intermedia* Matsumura, 1917
*lazarewi* del Guercio, 1906 (*Abamalekia*) HRL
*mimifushi* Matsumura, 1917
*miyabei* Matsumura, 1917
*peitan* Tsai & Tang, 1946 (*Melaphis*)
*sinensis* Walker, 1852 (*Pemphigus*)
elongallis (Tsai & Tang, 1946) (Meitanaphis)
intermedia Matsumura, 1917 = ? chinensis (Bell, 1851)
mimifushi Matsumura, 1917 = chinensis (Bell, 1851)
miyabei Matsumura, 1917 = chinensis (Bell, 1851)

SCHOUTEDENIA Mordvilko, 1921 nec Rübsaamen, 1905
Izv. Sev. Oblast. Sta. Zashch. Rast. Vredit. 3: 63
Type species Geoica cyperi Schouteden, 1902
= Geoica Hart, 1894

SCHOUTEDENIA Rübsaamen, 1905
Marcellia 4: 19
Type species Schoutedenia ralumensis Rübsaamen, 1905
*Cerciaphis* Theobald, 1920
*Setaphidia* Strand, 1942
*Setaphis* van der Goot, 1917 nec Simon, 1893

S. emblica (Patel & Kulkarni, 1953) (Cerciaphis) types BM HRL
emblica susbp. andhraka David & Hille Ris Lambers, 1956 BM HRL
lutea (van der Goot, 1917) (Setaphis) BM HRL
*bougainvilleae* Theobald, 1920 (*Cerciaphis*) BM HRL

S. ralumensis Rübsaamen, 1905
viridis (van der Goot, 1917) (Setaphis) BM HRL
*formosanus* Takahashi, 1929 (*Setaphis*)

SCHOUTEDENUM Mordvilko, 1928
In Philipjev, I.N., Keys for the identification of Russian Insects, Moscow, p. 173
Type species Geoica cyperi Schouteden, 1902
= Geoica Hart, 1894

SCIAMYZUS Stroyan, 1954
Proc. R. Ent. Soc. Lond. (B) 23: 10, described as subg. of Myzus Passerini, 1860
Type species Myzus (Sciamyzus) cymbalariae Stroyan, 1954
Here used as subg. of Myzus Passerini, 1860

SEMIAPHIS van der Goot, 1913
Tijdschr. Ent. 56: 105
Type species Aphis carotae Koch, 1954 = dauci Fabricius, 1775

S. anthrisci (Kaltenbach, 1843) (Aphis) BM HRL
cervariae (Börner, 1932) (Brachycolus) BM HRL
coryspermi Mamontova ex Shaposhnikov, 1964 HRL
dauci (Fabricius, 1775) (Aphis) BM HRL
*carotae* Koch, 1854 (*Aphis*)
dauci subsp. seselii Börner, 1952 HRL
*seselii* Borner, 1952
escherichi Börner, 1931 = pimpinellae (Kaltenbach, 1843)
heraclei (Takahashi, 1921) (Brachycolus) BM HRL
*lonicerae* Shinji, 1939 (*Brachycolus*)
horvathi Szelegiewicz, 1967 paratypes BM HRL
longissima Narzikulov, 1965
lonicerina Shaposhnikov, 1952 HRL
moiwaensis Takahashi, 1965 BM
montana van der Goot, 1917 to Brachysiphoniella Takahashi, 1921
nolitangere Aizenberg, 1954 BM
pastinacae Börner, 1950 HRL
peucedani (Nevsky, 1928) (Xerophilaphis)
pimpinellae (Kaltenbach, 1843) (Aphis) BM HRL
*escherichi* Börner, 1931
seselii Börner, 1950 = dauci subsp. seselii Börner, 1952

SEMIAPHOIDES Rusanova (1942) 1943
Trudy Azerb. Gos. Univ. biol. 3 (1): 36
Type species Semiaphoides cannabiarum Rusanova, 1943 nomen nudum
Without described species

S. cannabiarum Rusanova, 1943 nomen nudum

SENECIOBIUM Remaudière, 1954
Revue Path. vég. Ent. agric. Fr. 33: 55
Type species Seneciobium balachowskyi Remaudière, 1954

S. balachowskyi Remaudière, 1954 BM HRL

SENSORIAPHIS Cottier, 1953
Bull. N.Z. Dep. scient. ind. Res. 106: 96
Type species Sensoriaphis nothofagi Cottier, 1953

S. furcifera Carver & White, 1974 paratypes BM HRL
nothofagi Cottier, 1953 BM HRL
tasmaniae Carver & Martyn, 1962 paratypes BM HRL

SERRATAPHIS van der Goot, 1917
Contrib. Faune Indes Néerl. 1 (3): 262
Type species Tetraneura lucifuga Zehntner, 1897
= Geoica Hart. 1894

SETAPHIDIA Strand, 1942
Folia zool. hydrobiol. 11: 393
Type species Setaphis luteus van der Goot, 1917
= Schoutedenia Rübsaamen, 1905

SETAPHIS van der Goot, 1917 nec Simon, 1893 (Arachnida)
Contrib. Faune Indes Néerl. 1 (3): 153
Type species Setaphis luteus van der Goot, 1917
= Schoutedenia Rübsaamen, 1905

S. formosanus Takahashi, 1929 = Schoutedenia viridis (van der Goot, 1917)
lutea van der Goot, 1917 (type of Setaphidia Strand, 1942) to Schoutedenia Rübsaamen, 1905
viridis van der Goot, 1917 to Schoutedenia Rübsaamen, 1905

SHAPOSHNIKOVIA Aizenberg, 1954
Diss. biol. Sci. Lenin Univ., Moscow p. 9
Without species

SHAPOSHNIKOVIELLA Mamontova, 1963
Pratsi Inst. Zool. Kyyiv 19: 15
Type species Shaposhnikoviella paradoxa Mamontova, 1963
= Neosappaphis Hille Ris Lambers, 1959

S. paradoxa Mamontova, 1963 to Neosappaphis Hille Ris Lambers, 1959

SHENAHWEUM Hottes & Frison, 1931
Bull. Ill. St. nat. Hist. Surv. 19: 267
Type species Drepanaphis minutus Davis, 1910
Here used as subg. of Drepanaphis del Guercio, 1909

SHINJIA Takahashi, 1938
Tenthredo 2: 26
Type species Microtarsus pterydifoliae Shinji, 1929 = Atarsos orientalis Mordvilko, 1929
*Microtarsus* Shinji, 1929 nec Eyton, 1839

S. orientalis (Mordvilko, 1929) (Atarsos) BM HRL
*pterydifoliae* Shinji, 1929 (*Microtarsus*)

SHIVAPHIS Das, 1918
Mem. Indian Mus. 6: 245
Type species Shivaphis celti Das, 1918

S. arundinariae Takahashi, 1940 to Cranaphis Takahashi, 1939
bambusicola (David, Rajasingh & Narayanan, 1971) (Cranaphis) HRL
celti Das, 1918 BM HRL
*celticolens* Essig & Kuwana, 1918 (*Chromaphis*)
celticola (Nevsky, 1929) (Therioaphis) BM HRL

SILENOBIUM Börner, 1939
Arb. physiol. angew. Ent. Berl. 6: 78
Type species Silenobium schusteri Börner, 1939
= Volutaphis Börner, 1939

S. appeli Börner, 1950 = Volutaphis centaureae (Börner, 1939)
schusteri Börner, 1939 to Volutaphis Börner, 1939

SINOCALLIS Tseng & Tao, 1938
Jl W. China Border Res. Soc. 10: 213
Type species Sinocallis mirabilis Tseng & Tao, 1938 = Glyphina rhusae Shinji, 1922
= Dasyaphis Takahashi, 1938

S. mirabilis Tseng & Tao, 1938 = Dasyaphis rhusae (Shinji, 1922)

SINOCHAITOPHORUS Takahashi, 1936
Lingnan Sci. J. 15: 197
Type species Sinochaitophorus maoi Takahashi, 1936

S. maoi Takahashi, 1936 BM HRL

SINOCOLOPHA Tao, 1970
Q. Jl Taiwan Mus. 23: 138
Type species Colopha graminis (Takahashi, 1930) = Truncaphis graminis Takahashi, 1930
= Colopha Monell, 1877

SINOLACHNUS Hille Ris Lambers, 1956
Z. angew. Ent. 39: 472
Type species Lachnus niitakayamensis Takahashi, 1925

S. niitakayamensis (Takahashi, 1925) (Lachnus) HRL

SINOMEGOURA Takahashi, 1960
Kontyû 28: 228
Type species Acyrthosiphon photinae Takahashi, 1936

S. citricola (van der Goot, 1917) (Macrosiphoniella) BM HRL
*camphorae* Shinji, 1922 (*Tuberosiphum*)
*jacobsoni* Mason, 1927 (*Megoura*)
elaeocarpi (Tao, 1963) (Acyrthosiphon)
photiniae (Takahashi, 1936) (Acyrthosiphon) BM HRL
rhododendri (Takahashi, 1937) (Acyrthosiphon)
? vagans (van der Goot, 1917) (Rhopalosiphum)

SINONIPPONAPHIS Tao, 1966
Q. Jl Taiwan Mus. 19: 175
Type species Astegopteryx formosana Takahashi, 1927

S. formosana (Takahashi, 1927) (Astegopteryx)
monzeni (Takahashi, 1958) (Nipponaphis) BM HRL

SINOSIPHONIELLA Tao, 1963
Pl. Prot. Bull., Taiwan 5: 200
Type species Macrosiphoniella kuwayamai Takahashi, 1941
= Metopeurum Mordvilko, 1914

SIPHA Passerini, 1860
Gli Afidi, Parma, p. 29
Type species Aphis glyceriae Kaltenbach, 1843

S. agropyrella Hille Ris Lambers, 1939 = Sipha (Rungsia) elegans del Guercio, 1905
agropyronensis (Gillette, 1911) (Chaitophorus) HRL
arenariae Mordvilko, 1928 emendation of arenarii Mordvilko, 1921
arenarii Mordvilko, 1921 to Sipha (Rungsia)
avenae del Guercio, 1900 = Sipha (Rungsia) maydis Passerini, 1860
berlesei del Guercio, 1905 type of Chaetosiphella Hille Ris Lambers, 1939 q.v.
brunnea Nevsky, 1951 = Sipha (Rungsia) maydis Passerini, 1860
burakowskii Holman & Szelegiewicz, 1974 to Sipha (Rungsia)
carrerai E.E. Blanchard, 1939 = flava (Forbes, 1884)
elegans del Guercio, 1905 to Sipha (Rungsia)
euphorbiae Macchiati, 1881 = Aphis tirucallis Hille Ris Lambers, 1954
flava (Forbes, 1884) (Chaitophorus) BM HRL
*carrerai* E.E. Blanchard, 1939
glyceriae (Kaltenbach, 1843) (Aphis) BM HRL
*glyceriae* var. *italica* del Guercio, 1905
*schoutedeni* del Guercio, 1905
glyceriae var. italica del Guercio, 1905 = glyceriae (Kaltenbach, 1843)
graminis Kaltenbach, 1864 (type of Siphonella Borner, 1939) = Sipha (Rungsia) maydis Passerini, 1860
kurdjumovi Mordvilko, 1921 = Sipha (Rungsia) elegans del Guercio, 1905
littoralis (Walker, 1848) (Aphis) lectotype BM HRL
maydis Passerini, 1860 to Sipha (Rungsia)
maydis var. avenae del Guercio, 1905 = Sipha (Rungsia) maydis Passerini, 1860
minuta Tissot, 1932 to Chaitophorus Koch, 1854
paradoxa Theobald, 1918 = Atheroides serrulatus Haliday, 1839
polygoni Schouteden, 1907 = Aspidaphis adjuvans (Walker, 1848)
rubifolii Thomas, 1879 to Aphis Linnaeus, 1758
schoutedeni del Guercio, 1905 = glyceriae (Kaltenbach, 1843)
tshernavini Mordvilko, 1921 to Chaetosiphella Hille Ris Lambers, 1939
uvarovi Mordvilko, 1921 to Sipha (Rungsia)

Subgen. Rungsia Mimeur, 1933

S. (Rungsia) aegilopis (Bozhko, 1961) (Rungisa)
(Rungsia) arenarii Mordvilko, 1921 (Sipha) BM HRL
(Rungsia) burakowskii (Holman & Szelegiewicz, 1972) (Sipha) BM HRL
(Rungsia) elegans del Guercio, 1905 (Sipha) BM HRL
*agropyrella* Hille Ris Lambers, 1939 (*Sipha*) cotypes BM HRL
*kurdjumovi* Mordvilko, 1921 (*Sipha*)
*nemaydis* Narzikulov, 1963 (*Rungsia*) HRL
(Rungsia) maydis Passerini, 1860 (Sipha) BM HRL
*avenae* del Guercio, 1900 (*Sipha*)
*brunnea* Nevsky, 1951 (*Sipha*)
*graminis* Kaltenbach, (*Sipha*) 1864
*graminis* Mimeur, 1933 (*Rungsia*) BM
*maydis* var. *avenae* del Guercio, 1905 (*Sipha*)
(Rungsia) praecocis (Bozhko, 1959) (Rungsia)
(Rungsia) taurica (Mamontova, 1959) (Rungsia) HRL
(Rungsia) uvarovi Mordvilko, 1921 (Sipha)

SIPHOCORYNE Passerini, 1860
Gli Afidi, Parma, p. 28
Type species Aphis nymphaeae Linnaeus, 1761
= Rhopalosiphum Koch, 1854

S. acericola Matsumura, 1917 nomen dubium
alboapicalis Theobald, 1916 to Cryptomyzus Oestlund, 1922
angelicae del Guercio, 1911 = Coloradoa absinthiella Ossiannilsson, 1962
aquatica Gillette & Bragg, 1916 to Cavariella (Cavaraiellia)
archangelicae Oestlund, 1886 = Cavariella konoi Takahashi, 1939
artemisiae del Guercio, 1913 to Coloradoa Wilson, 1910
bicaudata Essig & Kuwana, 1918 = Cavariella salicicola (Matsumura, 1917)
cacaliae Matsumura, 1918 = Brachycaudus helichrysi (Kaltenbach, 1843)
conii Davidson, 1909 = Hyadaphis foeniculi (Passerini, 1860)
cornicolum Matsumura, 1918 to ? Brachycolus Buckton, 1879
donarium Matsumura, 1918 = Rhopalosiphum nymphaeae (Linnaeus, 1761)
essigi Gillette & Bragg, 1918 = Cavariella pastinacae (Linnaeus, 1758)
foeniculi Passerini, 1860 to Hyadaphis Kirkaldy, 1904
fraxinicola Matsumura, 1917 nomen dubium
indobrassicae Davis, 1918 = Lipaphis erysimi (Kaltenbach, 1843)
japonica Essig & Kuwana, 1918 to Cavariella del Guercio, 1911
passerinii del Guercio, 1911 to Hyadaphis Kirkaldy, 1904

S. polygoni Shinji, 1922 (Siphonoeoryne)
salicis Monell, 1879 to Cavariella del Guercio, 1911
splendens Theobald, 1915 = Rhopalosiphum rufiabdominalis (Sasaki, 1899)
xanthii Oestlund, 1886 to Capitophorus van der Goot, 1913

SIPHOCORYNE Passerini, 1863 nec 1860
Archo Zool. Anat. Fisiol. 2: 175
Type species Aphis xylostei Schrank, 1801
= Hyadaphis Kirkaldy, 1904

SIPHONAPHIS van der Goot, 1915
Beiträge zur Kenntnis der Holländischen Blattläuse, Haarlem-Berlin, p. 238
Type species Aphis nymphaeae Linnaeus, 1761
= Rhopalosiphum Koch, 1854

SIPHONATROPHIA Swain, 1918
Ent. News 29: 363
Type species Cerosipha cupressi Swain, 1918
*Minuticornis* Knowlton, 1928

S. cupressi (Swain, 1918) (Cerosipha) BM HRL
gravida (Knowlton, 1928) (Minuticornis) BM HRL

SIPHONELLA Börner, 1939 nec Macquart, 1835; Hagenow ex Geinitz, 1850; Issel, 1869; Verril, 1879; Ledenfeld, 1887.
Arb. physiol. angew. Ent. Berl. 6: 77
Type species Sipha graminis Kaltenbach, 1864 = Sipha maydis Passerini, 1860
= Rungsia Mimeur, 1933

SIPHONOCALLIS Del Guercio, 1914 (1913)
Redia 9: 293 (1913), pp. 227-297 were published on 23 February, 1914
Type species Callipterus betulaecolens Fitch, 1851
= Calaphis Walsh, 1863

SIPHONOEORYNE Shinji, 1922 lapsus pro Siphocoryne Passerini, 1860

SIPHONOPHORA Koch, 1855 nec Fischer, 1823; Brandt, 1837
Die Pflanzenläuse Aphiden, Nürnberg, p. 150
Type species Aphis rosae Linnaeus, 1758
= Macrosiphum Passerini, 1860

S. acerifoliae Thomas, 1878 type of Drepanaphis del Guercio, 1909 q.v.
achilleae Koch, 1855 to Uroleucon Mordvilko, 1914
achyrantes Monell, 1879 = Myzus (Necarosiphon) persicae (Sulzer, 1776)
adianthi Oestlund, 1886 to Sitobion Mordvilko, 1914
alliariae Koch, 1855 p. 160 = Uroleucon sonchi (Linnaeus, 1767)
alliariae Koch, 1855 p. 177 = Nasonovia ribisnigri (Mosley, 1841)
ambrosiae Thomas, 1878 to Uroleucon Mordvilko, 1914
antherrinii Macchiati, 1883 emended to antirrhinii
antirrhinii Macchiati, 1883 = Myzus (Nectarosiphon) persicae Sulzer, 1776
artemisiae Koch, 1855 to Titanosiphon Nevsky, 1928
artemisicola Williams, 1911 to Pleotrichophorus Börner, 1930
artocarpi Westwood, 1890 type of Greenidea Schouteden, 1905 q.v.
asclepiadifolii Thomas, 1878 = Macrosiphum euphorbiae (Thomas, 1878)
atra Ferrari, 1872 type of Macrosiphoniella del Guercio, 1911 q.v.
atropae Mordvilko, 1895 = Aulacorthum solani (Kaltenbach, 1843)
bifrontis Passerini, 1879 nomen dubium
caianensis del Guercio, 1900 = Sitobion avenae (Fabricius, 1775)
calendulae Monell, 1879 = Uroleucon chrysanthemi (Oestlund, 1886)
calendulella Monell, 1879 = Myzus (Nectarosiphon) persicae (Sulzer, 1776)
callae Henrich, 1909 = Aulacorthum (Neomyzus) circumflexum (Buckton, 1876)
caraganae Cholodkovsky, 1907 to Acyrthosiphon Mordvilko, 1914
carnosa Buckton, 1876 to Microlophium Mordvilko, 1914
carnosa var. impatientis Williams, 1911 incertae sedis
cholodkovskyi Mordvilko, 1909 to Macrosiphum Passerini, 1860
chrysanthemi Oestlund, 1886 to Uroleucon Mordvilko, 1914
chrysanthemicolens Williams, 1911 = Macrosiphoniella sanborni (Gillette, 1908)
cichorii Koch, 1855 to Uroleucon Mordvilko, 1914
circumflexa Buckton, 1876 (type of Neomyzus van der Goot, 1915) to Aulacorthum (Neomyzus)
citrifolii Ashmead, 1880 = Macrosiphum euphorbiae (Thomas, 1878)
coreopsidis Thomas, 1878 to Aphis Linnaeus, 1758
corydalis Oestlund, 1886 = Acyrthosiphon pisum (Harris, 1776)
crataegi Monell, 1879 to Utamphorophora Knowlton, 1947
cucurbitae Middleton ex Thomas, 1878 = Macrosiphum euphorbiae (Thomas, 1878)
cyparissiae Koch, 1855 (type of Mirotarsus Börner, 1939) to Acyrthosiphon Mordvilko, 1914
desmodii Williams, 1911 to Microparsus Patch, 1919
diplanterae Koch, 1855 = Aulacorthum solani (Kaltenbach, 1843)
dubia Ferrari, 1872 to ? Macrosiphoniella del Guercio, 1911 ?

S. erigeronensis Thomas, 1878 (type of Lambersius Olive, 1965) to Uroleucon (Lambersius)
eupatorii Williams, 1911 to Macrosiphum Passerini, 1860
euphorbiae Thomas, 1878 to Macrosiphum Passerini, 1860
euphorbicola Thomas, 1878 = Macrosiphum euphorbiae (Thomas, 1878)
filaginis Lichtenstein, 1884 nomen nudum
fragariae Koch, 1855 = Macrosiphum rosae (Linnaeus, 1758)
fragariella Theobald, 1905 = Amphorophora rubi (Kaltenbach, 1843)
frigidae Oestlund, 1886 to Obtusicauda Soliman, 1927
funesta Macchiati, 1885 to Macrosiphum Passerini, 1860
gaurae Williams, 1911 to Macrosiphum Passerini, 1860
gaurina Williams, 1911 = Macrosiphum gaurae (Williams, 1911)
gei Koch, 1855 to Macrosiphum Passerini, 1860
gerardiae Thomas, 1879 to Aphis Linnaeus, 1758
glaucii Lichtenstein, 1884 nomen nudum
gnidii Lichtenstein, 1884 nomen nudum
grindeliae Williams, 1911 to Illinoia Wilson, 1910
heucherae Thomas, 1879 to Kakimia Hottes & Frison, 1931
inulae Ferrari, 1872 to Uroleucon (Belochilum)
kochii Ferrari, 1872 = Titanosiphon artemisiae (Koch, 1855)
lactucae Passerini, 1860 to Acyrthosiphon Mordvilko, 1914
leptadeniae Vuillet & Vuillet, 1914 = Aphis nerii Boyer de Fonscolombe, 1841
leucanthemi Ferrari, 1872 to Macrosiphoniella del Guercio, 1911
lilacina Ferrari, 1872 = Macrosiphoniella tanacetaria (Kaltenbach, 1843)
lilii Monell, 1879 to Macrosiphum Passerini, 1860
linariae Koch, 1855 to Macrosiphoniella (Asterobium)
liriodendri Monell, 1879 type of Illinoia Wilson, 1910 q.v.
longipennis Buckton, 1876 = Metopolophium dirhodum (Walker, 1849)
ludovicianae Oestlund, 1886 to Macrosiphoniella del Guercio, 1911
lutea Buckton, 1876 to Sitobion Mordvilko, 1914
luteola Williams, 1911 to Uroleucon (Lambersius)
menthae Buckton, 1876 = Aulacorthum solani (Kaltenbach, 1843)
menthae Buckton of Shaposhnikov, 1964 (type of Ovatophorodon Aizenberg, 1966) = Ovatus mentharius (van der Goot, 1913)
minor Forbes, 1884 to Chaetosiphon (Pentatrichopus)
muralis Buckton, 1876 to Uroleucon Mordvilko, 1914
nasturtii Koch, 1855 = Myzus (Nectarosiphon) persicae (Sulzer, 1776)
oblonga Mordvilko, 1901 (type of Phalangomyzus Börner, 1939) to Macrosiphoniella del Guercio, 1911
obscura Koch, 1855 to Uroleucon Mordvilko, 1914

S. olivata Buckton, 1876 = Uroleucon cirsii (Linnaeus, 1758)
ononis Ferrari, 1872 = Acyrthosiphon pisum subsp. ononis (Koch, 1855)
ononis Koch, 1855 = Acyrthosiphon pisum subsp. ononis (Koch, 1855)
orobanches Lichtenstein, 1884 nomen nudum
panicola Thomas, 1878 = Hysteroneura setariae (Thomas, 1878)
poae Macchiati, 1885 = Sitobion fragariae (Walker, 1848)
polygoni Buckton, 1876 = Nasonovia ribisnigri (Mosley, 1841)
? portschinskyi Mordvilko, 1909 = Macrosiphum cholodkovskyi (Mordvilko, 1909)
rosae (Linnaeus, 1758) (Aphis) type of Macrosiphum Passerini, 1860 q.v.
rosae var. floridae Ashmead, 1882 = Macrosiphum floridae (Ashmead, 1882)
rosae var. glauca Buckton, 1876 = Macrosiphum rosae (Linnaeus, 1758)
rosaecola Passerini, 1871 = Macrosiphum rosae (Linnaeus, 1758)
rosarum Koch, 1855 = Chaetosiphon (Pentatrichopus) tetrarhodum (Walker, 1849)
rubi var. rufa Buckton, 1883 = Macrosiphum funestum (Macchiati, 1885)
salicicola Thomas, 1878 = Aphis farinosa Gmelin, 1790
scrophulariae Buckton, 1876 = Cryptomyzus galeopsidis (Kaltenbach, 1843)
setariae Thomas, 1878 type of Hysteroneura Davis, 1919 q.v.
sisymbrii Buckton, 1876 = Uroleucon cichorii (Koch, 1855)
solanifolii Ashmead, 1882 = Macrosiphum euphorbiae (Thomas, 1878)
sonchella Monell, 1879 to Uroleucon Mordvilko, 1914
spartii Koch, 1855 = Acyrthosiphon pisum (Harris, 1776)
subterranea Koch, 1855 to Macrosiphoniella del Guercio, 1911
tiliae Monell, 1879 to Macrosiphum Passerini, 1860
tulipae Monell, 1879 = Macrosiphum euphorbiae (Thomas, 1878)
verbenae Thomas, 1878 to Macrosiphum Passerini, 1860
viticola Thomas, 1878 = Aphis illinoisensis Shimer, 1866

SIPHONOPHOROIDES Buckton, 1883
Monograph of the British Aphides, London 4: 176
Type species Siphonophoroides antiqua Buckton, 1883
*Amalancon* Scudder, 1890
*Archilachnus* Buckton, 1883
*Cataneura* Scudder, 1890
*? Oryctaphis* Scudder, 1890
*Tephraphis* Scudder, 1890
Fossil

SITOBION Mordvilko, 1914
Faune Russie, Ins. Hém., 1 (1): 337
Type species Aphis granaria Kirby, 1798 = Aphis avenae Fabricius, 1775
*Anameson* Mordvilko, 1914
*Aphidiella* Theobald, 1923
*Sitobium* Mordvilko, 1914, without species, errore pro Sitobion

S. adgnatum (F.P. Müller, 1959) (Macrosiphum (Sitobion)) paratypes BM HRL
adianti (Oestlund, 1886) (Siphonophora) BM HRL
africanum (Hille Ris Lambers, 1954) (Macrosiphum (Sitobion)) BM HRL
africanum subsp. matatum (Eastop, 1959) (Macrosiphum (Sitobion)) types BM HRL
akebiae (Shinji, 1935 & 1941) (Macrosiphum) BM HRL
alopecuri (Takahashi, 1921) (Macrosiphum) HRL
alopecuri subsp. sylvesteri Hille Ris Lambers, 1966
*sylvesteri* Hille Ris Lambers, 1966
anselliae (Hall, 1932) (Myzus) types BM HRL
*schoelli* F.P. Müller, 1959 (*Macrosiphum (Sitobion)*) BM HRL
avenae (Fabricius, 1775) (Aphis) BM HRL
*adjuta* Walker, 1848 (*Aphis*) types BM
*alii* Jackson, 1918 (*Macrosiphum*) paratype BM
*caianensis* del Guercio, 1900 (*Siphonophora*)
*cerealis* Kaltenbach, 1843 (*Aphis*)
*consueta* Walker, 1848 (*Aphis*) lectotype BM
*gnaphalii* Walker, 1849 (*Aphis*) type BM
*granaria* Kirby, 1798 (*Aphis*)
*hordei* Kyber, 1815 (*Aphis*)
*lycopsidis* Walker, 1848 (*Aphis*) lectotype BM
*oljatae* Hottes, 1950 (*Macrosiphum*) BM
*secretocauda* Theobald, 1923 (*Aphidiella*) type BM
*triglochiniella* Theobald, 1928 (*Macrosiphoniella*) type BM
avenae subsp. graminis Takahashi, 1950 = graminis Takahashi, 1950
avenae var. longiarticulatus Rusanova, 1943 nomen nudum
avenae var. longisiphon Rusanova, 1962
bamendae (Eastop, 1959) (Macrosiphum (Sitobion)) types BM HRL
berchemiae (Takahashi, 1938) (Macrosiphum)
caricis (Glendenning, 1926) (Macrosiphum)
chanikiwiti (Eastop, 1959) (Macrosiphum (Sitobion)) types BM HRL
cissi (Theobald, 1920) (Macrosiphum) types BM HRL

S. colei (Eastop, 1959) (Macrosiphum (Sitobion)) types BM HRL
congolense (Hille Ris Lambers & Doncaster, 1956) (Macrosiphum (Sitobion)) types BM HRL
cuscutae (Holman, 1974) (Macrosiphum) HRL
cystopteridis Robinson, 1966 (Macrosiphum) paratypes BM HRL
dorsatum (Richards, 1967) (Aulacorthum) paratypes BM HRL
dryopteridis (Holman, 1959) (Aulacorthum) BM HRL
equiseti Holman, 1961 paratypes BM HRL
fragariae (Walker, 1848) (Aphis) lectotype BM HRL
*avenivorum* Kirkaldy, 1905 (*Macrosiphum*)
*dallmani* Theobald, 1924 (*Aphis*) type BM HRL
*harpagorubus* Knowlton, 1935 (*Macrosiphum*)
*laricellus* Theobald, 1927 (*Myzus*)
*molluginellus* Theobald, 1924 (*Myzus*)
*poae* Macchiati, 1885 (*Siphonophora*)
*rubiellum* Theobald, 1913 (*Macrosiphum*)
graminearum (Mordvilko, 1919) (Acyrthosiphon (Metopolophium))
graminis Takahashi, 1950 types BM HRL
*avenae* subsp. *graminis* Takahashi, 1950
*gathaka* Eastop, 1955 (*Macrosiphum (Sitobion)*) types BM HRL
gravelii van der Goot, 1917 (Macrosiphum) HRL
*spinotibium* M.R. Ghosh, A.K. Ghosh & RayChaudhuri, 1971 (*Macrosiphum*) HRL
halli (Eastop, 1959) (Macrosiphum (Sitobion)) types BM HRL
hirsutirostris (Eastop, 1959) (Macrosiphum (Sitobion)) types BM HRL
howlandae (Eastop, 1959) (Macrosiphum (Sitobion)) types BM HRL
ibarae (Matsumura, 1917) (Macrosiphum) BM HRL
*sumatraensis* Mason, 1927 (*Macrosiphum*)
indicum A.N. Basu, 1964 HRL
insuralis (Pergande, 1900) (Nectarophora)
insulare subsp. yagasogae (Hottes, 1948) (Macrosiphum) BM HRL
*yagasogae* Hottes, 1948 (Macrosiphum)
kamtshaticum (Mordvilko, 1919) (Anameson) HRL
*rubiphilum* Takahashi, 1964 (*Macrosiphum (Sitobion)*) HRL
krahi (Eastop, 1959) (Macrosiphum (Sitobion)) types BM HRL
lambersi David, 1956 BM HRL
leelamaniae (David, 1958) (Macrosiphum (Sitobion)) cotypes BM HRL
longisiphon Rusanova, (1943, 1960 nomina nuda) 1962
luteum (Buckton, 1876) (Siphonophora) types BM HRL
*aurantiaca* del Guercio, 1913 (*Macrosiphoniella*)
*polystachyae* Franssen & Tiggelovend, 1935 (*Macrosiphum*)

S. manitobense (Robinson, 1965) (Macrosiphum (Sitobion)) paratypes BM HRL
martorelli (Smith, 1960) (Macrosiphum) BM HRL
microspinulosum (David, Narayanan & Rajasingh, 1972) (Macrosiphum (Sitobion)) HRL
miscanthi (Takahashi, 1921) (Macrosiphum) BM HRL
*eleusines* Theobald, 1929 (*Macrosiphum*)
mucatha (Eastop, 1955) (Macrosiphum (Sitobion)) types BM HRL
neusi (Eastop, 1958) types BM HRL
*africanum* subsp. *neusi* Eastop, 1958 (*Macrosiphum (Sitobion)*)
nigeriense (Eastop, 1959) (Macrosiphum (Sitobion)) types BM HRL
nigrinectarium (Theobald, 1915) (Macrosiphum) types BM HRL
*brachytarsus* Hille Ris Lambers, 1933 (*Macrosiphum*) HRL
niwanistum Hottes, 1933 (Adactynus) BM HRL
ochnearum (Eastop, 1959) (Macrosiphum (Sitobion)) types BM HRL
orchidacearum Franssen & Tiggelovend, 1935 (Macrosiphum)
pauliani Remaudière, 1957 BM HRL
phyllanthi (Takahashi, 1937) (Amphorophora) BM HRL
plectranthi (M.R. Ghosh, A.K. Ghosh & RayChaudhuri, 1971) (Macrosiphum (Sitobion)) HRL
ptericolens (Patch, 1919) (Macrosiphum) BM HRL
pteridis (Wilson, 1915) (Macrosiphum) BM HRL
rhamni (Clarke, 1903) (Nectarophora) BM HRL
rosaeiformis (Das, 1918) (Macrosiphum) BM HRL
salicicornii (Richards, 1963) (Macrosiphum (Sitobion)) paratypes BM HRL
salviae (Bartholomew, 1932) (Macrosiphum) BM HRL
*hyptis* E.E. Blanchard, 1944 (*Macrosiphum*)
*mesosphaeri* Tissot, 1934 (*Macrosiphum*) HRL
scoticum (Stroyan, 1969) (Macrosiphum (Sitobion)) paratypes BM HRL
smilaceti (Takahashi, 1924) (Macrosiphum) or = ? Impatientinum impatiens (Shinji, 1922)
smilacicola (Takahashi, 1924) (Macrosiphum)
smilacicola subsp. sikkimense (A.K. Ghosh & RayChaudhuri, 1968) (Macrosiphum (Sitobion))
smilacifoliae (Takahashi, 1921) (Macrosiphum) BM HRL
sylvesteri Hille Ris Lambers, 1966 = alopecuri subsp. sylvesteri Hille Ris Lambers, 1966
takahashii (Eastop, 1959) (Macrosiphum (Sitobion)) BM HRL
*phyllanthi* Takahashi, 1939 (*Macrosiphum*)
wikstroemiae (Mamet, 1939) (Amphorophora) BM HRL
yakini (Eastop, 1959) (Macrosiphum (Sitobion)) types BM HRL

S. yasumatsui (Moritsu, 1958) (Macrosiphum)
yongyooti (Robinson, 1972) (Macrosiphum (Sitobion)) paratype BM HRL

SITOBIUM Mordvilko, 1914
Faune Russie, Ins. Hém. 1: 65, in key
Without species
Error pro Sitobion Mordvilko, 1914

SITOMYZUS Hille Ris Lambers, 1952
Meddr Grønland 136: 9
Type species Sitomyzus vibei Hille Ris Lambers, 1952

S. columbiae Richards, 1960 = Utamphorophora humboldti (Essig, 1941)
japonicus Takahashi, 1963 = Juncomyzus rhois (Takahashi, 1924)
vibei Hille Ris Lambers, 1952 BM HRL

SLAVUM Mordvilko, 1927
Mém. Soc. zoöl. Fr. 28: 73
Type species Slavum lentiscoides Mordivlko, 1927

S. esfandiarii Davatchi & Remaudière, 1957 HRL
lentiscoides Mordvilko, 1927 HRL
mordvilkoi Kreutzberg, 1953
wertheimae Hille Ris Lambers, 1957 HRL

SMIELA Mordvilko, (1929) 1948
Trudȳ prikl. Ent. 14: 50 nomen nudum (1929)
In Tarbinsky & Plaviltchikov, 1948, Keys to Insects of European Part of USSR, p. 226
Type species Smiela fusca Mordvilko, (1929) 1948
*Brevicorynaphis* Hille Ris Lambers, 1956

S. fusca Mordvilko, (1929 nomen nudum) 1948 BM HRL
mongolica Holman & Szelegiewicz, 1975
schneideri (Hille Ris Lambers, 1956) (Brevicorynaphis) HRL
schneideri subsp. alyssi Mamontova, 1962 HRL
syreniae Bozhko, 1963 type BM HRL

SMINTHURAPHIS Quednau, 1953
Zool. Anz. 150: 224
Type species Sminthuraphis ulrichi Quednau, 1953

S. ulrichi Quednau, 1953 cotypes BM HRL

SMINTHURODES Schaum, 1850 irregular rectification of Smynthurodes Westwood, 1849

SMYNTHURODES Westwood, 1849
Gardners' Chronicle 1849: 420
Type species Smynthurodes betae Westwood, 1849
*Sminthurodes* Schaum, 1850
*Trifidaphis* del Guercio, 1909
*Tullgrenia* van der Goot, 1912
*Tychea* Passerini, 1860 nec Koch, 1857

S. betae Westwood, 1849 BM HRL
*albicornis* Koch, 1857 (*Amycla*)
*globosus* Walker, 1852 (*Pemphigus*)
*gossypii* Kulkarni, 1956 (*Trifidaphis*)
? *karschii* Lichtenstein, 1885 (*Schizoneura*)
? *myrmicaria* Boisduval, 1867 (*Aphis*)
*natalensis* Theobald, 1920 (*Forda*) BM
*perniciosus* Nevsky, 1929 (*Trifidaphis*)
*phaseoli* Passerini, 1860 (*Tychea*)
*radicicola* Essig, 1909 (*Pemphigus*)
*silvestrii* Mordvilko, 1935 (*Trifidaphis*)
*trifolii* del Guercio, 1915 (*Pemphigus*)

SOGDIANELLA Mukhamediev, 1965 nec Schurenkova, 1939
Dokl. Acad. Nauk. tadzhik. SSR 8: 35
Type species Sogdianella lonicericola Mukhamediev, 1965
= Ferganaphis Mukhamediev ex Eastop & Hille Ris Lambers, 1976

S. lonicericola Mukhamediev, 1965 to Ferganaphis Mukhamediev ex Eastop & Hille Ris Lambers, 1976

SOMAPHIS Shinji, 1929
Lansania 1: 110
Type species Aphis somei Essig & Kuwana, 1918 = Longiunguis odinae van der Goot, 1917
= Toxoptera Koch, 1856

SONCHIFEX Amyot, 1847
Annls Soc. ent. Fr. (2) 5: 475
Invalid

SORBAPHIS Shaposhnikov, 1950
Ént. Obozr. 31: 224
Type species Sorbaphis chaetosiphon Shaposhnikov, 1950

S. chaetosiphon Shaposhnikov, 1950 BM HRL
kurilensis Ivanovskaja, 1966

SORBOBIUM MacGillivray & Bradley, 1961
Can. Ent. 93: 1000, described as subg. of Toxopterella Hille Ris Lambers, 1960
Type species Toxopterella (Sorbobium) drepanosiphoides MacGillivray & Bradley, 1961
Here used as subg. of Toxopterella Hille Ris Lambers, 1960

SPATULOPHORUS F.P. Müller, 1958
Beitr. Ent. 8: 84
Type species Spatulophorus incanae F.P. Müller, 1958

S. alyssi Holman, 1965 BM HRL
incanae F.P. Müller, 1958 paratypes BM HRL

SPHAEROCOCCUS (Coccoidea)

S. populi Maskell, 1898 to Doraphis Matsumura & Hori ex Hori, 1929

SPICAPHIS Essig, 1953
Proc. Calif. Acad. Sci. 28: 69
Type species Spicaphis michelbacheri Essig, 1953
= Neuquenaphis E.E. Blanchard, 1939

S. michelbacheri Essig, 1953 = Neuquenaphis essigi Hille Ris Lambers, 1968

SPINASPIDAPHIS Heinze, 1961
Beitr. Ent. 11: 26
Type species Spinaspidaphis droserae Heinze, 1961 = Aphis lythri Schrank, 1801
= Myzus Passerini, 1860

S. droserae Heinze, 1961 = Myzus lythri (Schrank, 1801)

STAEGERIELLA Hille Ris Lambers, 1947
Mitt. schweiz. ent. Ges. 20: 654
Type species Hydaphias necopinata Börner, 1939

S. asperulae Bozhko, 1959
necopinata (Börner, 1939) (Hydaphias) BM HRL

STAGONA Koch, 1857
Die Pflanzenläuse Aphiden, Nürnberg, p. 284
Type species Aphis xylostei De Geer, 1773
Here used as subg. of Prociphilus Koch, 1856

S. crataegistrobi Smith, 1969 to Prociphilus (Stagona)
piceaerubensis Smith, 1969 to Prociphilus (Stagona)
vesicalis Rudow, 1875 to Prociphilus (Stagona)

STATICOBIUM Mordvilko, 1914
Faune Russie, Ins. Hém. 1: 66
Type species Staticobium otolepidis Nevsky, 1928, indicated by Börner & Schilder, 1930: 192

S. caspicum Bozhko, 1957 HRL
*latifoliae* subsp. *caspicum* Bozhko, 1961 HRL
gmelini Bozhko, 1953 HRL
insularum Bozhko, 1959
latifoliae Bozhko, 1950 HRL
latifoliae subsp. caspicum Bozhko, 1961 = caspicum Bozhko, 1957
limonii (Contarini, 1847) (Aphis) BM HRL
limonii subsp. caucasicum Bozhko, 1961 HRL
loochooense Sorin, 1967 = loochooense (Takahashi, 1939)
loochooense (Takahashi, 1939) (Aulacorthum) HRL
*loochooense* Sorin, 1967 HRL
nevskyi Hille Ris Lambers, 1939 HRL
*staticis* Nevsky, 1928 nec Macrosiphoniella staticis Theobald, 1923
otolepidis Nevsky, 1928 HRL
staticis Nevsky, 1928 nec Macrosiphoniella staticis Theobald, 1923 = nevskyi Hille Ris Lambers, 1939
staticis (Theobald, 1923) (Macrosiphoniella) BM HRL
*limonii* Walker, 1852 (*Aphis*) nec Contarini, 1847 lectotype BM
strongilisiphon mis-spelling of strongilosiphon
strongilosiphon Bozhko, 1961
tauricum Bozhko, (1957) 1961 HRL

STAUROCERAS Börner, 1940
Neue Blattläuse aus Mitteleuropa, privately published, p. 2
Type species Stauroceras chaetosiphon Börner, 1940
= Pterocomma Buckton, 1879

S. chaetosiphon Börner, 1940 to Pterocomma Buckton, 1879

STEGOPHYLLA Oestlund, 1922
Rep. Minn. St. Ent. 19: 146
Type species Phyllaphis quercicola Baker, 1916 = Eriosoma querci Fitch, 1859

S. essigi Hille Ris Lambers, 1966 BM HRL
mordvilkoi Aizenberg, 1932 to Diphyllaphis Takahashi, 1960
querci (Fitch, 1859) (Eriosoma) BM HRL
*quercicola* Baker, 1916 (*Phyllaphis*) nec Callipterus quercicola Monell, 1879
quercicola (Monell, 1879) (? Callipterus) HRL
quercifoliae (Gillette, 1914) (Phyllaphis) BM HRL
quercina Quednau, 1966 HRL

STELLARIOPSIS Szelegiewicz, 1969
Annls zool., Warsz. 27: 173
Type species Stellariopsis songini Szelegiewicz, 1969

S. songini Szelegiewicz, 1969 HRL

STENAPHIS del Guercio, 1913
Redia 9: 185
Type species Stenaphis monticellii del Guercio, 1913 = Aphis maidis Fitch, 1856
= Rhopalosiphum Koch, 1854

S. monticellii del Guercio, 1913 = Rhopalosiphum maidis (Fitch, 1856)

STENAPHIS Quednau, 1954 nec del Guercio, 1913
Mitt. biol. BundAnst. Ld-u. Forstw. 78: 40
Type species Saltusaphis elongata Baker, 1917
= Saltusaphis Theobald, 1915

STEPHENSONIA Das, 1918
Mem. Indian Mus. 6: 175
Type species Stephensonia lahorensis Das, 1918 = Aphis rufomaculata Wilson, 1908
= Coloradoa Wilson, 1910

S. lahorensis Das, 1918 = Coloradoa rufomaculata (Wilson, 1908)

STERNAPHIS Heie, 1972
Steenstrupia 2: 257
Type species Sternaphis electricola Heie, 1972
Fossil, baltic amber

STOMACHIS Mordvilko, 1901 lapsus pro Stomaphis Walker, 1870

STOMAPHIS Walker, 1870
Zoologist 5: 2000
Type species Aphis quercus Linnaeus, 1758
*Macrhynchus* Haupt, 1913
*Neostomaphis* Takahashi, 1960
*Parastomaphis* Pašek, 1953
*Rhynchocles* Altum, 1882

S. acerina Mamontova, 1963
*graffii* subsp. *acerinus* Mamontova, 1963
aceris Takahashi,1960 HRL
aesculi Takahashi, 1960
*yanonis* subsp. *aesculi* Takahashi, 1960
alni Sorin, 1965 HRL
asiphon Szelegiewicz, (1974) 1975 HRL
betulae Mamontova, 1969
bobretzkyi Mordvilko, 1901 = longirostris (Fabricius, 1787)
carpini Sorin, 1965
cupressi (Pintera, 1965) (Parastomaphis) types BM HRL
fagi Takahashi, 1960 (Stomaphis (Neostomaphis)) HRL
graffii Cholodkovsky, 1894 type of Parastomaphis Pašek, 1953 BM HRL
graffii subsp. acerinus Mamontova, 1963 = acerina Mamontova, 1963
japonica Takahashi, 1960 HRL
*quercus* subsp. *japonica* Takahashi, 1960 HRL
liquidambarus Takahashi, 1925
longirostris (Fabricius, 1787) (Aphis) BM HRL
*bobretzkyi* Mordvilko, 1901
macrorhyncha Cholodkoskvy, 1894 = quercus (Linnaeus, 1758)
mordvilkoi Hille Ris Lambers, 1933 cotypes BM HRL
pini Takahashi, 1920
quercus (Linnaeus, 1758) (Aphis) BM HRL
*fusca* Geoffroy, 1762 (*Aphis*) invalid ?
*longirostris* Altum, 1882 (*Rhynchocles*)
*longirostris* Boyer de Fonscolombe, 1841 (*Phylloxera*)
*macrorhyncha* Cholodkovsky, 1894
*pini* Haupt, 1913 (*Macrhynchus*)
quercus subsp. japonica Takahashi, 1960 = japonica Takahashi, 1960
radicicola Hille Ris Lambers, 1947 cotypes BM HRL

S. takahashii Sorin, 1965 HRL
ulmicola Inouye, 1938
yanonis Takahashi, 1918 BM HRL
yanonis subsp. aesculi Takahashi, 1960 = aesculi Takahashi, 1960

Subgen. Neostomaphis Takahashi, 1960

S. (Neostomaphis) fagi Takahashi, 1960 (type of Neostomaphis Takahashi, 1960) to Stomaphis Walker, 1870

STYLETAPHIS Knowlton, 1972
Proc. Utah Acad. Sc. 49: 29-32
Invalid: In list of species, without description, without type citation.

SUBACYRTHOSIPHON Hille Ris Lambers, 1947
Temminckia 7: 260
Type species Subacyrthosiphon cryptobium Hille Ris Lambers, 1947

S. cryptobium Hille Ris Lambers, 1947 BM HRL

SUBANOECIA Börner, 1952
Mitt. thüring. bot. Ges. 4 (3): 180, 259, described as subg. of Anoecia Koch, 1857
Type species Schizoneura vagans Koch, 1856
= Anoecia Koch, 1857

SUBCALLIPTERUS Mordvilko, 1894
Rab. Lab. zool. Kab. imp. varsch. Univ. 8: 53
Type species Aphis alni Fabricius, 1781 = Aphis alni de Geer, 1773
= Pterocallis Passerini, 1860

SUBCINARA Borner, 1949
Beitr. tax. Zool. 1: 59, described as subg. of Cinara Curtis, 1835
Type species Cinara brauni Börner, 1940
= Cinara Curtis, 1835

SUBLACHNOBIUS Heinze, 1962
Dt. ent. Z. (n.s.) 9: 186
Type species Lachnus wichmanni Hille Ris Lambers, 1956
= Lachnus Burmeister, 1835

SUBMACROSIPHUM Hille Ris Lambers, 1931
Memorie Mus. Stor. nat. Venezia trident. 1: 22
Type species Aphis hieracii Kaltenbach, 1843 nec Schrank, 1801
= Nasonovia Mordvilko, 1914

S. brevipes Borner, 1950 to Nasonovia (Neokakimia)
hieracii subsp. nigrum Hille Ris Lambers, 1931 = Nasonovia nigra (Hille Ris Lambers, 1931)
hieracii subsp. teriolanum Hille Ris Lambers, 1931 = Nasonovia ribisnigri (Mosley, 1841)
nivalis Börner, 1950 to Nasonovia Mordvilko, 1914

SUBMEGOURA Hille Ris Lambers, 1953
Temminckia 9: 3, described as subg. of Rhopalosiphoninus Baker, 1920
Type species Myzotoxoptera heikinheimoi Börner, 1952
Here used as subg. of Rhopalosiphoninus Baker, 1920

SUBOVATOMYZUS A.N. Basu, 1964
J. Linn. Soc. (Zool.) 45: 241
Type species Subovatomyzus leucosceptri A.N. Basu, 1964

S. leucosceptri A.N. Basu, 1964 BM HRL
*vernoniae* A.K. Ghosh & RayChaudhuri, 1968 (*Capitophorus*)

SUBSALTUSAPHIS Quednau, 1953
In Börner, C., 1952, Mitt. thüring. bot. Ges. 4 (3): 465 (1952) nomen nudum; Zool. Anz. 130: 224 (1953)
Type species Saltusaphis intermedia Hille Ris Lambers, 1939
*Bacillaphis* Quednau, 1954

S. aquatilis (Ossiannilsson, 1959) (Bacillaphis) HRL
canadensis Richards, 1971
flava (Hille Ris Lambers, 1939) (Saltusaphis) BM HRL
intermedia (Hille Ris Lambers, 1939) (Saltusaphis)) BM HRL
lambersi (Quednau, 1954) (Bacillaphis) BM HRL
maritima (Hille Ris Lambers, 1956) (Bacillaphis) BM HRL
ornata (Theobald, 1927) (Saltusaphis) type BM HRL
*quadrilineata* Börner, 1939 (*Saltusaphis*)
pallida (Hille Ris Lambers, 1939) (Saltusaphis) BM HRL
*granulata* Börner, 1940 (*Saltusaphis*)
? pallida subsp. taurica (Bozhko, 1959) (Bacylaphis)
paniceae (Quednau, 1954) (Bacillaphis) BM HRL

S. picta (Hille Ris Lambers, 1939) (Saltusaphis) BM HRL
    *cinerea* Börner, 1940 (*Saltusaphis*)
    rossneri (Börner, 1940) (Saltusaphis) BM HRL
    saracola Higuchi, 1972 HRL
    virginica (Baker, 1917) (Saltusaphis) BM HRL
    wanica (Hottes & Frison, 1931) (Saltusaphis) BM HRL

SUBTAKECALLIS RayChaudhuri & Pal, 1974
    Orient. Insects 8: 96
    Type species Cranaphis pilosa David, Rajasingh & Narayanan, 1970

C. brevisetosus RayChaudhuri & Pal, 1974
    pilosus (David, Rajasingh & Narayanan, 1970) (Cranaphis) HRL

SUCCAPHIS Heie, 1967
    Spolia zool. Mus. haun. 26: 110
    Type species Succaphis helgeri Heie, 1967
        Fossil, baltic amber

SUCCINAPHIS Heie, 1967
    Spolia zool. Mus. haun. 26: 173
    Type species Succinaphis flauensgaardi Heie, 1967
        Fossil, baltic amber

SUMATRAPHIS Takahashi, 1933
    Miscnea zool. sumatr. 97: 2
    Type species Sumatraphis celti Takahashi, 1935

S. celti Takahashi, 1935 BM HRL?

SUMOIA Tao, 1963
    Pl. Prot. Bull., Taiwan 5: 176
    Type species Macrosiphum taiwanum Takahashi, 1923
    = Neomyzus van der Goot, 1915

SWIRSKIAPHIS Hille Ris Lambers, 1966
    Ktavim 16: 25
    Type species Swirskiaphis polychaeta Hille Ris Lambers, 1966

S. polychaeta Hille Ris Lambers, 1966 BM HRL

SYCHNOBROCHUS Scudder, 1890
U.S. Geol. Survey Territories 13: 244, 248
Type species Sychnobrochus reviviscens Scudder, 1890 = Mindarus scudderi Buckton, 1883
Fossil
= Mindarus Koch, 1857

SYMDOBIUS mis-spelling of Symydobius Mordvilko, 1894

SYMYDOBIUS Mordvilko, 1894
Rab. Lab. zool. Kab. imp. varsch. Univ. 8: 65
Type species Aphis oblonga von Heyden, 1837
*Yezocallis* Matsumura, 1917

S. agrifoliae Essig, 1917 to Neosymydobius Baker, 1920
albasiphus Davis, 1914 type of Neosymydobius Baker, 1920 q.v.
alniarius (Matsumura, 1917) (Yezocallis) BM HRL
*japonicus* Essig & Kuwana, 1918 (*Euceraphis*)
alter Mordvilko, 1909 = oblongus (von Heyden, 1837)
americanus Baker, 1918 BM HRL
chrysolepis Swain, 1918 to Neosymydobius Baker, 1920
intermedius Gillette & Palmer, 1930 BM HRL
kabae (Matsumura, 1917) (Yezocallis) BM HRL
macrostachyae Essig, 1912 to Chaitophorus Koch, 1854
oblongus (von Heyden, 1837) (Aphis) BM HRL
*alter* Mordvilko, 1909
*fuscipennis* Zetterstedt, 1840 (*Aphis*)
*piceus* Börner, 1950
*tenuinervis* Zetterstedt, 1840 (*Aphis*)
piceus Börner, 1950 = oblongus (von Heyden, 1837)
salicicorticis Ensig, 1912 to Chaitophorus Koch, 1854

SYNTHRIPAPHIS Quednau, 1954
Mitt. biol. Reichsanst. Ld-u. Forstw. 78: 38
Type species Thripsaphis cyperi (Walker) of Börner, 1952 nec Walker, 1848
= Callaphis caricicola Mordvilko, 1914
= Thripsaphis Gillette, 1917

SYPHA (Passerini) Ferrari, 1872 lapsus or emendation of Sipha Passerini, 1860

SYPHOCORYNE lapsus or emendation of Siphocoryne Passerini, 1860

SZELEGIEWICZIELLA Holman, 1974
Acta ent. bohemoslovica 71: 239
Type species Szelegiewicziella chamaerhodi Holman, 1974

S. chamaerhodi Holman, 1974

TACTILOTRAMA Börner, 1952
Mitt. thüring. bot. Ges. 4(3): 49, 242, described as subg. of Protrama Baker, 1920
Type species Trama antennata Mordvilko,1935

T. antennata (Mordvilko, 1935) (Trama)

TAIWANAPHIS Takahashi, 1934
Stylops 3: 56
Type species Taiwanaphis decaspermi Takahashi, 1934

T. decaspermi Takahashi, 1934 HRL
randiae A.K. Ghosh, H. Banerjee & RayChaudhuri, 1971

TAIWANOMYZUS Tao, 1963
Pl. Prot. Bull., Taiwan 5: 179
Type species Myzus montanus Takahashi, 1925
= Utamphorophora Knowlton, 1947

T. chrysosplenii Miyazaki, 1971 to Utamphorophora Knowlton, 1947

TAKECALLIS Matsumura, 1917
J. Coll. Agric. Tohoku imp. Univ. 7: 373
Type species Takecallis bambusae Matsumura, 1917 = Callipterus arundicolens Clarke, 1903

T. arundicolens (Clarke, 1903) (Callipterus) BM HRL
*bambusae* Matsumura, 1917
arundinariae (Essig, 1917) (Myzocallis) BM HRL
*bambusifoliae* Takahashi, 1921 (*Myzocallis*)
bambusae Matsumura, 1917 = arundicolens (Clarke, 1903)
sasae (Matsumura, 1917) (Myzocallis) BM HRL
*sasacola* Shinji, 1935 (*Agrioaphis*)
taiwanus (Takahashi, 1926) (Myzocallis) BM HRL
*tectae* Tissot, 1932 (*Therioaphis*) HRL

TAMALIA Baker, 1920
Bull. U.S. Dep. Agric. 826: 24
Type species Pemphigus coweni Cockerell, 1905

T. coweni (Cockerell, 1905 )(Pemphigus) BM HRL
*? tahoense* Davidson, 1911 (*Cryptosiphum*)
keltoni Richards, 1967
pallida Richards, 1967
tahoensis (Davidson, 1911) (Cryptosiphum) = ? coweni (Cockerell, 1905)

TAOIA Quednau, 1973
Can. Ent. 105: 217
Type species Euceraphis chuansiensis Tao, 1964
*Jaria* Quednau m.s. ex A.K. Ghosh & RayChaudhuri, 1972

T. chuansiensis (Tao, 1964) (Euceraphis) HRL
exotica Quednau, 1973 = indica (A.K. Ghosh & RayChaudhuri, 1972)
indica (A.K. Ghosh & RayChaudhuri, 1972) (Euceraphis) HRL
*exotica* Quednau, 1973

TAVARESIELLA del Guercio, 1911
Redia 7: 299
Type species Tavaresiella suberi del Guercio, 1911
= Thelaxes Westwood, 1840

T. suberi del Guercio, 1911 to Thelaxes Westwood, 1840

TELOCALLIS Shinji, 1922
Zool. Mag., Tokyo 34: 731
Type species Telocallis alnifoliae Shinji, 1922 = Sappocallis ulmicola Matsumura, 1919
= Sappocallis Matsumura, 1919

T. alnifoliae Shinji, 1922 = Sappocallis ulmicola Matsumura, 1919

TENUISIPHON Mordvilko, 1948
In Tarbinski & Plaviltshikov, Keys to the insects of the European part of the USSR, p. 215, described as subg. of Acyrthosiphon Mordvilko, 1914
Type species Acyrthosiphon gossypii Mordvilko, 1914
= Acyrthosiphon Mordvilko, 1914

TEPHRAPHIS Scudder, 1890
U.S. Geol. Survey Territories 13: 259
Type species Siphonophoroides simplex Buckton, 1883
= Siphonophoroides Buckton, 1883
Fossil

TERMITAPHIS Wasmann, 1902
Tijdschr. Ent. 45: 105
Type species Termitaphis circumvallata Wasmann, 1902
Erected in Aphididae, now in Termitocoridae (Heteroptera)

TERTIAPHIS Heie, 1969
Mitt. geol.-paläont. Inst. Univ. Hamburg 38: 144
Type species Tertiaphis haeatzcheli Heie, 1969
Fossil, baltic amber

TESTATAPHIS Börner, 1952
Mitt. thüring. bot. Ges. 4(3): 120, 252
Type species Amphorophora ledi Wahlgren, 1938
= Neoamphorophora Mason, 1924

TETRANEMA Derbes, 1869 lapsus pro Tetraneura Hartig, 1841

TETRANEURA Hartig, 1841
Z. Ent. (Germar) 3: 366
Type species Aphis ulmi Linnaeus, 1758
*Amycla* Koch, 1857
*Byrsocrypta* Haliday, 1838
*Dryopeia* Kirkaldy, 1904
*Endeis* Koch, 1857 nec Philippi, 1847
*Tetranevra* Agassiz (1846) 1847

T. aegyptiaca Theobald, 1923 = caerulescens (Passerini, 1856)
africana van der Goot, 1912 BM HRL
*cynodontis* Theobald, 1923
agnesii del Guercio, 1920
akinire Sasaki, 1904 to Tetraneura (Tetraneurella)
? bordschomica (Abashidze, 1951) (? Byrsocrypta)
caerulescens (Passerini, 1856) (Pemphigus) BM HRL
*aegyptiaca* Theobald, 1923
*rubra* Lichtenstein, 1880

T. chinensis Mordvilko, 1924 to Tetraneura (Tetraneurella)
colophoidea Howard, 1908 nomen nudum of Monell
cornicularia Passerini, 1856 (errore pro Pemphigus cornicularius Passerini, 1856) (type of Pemphigella Tullgren, 1909) = Baizongia pistaciae (Linnaeus, 1767)
cynodonti subsp. coimbatorensis George, 1928 = Tetraneura (Tetraneurella) javensis van der Goot, 1917
cynodontis Theobald, 1923 = africana van der Goot, 1912
flavescens del Guercio, 1920 to Geoica Hart, 1894
fusiformis Matsumura, 1917 = Tetraneura (Tetraneurella) nigriabdominalis (Sasaki, 1899)
graminis Monell, 1882 = Colopha ulmicola (Fitch, 1859)
heterohirsuta Carver & A.N. Basu, 1961 = radicicola Strand, 1929 + yezoensis Matsumura, 1917
javensis van der Goot, 1917 to Tetraneura (Tetraneurella)
lentisci Passerini, 1856 type of Aploneura Passerini, 1863 q.v.
longisetosa (Dahl, 1912) (Tycheoides)
lucifuga Zehntner, 1897 (type of Serrataphis van der Goot, 1917) to Geoica Hart, 1894
moriokaensis Monzen, 1927 to Colopha Monell, 1877
oligocenica N. Theobald, 1937 fossil, type of Aixaphis Heie, 1970
oryzae van der Goot ex van Heurn, 1923 = Tetraneura (Tetraneurella) nigriabdominalis (Sasaki, 1899)
paiki Hille Ris Lambers, 1970 HRL
provincialis N. Theobald, 1937 fossil
pruni Taschenberg ex Schmidt, 1869, name for the imaginary agent of a fungus gall of Prunus
radicicola Strand, 1929 BM HRL
*heterohirsuta* Carver & A.N. Basu, 1961 partim, Indian specimens
*takahashii* Mordvilko, 1930
reticulata del Guercio, 1920
rubra Lichtenstein, 1880 = caerulescens (Passerini, 1856)
rugicornis Hartig, 1841 nomen nudum
takahashii Mordvilko, 1930 = radicicola Strand, 1929
theobaldi Zwölfer, 1958 = ulmi (Linnaeus, 1758)
ulmi (Linnaeus, 1758) (Aphis) BM HRL
*bella* Koch, 1857 (*Endeis*)
*boyeri* Passerini, 1856 (*Pemphigus*)
*boyeri* var. *saccharata* del Guercio, 1900 (*Pemphigus*)
*fuscifrons* Koch, 1857 (*Amycla*)

*fuscifrons* var. *saccharata* del Guercio, 1895 (*Pemphigus*)
*gallarum* Gmelin, 1790 (*Aphis*)
*gallarum ulmi* de Geer, 1773 (*Aphis*)
*graminis* Koch, 1857 (*Tychea*) sensu Schouteden, 1906
*personata* Börner, 1950 (*Byrsocrypta*)
*radicum* Boyer de Fonscolombe, 1841 (*Aphis*) nec Kirby, 1818
*rosea* Koch, 1857 (*Endeis*)
*theobaldi* Zwölfer, 1958
*ulmifoliae* Baker, 1920
*ulmi-gallarum* Haliday, 1838 (*Eriosoma*)
*ulmisacculi* Patch, 1910
*zeae-maydis* Dufour, 1824 (*Coccus*)
T. ulmifoliae Baker, 1920 = ulmi (Linnaeus, 1758)
ulmisacculi Patch, 1910 = ulmi (Linnaeus, 1758)
? ulmoides Lichtenstein, 1880
? walshii, Williams, 1911 (Pemphigus)
yezoensis Matsumura, 1917 HRL
*heterohirsuta* Carver & A.N Basu, 1961 partim, Australian specimens

Subgen. Tetraneurella Hille Ris Lambers, 1970

T. (Tetraneurella) akinire Sasaki, 1904 (Tetraneura) BM HRL
(Tetraneurella) basui Hille Ris Lambers, 1970 HRL
(Tetraneurella) chinensis Mordvilko, 1924 (Tetraneura) HRL
(Tetraneurella) javensis van der Goot, 1917 (Tetraneura) HRL
*cynodonti* subsp. *coimbatorensis* George, 1928 (*Tetraneura*)
(Tetraneurella) nigriabdominalis (Sasaki, 1899) (Schizoneura) BM HRL
*argrimoniae* Shinji, 1924 (*Pemphigus*)
*fusiformis* Matsumura, 1917 (*Tetraneura*)
*hirsuta* Baker, 1921 (*Dryopeia*)
*oryzae* van der Goot ex van Heurn, 1923 (*Tetraneura*)
(Tetraneurella) nigriabdominalis subsp. bispina Hille Ris Lambers, 1970 BM HRL
(Tetraneurella) polychaeta Hille Ris Lambers, 1970 BM HRL
(Tetraneurella) sorini Hille Ris Lambers, 1970 HRL

TETRANEURELLA Hille Ris Lambers, 1970
Boll. Zool. agr. Bachic. 9: 24, described as subg. of Tetraneura Hartig, 1841
Type species Dryopeia hirsuta Baker, 1921 = Schizoneura nigriabdominalis Sasaki, 1899
Here used as subg. of Tetraneura Hartig, 1841

TETRANEVRA Agassiz, 1847 (1846)
Nomenclator zoologici index universalis, Solothurn, p. 366 emendation of Tetraneura Hartig, 1841
= Tetraneura Hartig, 1841

TETRANYCHUS (Acarina)

T. longirostris Mola, 1907 to ? Aploneura Passerini, 1863

TETRAPHIS Horvath, 1896
Wien. Ent. Ztg 15: 6
Type species Tetraphis betulina Horvath, 1896
= Hamamelistes Shimer, 1867

T. betulina Horvath, 1896 to Hamamelistes Shimer, 1867

TETRENEMA Derbes, 1869 errore pro Tetraneura Hartig, 1841

THARGELIA Oestlund, 1922 nec Puengelev, 1899
Rep. Minn. St. Ent. 19: 127
Type species Aphis albipes Oestlund, 1887 = Aphis symphoricarpi Thomas, 1878
= Aphthargelia Hottes, 1958

THECABIUS Koch, 1857
Die Pflanzenläuse Aphiden, Nürnberg, p. 294
Type species Thecabius populneus Koch, 1857 = Pemphigus affinis Kaltenbach, 1843
*Bucktonia* Lichtenstein, 1896

T. affinis (Kaltenbach, 1843) (Pemphigus) BM HRL
*agnotus* Börner & Blunck, 1916
*populneus* Koch, 1857
*ranunculi* Kaltenbach, 1843 (*Pemphigus*)
affinis subsp. orientalis Mordvilko, 1935 = orientalis Mordvilko, 1935
agnotus Börner & Blunck, 1916 = affinis (Kaltenbach, 1843)
altaicus Dolgova, 1971 = Thecabius (Parathecabius) latisensorius Hori, 1938
anemoni (Shinji, 1922) (Prociphilus)
kelloggi (Takahashi, 1939) (Pemphiginus) BM
latisensoria Hori, 1938 to Thecabius (Parathecabius)
ligustrifoliae Tseng & Tao, 1938 to Prociphilus Koch, 1857
lysimachiae Börner, 1916 (type of Parathecabius Börner, 1950) to Thecabius (Parathecabius)

T. orientalis Mordvilko, 1935 HRL
*affinis* subsp. *orientalis* Mordvilko, 1935
*ranunculi* Shinji, 1922 nec Kaltenbach, 1843; Davidson, 1911 (*Pemphigus*)
patchiae Gillette, 1914 errore pro patchii Gillette, 1914
patchii Gillette, 1914 BM
populiconduplifolius (Cowen, 1895) (Pemphigus) BM HRL
*californicus* Davidson, 1911 (*Pemphigus*)
*ranunculi* Davidson, 1910 (*Pemphigus*)
populimonilis (Riley, 1879) (Pemphigus) BM HRL
populneus Koch, 1857 = affinis (Kaltenbach, 1843)

Subgen. Parathecabius Börner, 1950

T. (Parathecabius) auriculae (Murray, 1877) (Trama) BM HRL
*pseudoaruriculae* Theobald, 1929 (*Pemphigus*) BM
(Parathecabius) cerastii (Börner, 1950) (Parathecabius)
(Parathecabius) gravicornis (Patch, 1913) (Pemphigus) HRL
(Parathecabius) latisensorius Hori, 1938 (Thecabius) HRL
*altaicus* Dolgova, 1971 (*Thecabius*) HRL
(Parathecabius) luppovae Narzikulov, 1963 (Pemphigus)
(Parathecabius) lysimachiae Börner, 1916 (Thecabius) BM HRL
*? glebae* Jackson, 1918 (*Pemphigus*)
(Parathecabius) ? stammeri (Zwölfer, 1957) (Parathecabius) HRL

THELAXES Westwood, 1840
An Introduction to the modern classification of Insects, ... London vol. 2, Synopsis of the genera of British Insects, p. 118 (separately paginated from rest of volume)
Type species Thelaxes quercicola Westwood, 1840 = Aphis dryophila Schrank, 1801
*Tavaresiella* del Guercio, 1911

T. albipes Richards, 1966 = californica (Davidson, 1919)
betulina Buckton, 1886 = Glyphina betulae (Linnaeus, 1758)
californica (Davidson, 1919) (Vacuna) BM HRL
*albipes* Richards, 1966
cerridis Börner, 1942 = ? suberi (del Guercio, 1911)
confertae Börner, 1942 = suberi (del Guercio, 1911)
dewildei Hille Ris Lambers, nomen nudum ex Halmagyi, 1974
dryophila (Schrank, 1801) (Aphis) BM HRL
*? pedunculi* Hartig, 1841 (*Pemphigus*)
*quercicola* Westwood, 1840
*quercus* Mosley, 1841 (*Cinara*)

T. quercicola Westwood, 1840 = dryophila (Schrank, 1801)
subcri (del Guercio, 1911) (Tavaresiella) BM HRL
*carlucciana* del Guercio, 1913 (*Vacuna*)
*castaneae* del Guercio, 1913 (*Vacuna*)
*? cerridis* Börner, 1942
*confertae* Börner, 1942
ulmicola Walsh, 1867 is reference to Byrsocrypta ulmicola Fitch, 1856

THERIOAPHIS Walker, 1870
Zoologist (2) 5: 1999
Type species Aphis ononidis Kaltenbach, 1846
*Myzocallidium* Börner, 1949
*Pterocallidium* Börner, 1949
*Triphyllaphis* Börner, (1944) 1949

T. aizenbergi Ivanovskaja & Tomilova, 1968 HRL
alatina Hille Ris Lambers & van den Bosch, 1964 types BM HRL
astragali (Dzhibladze, 1959) (Myzocallidium) BM HRL
astragalica Narzikulov & Umarov, 1971
bambusae Abashidze, 1951 nomen nudum
bonjeaneae Hille Ris Lambers & van den Bosch, 1964 types BM HRL
celticola Nevsky, 1929 to Shivaphis Das, 1918
collina Börner, 1942 = trifolii (Monell, 1882)
dorycnii (Pintera, 1956) (Myzocallidium) BM HRL
hillerislambersi Szelegiewicz, 1968 HRL
hungarica Szelegiewicz, 1968 HRL
japonica Shinji, 1933 = Tiliaphis shinae (Shinji, 1924)
litoralis Hille Ris Lambers & van den Bosch, 1964 types BM HRL
loti Hille Ris Lambers & van den Bosch, 1964 BM HRL
luteola (Börner, 1949) (Triphyllaphis) BM HRL
maculata (Buckton, 1899) (Chaitophorus) here regarded as form of trifolii (Monell, 1882)
natricis Hille Ris Lambers & van den Bosch, 1964 HRL
obscura Hille Ris Lambers & van den Bosch, 1964 types BM HRL
ononidis (Kaltenbach, 1846) (Aphis) BM HRL
*spinulosus* Koch in litteris 1854 (*Chaitophorus*)
riehmi (Börner, 1949) (Myzocallidium) BM HRL
shinae Shinji, 1924 type of Tiliaphis Takahashi, 1961 q.v.
subalba Börner, 1949 BM HRL

T. tectae Tissot, 1932 = Takecallis taiwanus (Takahashi, 1926)
tenera (Aizenberg, 1956) (Myzocallidium) HRL
tilicola Shinji, 1933 to Eucallipterus Schouteden, 1906
trifolii (Monell, 1882) (Callipterus) BM HRL
*collina* Börner, 1942
*genevii* Sanborn, 1904 (*Callipterus*)
*lydiae* Börner, 1949 (*Pterocallidium*)
*propinquum* Borner, 1949 (*Pterocallidium*)
trifolii forma maculata (Buckton, 1899) (Chaitophorus) BM HRL
trifolii subsp. albae Bozhko, 1959 (Therioaphis (Pterocallidium))
trifolii subsp. brevipilosa Hille Ris Lambers & van den Bosch, 1964 types BM HRL
trifolii subsp. ventromaculata F.P. Müller, (1968 nomen nudum) 1968 paratypes BM HRL

Subgen. Pterocallidium Borner, 1949

T. (Pterocallidium) maculatum subsp. albae Bozhko, 1959 = Therioaphis trifolii subsp. albae Bozhko, 1959

Subgen. Rhizoberlesia del Guercio, 1915

T. (Rhizoberlesia) brachytricha Hille Ris Lambers & van den Bosch, 1964 types BM HRL
*trifolii* del Guercio, 1915 (*Rhizoberlesia*) nec Callipterus trifolii Monell, 1882
(Rhizoberlesia) langloisi Remaudière & Leclant, 1968 HRL

THOMASIA Wilson, 1910 nec Poche, 1908
Can. Ent. 42: 38
Type species Chaitophorus populicola Thomas, 1878
= Chaitophorus Koch, 1854

T. californiensis Shinji, 1917 to Periphyllus van der Hoeven, 1863
crucis Essig, 1912 to Chaitophorus Koch, 1854
sansho Shinji, 1922 = Toxoptera odinae (van der Goot, 1917)
shidae Shinji, 1922 nomen dubium
tomitae Shinji, 1922 nomen dubium
zanthoxylum Shinji, 1922 nomen dubium

THOMASINIELLULA Strand, 1917
Arch. Naturgesch. 82 A 5: 82
Type species Chaitophorus populicola Thomas, 1878
= Chaitophorus Koch, 1854

THORACAPHIS van der Goot, 1917
Contrib. Faune Indes Néerl. 1 (3): 242
Type species Thoracaphis arboris van der Goot, 1917

T. arboris van der Goot, 1917 cotypes BM HRL
betulifoliae Shinji, 1924 = Hormaphis ? betulae Mordvilko, 1907 or Hamamelistes ? betulina (Horvath, 1896)
cheni Takahaish, 1936, to Paratoracaphis Takahashi, 1958
cinnamomi van der Goot, 1918 nomen dubium?
cinnamoniae Shinji, 1941 to Euthoracaphis Takahashi, 1938
coreanus Paik, 1965 to Nipponaphis Pergande, 1906
crematogastri Takahashi, 1941
depressus Takahashi, 1933 (type of Microunguis Tao, 1966) to Neothoracaphis Takahashi, 1958
dorocola Shinji, 1941 = Doraphis populi (Maskell, 1898)
elongatus Takahashi, 1941 to Parathoracaphis Takahashi, 1958
fici var. foveolatae Takahashi, 1935 = Reticulaphis distylii subsp. foveolatae (Takahashi, 1935)
fici var. mirabilis Takahashi, 1939 = Reticulaphis mirabilis (Takahashi, 1939)
flava Takahashi, 1950 types BM
gooti Takahashi, 1950 to Parathoracaphis Takahashi, 1958
hongkongensis van der Goot, 1918 = Schizoneuraphis gallarum (van der Goot, 1917)
jici Swezey, 1921
kashiwae Takahashi, 1932 (type of Xenothoracaphis Takahashi, 1958) = Xenothoracaphis kashifoliae (Uye, 1924)
kayashimai Takahashi, 1950 to Parathoracaphis Takahashi, 1958
linderae Shinji, 1926
lithocarpicola Takahashi, 1933 to Metanipponaphis Takahashi, 1959
machili Takahashi, 1923 to Nipponaphis Pergande, 1906
machilicola Shinji, 1941 to Nipponaphis Pergande, 1906
malayana Takahashi, 1950 to Schizoneuraphis van der Goot, 1917
piyananensis Takahashi, 1935 type of Allothoracaphis Takahashi, 1958 q.v.
rappardi Hille Ris Lambers & Takahashi, 1959 cotypes BM HRL
saramaoensis Takahashi, 1935 to Neothoracaphis Takahashi, 1958
setigera Takahashi, 1932 type of Parathoracaphis Takahashi, 1958 q.v.
siamensis Takahashi, 1941 to Nipponaphis (Pseudonipponaphis)
silvestrii Takahashi, 1935 to ? Metanipponaphis Takahashi, 1959
sutepensis Takahashi, 1941 to Neothoracaphis Takahashi, 1958
tarakoensis Takahashi, 1937 to Neothoracaphis Takahashi, 1958 q.v.

T. tuberculatus (Takahashi, 1933) (Hamamelistes) type of Quernaphis Takahashi, 1958 q.v.
umbellulariae Essig, 1932 type of Euthoracaphis Takahashi, 1938 q.v.
yamanarashii Shini, 1941 = Doraphis populi (Maskell, 1898)

THRIPSAPHIS Gilette, 1917
Can. Ent. 49: 193
Type species Brachycolus ballii Gillette, 1908
*Allaphis* Mordvilko, 1921
*Callaphis* Mordvilko, (1909) 1914 nec Walker, 1870
*Synthripaphis* Quednau, 1954
ballii (Gillette, 1908) (Brachycolus) BM HRL
ballii subsp. caespitosae Ossiannilsson, 1954 HRL
*caespitosae* Ossiannilsson, 1954 HRL
ballii subsp. longisetis Richards, 1971 HRL
*longisetis* Richards, 1971
brevicornis subsp. carpaticae Mamontova, 1963 to Thripsaphis (Larvaphis)
caespitosae Ossiannilsson, 1954 = ballii subsp. caespitosae Ossiannilsson, 1954
caricicola (Mordvilko, 1914) (Callaphis) BM HRL
*cyperi* Walker of Börner, 1952 nec Walker, 1848
*gelrica* Hille Ris Lambers, 1955 HRL
caricicola Swain, 1918 nec caricicola (Mordvilko, 1914)
cyperi (Walker, 1848) (Aphis) to Thripsaphis (Trichocallis)
cyperi (Walker) of Börner, 1952 nec (Walker, 1848) (type of Synthripaphis Quednau, 1954) = caricicola (Mordvilko, 1914)
foxtonensis Cottier, 1953 to Thripsaphis (Trichocallis)
gelrica Hille Ris Lambers, 1955 = caricicola (Mordvilko, 1914)
leporinae Börner, 1939 = Thripsaphis (Trichocallis) verrucosa Gillette, 1917
longisetis Richards, 1971 = ballii subsp. longisetis Richards, 1971
ossiannilssoni Hille Ris Lambers, 1952 to Thripsaphis (Trichocallis)
producta Gillette, 1917 to Thripsaphis (Trichocallis)
sensoriata Hille Ris Lambers, 1952
striata Bozhko, 1961 to Neobacillaphis Huculak, 1968
thripsoides Hille Ris Lambers, 1939 = Thripsaphis (Trichocallis) caricis (Mordvilko, 1921)
utahensis Knowlton & Hall, 1950 to Thripsaphis (Trichocallis)
verrucosa Gillette, 1917 to Thripsaphis (Trichocallis)
vibei Hille Ris Lambers, 1952 to Thripsaphis (Trichocallis)
vibei subsp. arctica Hille Ris Lambers, 1960 = Thripsaphis (Trichocallis) vibei subsp. arctica Hille Ris Lambers, 1955

Subgen. Larvaphis Ossiannilsson, 1953

T. (Larvaphis) brevicornis Ossiannilsson, 1953 HRL (Larvaphis)
(Larvaphis) brevicornis subsp. carpaticae Mamontova, 1963 (Thripsaphis)

Subgen. Peltaphis Frison & Ross, 1933

T. (Peltaphis) hottesi Frison & Ross, 1933 type of Peltaphis Frison & Ross, 1933; and of Zabapaphis Richards, 1971 BM HRL

Subgen. Trichocallis Börner, 1930

T. (Trichocallis) californica Hille Ris Lambers, 1974 HRL
(Trichocallis) caricis (Mordvilko, 1921) (Allaphis) BM HRL
*thripsoides* Hille Ris Lambers, 1939 (*Thripsaphis*) HRL
(Trichocallis) caricis subsp. amurensis (Mordvilko, 1921) (Allaphis)
(Trichocallis) cyperi (Walker, 1848) (Aphis) lectotype BM HRL
(Trichocallis) cyperi subsp. arctica Hille Ris Lambers, 1955 = vibei subsp. arctica Hille Ris Lambers, 1955
(Trichocallis) daviaulti (Quednau, 1966) (Heterocallis) HRL
(Trichocallis) foxtonensis Cottier, 1953 (Thripsaphis) BM HRL
(Trichocallis) hybrida Hille Ris Lambers, 1974 HRL
(Trichocallis) ossiannilssoni Hille Ris Lambers, 1952 (Thripsaphis) BM HRL
(Trichocallis) ossiannilssoni subsp. pacifica Hille Ris Lambers, 1974 HRL
(Trichocallis) producta Gillette, 1917 (Thripsaphis) BM HRL
(Trichocallis) scabra Hille Ris Lambers, 1974 HRL
(Trichocallis) utahensis Knowlton & Hall, 1950 (Thripsaphis) BM HRL
(Trichocallis) verrucosa Gillette, 1917 (Thripsaphis) BM HRL
*leporinae* Borner, 1939 (*Thripsaphis*)
(Trichocallis) verrucosa subsp. nodulosa Hille Ris Lambers, 1974 HRL
(Trichocallis) verrucosa subsp. subverrucosa Hille Ris Lambers, 1974 BM HRL
(Trichocallis) vibei Hille Ris Lambers, 1952 (Thrisaphis) BM HRL
(Trichocallis) vibei subsp. arctica Hille Ris Lambers, 1955, 1960 (Thripsaphis) HRL
*cyperi* subsp. *arctica* Hille Ris Lambers, 1955
*vibei* subsp. *arctica* Hille Ris Lambers, 1960

THULEAPHIS Hille Ris Lambers, 1960
Meddr Grønland 159(5): 7
Type species Thuleaphis acaudata Hille Ris Lambers, 1960
*Brevicaudus* Shaposhnikov, 1964
Here used as subg. of Brachycaudus van der Goot, 1913

T. acaudata Hille Ris Lambers, 1960 to Brachycaudus (Thuleaphis)
sedi Jacob, 1964 to Brachycaudus (Thuleaphis)

TILIAPHIS Takahashi, 1961
Kontyû 29: 251
Type species Therioaphis shinae Shinji, 1924

T. shinae (Shinji, 1924) (Therioaphis) BM HRL
*japonica* Shinji, 1933 (*Therioaphis*)
shinjii Higuchi, 1972 HRL

TILIPHAGUS Smith, 1965
Ann. ent. Soc. Am. 58: 782
Type species Tiliphagus lycoposugus Smith, 1965

T. lycoposugus Smith, 1965 paratype BM HRL

TINOCALLIS Matsumura, 1919
Trans. Sapporo Nat. Hist. Soc. 7: 100
Type species Tinocallis ulmiparvifoliae Matsumura, 1919
*Lutaphis* Shinji, 1924
*Tuberocallis* Nevsky, 1929

T. coreanus (Paik, 1965) (Chromaphis) BM HRL
distinctus M.R. Ghosh, A.K. Ghosh & RayChaudhuri, 1971 HRL
himalayensis A.K. Ghosh, M.R. Ghosh & RayChaudhuri, 1971
insularis (Takahashi, 1927) (Myzocallis) BM HRL
magnoliae A.K. Ghosh & RayChaudhuri, 1972
mushensis (Takahashi, 1925) (Myzocallis)
nikkoensis Higuchi, 1972
nirecola (Shinji, 1924) (Lutaphis) BM HRL
*zelkowae* auctt. nec Takahashi, 1919*
platani (Kaltenbach, 1843) (Lachnus) BM HRL
*elegans* Koch, 1855 (*Callipterus*)
*pulchellus* Glendenning, 1929 (*Myzocallis*)

---

* Takahashi (1919) described his zelkowae from Zelkowa and "cultivated beans", Tokyo, Japan. His decription (25 rhinaria on ant. segment III) and figure (26 rhinaria) suggest that he had strays of Tinocallis ulmiparvifoliae (Matsumura, 1919) from beans rather than the species with 14-18 (Higuchi, 1972) rhinaria (or up to 20 in fundatrices) from Japanese and Korean Zelkowa. For the latter the name Lutaphis nirecola Shinji, 1924 = Tinocallis nirecola (Shinji, 1924) is available. Records of zelkowae from Formosa (Taiwan) relate to another species.

T. saltans (Nevsky, 1929) (Tuberocallis) BM HRL
sapporoensis Higuchi, 1972
takachihoensis Higuchi, 1972
ulmifolii (Monell, 1879) (Callipterus) BM HRL
? *subulmifolii* Miller, 1936 (*Myzocallis*)
ulmiparvifoliae Matsumura, 1919 BM HRL
*zelkowae* Takahashi, 1919 (*Myzocallis*)*
viridis (Takahashi, 1929) (Myzocallis) HRL
zelkovae Dzhibladze, 1957 BM HRL
zelkowae auctt. including Higuchi, 1973 nec Takahashi, 1919 = nirecola (Shinji, 1924*)

TINOCALLOIDES A.N. Basu, 1970 (1969)
Orient. Insects 3: 367
Type species Tinocalloides montanus A.N. Basu, 1970 (1969)
*Tuberdefectus* Kumar & Lavigne, 1970

T. montanus A.N. Basu, 1970 (1969) BM HRL
*eastopi* Kumar & Lavigne, 1970 (*Tuberdefectus*) paratypes BM

TITANOSIPHON Nevsky, 1928
Ent. Mitt. 17: 189
Type species Titanosiphon bellicosum Nevsky, 1928

T. artemisiae (Koch, 1855) (Siphonophora) BM HRL
*benoisti* Balachowsky, 1933
*kochii* Ferrari, 1972 (*Siphonophora*)
*minkiewiczi* Judenko, 1931 HRL
bellicosum Nevsky, 1928 BM HRL
benoisti Balachowsky, 1933 = artemisiae (Koch, 1855)
dracunculi Nevsky, 1928 HRL
intermedium Narzikulov & Umarov, 1972
minkiewiczi Judenko, 1931 = artemisiae (Koch, 1855)
neoartemisiae (Takahashi, 1921) (Macrosiphum)

TLJA Mordvilko, 1914
Faune Russie, Ins. Hém. 1: 72, in key, without species
Type species Macrosiphum lactucae (Passerini, 1860) = Siphonophora lactucae Passerini, 1860 included by Mordvilko (1932)
= Acyrthosiphon Mordvilko, 1914

* vide footnote on p. 428

TODOLACHNUS Matsumura, 1917
Jl Coll. Agric. Tohoku Imp. Univ. 7: 381
Type species Todolachnus abietis Matsumura, 1917 = Cinara matsumurana Hille Ris Lambers, 1966
= Cinara Curtis, 1835

T. abieticola subsp. bulgarica Pintera, 1959 = Cinara confinis (Koch, 1856)
abietis Matsumura, 1917 = Cinara matusumurana Hille Ris Lambers, 1966

TOXOPTERA Koch, 1856
Die Pflanzenläuse Aphiden, Nürnberg, p. 253
Type species Toxoptera aurantiae Koch, 1856 = Aphis aurantii Boyer de Fonscolombe, 1841
*Arimakia* Matsumura, 1917
*Ceylonia* Buckton, 1891
*Paratoxoptera* E.E. Blanchard, 1944
*Somaphis* Shinji, 1929

T. acori Shinji, 1922 = Schizaphis rotundiventris (Signoret, 1860)
argrimoniae Shinji, 1941*
alaterna del Guercio, 1909 = aurantii (Boyer de Fonscolombe, 1841)
aphoides van der Goot, 1917 = citricidus (Kirkaldy, 1907)
aurantiae Koch, 1856 = aurantii (Boyer de Fonscolombe, 1841)
aurantii (Boyer de Fonscolombe, 1841) (Aphis) BM HRL
*alaterna* del Guercio, 1909
*aurantiae* Koch,1856
*camelliae* Kaltenbach, 1843 (*Aphis*)
*citrifoliae* Maki, 1913
*clematidis* del Guercio, 1900
*coffeae* Nietner, 1861 (*Aphis*)
*coffeae* subsp. *thomensis* Seabra, 1921
*djarani* van der Goot, 1917
*papaveris* var. *buxi* del Guercio, 1919 (*Aphis*)
*theaecola* Buckton, 1891 (*Ceylonia*)
*theobromae* Schouteden, 1906
*variegata* del Guercio, 1909

*** Shinji, 1941: 683-687 describes an Aphis-like species under this name and quotes as the original description: Shinji, 1924: 371 which refers not to Toxoptera argrimoniae but to Pemphigus arrigmoniae, spelled argrimoniae in p. 352 in the legend to the plate where an antenna of Tetraneura (Tetraneurella) nigriabdominalis (Sasaki, 1899) is figured.**

T. aurantii var. limonii del Guercio, 1917 = Aphis gossypii Glover, 1877
aurantii var. soyogo (Uye, 1923) (Aphis) BM HRL
caricis Fullaway, 1909 type of Vesiculaphis del Guercio, 1911 q.v.
celtis (Shinji, 1922) (*Aphis*)
citricidus (Kirkaldy, 1907) (Myzus) BM HRL
*aeglis* Shinji, 1922 (*Aphis*)
*aphoides* van der Goot, 1917
*argentiniensis* E.E. Blanchard, 1944 (*Paratoxoptera*)
*nigricans* van der Goot, 1917 (*Aphis*)
*tavaresi* del Guercio, 1908 (*Aphis*)
citrifoliae Maki, 1913 = aurantii (Boyer de Fonscolombe, 1841)
clematidis del Guercio, 1900 = aurantii (Boyer de Fonscolombe, 1841)
coffeae subsp. thomensis Seabra, 1921 = aurantii (Boyer de Fonscolombe, 1841)
cyperi van der Goot, 1917 = Schizaphis rotundiventris (Signoret, 1860)
djarani van der Goot, 1917 = aurantii (Boyer de Fonscolombe, 1841)
fusca Macchiati, 1881 = Cavariella archangelicae (Scopoli, 1763)
graminum (Rondani, 1852) (Aphis) to Schizaphis Börner, 1931
jaroslavi Mordvilko, 1921 to Schizaphis Börner, 1951
leonuri Takahashi, 1921 = Aphis gossypii Glover, 1877
mehijiwae Uye, 1924 to Schizaphis Börner, 1931
minuta van der Goot, 1917 to Schizaphis Börner, 1931
muhlenbergiae Phillips & Davis, 1912 to Schizaphis Börner, 1931
nigerrima Hille Ris Lambers, 1931 to Schizaphis Börner, 1931
nigra Baker, 1918 to Schizaphis (Paraschizaphis)
odinae (van der Goot, 1917) (Longiunguis) BM HRL
*adivae* Shiraki, 1952 (*Aphis*)
*araliae* Matsumura, 1917 (*Arimakia*)
*ficicola* Takahashi, 1921 (*Aphis*)
*hameliae* Theobald, 1929 (*Longicaudus*)
*mokulen* Shinji, 1922 (*Aphis*)
*rutae* Shinji, 1922 (*Aphis*)
*sansho* Shinji, 1922 (*Thomasia*)
*somei* Essig & Kuwana, 1918 (*Aphis*)
*spathodeae* van der Goot, 1918 (*Longiunguis*)
*taranbonis* Matsumura, 1917 (*Arimakia*)
piricola Matsumura, 1917 to Schizaphis Börner, 1931
punjabipyri Das, 1918 to Schizaphis Börner, 1931
rhododendri Uye, 1923 = ? Vesiculaphis caricis (Fullaway, 1909)

T. rufiabdominalis Sasaki, 1899 to Rhopalosiphum Koch, 1854
schlingeri Tao, 1961
scirpi Passerini, 1874 to Schizaphis (Paraschizaphis)
setariae Rusanova, 1943 nomen nudum
shichito Uye, 1923 = Vesiculaphis caricis (Fullaway, 1909)
theobromae Schouteden, 1906 = aurantii (Boyer de Fonscolombe, 1841)
typhae Laing, 1923 (type of Paraschizaphis Hille Ris Lambers, 1947) = Schizaphis (Paraschizaphis) scirpi (Passerini, 1874)
vandergooti Börner, 1939, type of Toxopterina Börner, 1940 q.v.
variegata del Guercio, 1909 = aurantii (Boyer de Fonscolombe, 1841)
viridi-rubra Gillette & Palmer, 1932 to Schizaphis Börner, 1931

Subgen. Schizaphis Börner, 1931

T. (Schizaphis) setariae Rusanova, 1962: 76 = Rhopalosiphum maidis (Fitch, 1856)

TOXOPTERELLA Hille Ris Lambers, 1960
Can. Ent. 92: 263
Type species Toxopterella canadensis Hille Ris Lambers, 1960

T. canadensis Hille Ris Lambers, 1960 BM HRL
smithi Hille Ris Lambers, 1962 BM HRL

Subgen. Sorbobium MacGillivray & Bradley, 1961

T. (Sorbobium) drepanosiphoides MacGillivray & Bradley, 1961 type of Sorbobium MacGillivray & Bradley, 1961 BM HRL
(Sorbobium) drepanosiphoides subsp. irae Shaposhnikov, 1963

TOXOPTERINA Börner, 1940
Neue Blattläuse aus Mitteleuropa, privately published, p. 3
Type species Toxoptera vandergooti Börner, 1939

T. vandergooti (Börner, 1939) (Toxoptera) BM HRL
*verae* Shaposhnikov, 1956 (*Chomaphis*)

Subgen. Tuberculaphis Börner, 1952

T. (Tuberculaphis) grosmannae Börner, 1952 to Aphis Linnaeus, 1758
(Tuberculaphis) parietariella Börner, 1952* = parietariae Theobald, 1922

*** nomen novum pro Aphis parietariae Theobald, 1922 nec Lichtenstein, 1884; but parietariae Lichtenstein as nomen nudum has no nomenclatural standing.**

TRAMA von Heyden, 1938
Ent. Beitr. Mus. Senckenb. 2: 298
Type species Trama troglodytes von Heyden, 1837

T. afghanica (Narzikulov ex Narzikulov & Umarov, 1971 nomen nudum) Narzikulov, 1973
antennata Mordvilko, 1935 type of Tactilotrama Börner, 1952 q.v.
auriculae Murray, 1877 to Thecabius (Parathecabius)
bazarovi Narzikulov, 1966
caudata del Guercio, 1909 to Neotrama Baker, 1920
centaureae Borner, 1940 BM ?
donisthorpei Theobald, 1913 = Protrama flavescens (Koch, 1856)
eastopi Heinze, 1962 BM ?
euphorbiae Juchnevitch & Kan, 1971
flavescens Koch, 1856 to Protrama Baker, 1920
helianthemi (Westwood, 1843) (Rhizobius)
horvathi del Guercio, 1909 = Neotrama caudata (del Guercio, 1909)
kulinitschae Narzikulov, 1973
lomovae Narzikulov, 1969
mordvilkoi Börner, 1940
oculata Gillette & Palmer, 1930 = rara Mordvilko, 1908
pubescens Koch, 1857 BM HRL ?
radicis Kaltenbach, 1843 type of Protrama Baker, 1920 q.v.
radicis subsp. asiatica Narzikulov, 1973 to Protrama Baker, 1920
ranunculi del Guercio, 1909 to Protrama Baker, 1920
rara Mordvilko, 1908 BM HRL
*oculata* Gillette & Palmer, 1930
taraxaci Shinji, 1929 (type of Longitarsus Shinji, 1929/30 nec Berthold ex Latreille, 1827, Latreille ex Cuvier, 1829) to Neotrama Baker, 1920
troglodytes von Heyden, 1837 BM HRL
*radicum* Goureau, 1861 (*Aphis (Forda)*)
troglodytes subsp. sibirica Mordvilko, 1935
voigti Börner, 1940 BM ?
xerophilaphica Juchnevitch & Kan, 1971

TRAMIA Amyot, 1847
Annls Soc. ent. Fr. (2) 5: 487
Invalid

TRANAPHIS Walker, 1870
Zoologist (2) 5: 1999
Type species Aphis salicivora Walker, 1848 = Aphis capreae Mosley, 1841 = Chaitophorus Koch, 1854

T. beuthani Börner, 1950 = Chaitophorus horii subsp. beuthani (Börner, 1950)
palaestina Börner ex Bodenheimer, 1930; in litteris synonym of syriaca Börner ex Bodenheimer ex Swirskii, 1957 nomen nudum
syriaca Börner ex Bodenheimer, 1937 nomen nudum

TRIASSOAPHIS Evans, 1956
Aust. J. Zool. 4: 238
Type species Triassoaphis cubitus Evans, 1956
Fossil

TRICAUDATUS Narzikulov, 1957
Ént. Obozr. 36: 683
Type species Rhopalosiphoninus polygoni Narzikulov, 1957

T. indicus A.K. Ghosh, R.C. Basu & RayChaudhuri, 1969 = polygoni (Narzikulov, 1957)
polygoni (Narzikulov, 1957) (Rhopalosiphoninus) HRL
*indicus* A.K. Ghosh, R.C. Basu & RayChaudhuri, 1969
*polygoni* subsp. *tuberculatus* Hille Ris Lambers & A.N. Basu, 1966 HRL
polygoni subsp. tuberculatus Hille Ris Lambers & A.N. Basu, 1966 = polygoni (Narzikulov, 1957)

TRICHAITOPHORUS Takahashi, 1937
Annotnes zool. jap. 16: 17
Type species Trichaitophorus aceris Takahashi, 1937

T. aceris Takahashi, 1937 BM HRL
albus Takahashi, 1961 type of Yamatochaitophorus Higuchi, 1972 q.v.
koyaensis Takahashi, 1961 BM HRL
recurvispinus Hille Ris Lambers & A.N. Basu, 1966 HRL

TRICHOCALLIS Börner, 1930
Arch. klassif. phylogen. Ent. 1: 127
Type species Allaphis caricis Mordvilko, 1921
*Heterocallis* Quednau, 1966
Here used as subg. of Thripsaphis Gillette, 1917

TRICHONAPHIS Shinji, 1930 lapsus pro Trichosiphonaphis Takahashi, 1922

TRICHONOSIPHONIELLA Shinji, lapsus pro Trichosiphoniella Shinji, 1929

TRICHOREGMA Takahshi, 1929
Trans. nat. Hist. Soc. Formosa 19: 252
Type species Oregma bambusifoliae Takahashi, 1921
= Astegopteryx Karsch, 1890

T. bambusae Takahashi, 1935 to Astegopteryx Karsch, 1890
esakii Takahashi, 1941 to Astegopteryx Karsch, 1890
flava Takahashi, 1950 to Astegopteryx Karsch, 1890
malaccensis Takahashi, 1950 to Astegopteryx Karsch, 1890
musae Takahashi, 1925 to Astegopteryx Karsch, 1890
pandani Takahashi, 1935 to Astegopteryx Karsch, 1890

TRICHOSIPHON Pergande lapsus pro Trichosiphum Pergande, 1906

TRICHOSIPHONAPHIS Takahashi, 1922
Proc. ent. Soc. Wash. 24: 205
Type species Myzus polygoniformosanus Takahashi, 1921

T. cornuta Miyazaki, 1971 to Trichosiphonaphis (Xenomyzus)
gerberae A.K. Ghosh & RayChaudhuri, 1972
horii Miyazaki, 1971 to Trichosiphonaphis (Xenomyzus)
lonicerae (Uye, 1923) (Macrosiphum) BM HRL
polygoniformosanus (Takahashi, 1921) (Myzus) BM HRL

Subgen. Xenomyzus Aizenberg, 1935

T. (Xenomyzus) alpestris (Hille Ris Lambers, 1969) (Xenomyzus) HRL
(Xenomyzus) cornuta Miyazaki, 1971 (Trichosiphonaphis)
(Xenomyzus) corticis (Aizenberg, 1935) (Xenomyzus) BM HRL
*carpathica* Knechtel & Manolache, 1941 (*Alphitoaphis*) HRL
(Xenomyzus) horii Miyazaki, 1971 (Trichosiphonaphis)
*lonicerae* Hori, 1938 (*Aulacorthum*)
(Xenomyzus) ishimikawae (Shinji, 1941) (Phorodon) BM HRL
(Xenomyzus) polygoni (van der Goot, 1917) (Phorodon) BM HRL
(Xenomyzus) polygonifoliae (Shinji, 1944) (Myzus) HRL
(Xenomyzus) tade (Shinji, 1927) (Carolinaia) BM HRL
*polygoni-japonica* Shinji, 1941 (*Macrosiphum*)
*polygoniphaga* Takahashi, 1961 (*Aphorodon*) HRL

TRICHOSIPHONIELLA Shinji, 1929
Lansania 1: 46
Type species Aphis spinosula Essig & Kuwana, 1918 = Myzus sakurae Matsumura, 1917
= Tuberocephalus Shinji, 1929

T. formosana Hille Ris Lambers, 1965 = Tuberocephalus momonis (Matsumura, 1917)

TRICHOSIPHUM Pergande, 1906
Ent. News 17: 206
Type species Trichosiphum anonae Pergande, 1906
Here used as subg. of Greenidea Schouteden, 1905

T. anonae Pergande, 1906 to Greenidea (Trichosiphum)
dubium van der Goot, 1918 to Eutrichosiphum Essig & Kuwana, 1918
formosanum Maki, 1917 to Greenidea (Trichosiphum)
kashicola Kurisaki, 1920 type of Allotrichosiphon Takahashi, 1962 q.v.
kuwanae Pergande, 1906 emended to kuwanai by Takahashi, 1919
kuwanai Pergande, 1906 to Greenidea (Trichosiphum)
lithocarpae Maki, 1918 = Eutrichosiphum pasaniae (Okajima, 1908)
minutum van der Goot, 1917 to Eutrichosiphum Essig & Kuwana, 1918
montanum van der Goot, 1918 to Mollitrichosiphum (Metatrichosiphon)
nigra Maki, 1917 to Greenidea (Trichosiphum)
nigrofasciatum Maki, 1917 (type of Metatrichosiphon RayChaudhuri, 1956) to Mollitrichosiphum (Metatrichosiphon)
pasaniae Okajima, 1908 type of Eutrichosiphum Essig & Kuwana, 1918 q.v.
roepkei van der Goot, 1918 to Eutrichosiphum Essig & Kuwana, 1918
tenuicorpus Okajima, 1908 (type of Neotrichosiphon RayChaudhuri, 1956) to Mollitrichosiphum Suenaga, 1934
vandergooti Franssen, 1930 to Eutrichosiphum Essig & Kuwana, 1918

TRIFIDAPHIS del Guercio, 1909
Pomona Coll. J. Ent. 1: 75
Type species Pemphigus radicicola Essig, 1909 = Smynthurodes betae Westwood, 1849
= Smynthurodes Westwood, 1849

T. gossypii Kulkarni, 1956 = Smynthurodes betae Westwood, 1849
perniciosus Nevsky, 1929 = Smynthurodes betae Westwood, 1849
silvestrii Mordvilko, 1935 = Smynthurodes betae Westwood, 1849

TRILOBAPHIS Theobald, 1922
Entomologist's mon. Mag. 58: 138
Type species Trilobaphis caricis Theobald, 1922 = Vesiculaphis theobaldi Takahashi, 1930
= Vesiculaphis del Guercio, 1911

T. caricis Theobald, 1922 = Vesiculaphis theobaldi Takahashi, 1930
langei Börner, 1933 (type of Galiobium Börner, 1933) to Myzus (Galiobium)
rhodolestes Wood-Baker, 1943 = Myzaphis rosarum (Kaltenbach, 1843)

Subgen. Galiobium Börner, 1933

T. (Galiobium) langei, 1933 to Myzus (Galiobium)

TRINACRIELLA del Guercio, 1913
Redia 9: 169
Type species Trinacriella magnifica del Guercio, 1913
= Geoica Hart, 1894

T. magnifica del Guercio, 1913 to Geoica Hart, 1894

TRIOCULA Rusanova, 1943 (1942)
Trudy̅ Azerb. Gos. Univ. (Biol.) 3 (1): 36
Type species Triocula distorta Rusanova, 1943 nomen nudum
Without described species

T. distorta Rusanova, 1943 nomen nudum

TRIPHYLLAPHIS Börner, (1944) 1949
In Brohmer, P., Fauna von Deutschland, ed. 5: 214, in key, without species (1944)
Beitr. tax. zool. 1: 214 (1949)
Type species Triphyllaphis luteola Börner, 1949
= Therioaphis Walker, 1870

T. luteola Börner, 1949 to Therioaphis Walker, 1870

TRITOGENAPHIS Oestlund, 1922
Rep. Minn. St. Ent. 19: 142
Type species Aphis rudbeckiae Fitch, 1851
=Uroleucon Mordvilko, 1914

T. eupatorifoliae Tissot, 1934 to Uroleucon (Uromelan) HRL
gutierrezia Pack & Knowlton, 1929 = Uroleucon (Lambersius) escalantii (Knowlton, 1928)
kosacaudis Knowlton, 1928 = Uroleucon ambrosiae (Thomas, 1877)
utahensis Pack & Knowlton, 1929 = Uroleucon (Lambersius) escalantii (Knowlton, 1928)

TRITRICHOSIPHUM Robinson, 1972
Can. Ent. 104: 606
Type species Tritrichosiphum thailandicum Robinson, 1972

T. thailandicum Robinson, 1972 paratype BM HRL

TROITZKYA Börner, 1930
Arch. klassif. phylogen. Ent. 1: 160
Type species Phylloxera caryaesemen Walsh, 1867
Phylloxeridae

TRUNCAPHIS Theobald, 1918
Entomologist 51: 25
Type species Truncaphis newsteadi Theobald, 1918 = Byrsocrypta rhois Fitch, 1866
= Melaphis Walsh, 1867

T. graminis Takahashi, 1930 to Colopha Monell, 1877
newsteadi Theobald, 1918 = Melaphis rhois (Fitch, 1866)

TSHERNOVAIA Holman & Szelegiewicz, 1964
Bull. Acad. pol. Sci. (Cl. II Ser. Sci. biol.) 12: 351
Type species Tshernovaia adenophorae Holman & Szelegiewicz, 1964

T. adenophorae Holman & Szelegiewicz, 1964 HRL

TSUGAPHIS Takahashi, 1967
Proc. R. ent. Soc. Lond. (B) 26: 107
Type species Tsugaphis sorini Takahashi, 1957

T. sorini Takahashi, 1957 HRL

TUBAPHIS Hille Ris Lambers, 1947
Temminckia 7: 312
Type species Aphis ranunculina Walker, 1852
*Ovatopsis* Aizenberg, 1954

T. clematophilus (Takahashi, 1965) (Myzus (Tubaphis)) HRL
ranunculina (Walker, 1852) (Aphis) lectotype BM HRL
*ranunculi* Aizenberg, 1954 (*Ovatopsis*)
*ranunculi* del Guercio, 1900 (*Myzus*)

TUBERAPHIS Takahashi, 1933
Stylops 2: 27
Type species Tuberaphis coreanus Takahashi, 1933

T. coreana Takahashi, 1933 BM HRL
indica A.K. Ghosh, M.R. Ghosh & RayChaudhuri, 1971 HRL
loranthi Tseng & Tao, 1938 (Astegopteryx) HRL
loranthi (van der Goot, 1917) (Oregma) HRL

TUBERCULAPHIS Börner, 1952 nec Theobald ex Das, 1918
Mitt. thüring. bot. Ges. 4 (3): 91, described as subg. of Toxopterina Börner, 1940
Type species Doralina taraxacicola Börner, 1940
= Aphis Linnaeus, 1758

T. grosmannae Börner, 1952 to Aphis Linnaeus, 1758

TUBERCULAPHIS Theobald ex Das, 1918
In Das, B., 1918, Mem. Indian Mus. 6: 174, in key, without species
Without species
= ? Cavariella del Guercio, 1911

TUBERCULATUS Mordvilko, 1894
Rab. Lab. zool. Kab. imp. varsh. Univ. 8 (3): 51
Type species Aphis quercea Kaltenbach, 1843

T. cornutus Richards, 1969 to Tuberculatus (Camelaphis)
eggleri Börner, 1950 to Tuberculatus (Tuberculoides)
flavus Mordvilko, 1929 nomen nudum
indicus L.K. Ghosh, 1972 to Tuberculatus (Acanthocallis)
japonicus Higuchi, 1969 to Tuberculatus (Acanthocallis)
multituberculatus Mordvilko, 1929 nomen nudum
orientalis Richards, 1968 = Tuberculatus (Acanthocallis) stigmatus (Matsumura, 1917)
punctatellus of authors nec Fitch, 1855 = tuberculatus (Richards, 1965) nec 1968
querceus (Kaltenbach, 1843) (Aphis) BM HRL
spiculatus Richards, 1971
tuberculatus (Richards, 1965) (Myzocallis (Neomyzocallis)) BM HRL
*punctatellus* Fitch of Hottes & Frison, 1931 nec Fitch, 1855
tuberculatus Richards, 1968 = Tuberculatus (Orientuberculoides) fangi (Tseng & Tao, 1938)

Subgen. Acanthocallis Matsumura, 1917

T. (Acanthocallis) indicus L.K. Ghosh, 1972 (Tuberculatus) HRL
(Acanthocallis) japonicus Higuchi, 1969 (Tuberculatus) HRL
(Acanthocallis) pilosus (Takahashi, 1929) (Myzocallis) BM HRL
*lewi* Tao, 1964 (*Nippocallis*)
(Acanthocallis) quercicola (Matsumura, 1917) (Acanthocallis) BM HRL
*macrotuberculata* Essig & Kuwana, 1918 (*Myzocallis*)
*quercicola* Matsumura, 1919 (*Ptychodes*)
(Acanthocallis) stigmatus (Matsumura, 1917) (Arakawana) BM HRL
*nigra* Okamoto & Takahashi, 1927 (*Myzocallis*)
*orientalis* Richards, 1968 (*Tuberculatus*)

Subgen. Camelaphis Hille Ris Lambers, 1974

T. (Camelaphis) cornutus Richards, 1969 (Tuberculatus) HRL
(Camelaphis) maculipennis Hille Ris Lambers, 1974 BM HRL
(Camelaphis) pallescens Hille Ris Lambers, 1974 HRL

Subgen. Orientuberculoides Hille Ris Lambers, 1974

T. (Orientuberculoides) capitatus (Essig & Kuwana, 1918) (Myzocallis) HRL
(Orientuberculoides) capitatus subsp. intermedius Hille Ris Lambers, 1974 HRL
(Orientuberculoides) fangi (Tseng & Tao, 1938) (Tuberculoides) HRL
*tuberculatus* Richards, 1968 (*Tuberculatus*) nec 1965
(Orientuberculoides) higuchii Hille Ris Lambers, 1974 HRL
(Orientuberculoides) higuchii subsp. breviunguis Hille Ris Lambers, 1974 HRL
(Orientuberculoides) kashiwae (Matsumura, 1917) (Myzocallis) HRL
*naracola* Matsumura, 1918 (Myzocallis)
(Orientuberculoides) konaracola (Shinji, 1941) (Tuberculoides) HRL
(Orientuberculoides) kunugi (Shinji, 1924) (Myzocallis)
(Orientuberculoides) paranaracola Hille Ris Lambers, 1974 HRL
(Orientuberculoides) paranaracola subsp. hemitrichus Hille Ris Lambers, 1974 HRL
(Orientuberculoides) paiki Hille Ris Lambers, 1974 HRL
(Orientuberculoides) querciformosanus (Takahashi, 1921) (Myzocallis) HRL
(Orientuberculoides) yokoyamai (Takahashi, 1923) (Myzocallis) HRL

Subgen. Pacificallis Richards, 1965

T. (Pacificallis) californicus (Baker, 1917) (Myzocallis) BM HRL
(Pacificallis) columbiae Richards, 1965 paratype BM HRL
(Pacificallis) kiowanicus (Hottes, 1933) (Myzocallis) paratype BM HRL
*tonkawa* Hottes, 1949 (*Myzocallis*)
*youngii* Knowlton & Hall, 1950 (*Myzocallis*)
(Pacificallis) maureri (Swain, 1918) (Myzocallis) BM HRL
(Pacificallis) pallidus (Davidson, 1919) (Myzocallis) BM HRL
*californica* var. *pallida* Davidson, 1919 (*Myzocallis*)
(Pacificallis) pasaniae (Davidson, 1915) (Myzocallis) HRL
(Pacificallis) quercifolii (Davidson, 1919) (Myzocallis) HRL

Subgen. Tuberculoides van der Goot, 1915

T. (Tuberculoides) africanus Hille Ris Lambers, 1974 HRL
(Tuberculoides) albosiphonatus Hille Ris Lambers, 1974 BM HRL
(Tuberculoides) annulatus (Hartig, 1841) (Aphis) BM HRL
*elegans* E.E. Blanchard, 1939 (*Tuberculoides*)
*essigi* Shinji, 1917 (*Myzocallis*)
*querciplatensis* E.E. Blanchard, 1926 (*Myzocallis*)
*quercus* Kaltenbach, 1843 (*Aphis*)
(Tuberculoides) borealis (Krzywiec, 1971) (Tuberculoides) paratypes BM HRL
(Tuberculoides) eggleri Börner, 1950 (Tuberculatus) BM HRL
*quercus* var. *insignis* Ferrari, 1872 (*Mizocallis*)
(Tuberculoides) fulviabdominalis (Shinji, 1941) (Tuberculoides) nomen dubium
(Tuberculoides) maximus Hille Ris Lambers, 1974 HRL
(Tuberculoides) moerickei Hille Ris Lambers, 1974 BM HRL
(Tuberculoides moerickei subsp. galatensis Hille Ris Lambers, 1974 HRL
(Tuberculoides) naganoe (Shinji, 1941) (Tuberculoides) nomen dubium
(Tuberculoides) neglectus (Krzywiec, 1966) (Tuberculoides) paratypes BM HRL
(Tuberculoides) remaudierei (Nieto Nafria, 1974) (Tuberculoides)

TUBERCULOIDES Van der Goot, 1915
Beiträge zur Kenntnis der Holländischen Blattläuse, Haarlem - Berlin, p. 312
Type species Aphis quercus Kaltenbach, 1843 = Aphis annulata Hartig, 1841
*Neotuberculatus* van der Goot, 1915
Here used as subg. of Tuberculatus Mordvilko, 1894

T. alnifoliae Shinji, 1941 = Pterocallis (Recticallis) alnijaponicae (Matsumura, 1917)
borealis Krzywiec, 1917 to Tuberculatus (Tuberculoides)

T. elegans E.E. Blanchard, 1939 = Tuberculatus (Tuberculoides) annulatus (Hartig, 1841)
fangi Tseng & Tao, 1938 = Tuberculatus (Orientuberculoides)
fulviabdominalis Shinji, 1941 nomen dubium
grodsinskyi E.E. Blanchard, 1939 = Hoplocallis pictus (Ferrari, 1872)
konaracola Shinji, 1941 to Tuberculatus (Orientuberculoides)
naganoe Shinji, 1941 nomen dubium
neglectus Krzywiec, 1966 to Tuberculatus (Tuberculoides)
nigrostriata Shinji, 1941 to Pterocallis (Recticallis)
petraeae Bozhko, 1957 = Myzocallis castanicola Baker, 1917
remaudierei Nieto Nafria, 1974 to Tuberculatus (Tuberculoides)

TUBERDEFECTUS Kumar & Lavigne, 1970
Pan-Pacif. Ent. 46: 120
Type species Tuberdefectus eastopi Kumar & Lavigne, 1970 = Tinocalloides montanus A.N. Basu 1970 (1969)
= Tinocalloides A.N. Basu, 1970 (1969)

T. eastopi Kumar & Lavigne, 1970 = Tinocalloides montanus A.N. Basu 1970 (1969)

TUBEROAPHIS Tseng & Tao, 1938
Jl W. China Border Res. Soc. 8: 207
Type species Tuberoaphis hydrangeae (hydranglee, errore ) Tseng & Tao, 1938

T. hydrangeae Tseng & Tao, 1938 HRL
hydrangeae subsp. digitata Hille Ris Lambers & A.N. Basu, 1966 BM HRL
*indica* A.K. Ghosh, M.R. Ghosh & RayChaudhuri, 1971 (*Matsumuraja*)

TUBEROCALLIS Nevsky, 1929
Zool. Anz. 82: 221
Type species Tuberocallis saltans Nevsky, 1929
= Tinocallis Matsumura, 1919

T. saltans Nevsky, 1929 to Tinocallis Matsumura, 1919

TUBEROCEPHALUS Shinji, 1929
Lansania 1: 39
Rep. Jap. Ass. Advmt sci. 5: 189 (1930)
Type species Tuberocephalus artemisiae Shinji, 1929
*Heterogenaphis* Ivanovskaja, 1966
*Trichosiphoniella* Shinji, 1929

T. artemisiae Shinji, 1929
higansakurae (Monzen, 1927) (Myzus) HRL
momonis (Matsumura, 1917) (Myzus) BM HRL
*formosana* Hille Ris Lambers, 1965 (*Trichosiphoniella*) HRL
sakurae (Matsumura, 1917) (*Myzus*) BM HRL
*rarus* (Matsumura, 1927) (*Myzus*)
*spinosula* Essig & Kuwana, 1918 (*Aphis*)
sasakii (Matsumura, 1917) (Myzus) HRL
*kunashyri* Ivanovskaja, 1966 (*Heterogenaphis*)
*tsengi* Tao, 1963 (*Myzus*)

TUBEROCORPUS Shinji, 1929 nec 1932
Lansania 1: 48
Type species Anoecia onigurumi Shinjii, 1923
= Kurisakia Takahashi, 1924

TUBEROCORPUS Shinji, 1932 nec 1929
Oyo-Dobuts. Zasshi 4 (3): 120
Type species Tuberocorpus onigurumi Shinji, 1932 = Glyphina rhusae Shinji, 1922
= Dasyaphis Takahashi, 1938

T. coreanus Paik, 1965 = Dasyaphis rhusae (Shinji, 1922)
onigurumi Shinji, 1932 (type of Dasyaphis Takahashi, 1938) = Dasyaphis rhusae (Shinji, 1922)

TUBERODRYOBIUS Das, 1918
Mem. Indian Mus. 6: 259
Type species Lachnus persicae Cholodkovsky, 1899
= Pterochloroides Mordvilko, 1914

TUBEROLACHNIELLA Hille Ris Lambers & A.N. Basu, 1966
Ent. Ber., Amst. 26: 34, described as subg. of Tuberolachnus Mordvilko, 1909
Type species Tuberolachnus (Tuberolachniella) sclerata Hille Ris Lambers & A.N. Basu, 1966
Here used as subg. of Tuberolachnus Mordvilko, 1909

TUBEROLACHNUS Mordvilko, 1909
Ezheg. zool. Muz. 13: 374
Type species Aphis viminalis Boyer de Fonscolombe, 1841 = Aphis saligna Gmelin, 1790

T. salignus (Gmelin, 1790) (Aphis) BM HRL
*dentatus* Le Baron, 1872 (*Lachnus*)
*fuliginosus* Buckton, 1891 (*Lachnus*)
*nigripes* Takahashi, 1932 (*Lachnus*)
*punctatus* Burmeister, 1835 (*Lachnus*)
*riparius* Snellen van Vollenhoven, 1862 (*Dryobius*)
*salicina* Zetterstedt, 1840 (*Aphis*)
*salicis* Sulzer, 1776 (*Aphis*)
? *tatakaensis* Takahashi, 1937 (*Lachnus*)
*viminalis* Boyer de Fonscolombe, 1841 (*Aphis*)
*vitellinae* Hartig, 1841 (*Aphis*)
todocola Inouye, 1936 to Cinara Curtis, 1835

Subgen. Tuberolachniella Hille Ris Lambers & A.N. Basu, 1966

T. (Tuberolachniella) scleratus Hille Ris Lambers & A.N. Basu, 1966 type of Tuberolachniella Hille Ris Lambers & A.N. Basu, 1966

TUBEROSIPHUM Shinji, 1922
Zool. Mag., Tokyo, 34: 789
Type species Tuberosiphum impatiens Shinji, 1922
= Impatientinum Mordvilko, 1914

T. camphorae Shinji, 1922 = Sinomegoura citricola (van der Goot, 1917)
impatiens Shinji, 1922 to Impatientinum Mordvilko, 1914

TULLGRENIA van der Goot, 1912
Tijdschr. Ent. 15: 96
Type species Tychea phaseoli Passerini, 1860 = Smynthurodes betae Westwood, 1849
= Smynthurodes Westwood, 1849

TURANAPHIS Mordvilko, 1914
Faune Russie, Ins. Hém. 1: 66
Without species

TYCHEA Koch, 1857
Die Pflanzenläuse Aphiden, Nürnberg, p. 296
Type species Tychea graminis Koch, 1857, a coccid
Coccoidea, Pseudococcidae

T. amycli Koch, 1857 nomen dubium
brevicornis Forbes of Mordvilko, 1921: 72 species dubium
brevicornis Hart, 1894 to Pemphigus Hartig, 1839
crassa Cockerell, 1903 nomen dubium
eragrostidis Passerini, 1860 = Geoica utricularia (Passerini, 1856)
eragrostidis Passerini of Schouteden, 1906 nec Passerini, 1860 (type of Tycheoides Schouteden, 1906) = Aploneura lentisci (Passerini, 1856)
erigeronensis Thomas, 1879 to Prociphilus Koch, 1857
graminis Koch, 1857 to Pseudococcidae, Coccoidea
graminis Koch, 1857 sensu Schouteden, 1906 = Tetraneura ulmi (Linnaeus, 1758)
groenlandica Rübsaamen, 1898 to Pemphigus Hartig, 1839
lasii Cockerell, 1903 to Pemphigus Hartig, 1839
pallidula Cockerell, 1903 nomen dubium
panici Thomas, 1878 to ? Anoecia Koch, 1857
phaseoli Passerini, 1860 (type of Tullgrenia van der Goot, 1912; and of Tychea Passerini, 1860 nec Koch, 1857) = Smynthurodes betae Westwood, 1849
radicola Oestlund, 1886 = Prociphilus erigeronensis (Thomas, 1879)
setariae Passerini, 1860 to Geoica Hart, 1894
setulosa Passerini, 1860 to Geoica Hart, 1894
silvestrii Mordvilko, 1921 nomen dubium
trivialis Passerini, 1860 = Forda marginata (Koch, 1857)

TYCHEA Passerini, 1860
Gli Afidi, Parma, p. 30, ascribed to Koch
Type species Tychea phaseoli Passerini, 1860 = Smynthurodes betae Westwood, 1849
= Smynthurodes Westwood, 1849

TYCHEIDES Dahl, 1912 lapsus pro Tycheoides Schouteden, 1906

TYCHEOIDES Schouteden, 1906
Mém. Soc. r. ent. Belg. 12: 194
Type species Tychea eragrostidis Passerini of Schouteden nec Passerini, 1860 = Tetraneura lentisci Passerini, 1856
= Aploneura Passerini, 1863

T. longisetosa Dahl, 1912 to Tetraneura Hartig, 1841

UHLMANNIA Börner, 1952
Mitt. thüring. bot. Ges. 4 (3): 109, 251
Type species Brachycolus singularis Börner, 1950

U. singularis (Börner, 1950) (Brachycolus) BM HRL

UMBELLIFERARIA Shaposhnikov, 1964
In Bei-Bienko, G.Y., Keys to the Insects of the European part of the USSR, 1: 582, described as subg. of Dysaphis Börner, 1931
Type species Dysaphis aizenbergi Shaposhnikov, 1949 = Yezabura aizenbergi Shaposhnikov, 1949
= Dysaphis Börner, 1931

UNILACHNUS Wilson, 1919
Ent. News 30: 5
Type species Unilachnus parvus Wilson, 1919 = Lachnus parvus Wilson, 1915
= Schizolachnus Mordvilko, 1909 (1908)

U. orientalis Takahashi, 1924 to Schizolachnus Mordvilko, 1909 (1908)

UNIPTERUS Hall, 1932
Stylops 1: 50
Type species Unipterus terminaliae Hall, 1932
= Paoliella Theobald, 1928

U. ayari Eastop, 1955 to Paoliella Theobald, 1928
commiphorae Doncaster, 1954 to Paoliella Theobald, 1928
commiphorae subsp. persimilis Eastop, 1955 to Paoliella Theobald, 1928
delottoi Hille Ris Lambers, 1954 to Paoliella Theobald, 1928
nachensis Eastop, 1955 to Paoliella Theobald, 1928
papillata Hall, 1932 to Paoliella Theobald, 1928
sawus Eastop, 1955 to Paoliella Theobald, 1928
terminaliae Hall, 1932 to Paoliella Theobald, 1928
terminaliae subsp. kenyensis Eastop, 1956 = Paoliella kenyensis (Eastop, 1956)
ufuasi Eastop, 1955 to Paoliella Theobald, 1928

Subgen. Paoliella Theobald, 1928

U. (Paoliella) nirmalae David, 1969 to Paoliella Theobald, 1928

UNISITOBION Takahashi, 1961
Insecta matsum. 24: 104
Type species Macrosiphum sorbi Matsumura, 1918
Here used as subg. of Macrosiphum Passerini, 1860

URAPHIS del Guercio, 1907
Redia 4: 191, 192
Type species Aphis genistae Kaltenbach, 1843
= Aphis Linnaeus, 1758

UROLEUCON Mordvilko, 1914
Faune Russie, Ins. Hém. 1: 64, described as subg. of Macrosiphum Passerini, 1860
Type species Aphis sonchi Linnaeus, 1767 designated by Borner, 1930: 141
*Dactynotus* Rafinesque, 1818 invalid
*Eurythaphis* Mordvilko, 1914
*Megalosiphum* Mordvilko, 1919
*Tritogenaphis* Oestlund, 1922

U. aaroni (Knowlton, 1949) (Macrosiphum) BM
achilleae (Koch, 1855) (Siphonophora) BM HRL
adenocaulonae (Essig, 1936) (Macrosiphum) BM HRL
ambrosiae (Thomas, 1878) (Siphonophora) BM HRL
*kosacaudis* Knowlton, 1928 (*Tritogenaphis*)
astronomus (Hille Ris Lambers, 1962) (Dactynotus) BM HRL
atripes (Gillette & Palmer, 1933) (Macrosiphum) BM HRL
bicolor (Nevsky, 1929) (Macrosiphum)
bielawskii (Szelegiewicz, 1962) (Dactynotus) paratypes BM
brevirostre Holman, 1975
chilensis (Essig, 1953) (Macrosiphum) paratypes BM
chondrillae (Nevsky, 1929) (Macrosiphum) BM HRL
*margarithae* Hille Ris Lambers, 1950 (*Dactynotus*) cotypes BM HRL
chrysanthemi (Oestlund, 1886) (Siphonophora) BM HRL
*calendulae* Monell, 1879 (*Siphonophora*)
chrysopsidicola (Olive, 1963) (Dactynotus) BM HRL
cichorii (Koch, 1855) (Siphonophora) BM HRL
*phillipsii* Theobald, 1925 (*Macrosiphum*) BM
*sisymbrii* Buckton, 1876 (*Siphonophora*)
cichorii subsp. carduicola (Hille Ris Lambers, 1939) (Dactynotus) HRL
cichorii subsp. grossum (Hille Ris Lambers, 1939) (Dactynotus) BM HRL
*picridicola* Hille Ris Lambers, 1939 (*Dactynotus*)
cichorii subsp. leontodontis (Hille Ris Lambers, 1939) (Dactynotus) BM HRL
ciefi (Olive, 1963) (Dactynotus) HRL
cirsicola (Holman, 1962) (Dactynotus) BM HRL
cirsii (Linnaeus, 1758) (Aphis) BM HRL
*marcatus* Hille Ris Lambers, 1931 (*Dactynotus*) HRL
*olivata* Buckton, 1876 (*Siphonophora*) lectotype BM
*serratulae* Kaltenbach, 1843 (*Aphis*)

U. debilis (Takahashi, 1923) (Macrosiphum)
ensifoliae (Holman, 1965) (Dactynotus) BM HRL
epilobii (Pergande, 1900) (Nectarophora)
eupatoricolens (Patch, 1919) (Macrosiphum) BM HRL
fagopyri (Chowdhuri, R.C. Basu, Chakrabarti & RayChaudhuri, 1969) (Dactynotus) BM HRL
fallacis (Nevsky, 1929) (Macrosiphum)
formosanum (Takahashi, 1921) (Macrosiphum) BM HRL
formosanum subsp. crepidis A.K. Ghosh, M.R. Ghosh & RayChaudhuri, 1971
fuchuensis (Shinji, 1942) (Macrosiphum) HRL
gulbenkiani Ilharco, 1968 = Macrosiphoniella tapuskae (Hottes & Frison, 1931)
hieracicola (Hille Ris Lambers, 1962) (Dactynotus) HRL
*pieloui* Richards, 1972 (*Dactynotus*)
hypochoeridis (Fabricius, 1779) (Aphis) BM HRL
*hypochoeridis* Hille Ris Lambers, 1939 (*Dactynotus*) HRL
hyssopii (Narzikulov & Daniarova, 1973) (Dactynotus)
impatiensicolens (Patch, 1919) (Macrosiphum) HRL
inulicola (Hille Ris Lambers, 1939) (Dactynotus) BM HRL
inulicola subsp. hirticola (Ossiannilsson, 1959) (Dactynotus) BM HRL
jaceicola (Hille Ris Lambers, 1939) (Dactynotus) BM HRL
jaceicola subsp. pasqualei (Hille Ris Lambers & Stroyan, 1959)(Dactynotus) HRL
kashmiricum (Verma, 1966) (Dactynotus) BM HRL
kikioense (Shinji, 1942) (Macrosiphum) HRL
*torajicolus* Paik, 1965 (*Dactynotus*) HRL
kumaoni H. Banerjee, A.K. Ghosh & RayChaudhuri, 1969
lanceolatum (Patch, 1919) (Macrosiphum) BM HRL
leonardi (Olive, 1965) (Dactynotus) HRL
lizerianum (E.E. Blanchard, 1922) (Macrosiphum)
longisetosum Chakrabarti & Verma, 1975 HRL
mongolicum Holman 1975 HRL
monticola (Takahashi, 1935) (Macrosiphum) BM HRL
muermosum (Essig, 1953) (Macrosiphum) paratypes BM
muralis (Buckton, 1876) (Siphonophora) lectotype BM HRL
narzykulovi Holman, 1974 BM HRL
nickeli (Essig, 1956) (Macrosiphum)
nigrotibium (Olive, 1963) (Dactynotus) BM HRL
nigrotuberculatum (Olive, 1963) (Dactynotus) BM HRL
obscuricaudatum (Olive, 1965) (Dactynotus) HRL
obscurum (Koch, 1855) (Siphonophora) BM HRL

U. ochropus (Hille Ris Lambers, 1939) (Dactynotus) BM HRL
paucosensoriatum (Hille Ris Lambers, 1960) (Dactynotus) BM HRL
pepperi (Olive, 1965) (Dactynotus) HRL
phyteumatis (Bozhko, date ?) (genus ?)
picridiphagum (Takahashi, 1962) (Dactynotus) HRL
picridis (Fabricius, 1775) (Aphis) BM HRL
pilosellae (Börner, 1933) (Dactynotus) BM HRL
pseudambrosiae (Olive, 1963) (Dactynotus) BM HRL
pseudobscurum (Hille Ris Lambers, 1967) (Dactynotus) HRL
pseudochrysanthemi (Olive, 1967) (Dactynotus)
pseudotanaceti (Verma, 1969) (Dactynotus) HRL
*tanaceti* subsp. *indica* L.K. Ghosh, 1971 HRL
pulicariae (Hille Ris Lambers, 1939) (Dactynotus) HRL
pyrethri Holman, 1974 BM HRL
reynoldense (Olive, 1965) (Dactynotus) HRL
? rubecula (Haldeman, 1844) (Aphis)
rudbeckiae (Fitch, 1851) (Aphis) BM HRL
russellae (Hille Ris Lambers, 1960) (Dactynotus) BM HRL
saussureae (Takahashi, 1962) (Dactynotus) BM HRL
seneciocola (Paik, 1965) (Macrosiphum) HRL
sijpkensi Hille Ris Lambers, 1974 HRL
simlaense Chakrabarti, A.K. Ghosh & RayChaudhuri, 1971 HRL
solirostratum (Richards, 1972) (Dactynotus)
sonchellum (Monell, 1879) (Siphonophora) BM HRL
*williamsi* Gillette & Palmer, 1929 (*Macrosiphum*)
sonchi (Linnaeus, 1767) (Aphis) BM HRL
*alliariae* Koch, 1855 p. 160 (*Siphonophora*)
*sonchi* Geoffroy, 1762 (*Aphis*) invalid
*sonchicola* Matsumura, 1917 (*Macrosiphum*)
sonchi subsp. afghanicum (Narzikulov ex Narzikulov & Umarov, 1972) (Dactynotus)
tanaceti (Linnaeus, 1758) (Aphis) BM HRL
*tanaceticola* Kaltenbach, 1843 (*Aphis*)
tanaceti subsp. indica L.K. Ghosh, 1971 = pseudotanaceti (Verma, 1969)
telekiae (Holman, 1965) (Dactynotus) BM
tenuirostris L.K. Ghosh, 1970 = Paczoskia budhium H. Banerjee, A.K. Ghosh & RayChaudhuri, 1969
tucumani (Essig, 1953) (Macrosiphum) BM HRL
tussilaginis (Walker, 1850) (Aphis) type BM HRL
*tussilaginis* subsp. *petasitis* Bozhko, 1959 (*Eurythaphis*)

U. vernonicola (Holman, 1974) (Dactynotus) HRL
zinzalae (Hottes & Frison, 1931) (Macrosiphum) BM

Subgen. Belochilum Börner, 1932

U. (Belochilum) inulae (Ferrari, 1872) (Siphonophora) BM HRL
*inulae* Börner, 1932 errore Passerini, 1860 (*Belochilum*)
*inulifoliae* Mimeur, 1934 (*Macrosiphum*)

Subgen. Lambersius Olive, 1965

U. (Lambersius) anomalae (Hottes & Frison, 1931) (Macrosiphum) BM HRL
(Lambersius) baccharidis (Clarke, 1903) (Nectaropora) HRL
(Lambersius) bereticum (E.E. Blanchard, 1922) (Macrosiphoniella)
(Lambersius) bradburyi (Olive, 1965) (Dactynotus (Lambersius))
(Lambersius) breviscriptum Palmer, 1936 (Macrosiphum) BM HRL
(Lambersius) caligatum (Richards, 1966) (Dactynotus) BM HRL
(Lambersius) canadense (Richards, 1972) (Dactynotus) HRL
(Lambersius) chilense (Essig, 1953) (Macrosiphum) paratypes BM HRL
(Lambersius) cocoense E.E. Blanchard, 1932) (Macrosiphum)
(Lambersius) cordobense (E.E. Blanchard, 1932) (Macrosiphum)
(Lambersius) crepusisiphon (Olive, 1965) (Dactynotus (Lambersius))
(Lambersius) dubium Holman, 1975
(Lambersius) erigeronensis (Thomas, 1878) (Siphonophora) BM HRL
*erigeronella* Soliman, 1927 (*Macrosiphum*)
(Lambersius) escalantii (Knowlton, 1928) (Macrosiphum) BM HRL
*gutierrezia* Pack & Knowlton, 1929 (*Tritogenaphis*)
*utahensis* Pack & Knowlton, 1929 (*Tritogenaphis*)
(Lambersius) gravicorne (Patch, 1919) (Macrosiphum) BM HRL
(Lambersius) helianthi (Tao, 1963) (Dactynotus)
(Lambersius) huantanum (Essig, 1953) (Macrosiphum) paratypes BM HRL
(Lambersius) idahoense (Miller, 1933) (Macrosiphum)
(Lambersius) katonkae (Hottes, 1933) (Adactynus) BM HRL
(Lambersius) littorale (E.E. Blanchard, 1939) (Macrosiphum)
(Lambersius) longirostre Gillette & Palmer, 1933 (Macrosiphum) BM HRL
(Lambersius) luteolum (Williams, 1911) (Siphonophora)
(Lambersius) macgillivrayae (Olive, 1967) (Dactynotus (Lambersius))
(Lambersius) madia (Swain, 1919) (Macrosiphum)
*rudbeckiae* var. *madia* Swain, 1919 (*Macrosiphum*)
(Lambersius) nodulum (Richards, 1972) (Dactynotus)
(Lambersius) nuble (Essig, 1953) (Macrosiphum)

U. (Lambersius) richardsi (Robinson, 1964) (Dactynotus) paratypes BM HRL
(Lambersius) tenuitarsum (Gillette & Palmer, 1933) (Macrosiphum) HRL
(Lambersius) tissoti (Boudreaux, (1948) 1949) (Macrosiphum) BM
(Lambersius) zayasi Holman, 1974 (Dactynotus (Lambersius)) HRL
(Lambersius) zymozionense (Knowlton, 1946) (Macrosiphum) BMHRL
*brevitarsus* Robinson, 1974 (*Dactynotus*) paratypes BM HRL

Subgen. Satula Olive, 1963

U. (Satula) brachychaetum (Olive, 1963) (Satula) BM HRL

Subgen. Uromelan Mordvilko, 1914

U. (Uromelan) adenophorae (Matsumura, 1918) (Macrosiphum) BM HRL
*triphyllae* Miyazaki, 1966 (*Dactynotus (Uromelan)*) HRL
(Uromelan) adenophoricola Holman, 1975 HRL
(Uromelan) amamianum (Takahashi, 1930) (Macrosiphum) BM HRL
(Uromelan) asyneumatis (Holman, 1969) (Dactynotus) BM HRL
(Uromelan) bonitum (Hottes, 1950) (Macrosiphum) BM HRL
(Uromelan) cameronense (Takahashi, 1950) (Dactynotus) BM
(Uromelan) campanulae (Kaltenbach, 1843) (Aphis) BM HRL
(Uromelan) campanulae subsp. longius (Börner, 1950) (Dactynotus) HRL
(Uromelan) carlinae (Börner, 1933) (Dactynotus) HRL
(Uromelan) carthami (Hille Ris Lambers, 1948) (Dactynotus) BM HRL
(Uromelan) cephalonopli (Takahashi, 1962) (Dactynotus (Uromelan)) BM
(Uromelan) compositae (Theobald, 1915) (Macrosiphum) types BM HRL
*dahliafolii* Theobald, 1918 (*Macrosiphum*)
*sonchi* var. *flavomarginata* del Guercio, 1916 (*Macrosiphum*)
*vernoniae* van der Goot, 1915 (*Macrosiphum*)
(Uromelan) compositae subsp. evansi (Eastop, 1956) (Dactynotus (Uromelan)) types BM HRL
(Uromelan) doronici (Börner, 1942) (Dactynotus) HRL
(Uromelan) eoessigi (Knowlton, 1947) (Macrosiphum) BM HRL
(Uromelan) erigeronis (Nevsky, 1928) (Macrosiphoniella)
(Uromelan) eupatorifoliae (Tissot, 1934) (Tritogenaphis) BM HRL
(Uromelan) giganteum (Matsumura, 1918) (Macrosiphum) BM HRL
(Uromelan) gobonis (Matsumura, 1917) (Macrosiphum) BM HRL
(Uromelan) griersoni (E.E. Blanchard, 1932) (Macrosiphum)
(Uromelan) helenae (Hille Ris Lambers, 1950) (Dactynotus (Uromelan)) BM HRL
(Uromelan) helianthicola (Olive, 1963) (Dactynotus) BM HRL
(Uromelan) hieracioides (Bozhko, (1957) 1961) (Dactynotus (Uromelan))

U. (Uromelan) himachali L.K. Ghosh, 1975 HRL
(Uromelan) illini (Hottes & Frison, 1931) (Macrosiphum)
(Uromelan) illini var. crudae (Hottes & Frison, 1931) (Macrosiphum)
(Uromelan) illini var. sangamonense (Hottes & Frison, 1931) (Macrosiphum)
(Uromelan) jaceae (Linnaeus, 1758) (Aphis) BM HRL
(Uromelan) jaceae subsp. aeneum (Hille Ris Lambers, 1939) (Dactynotus (Uromelan)) BM HRL
(Uromelan) jaceae subsp. henrichi (Börner, 1950) (Dactynotus) BM HRL
*jaceae* subsp. *scabiosae* Börner ex Franz, 1949 (*Dactynotus*)
(Uromelan) jaceae subsp. macrosiphon (Hille Ris Lambers, 1939) (Dactynotus (Uromelan)) HRL
(Uromelan) jaceae subsp. reticulatum (Hille Ris Lambers, 1939) (Dactynotus (Uromelan)) HRL
(Uromelan) lactucicola (Strand, 1928) (Macrosiphum) BM HRL
*nipponicum* Shinji, 1942 (*Macrosiphum*)
(Uromelan) minus (Borner, 1940) (Dactynotus) BM HRL
(Uromelan) minutum (van der Goot, 1916) (Macrosiphum) HRL
*dravidiana* David, 1956 (*Dactynotus (Uromelan)*) HRL
(Uromelan) montanivorum (Mosbacher, 1959) (Dactynotus (Uromelan)) HRL
(Uromelan) neocampanulae (Takahashi, 1962) (Dactynotus) HRL
(Uromelan) nigrocampanulae (Theobald, 1928) (Macrosiphum) BM HRL
*glomeratae* Börner, 1950 (*Dactynotus*) HRL
*trachelli* Börner, 1939 (*Dactynotus*) HRL
(Uromelan) omeishanense (Tao, 1963) (Uromelan)
(Uromelan) orientale (van der Goot, 1912) (Macrosiphum) BM HRL
(Uromelan) pachysiphon (Börner, 1950) (Dactynotus)
(Uromelan) parvotuberculatum (Olive, 1963) (Dactynotus)
(Uromelan) rapunculoidis (Börner, 1939) (Dactynotus) BM HRL
(Uromelan) riparium (Stroyan, 1955) (Dactynotus (Uromelan)) BM HRL
(Uromelan) rurale (Hottes & Frison, 1931) (Macrosiphum) BM HRL
(Uromelan) scorzonerae Danielsson, 1974 paratypes BM HRL
(Uromelan) simile (Hille Ris Lambers, 1935) (Dactynotus) BM HRL
(Uromelan) solidaginis (Fabricius, 1779) (Aphis) BM HRL
(Uromelan) squarrosum (Sanborn, 1904) (Macrosiphum)
(Uromelan) stachydis ((Bozhko, (1957) 1961)) (Dactynotus (Uromelan))
(Uromelan) taraxaci (Kaltenbach, 1843) (Aphis) BM HRL
(Uromelan) tardae (Hottes & Frison, 1931) (Macrosiphum) paratypes BM
(Uromelan) tripartitum (Ivanovskaja, 1971) (Dactynotus)
(Uromelon) tuataiae (Olive, 1963) (Dactynotus) BM HRL
(Uromelan) verbesinae (Boudreaux, 1949) (Macrosiphum) HRL

UROMELAN Mordvilko, 1914
Faune Russie, Ins. Hém., 1: 64, in key, as subg. of Macrosiphum Passerini, 1860
Type species Aphis jaceae Linnaeus, 1758, designated by Hille Ris Lambers 1939: 3
Here used as subg. of Uroleucon Mordvilko, 1914

U. omeishanensis Tao, 1963 to Uroleucon (Uromelan)

USSURAPHIS Mordvilko, 1924
Dokl. Acad. Nauk USSR 1924: 49
Type species Ussuraphis echinacea Mordvilko, 1924 nomen nudum
Without described species

U. echinacea Mordvilko, 1924 nomen nudum

UTAMPHOROPHORA Knowlton, 1947
Gt Basin Nat. &: 1
Type species Utamphorophora timpanagos Knowlton, 1947
*Taiwanomyzus* Tao, 1963

U. alpicola Hille Ris Lambers, 1969 HRL
bossekiae (Gillette & Palmer, 1929) (Rhopalosiphum) paratypes BM HRL
chrysosplenii (Miyazaki, 1971) (Taiwanomyzus) HRL
commelinensis (Smith, 1960) (Amphorophora) paratypes BMHRL
crataegi (Monell, 1879) (Siphonophora) BM HRL
*crataegi* Tissot, 1932 (*Amphorophora*)
filicis Miyazaki, 1968 HRL
humboldti (Essig, 1941) (Myzus) paratype BM HRL
*columbiae* Richards, 1960 (*Sitomyzus*)
*physocarpi* Pepper, 1950 (*Myzus*) BM HRL
montanus (Takahashi, 1925) (Myzus)
? prokovskajae (Mordvilko, 1914) (Myzodes)
singularis (Hottes & Frison, 1931) (Amphorophora)
timpanagos Knowlton, 1947 HRL

UZBEKAPHIS Mordvilko, 1929
Trudy̅ prikl. Ent. 14: 83
Type species Uzbekaphis carthami Nevsky ex Mordvilko, 1929

U. carthami Nevsky ex Mordvilko, 1929 ms name ?

VACUNA von Heyden, 1837
Ent. Beitr. Mus. Senckenb. 2: 289
Type species Vacuna coccinea von Heyden 1837*
= Phylloxera Boyer de Fonscolombe, 1834
Phylloxeridae

V. betulae Kaltenbach, 1843 (type of Glyphina Koch, 1856) = Glyphina betulae (Linnaeus, 1758)
californica Davidson, 1919 to Thelaxes Westwood, 1840
carlucciana del Guercio, 1913 = Thelaxes suberi (del Guercio, 1911)
castaneae del Guercio, 1913 = Thelaxes suberi (del Guercio, 1911)
coccinea von Heyden, 1837 to Phylloxera Boyer de Fonscolombe, 1834
elegantula Dahlbom, 1851 nomen dubium
ferulae Lichtenstein, 1885 nomen nudum
glabra von Heyden, 1837 to Phylloxera Boyer de Fonscolombe, 1834

VESICULAPHIS del Guercio, 1911
Redia 7: 464
Type species Toxoptera caricis Fullaway, 1909
*Pentaceratinaphis* Ivanovskaja-Shubina, 1965
*Trilobaphis* Theobald, 1922

V. caricis (Fullaway, 1909) (Toxoptera) BM HRL
*? rhododendri* Uye, 1923 (*Toxoptera*)
*shichito* Uye, 1923 (*Toxoptera*)
cephalata Miyazaki, 1971 BM HRL
congoensis Takahashi, 1965
grandis A.N. Basu, 1964 HRL
kuwanais A.K. Ghosh, R.C. Basu & RayChaudhuri, 1970
pieridis A.N. Basu, 1964 BM HRL
rhododendri A.K. Ghosh & RayChaudhuri, 1972
samagaltaica (Ivanovskaja-Shubina, 1965) (Pentaceratinaphis)
theobaldi Takahashi, 1930 BM HRL
*caricis* Theobald, 1922 (*Trilobaphis*) types BM
verbasci Chowdhuri, R.C. Basu, Chakrabarti & RayChaudhuri, 1969 to Myzakkaia A.N. Basu, 1969

VITEUS Shimer, 1867
Proc. Acad. nat. Sci. Philad. 19: 6, footnote
Type species Pemphigus vitifoliae Fitch, 1855
*? Pergandea* Börner, 1909 nec Ashmead, 1905
*Peritymbia* Westwood, 1869

*** Passerini 1860, p. 30 indicated Aphis dryophila Schrank, 1801 as type species of Vacuna with the result that for the following seventy years Vacuna was used in the sense of Thelaxes Westwood, 1840**

*Rhizaphis* Planchon, 1868
*Rhizocera* Kirkaldy, 1897
*Rizaphis* Planchon, 1868
*Xerampelus* del Guercio, 1900
Phylloxeridae

VOLUTAPHIS Börner, 1939
Arb. physiol. angew. Ent. Berl. 6: 80, described as subg. of Lipaphis Mordvilko, 1928
Type species Lipaphis (Volutaphis) centaureae Börner, 1939
*Silenobium* Börner, 1939

V. alpinae Prior, 1970 paratypes BM HRL
centaureae (Börner, 1939) (Lipaphis (Volutaphis)) BM HRL
*appeli* Börner, 1950 (*Silenobium*)
schusteri (Börner, 1939) (Silenobium) BM HRL

WAHLGRENIELLA Hille Ris Lambers, 1949
Temminckia 8: 246
Type species Rhopalosiphum arbuti Davidson, 1910, here considered a subsp. of W. nervata (Gillette, 1908)

W. empetri Richards, 1963
neoempetri A.K. Ghosh, R.C. Basu & RayChaudhuri, 1971 = ? Liosomaphis atra Hille Ris Lambers, 1966
nervata (Gillette, 1908) (Rhopalosiphum) BM HRL
*cicutae* Shinji, 1917 (*Amphorophora*)
*clavicornis* Richards, 1972 (*Aulacorthum*)
*halli* Knowlton, 1927 (*Amphorophora*)
*janesi* Knowlton, 1938 (*Amphorophora*)
nervata subsp. arbuti (Davidson, 1910) (Rhopalosiphum) BM HRL
*henryi* Balachowsky & Cairaschi, 1941 (*Amphorophora*)
ossiannilssoni Hille Ris Lambers, 1949 BM HRL
vaccinii Theobald, 1924 (Myzus) BM HRL
viburni (Takahashi, 1925) (Amphorophora)

WAPUNA Hottes & Wehrle, 1951
Proc. biol. Soc. Wash. 64: 57, described as subg. of Aphis Linaeus, 1758
Type species Aphis (Wapuna) tahosalea Hottes & Wehrle, 1951
= Aphis Linnaeus, 1758

WATABURA Matsumura, 1917
In Nagano, K., A Collection of Essays for Mr. Yasushi Nawa, Gifu, 3: 88
Type species Watabura nishiyae Matsumura, 1917
*Aphidounguis* Takahashi, 1963

W. nishiyae Matsumura, 1917 BM HRL
*gramineus* Shinji, 1923 (*Pemphigus*)
*mali* Takahashi, 1963 (*Aphidounguis*) HRL
*watamushie* Matsumura ex Nishiya, 1917 (*Pemphigus*)

WATAMUSHIA Matsumura, 1917
In Nagano, K., A Collection of Essays for Mr. Yasushi Nawa, Gifu 3: 93
Lapsus pro Watabura Matsumura, 1917

WILSONIA Baker, 1919 nec Bonaparte, 1838
Can. Ent. 51: 212
Type species Lachniella gracilis Wilson, 1919
= Cinara Curtis, 1835

XANTHOMYZUS Narzikulov, 1966
Ént. Obozr. 45: 575
Type species Xanthomyzus glaucii Narzikulov, 1966
Here used as subg. of Acyrthosiphon Mordvilko, 1914

X. glaucii Narzikulov, 1966 to Acyrthosiphon (Xanthomyzus)

XENOMYZUS Aizenberg, 1935
Zap. biol. Sta. Bolcheva 7-8: 152
Type species Xenomyzus corticis Aizenberg, 1935
*Acanthulipes* Börner, 1952
*Aphorodon* Takahashi, 1961
*Metaphorodon* Takahashi, 1961
Here used as subg. of Trichosiphonaphis Takahashi, 1922

X. alpestris Hille Ris Lambers, 1969 to Trichosiphonaphis (Xenomyzus)
corticis Aizenberg, 1935 to Trichosiphonaphis (Xenomyzus)
foliotus Shaposhnikov ex Juchnevitch, 1968

XENOPTERYGUS Smith, 1948
Fla Ent. 31: 24
Type species Xenopterygus ipomoiae Smith, 1948 = Geoica floccosa Moreira, 1925
= Geopemphigus Hille Ris Lambers, 1933

X. ipomoiae Smith, 1948 = Geopemphigus floccosus (Moreira, 1925)

XENOSIPHONAPHIS Takahashi, 1961
Bull. Univ. Osaka Pref. (B) 11: 6
Type species Xenosiphonaphis conandri Takahashi, 1961

X. conandri Takahashi, 1961 HRL
folisacculata Kumar & Burkhardt, 1971 paratype BM
japonica Takahashi, 1961 BM HRL

XENOTHORACAPHIS Takahashi, 1958
Kontyû 26: 184
Type species Thoracaphis kashiwae Uye, 1924 = Astegopteryx kashifoliae Uye, 1924

X. kashifoliae (Uye, 1924) (Astegopteryx) BM HRL
*kashiwae* Takahashi, 1932 (*Thoracaphis*)*

XERAMPELUS del Guercio, 1900
Nuove Relaz. R. Staz. Ent. agr. 2: 77, 80
Type species Xerampelus vastator without author*? Lapsus pro Rhizaphis vastatrix Planchon, 1868 = Pemphigus vitifoliae Fitch, 1855
= Viteus Shimer, 1867
Phylloxeridae

XEROBION Nevsky, 1928
Acta Univ. Asiae mediae (series VIII-a, Zool.) 3: 22
Type species Xerobion eriosomatinum Nevsky, 1928

X. eriosomatinum Nevsky, 1929 BM HRL
*camphorosmae* Tashev, 1964 (*Brachyunguis*) HRL

XEROPHILAPHIS Nevsky, 1928
Acta Univ. Asiae mediae (series VIII-a, Zool.) 3: 4
Type species Xerophilaphis saxaulica Nevsky, 1928
*Clypeoaphis* Soliman, 1937

* Vide footnote with Astegopteryx kashiwae

X. alexandrae Nevsky, 1928 to Aphis (Protaphis)
anthemiae Rusanova, 1943 nomen nudum
anuraphoides Nevsky, 1928 to Aphis (Protaphis)
artemisiae Narzikulov, 1949 to Aphis (Protaphis)
atraphaxidis Nevsky, 1928
berezhkovi Ivanovskaja, 1959
calligoni Nevsky, 1928
chondrillae (Mordvilko, 1932 nomen nudum), Mordvilko ex Tarbinsky & Plaviltchikov, 1948
cuscutae Nevsky, 1928 to Brachyunguis Das, 1918
cynanchi Nevsky, 1928 to Brachyunguis Das, 1918
deformans Nevsky, 1929 to Aphis (Protaphis)
elatior Nevsky, 1928 to Aphis (Protaphis)
elongata Nevsky, 1928 to Aphis (Protaphis)
hamatus lanattus Linnaeus ex Rusanova, 1947, p. 173 (incomprehensible)
lycii Nevsky, 1928 to Brachyunguis Das, 1918
nevskyi Kreuzberg, 1953
ninae Rusanova, 1943 nomen nudum
peucedani Nevsky, 1928 to Semiaphis van der Goot, 1913 or Brachyunguis Das, 1918
plotnikovi Nevsky, 1928 to Brachyunguis Das, 1918
rhei Nevsky, 1951
rutae Nevsky, 1928 to Aphis (Protaphis)
salsolacearum Nevsky, 1928
saxaulica Nevsky, 1928
scorzonerae Mordvilko, 1937 to Aphis (Protaphis)
scorzonerae Mordvilko ex Tarbinsky & Plaviltchikov, 1948 = Aphis (Protaphis) scorzonerae (Mordvilko, 1937)
stavropolensis (Ivanovskaja, 1960) (Clypeoaphis)
suaedae (Mimeur, 1934) (Longicaudus) BM HRL
*suaedae* Soliman, 1937 (*Clypeoaphis*) BM
*suaedicola* Hille Ris Lambers, 1939 (*Xerophyllaphis*) HRL
*suaedus* Paik, 1965 (*Hyalopterus*) HRL
suaedicola Hille Ris Lambers, 1939 (Xerophyllaphis errore) = suaedae (Mimeur, 1934)
tamariciarum Rusanova, 1943 nomen nudum
tamaricivorum Narzikulov, 1954 to Brachyunguis Das, 1918
tamaricophila Nevsky, 1928 to Brachyunguis Das, 1918

X. tetrapteralis (Cockerell, 1902) (Aphis) BM HRL
*cahuille* Dickson, 1940 (*Pergandeida*)
*piutapa* Hottes & Wehrle, 1951 (*Aphis*)
zawadovskii, Nevsky 1928
zygophylli Nevsky, 1928 to Brachyunguis Das, 1918

XEROPHYLLA Walsh, 1867
Proc. ent. Soc. Philad. 6: 283, footnote
Type species Xerophylla caryae-pilula Walsh, 1867
*Notabilia* Mordvilko, 1909
Phylloxeridae

XEROPHYLLAPHIS Hille Ris Lambers, 1939
Zool. Meded., Leiden 22: 117
Lapsus pro Xerophilaphis Nevsky, 1928

X. suaedicola Hille Ris Lambers, 1939 = Xerophilaphis suaedae (Mimeur, 1934)

YAMATAPHIS Matsumura, 1917
J. Coll. Agric. Tohoku imp. Univ. 7: 412
Type species Yamataphis oryzae Matsumura, 1917 = Toxoptera rufiabdominalis Sasaki, 1899
= Rhopalosiphum Koch, 1854

Y. oryzae Matsumura, 1917 = Rhopalosiphum rufiabdominalis (Sasaki, 1899)
papaveri Takahashi, 1921 = Rhopalosiphum rufiabdominalis (Sasaki, 1899)
rhodesiensis Hall, 1932 type of Hallaphis Doncaster, 1956 q.v.

YAMATOCALLIS Matsumura, 1917
J. Coll. Agric. Hokkaido imp. Univ. 7: 366 (566 errore)
Type species Yamatocallis hirayamae Matsumura, 1917
*Chaitophoraphis* Shinji, 1923
*Dichaitophorus* Shinji, 1927
*Megalocallis* Takahashi, 1963

Y. acericola Higuchi, 1974
hirayamae Matsumura, 1917 HRL
*acerifloris* Shinji, 1923 (*Chaitophoraphis*)
moriokae Shinji, 1933 = tokyoensis (Takahashi, 1923)
sauteri (Takahashi, 1927) (Drepanaphis)
takagii (Takahashi, 1963) (Megalocallis) HRL
tokyoensis (Takahashi, 1923) (Drepanaphis) HRL
*moriokae* Shinji, 1933

Subgen. Megalophyllaphis M.R. Ghosh, A.K. Ghosh & RayChaudhuri, 1971 (1970)

Y. (Megalophyllaphis) obscurus (M.R. Ghosh, A.K. Ghosh & RayChaudhuri, 1971 (1970) (Megalophyllaphis)

YAMATOCHAITOPHORUS Higuchi, 1972
Insecta matsum. 35: 98
Type species Trichaitophorus albus Takahashi, 1961

Y. albus (Takahashi, 1961) (Trichaitophorus) HRL

YEZABURA Matsumura, 1917
J. Coll. Agric. Hokkaido imp. Univ. 7: 392
Type species Yezabura sasae Matsumura, 1917 = Aphis bambusae Fullaway, 1910
= Melanaphis van der Goot, 1917

Y. aethusae Börner, 1950 = Dysaphis crataegi subsp. aethusae (Börner, 1950)
aizenbergi Shaposhnikov, 1949 to Dysaphis Börner, 1931
annulata Börner, 1950 to Dysaphis Börner, 1931
bonomii Hille Ris Lambers, 1935 to Dysaphis Börner, 1931
brancoi Börner, 1950 (type of Annaja Börner, 1952) to Dysaphis Börner, 1913
cephalaria Narzikulov, 1953 to Dysaphis Börner, 1931
chaerophylli Börner, 1940 to Dysaphis Börner, 1931
crataegi subsp. aegopodii Börner, 1950 = Dysaphis crataegi (Kaltenbach, 1843)
crataegi subsp. anthrisci Börner, 1950 = Dysaphis crataegi (Kaltenbach, 1843)
depilosa Nevsky, 1951 = Anuraphis subterranea (Walker, 1852)
eremuri Narzikulov, 1954 to Dysaphis Börner, 1931
hirsutissima Börner, 1940 to Dysaphis Börner, 1931
inculta subsp. petroselini Börner, 1950 = Dysaphis apiifolia subsp. petroselini (Börner, 1950)
inculta subsp. nudicaulium Börner, 1950 = Dysaphis apiifolia subsp. petroselini (Börner, 1950)
kunzei Börner, 1950 = Dysaphis crataegi subsp. kunzei Börner, 1950
laserpitii Börner, 1950 to Dysaphis Börner, 1931
lauberti Börner, 1940 to Dysaphis Börner, 1931
ligulariae Narzikulov, 1954 to Dysaphis Börner, 1931
marchali Börner, 1931 (type of Roepkea Hille Ris Lambers, 1935) = Roepkea phlomicola subsp. marchali (Börner, 1931)
mirabilis Nevsky, 1951 = ? Dysaphis ferulae (Nevsky, 1929)
papillata Nevsky, 1951 to Dysaphis Börner, 1931
photiniae Matsumura, 1918 = Melanaphis bambusae (Fullaway, 1910)
pomaria Shaposhnikov, 1950 = Dysaphis brancoi (Börner, 1950)

Y. ranunculi subsp. bulbosi Börner, 1950 = Dysaphis ranunculi (Kaltenbach, 1843)
sasae Matsumura, 1917 = Melanaphis bambusae (Fullaway, 1910)
sorbiarum Narzikulov, 1954 to Dysaphis Börner, 1931

YEZAPHIS Matsumura, 1917
J. Coll. Agric. Tohoku imp. Univ. 7: 411
Type species Yezaphis sasicola Matsumura, 1917

Y. sasicola Matsumura, 1917

YEZOCALLIS Matsumura, 1917
J. Coll. Agric. Tohoku imp. Univ. 7: 369
Type species Yezocallis kabae Matsumura, 1917
= Symydobius Mordvilko, 1894

Y. alniaria Matsumura, 1917 to Symydobius Mordvilko, 1894
kabae Matsumura, 1917 to Symydobius Mordvilko, 1894

YEZOSIPHUM Matsumura, 1919
Trans. Sapporo nat. Hist. Soc. 7: 7
Type species Yezosiphum thalictri Matsumura, 1919 = Aphis trirhoda Walker, 1849
= Longicaudus van der Goot, 1913

Y. thalictri Matsumura, 1919 = Longicaudus trirhodus (Walker, 1849)

ZABAPAPHIS Richards, 1971
Mem. ent. Soc. Can. 80: 93, legend to figure
Type species Thripsaphis (Peltaphis) hottesi Frison & Ross, 1933
= Peltaphis Frison & Ross, 1933

ZINIA Shaposhnikov, 1950
Ént. Obozr. 31: 221
Type species Zinia veronicae Shaposhnikov, 1950

Z. veronicae Shaposhnikov, 1950 BM HRL

ZYMUS Heie, 1972
Steenstrupia 2: 254
Type species Zymus succinicola Heie, 1972
Fossil, baltic amber

ZYXAPHIS Knowlton, 1947
Pan-Pacif. Ent. 23: 25
Type species Zyxaphis utahensis Knowlton, 1947
Here used as subg. of Aphis Linnaeus, 1758

Z. utahensis Knowlton, 1947 to Aphis (Zyxaphis)

# PART II. INDEX BY SPECIES-GROUP NAMES AND INFRA-SUBSPECIFIC NAMES

# STRUCTURE OF THE INDEX

This index was constructed to trace aphids known only by their specific name or by a subspecific or varietal name. Species group-names and infra-subspecific names are listed in one alphabetical sequence. Each of these names is followed by the rest of the combination(s) in which it occurs in Part I, again in alphabetical order. Page numbers are not needed as the genera are arranged alphabetically in part I.

Examples. After the specific name "rosae" one finds: *Anuraphis* (lapsus pro rosea), Aphiduromyzus, *Aphis, Endeis* (lapsus pro rosea), *Lachnus*, Macrosiphum, Maculolachnus, *Nectarophora, Neolachnus, Passerinia*, Pseudocercidus, *Siphonophora*. All the combinations in which an italicized name occurs are invalid for one reason or another. Combinations, not specifically mentioned as combinations in Part I, in which a species was used as type species for an invalid genus or a subgenus (Nectarophora, Neolachnus, Passerinia, and Siphonophora), are also included in the Index.

After the subspecific name "arenaria" one finds: Aphis coronillae subsp., *Aphis scaliae* subsp., of which the italicized part relates to the invalid combination Aphis scaliai subsp. arenaria.

After the name "glauca" one finds: Pseudoepameibaphis, *Siphonophora rosae* var., which indicates that Pseudoepameibaphis glauca is the valid name for a species while Siphonophora rosae subsp. glauca is not.

This method caused difficulties in adjectival latin or latinized species-group names and infra-subspecific names. The ending of such names usually follows the gender of the generic (not subgeneric) name. When more than one generic name follows an adjectival species-group name or infra-subspecific name, we wrote the masculine form. To avoid new difficulties, the feminine or neuter form is sometimes also given.

Example. "alba vide albus" is listed because 16 other names starting with "alb-" come between alba and albus. But "album" is not listed because it would immediately precede albus.

A table at the end of this chapter summarizes the latin adjectival endings occurring in aphid names.

N.B. Some names, like those ending on -cola and -cidus look like adjectives but are not. Compounds with -cola mean "inhabitant of" and are nouns. Citricidus can be a noun meaning killer of Citrus, or an adjective meaning the Citrus-killing (Toxoptera). We opted for the noun.

The gender of generic names is often a matter of dispute. When the original author clearly stated or demonstrated the gender of his generic name, we adhered to that. Koch (1854-1857) used generic names ending on -siphum as neuter and so did we. Mordvilko (1914-1919) left no doubt as to the neuter gender of his genera ending in -siphon. Generic names ending in -aphis we considered feminine, those in -callis masculine. Ipuka was stated to be neuter. Eriosoma and Pterocomma are Greek neuters. However, names like Uroleucon and Uromelan are Greek neuters, according to their endings, but then mean white urine and black urine, respectively, instead of white tail and black tail as undoubtedly intended by their author, Mordvilko. However, we accepted them as neutra rather than correcting or amending them.

Table of endings of latin adjectival names of aphids.

| masculine | feminine | neuter | examples |
|---|---|---|---|
| -us | -a | -um | albus, americanus, luteus, magnus |
| -er | -ra | -rum | ater, glaber, niger, ruber |
| or keeping the e | -era | -erum | alter, tener, laniger |
| -is | -is | -e | asterensis, bengalensis, nivalis |
| -or (comparatives) | -or | -us | longior, major, minor, elatior |

# INDEX

UNIV. COLL. N.W.
BANGOR